D0597055

MASSIVE STARS: THEIR LIVES IN THE INTERSTELLAR MEDIUM

A SERIES OF BOOKS ON RECENT DEVELOPMENTS IN ASTRONOMY AND ASTROPHYSICS

© Copyright 1993 Astronomical Society of the Pacific
390 Ashton Avenue, San Francisco, California 94112

Printed by BookCrafters, Inc.

First published 1993

Library of Congress Catalog Card Number: 92-75945
ISBN 0-937707-54-6

D. Harold McNamara, Managing Editor of Conference Series
408 ESC Brigham Young University
Provo, UT 84602
801-378-2298

A SERIES OF BOOKS ON RECENT DEVELOPMENTS IN ASTRONOMY AND ASTROPHYSICS

Vol. 22-Nonisotropic and Variable Outflows from Stars
ed. L. Drissen, C. Leitherer, and A. Nota   ISBN 0-937707-41-4

Vol. 23-Astronomical CCD Observing and Reduction Techniques
ed. S. B. Howell   ISBN 0-937707-42-4

Vol. 24-Cosmology and Large-Scale Structure in the Universe
ed. R. R. de Carvalho   ISBN 0-937707-43-0

Vol. 25-Astronomical Data Analysis Software and Systems I
ed. D. M. Worrall, C. Biemesderfer, and J. Barnes   ISBN 0-937707-44-9

Vol. 26-Cool Stars, Stellar Systems, and the Sun, Seventh Cambridge Workshop
ed. M. S. Giampapa and J. A. Bookbinder   ISBN 0-937707-45-7

Vol. 27-The Solar Cycle
ed. K. L. Harvey   ISBN 0-937707-46-5

Vol. 28-Automated Telescopes for Photometry and Imaging
ed. S. J. Adelman, R. J. Dukes, Jr., and C. J. Adelman   ISBN 0-937707-47-3

Vol. 29-Workshop on Cataclysmic Variable Stars
ed. N. Vogt   ISBN 0-937707-48-1

Vol. 30-Variable Stars and Galaxies, in honor of M. S. Feast on his retirement
ed. B. Warner   ISBN 0-937707-49-X

Vol. 31-Relationships Between Active Galactic Nuclei and Starburst Galaxies
ed. A. V. Filippenko   ISBN 0-937707-50-3

Vol. 32-Complementary Approaches to Double and Multiple Star Research, IAU Colloquim 135
ed. H. A. McAlister and W. I. Hartkopf   ISBN 0-937707-51-1

Vol. 33-Research Amateur Astronomy
ed. S. J. Edberg   ISBN 0-937707-52-X

Vol. 34-Robotic Telescopes in the 1990s
ed. A. V. Filippenko   ISBN 0-937707-53-8

Inquiries concerning these volumes should be directed to the:
Astronomical Society of the Pacific
CONFERENCE SERIES
390 Ashton Avenue
San Francisco, CA 94112-1722
415-337-1100

ASTRONOMICAL SOCIETY OF THE PACIFIC
CONFERENCE SERIES

Volume 35

MASSIVE STARS: THEIR LIVES IN THE INTERSTELLAR MEDIUM

Proceedings of a Symposium held as part of the 104th
Annual Meeting of the Astronomical Society of the Pacific,
at the University of Wisconsin, Madison, Wisconsin
23-25 June 1992

Edited by
Joseph P. Cassinelli and Edward B. Churchwell

Contents

SECTION 2. POST-MAIN-SEQUENCE MASSIVE STARS

SECTION 3. EFFECTS OF MASSIVE STAR WINDS AND RADIATION
ON THE ISM

PREFACE

Throughout their relatively short lives, massive stars are closely coupled to the interstellar medium. Each star spends about ten percent of its lifetime deeply embedded in the giant molecular cloud in which it formed. It greatly influences the surrounding medium because of its ultraviolet radiation and strong stellar wind. The large H II regions later produced are among the most spectacular objects seen in our Galaxy and clearly delineate the locations of the hot young stars in galaxies. The stellar wind of a single very massive star can replenish the interstellar medium with tens of solar masses of chemically enriched material while the star is in the form of a blue supergiant, a red supergiant, or a Wolf-Rayet star near the end of its life cycle. The winds not only contribute mass, but also make significant contributions to the momentum in the interstellar medium. At the end, the stars explode as supernovae, a most dramatic example of which is Supernova 1987A, the spectacular explosion of a blue supergiant star in the nearby Large Magellanic Cloud. Supernovae can compress the nearby molecular clouds, leading in turn to the start of another cycle of star formation. There is a steady heavy-element enrichment of the interstellar medium, and conditions in the galaxy continually change as a result of the massive stars that form. Many of these topics have become major areas of research over the past decade. The subject of massive star formation is a very lively new topic, there have been new types of massive stars recently classified, and of course SN 1987A has led to major improvements in our understanding of the final stages of stellar evolution.

Given the rapid development of studies of massive stars and associated interstellar phenomena, the faculty of the Astronomy Department at the University of Wisconsin chose this topic for the 104th annual meeting of the Astronomical Society of the Pacific held in Madison from June 23-25, 1992. The topic encompasses much of the astrophysics being pursued in the Astronomy and Physics Departments. Cassinelli chaired the scientific organizing committee because of his interest in winds and in the formation of very massive stars. In spite of the close relation between massive stars and the interstellar medium, specialists in the two research areas rarely have a common forum for communication. There had not been a meeting in several years that encompassed both topics. Hence this meeting was arranged from the beginning to be quite broad in scope.

The Scientific Organizing Committee, whose members are listed below, includes experts across the spectrum of topics associated with massive star phenomena and interstellar medium physics and chemistry. Because clusters of massive stars can affect the appearance and spectrum of an entire galaxy, we have also included a section on "broader issues". This includes topics such as starburst galaxies that are certainly related to the presence of massive stars, but are not generally considered in discussions of the evolution of individual stars. Finally, we wanted to make the meeting of great interest to graduate students and do whatever we could to make it possible for them to attend the meeting.

xii

Given these goals for the meeting, the Scientific Organizing Committee was able to attract an excellent group of speakers for each of the five half day sessions. The broad scope of this meeting brought together leading experts in stellar or interstellar medium astrophysics. Each invited speaker was asked to address, where possible, the "interactions" that exist in their topic between massive stars and the interstellar medium.

The scientific meeting was composed of five scientific sessions plus a related history session. This book starts with a paper from the history session by Arthur Code regarding the role of instrumentation in the development of our understanding of the distribution of massive and luminous stars in our Galaxy. The rest of the book follows the agenda of the scientific meeting.

The first section concerns the early formation phases in the life of the star. This is perhaps the newest topic in the field, and there remain great uncertainties and quite different pictures for what is occurring while the stars are still embedded in their natal cloud. The topic of star formation has until very recently been concentrated on low mass stars because they are clearly identifiable. In the case of massive stars, astronomers have only recently recognized ultracompact HII regions as manifestations of newly formed massive stars still embedded in molecular clouds.

Sections two and three deal with more mature massive stars, those which are evolving beyond the hydrogen burning main-sequence. In following their aging process we encounter several classes of the most luminous stars: blue and red supergiants, hypergiants of intermediate spectral type, luminous blue variables, and the Wolf-Rayet stars. The evolutionary connection between all of the classes remains a major topic in stellar interior and evolution theory. All of these stars interact strongly with their environment because of their powerful stellar winds, and in the case of the hot stars, by way of their large input of ionizing radiation. These interactions are the focus of section three.

Section four concerns the final explosive phase in the evolution of a massive star. In the case of supernova 1987A, the fact that it involved the explosion of a blue supergiant was in itself unexpected, since most evolutionary modeling had led to the conclusion that red supergiants or Wolf-Rayet phases were the likely Supernovae precursors. This section contains discussions of recent information derived from SN 1987A as well as summaries of our current understanding of the effects of supernovae shocks on the interstellar medium.

This book closes with section five which contains reviews of starburst galaxies, giant HII regions, and the new topic of Wolf-Rayet Galaxies. These are clearly phenomena which require the presence of very populous clusters of massive stars. Here we see a strong connection between a major field of extragalactic astronomy, and the subjects of massive star evolution and star/interstellar interactions that are discussed in the earlier sections of the book.

To encourage the participation of younger Astronomers we were able to obtain a meeting travel grant from the National Science Foundation. We are grateful to the NSF for this "seed money", which in combination with funds from their home institutions, allowed 19 graduate students to attend and present their research. We also wish to thank the Scientific Organizing Committee for helping us arrange a very stimulating meeting. We thank Robert Mathieu and the rest of the Local Organizing Committee for the smooth operation of the meeting. The work of Phyllis Fass in handling the travel funds is much appreciated. We thank Lizabeth Kinney and Barbara Brodie for their help in putting this book together. Sharon Pittman deserves special recognition for her help at all phases of the meeting, including assembling of the papers for this book.

Joseph P. Cassinelli & Edward B. Churchwell
Madison, Wisconsin

THE ORGANIZING COMMITTEES

Scientific Organizing Committee

Joseph Cassinelli (Chair, University of Wisconsin)
Roger Chevalier (University of Virginia)
Edward Churchwell (University of Wisconsin)
Peter Conti (University of Colorado)
Kris Davidson (University of Minnesota)
John Gallagher (University of Wisconsin)
Michael Jura (University of California, Los Angeles)
Nancy Morrison (University of Toledo)
Frank Shu (University of California, Berkeley)
Stanford Woosley (University of California, Santa Cruz)

Local Organizing Committee

Robert Mathieu (Chair, University of Wisconsin)
Robert Bless (University of Wisconsin)
Karen Bjorkman (University of Wisconsin)
Jane Breun (Madison Astronomical Society, Wisconsin)
Lowell Doherty (University of Wisconsin)
Alan Dyer (ASTRONOMY Magazine)
Kathy Stittleburg (University of Wisconsin)

LIST OF PARTICIPANTS

Fred C. Adams	University of Michigan
Bruce Altner	Applied Research Corporation
William G. Bagnuslo	Georgia State University
Peter Barnes	Center for Astrophysics
Mary Barsony	Center for Astrophysics
Martin Beech	University of Western Ontario
Robert A. Benjamin	University of Texas-Austin
Frank Bertoldi	Princeton University Observatory
Andriaan Blaauw	Kapteyn Laboratory, Groningen
John Blondin	University of North Carolina
Alexander Brown	Leiden University
Anthony G.A. Brown	Sterrewacht Leiden, The Netherlands
Douglas Brown	University of Washington & Bellevue C.C.
Kenneth Brownsberger	JILA, University of Colorado
Geoffrey Burks	CASA, University of Colorado
David Burrows	Pennsylvania State University
Murray F. Campbell	Colby College, Maine
Cristina Cappa de Nicolau	Instituto Argentino de Radioastronomia
John Carpenter	University of Massachusetts
Paola Caselli	Ohio State University
Brian Casey	University of Wisconsin
Joseph Cassinelli	University of Wisconsin
Hector Castaneda	Instituto de Astrofisica de Canarias
John Castor	Lawrence Livermore National Laboratory
Haiqi Chen	University of Western Ontario
Wan Chen	NASA Goddard Space Flight Center
Roger Chevalier	University of Virginia
You-Hua Chu	University of Illinois
Mark Chun	University of Chicago
Ed Churchwell	University of Wisconsin
Arthur Code	University of Wisconsin
Peter Conti	JILA, University of Colorado
Michael Corcoran	NASA Goddard Space Flight Center
George J. Corso	DePaul University
Rafael Costero	Observatorio Astronomico Nacional
Don Cox	University of Wisconsin
Kyle Cudworth	University of Chicago
Robert Cumming	Imperial College - London
Augusto Daminelli	Universidade de Sao Paulo, Brazil
Kris Davidson	University of Minnesota
Eugene J. De Geus	University of Maryland
Athanassios Diplas	University of Wisconsin
Linda Dressel	Applied Research Corporation
Laurent Drissen	Space Telescope Science Institute
Dennis Ebbets	Ball Aerospace Systems Group
Philippe Eenens	Royal Observatorio Astronomico Nacional
Joann Eisberg	University of Wisconsin
Jay Elias	Cerro Tololo Inter-American Observatory

Marco Fatuzzo	University of Michigan
Michael Fich	Smithsonian Astrophysical Observatory
Douglas Forbes	Grenfell College, Newfoundland
James Fowler	Apache Point Observatory
Geoffrey Fox	Glasgow University
David Friend	University of Montana
John S. Gallagher	University of Wisconsin
Guido Garay	Universidad de Chile
Guillermo Garcia-Segura	University of Illinois
Katy Garmany	University of Colorado
Steven Gibson	University of Wisconsin
Harry Guetter	US Naval Observatory
Margaret Hanson	CASA, University of Colorado
John Hester	Arizona State University
Lynne Hillenbrand	University of Massachusetts
Melvin Hoare	University of Oxford
Paul Hodge	University of Washington
Peter Hofner	University of Wisconsin
David Hollenbach	NASA-Ames Research Center
Jamie M. Howard	Yale University
Roberta Humphreys	University of Minnesota
Vera Jatenco Silva Pereira	Universidade de Sao Paulo, Brazil
Mike Jura	Univ. of California, Los Angeles
Michael Kaufman	Johns Hopkins University
William C. Keel	University of Alabama
Gillian Knapp	Princeton University
Gloria Koeningsberger	Instituto de Astronomia, Mexico
Susan Lamb	University of Illinois
Henny Lamers	SRON, Space Research Utrecht
Norbert Langer	University-Observatory, Gottingen
Matthew Lehnert	The Johns Hopkins University
David Leisawitz	Pennsylvania State University
Daniel F. Lester	University of Texas-Austin
Richard Lines	Lines Observatory
Felix J. Lockman	NRAO Charlottesville
Peter Lucke	Mount Union College
Joseph MacFarlane	University of Wisconsin
Murugesap Maheswaran	University of Wisconsin Centers
Crystal Martin	University of Arizona
Steve Martin	University of Chicago
Derck Massa	Applied Research Corporation
Phil Massey	Kitt Peak National Observatory
Lynn Matthews	University of Wisconsin
Robert Mathieu	University of Wisconsin
Mark McCaughrean	University of Arizona
Peter McCullough	University of California, Berkeley
P. Mitalas	University of Western Ontario
Partrick Morris	University of Colorado
Nancy Morrison	University of Toledo

Edward Murphy	University of Virginia
Phil Myers	Center for Astrophysics
Jaylene Naylor	University of Montana
Sally Oey	University of Arizona
Gordon Olson	Los Alamos National Laboratory
Michael Pahre	Center for Astrophysics
Nino Panagia	Space Telescope Science Institute
Joel Wm. Parker	CASA,University of Colorado
Tony Phillips	California Institute of Technology
Rene Plume	University of Texas-Austin
John Porter	University of Wisconsin
Ron Reynolds	University of Wisconsin
Carmelle Robert	Space Telescope Science Institute
V. Robledo-Rella	JILA, University of Colorado
Michael Rosa	European Southern observatory
Alexander Rudolph	NASA-Ames Research Center
Robert Ruotsalainen	Eastern Washington University
Eric M. Schlegel	NASA-Goddard Space Flight Ctr.
Regina Schulte-Ladbeck	University of Pittsburgh
Kenneth Sembach	Massachusetts Institute of Technology
Lea Shanley	University of Wisconsin
Yaron Sheffer	University of Texas
Debra Shepherd	University of Wisconsin
Steven M. Shore	NASA Goddard Space Flight Center
Michael Shull	JILA, Universtiy of Colorado
Ronald Snell	University of Massachusetts
Linda Sparke	University of Wisconsin
Meenakshi Srinivasan-Sahu	Kapteyn Laboratory, Groningen
Todd Tripp	University of Wisconsin
Ken Swanson	New Mexico State University
Anne B. Underhill	University of British Columbia
David Van Buren	IPAC, California Institute of Technology
Schuyler Van Dyk	Naval Research Laboratory
Dany Vanbeveren	Vrije Universiteit Brussels
Kim Venn	University of Texas
William Waller	NASA Goddard Space Flight Center
Lifan Wang	University of Manchester, UK
Alan Watson	University of Wisconsin
William J. Welch	University of California, Berkeley
Karl Wilk	University of Waterloo, Canada
Mark G. Wolfire	Center for Astrophysics
Douglas Wood	National Radio Astronomy Observatory
Michael Woodhams	Princeton University
Hui Yang	University of Minnesota
Harold Yorke	Institute for Astronomy & Astrophysics, University of Wuerzburg

HISTORICAL INTRODUCTION

Massive Stars: Their Lives in the Interstellar Medium
ASP Conference Series, Vol. 35, 1993
Joseph P. Cassinelli and Edward B. Churchwell (eds.)

PHOTOCELLS, HOT STARS AND SPIRAL ARMS

ARTHUR D. CODE
Department of Astronomy, University of Wisconsin-Madison,53706

ABSTRACT   This paper describes three different developments in astronomy which converged to provide valuable insight into the nature of massive early type stars and to the delineation of spiral structure in our galaxy. The first of these was the development of photoelectric techniques by Joel Stebbins at Washburn Observatory. The second was the pioneering work on two dimensional spectral classification by William Morgan at Yerkes observatory and the third was the design of the Henyey-Greenstein wide angle camera. The impact of these advances on stellar physics and galactic structure during the first half of this century is presented from the authors perspective as a participant in some of the later developments.

PHOTOCELLS

The history that I shall describe is both parochial and selective. Parochial because it focuses on Wisconsin astronomy and selective because it reflects the lore and experiences that I have been exposed to here at Washburn Observatory.

In the early days, Washburn Observatory followed the traditional research methods of the time and continued to do so until 1922 when Joel Stebbins arrived to take over the Directorship of the Observatory. The observatory then moved from visual observing directly to photoelectric astronomy establishing a tradition of instrumental innovation that has continued to this day.

Stebbins started his astronomical career at Washburn Observatory as a first year graduate student. and published his first paper on the photometry of Nova Persei 1901. He then finished his studies at Lick Observatory completing a spectroscopic thesis in 1903 (Stebbins 1903), following which he took an appointment at the University of Illinois.   It was there that he heard of the photosensitivity of selenium and embarked on a life-long occupation with "electric astronomy" (Stebbins 1915). The first successful measurements with a Selenium Cell were of the moon. In 1907 Stebbins and Brown (1907) published a paper giving accurate measurements of moon light throughout the lunation. They commented that as far as they could determine the only other use of selenium cells in astronomy were the measurement of a current produced by bright stars with a 24 inch reflector by G. M. Minchin in England, and use by E. Ruhner in observations of solar and lunar eclipses. Later that year with an improved cell they succeeded in getting a deflection from a first magnitude star, Aldebaran. By 1910 Stebbins had made significant improvements in his photometer and had obtained measurements of Halley's comet and of other bright stars including Algol (Stebbins 1910). The variability of Algol or β Persei was discovered by John Goodricke in 1782 and was in fact the first variable star to be systematically studied. Some two years later Goodricke discovered δ Cephei another star that was to be carefully studied by Stebbins. Goodricke had suggested that the light variation in Algol might be due to an eclipse; but since no secondary component was observed the idea was not immediately accepted. The importance of Stebbins observations were that for the first time the secondary minimum, which had been exhaustively searched for visually, was measured. This established the binary nature of Algol and demonstrated the superiority of photoelectric photometry over visual or photographic methods.

The Selenium cell was a photoconductive device which changed its resistance in response to temperature as well as to light and for this reason Stebbins had placed the cell in an ice chamber. An interesting side-light on the early investigations of selenium receivers comes for the aforementioned mentioned George Minchin (1880). It had been suggested to him that the measured current might be found to vary inversely as the distance of the luminous source. The argument went as follows ; the energy of the incident light varied as k/r^2, where k is a constant, whereas the energy of the photoelectric current was I^2 R. Thus the current I should vary as $1/r$. He had not yet been able to verify this prediction at the time of publication although the few experiments he had made roughly agreed with the above law. Were it so, our stellar distance measurements would be in need of serious revision.

In 1913 Stebbins was able to switch over to a far more sensitive and stable detector . An Illinois physics colleague Prof. Jacob Kunz had succeeded in making photo cells employing photoelectric emission. One of the early Kunz cells was number QK 99. Being the 99 th attempt at a Potassium Hydride photocathode in a quartz cell. The cell was filled with a He, Ne or Ar which permitted gas gains of 5 to 10 depending upon the applied voltage. Stebbins continued his association with Jacob Kunz until the latter's death in 1938. This early work concentrated on variable stars. and when in 1922 he was offered the Directorship here at Washburn Observatory he brought with him his photocells and innovative research.

The photoelectric current in these early photometers was measured by means of a string electrometer and in 1927 the substitution of a Lindeman electrometer was considered an important improvement in the sensitivity stability and speed of observing. In order to reduce leakage and stray charge the photometer and wiring were enclosed and purged with dry air. It was with such a photometer that the first work on galactic structure was undertaken. In 1930 during a visit to Lick Observatory Dr. Robert Trumpler suggested to Stebbins that he undertake a study of the colors of B-stars. Trumpler was convinced from his study of clusters that there was considerable absorption of light in interstellar space. Trumpler provided Stebbins with a list of a number of O and B stars near dark clouds and at higher latitudes for comparison. Not long after Stebbins and C.M. Huffer embarked on this program they realized that the interstellar extinction was far from uniform and would require observations of many more stars. The observing list grew to include all stars north of $-15°$ brighter than 7.5 and of classes B0 to B5 in the Draper Catalogue. In 1933 Stebbins and Huffer

(1934) published results on the space reddening in the galaxy based on the colors of 733 stars. The results showed the strong concentration to the galactic plane and to the zone of avoidance of extragalactic nebulae delineated by Edwin Hubble. They also determined the ratio of total to selective absorption for their color system and derived distances for each of the B stars. In summarizing this work Stebbins said "The results of the present study are that the reddening of B- stars confirms the conclusion that our galactic systems is filled near its median plane with a layer of dark material, which reddens all stars at sufficient distances and blots out everything behind it. The presence of this dark region is alone a good reason for concluding that our own galaxy is simply one of many similar systems which we can observe.

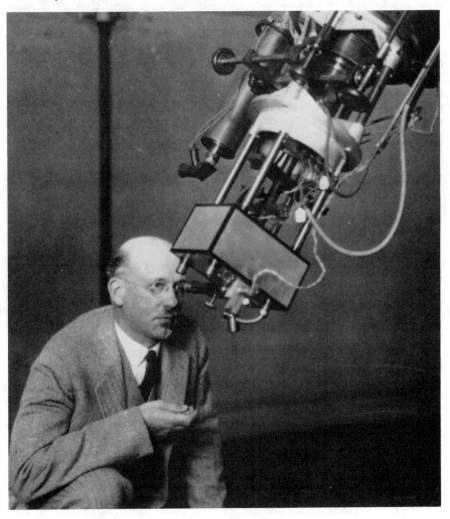

Fig. 1. Professor Joel Stebbins with an early photometer on the Washburn Observatory 15-inch refractor.

At this same time Albert Whitford (1932) had constructed a vacuum tube amplifier capable of amplifying the weak currents from the Kunz cells. This had been made possible by the development by General Electric of the FP-54 Thermionic amplifier, a very low grid current tetrode.Typically a 14th magnitude star would produce a current of the order of 10^{-16} amps with the Washburn 15-inch refractor. The amplifier could provide a gain of $2x10^6$ which meant that the signal could be measured with a galvanometer. Usually the galvanometer would be located in another area of the observatory and the observer would call to an assistant to make and record readings at the appropriate time. My first exposure to photoelectric photometry was as an observing assistant watching the dancing beam of light on the galvanometer scale. Whitford also found that an order of magnitude decrease in noise was achieved by evacuating the enclosure containing the amplifier tube, high resistance circuitry and photocell. The vacuum reduced leakage currents caused by moisture and reduced stray charges produced by cosmic rays. This photometer represented a great advance in photoelectric photometry. An experiment carried out in the 1930's with this instrument illustrates the direct approach Stebbins had to problem solving. I quote Joel Stebbins. "We used to say that this photometer even without a telescope, would detect a candle a mile away. We said this so often we felt constrained to try it at least once with a real candle at a real mile. When we did so by setting of a standard candle on Picnic Point a mile across the lake from the observatory, we found that with no optical aid except a black tube to eliminate stray light the photocell would not only give a conspicuous response but 1/5 candle would do."

HOT STARS

The distribution of B stars in our galaxy as found by Stebbins and Huffer suffered in two ways. Many of the spectral types were not well determined and the absolute magnitudes of B stars were at the time seriously underestimated, a result, of course, of not having properly accounted for interstellar extinction. Some 60 miles south of Madison at Yerkes Observatory, however, W.W. Morgan had started on a program that would remedy this situation. Using a small prism spectrograph on the 40-inch refractor with a linear dispersion of 120 A/mm near H_gamma he had found that it was possible to develop a two dimensional classification define by a grid of standard stars. The MKK Atlas of Stellar Spectra was published in 1942 and represented a lasting landmark in spectral classification (Morgan,Keenan and Kellman 1942). The spectral type and luminosity class of a star provides the means of determining it's normal color and absolute magnitude, while photoelectric photometry provide observed magnitude and color and therefore corrections for interstellar extinction . In a monumental paper in 1953 Harold Johnson, who had just joined the Yerkes staff coming from Washburn Observatory, and Morgan (Johnson and Morgan 1953) published the paper that defined the UBV stellar photometry system in terms of the spectral types on the revised system of the Yerkes Atlas. Harold Johnson had brought with him to Yerkes the advances in photoelectric photometry that he and Whitford had developed here at Washburn . In the 1930's Whitford had continued to improve the detection limits of the photocell and he and Stebbins introduced a multicolor photometric system. Thus by 1940 the six color photometric system was introduced defining band passes from 3500 Å to 12,500 Å. They had been observing extra galactic systems and had established the multi-color system in order to determine red shifts of faint galaxies from colors. In order to do so it would be necessary not only to observe nearby galaxies but for calibration purposes they started a series of stellar observations. It was nearly a decade later before they got back to the problem of galaxy red shifts. The first six color photometry paper (Stebbins and Whitford 1943) was on law of space reddening from the colors of O and B stars. Until this time it was generally thought that the wavelength variation would be Rayleigh like and vary as λ^{-4} . Stebbins and Whitford found, however, a λ^{-1} variation over this extended spectral region. Studies of interstellar extinction and the nature of the interstellar grains has remained an important area of investigation at the Washburn Observatory to this day. By 1947 Whitford (1947) had extended the determination of the wavelength dependence to 2.2 μ and replaced the wide band determinations to resolutions of the order of 10 Å using a photoelectric spectral scanner. The Whitford extinction curve became the standard for many years. Later on we extended the measurements into the vacuum ultraviolet.

SPIRAL ARMS

Just prior to World War II Whitford had obtained some developmental photomultiplier tubes which simplified the amplification of photoelectric signals and made it possible to drive chart recorders. During the war industry developed mass production techniques for photmultipliers for what seems to me a rather curious reason. Since the dark noise of a photomultiplier was random the tubes were used in a shielded container to generate random signals for the purpose of radar jamming. After the war the surplus 931-A's became a bonanza for photoelectric photometry. Ever since the days of George Ritchey Yerkes Observatory had an optical shop. During the war Jessie Greenstein and Louis Henyey did optical design for the Navy. One of the products, a projector to display constellations for training navigators employed a 12 inch concave mirror and an F/2 petzval portrait lens design to flatten the curved field. By virtue of the reciprocity theorem this would work equally well as an all sky camera (Struve 1951). Thus after World War II The Henyey-Greenstein camera became available and two Yerkes graduate students Stuart Sharpless and Don Osterbrock set about taking photographs of the Milky way. The 140 degree field of view provided an opportunity to view our galaxy in a manner more like that we had of extragalactic nebulae. Baade had been photographing nearby galaxies with an H alpha filter and found that in many spirals the HII regions delineated the spiral arms almost like beads on a string. In collaboration with W.W. Morgan, Sharpless and Osterbrock obtained H alpha images of the northern sky which were strikingly similar to Baade's pictures of M31. By identifying the exciting star and obtaining spectroscopic distances they constructed the first model of the spiral arms in our galaxy (Morgan,Sharpless and Osterbrock 1952). Meanwhile a collaborative program on the photometry and spectral classification of OB stars between Morgan, Whitford and myself was underway. In a paper in 1953 (Morgan Whitford and Code 1953) we identified 27 large scale clusterings or aggregates as we called the associations at that time.

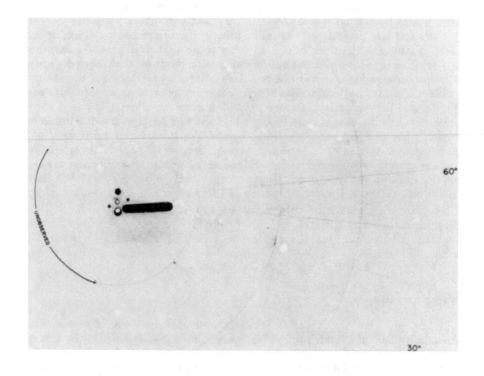

Fig. 2. The Morgan,Sharpless and Osterbrock model of the distribution of HII regions showing for the first time spiral arms in the neighboorhood of the sun.

The paper includes a list of the highest luminosity O-A stars in these associations. A plot of the spatial distribution, figure 4a, confirmed the identification of an inner arm suggested by Morgan Sharpless and Osterbrock. In order to extend this study to the southern milky way Houck and I went to South Africa where we carried out spectroscopic observations at the Radcliff Observatory and photometry at the Cape Observatory. We also took along the Henyey-Greenstein wide angle camera and set it up under the dark skies of the Harvard kopje near Bloemfontein. In addition to visual and H-alpha photographs we took advantage of the fine view of the galactic center to obtain near infra-red images. In these images the nature of our own galaxy is readily apparent. Becker (1956) had already pointed out that a characteristic of spiral structure is that independent of were an observer is located away from the center the surface brightness distribution in the galactic plane should exhibit an asymmetry with respect to the center and extend further along the trailing edge of the spiral. This appearance could of course be confused by the presence of interstellar extinction. In this infra-red picture where extinction is much reduced we see the trailing arm in the direction of Carina. That this is indeed a real change in surface bright is born out by the fact that photometric distance along the arm are extended and that the OB stars observed on either side of the inner arm are more distant and are relatively blue. While in the southern hemisphere we could not resist extending the observations to blue stars in the Magellanic clouds. One of our interests appears in a short paper Houck and myself on superluminous stars in the LMC and our own galaxy (Code and Houck 1958). In the discussion we evoked the Eddington limit but thought of it not as indicating a limit on the mass of stars but rather as a lower limit on the mass of these stars. The results of our studies of galactic structure were presented in a catalogue of 1270 OB stars published in 1955 (Morgan, Code and Whitford 1955). Our next speaker Adriaan Blauuw, for a time a Yerkes astronomer has made significant contributions to these studies of galactic structure and in particular the calibration of the absolute magnitudes of the MK luminosity classes.

Fig. 3. Henyey-Greenstein Wide Angle Camera image towards the Galactic Center photographed in the near infra-red.

Another direction that advances in photometry took here at the Washburn observatory was to try to provide a better link between theoretical predictions fom modeling of stellar interiors and stellar atmospheres. For this reason we started a program of "monochromatic" photometry by scanning the spectrum. Continuous spectral scans were obtained with a resolution of the order of 10 Angstroms for a diverse range of objects (Code 1952). Before discussing some of these other developments let me return to the comment I made earlier about the initial objective of the 6-color photometry being to determining the Hubble

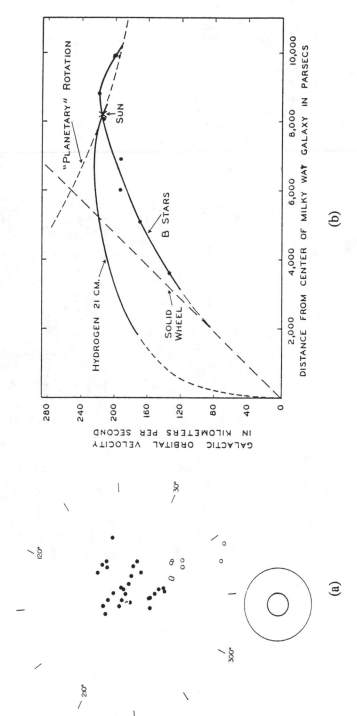

Figure 4 (a) Space distribution of associations according to Morgan, Whitford and Code (1953). Center of galaxy is at 9 kpc and the longitude index lines are 500 psc long. (b) Galactic rotation curve according to Bahng, Code and Whitford (1957). In this plot the galactic center is at 8.2 kpc redshift. When Stebbins and Whitford (1948) attempted to calibrate the color dependence of elliptical galaxies as a function of red shift they found an excess the galaxies were redder than expected. If due to intergalactic dust an increase in density by a factor of a thousand is implied. This became known as the Stebbins-Whitford effect and it's interpretation was troubling. If an evolution effect then galaxies evolved more rapidly than expected. It turned out, however that this was an artifact introduced by the fact that in the presence of intensity fall in the elliptical spectra shortward of the 3800 Å the band width of the filters was too large (Code 1958). When we determined spectral energy distributions of galaxies using the spectral scanner we found that we could account for the colors of galaxies nicely.

One of the features of spectral scanning was that it became possible to move astronomy away from the traditional magnitudes to measurements of real physical units, $ergs/cm^2/sec/\mathring{A}$ say. These were the kind of measurements that were need to make direct comparison with the stellar model calculations. Whitford and I attempted to determine the observed flux of stars over the optical and near infrared by comparison with standard lamps and by use of a sun reducer with the solar spectrum. Later at the Pine Bluff Observatory west of Madison where a new 36-inch reflector had been placed in operation Bob Bless and I started on a calibration program were we made direct comparison of spectral scans of stars with a standard lamp placed on a nearby tower. The lamp in turn was calibrated directly against a fundamental Platinum black body. A useful number to remember is that a V magnitude of zero corresponds to 1000 $photons/cm^2/sec/\mathring{A}$ at 5500 Å. Within the uncertainties introduced by the wide band nature of a V response this is as good a number today as it was when first determined (Code 1960).

The observed spectral energy distribution of the massive hot O B stars that we had been so interested in did not provide much basis for assessing the validity of the model calculations because most of the flux was radiated in the vacuum ultraviolet. When the first uv results came in they indicated that the flux was much less than predicted and that Spica at least seemed to have a uv halo around it. We ourselves had started up a program to carry out uv photometry from a satellite. Until such time as we had an operating space telescope, however, we started flying instruments in the X-15 rocket plane that went well above the ozone layer. The first photometer simply replace one of the movie camera usually carried for engineering purposes. Later we installed a pointed telescope. The results indicated that there were no halos and the flux was about what was expected. After the launch of OAO-2 in December 1968 we began to accumulated a large body of data on the spectral flux of stars as well as planets, comets and galaxies. The ultraviolet flux combined with ground based data covered most of the spectral distribution except for the hottest stars. In collaboration with John Davis and Hanbury Brown, who had been able to measure the angular diameter of hot stars, we were able for the first time to make an empirical measurement of the effective temperature of stars (Code, Davis, Bless and Brown 1976).

I will terminate my discussion at this point for the pace of astronomy has grown much too fast to review in anything less than great compendium. Even so I have had leave out many important contributions by the heroes of this presentation , Stebbins, Whitford and Morgan to whom I am beholden for providing much wise guidance.

I am not a historian. A fact that I may have adequately demonstrated during the last half hour. I still regard myself as somewhat of a futurist and I believe that the story that I have been telling has not yet reached the final chapter. Let me return to one example. The rotation of the galaxy is today derived primarily from radio data i.e. the interstellar gas. The motions of stars and gas do not appear to be identical. The 1957 rotation curve, shown in figure 3b, illustrates the point (Bahng, Code and Whitford 1957). Modern 21 cm curves would differ some from this presentation, but the difference would still exist. Now returning to an earlier plot of the stellar spiral arms, figure 3a, we see that the inner points on the rotation curve can not be moved much closer to the galactic center. The difference can not be due to errors in the stellar distance. Either the stellar velocities are wrong or one or both of the data sets does not represent simple galactic rotation. The answer is not in our past. Perhaps it is for new generations of astronomers to answer.

REFERENCES

Bahng, J.D. Code, A.D. and Whitford, A.E. 1957, Sky and Tel 16, 529
Becker, W. 1956, Vistas in Astronomy 2, 1515
Code, A.D. 1952, Observatory 72, 201
Code, A.D. 1958, PASP, 71, 118
Code, A.D. 1960, in Stellar Atmospheres, ed. J.L. Greenstein (Univ. of Chicago Press) 50
Code, A.D. and Houck, T.E. 1958, PASP, 70, 261
Code, A.D. Davis, J. Bless, R.C. and Brown, R.H. 1976, ApJ, 203, 417
Johnson, H.L. and Morgan, W.W. 1953, ApJ, 117, 313
Minchin, G.M. 1880, "An Account of Experiments in Photoelectricity" Univ. Press Dublin
Morgan, W.W. Keenan, P.C. and Kellman E. 1943, "An Atlas of Stellar Spectra" Univ. of Chicago Press
Morgan, W.W. Sharpless, S. and Osterbrock, D.E. 1952, Sky and Tel 11, 138
Morgan, W.W. Whitford, A.E. and Code, A.D. 1953, ApJ, 118, 318
Morgan, W.W. Code, A.D. and Whitford, A.E. 1955, ApJS, 2, 41
Stebbins, J. 1903, Lick Obs Pub 41
Stebbins, J. 1910, ApJ, 32, 185
Stebbins, J. 1915, Science 41, 809
Stebbins, J. and Brown, F.C. 1913, ApJ, 24, 326
Stebbins, J. and Huffer, C.M. 1934, Pub of Washburn Obs 15 part 5
Stebbins, J. and Whitford, A.E. 1943, ApJ, 98, 20
Stebbins, J. and Whitford, A.E. 1948, ApJ, 108, 413
Struve, O. 1951, Sky and Tel 10, 215
Whitford, A.E. 1932, ApJ, 76, 213
Whitford, A.E. 1947, ApJ, 107, 102

SECTION 1

FORMATION AND EARLY EVOLUTION OF MASSIVE STARS

Massive Stars: Their Lives in the Interstellar Medium
ASP Conference Series, Vol. 35, 1993
Joseph P. Cassinelli and Edward B. Churchwell (eds.)

FORMATION AND EARLY EVOLUTION OF YOUNG MASSIVE STARS

Wm. J. Welch
Radio Astronomy Laboratory, University of California, Berkeley 94720

ABSTRACT   The present status of long wavelength observations of massive star formation is briefly reviewed. O stars spend about 15 percent of their Main Sequence lives buried in clouds in which they form, which means that at this stage they must be studied in the infrared and radio wavelengths. There is recent direct evidence of disks around both very young and somewhat older (revealed) O stars. These primordial disks may provide the source for the dense ionized gas around imbedded O stars, the compact HII regions, for periods as long as a million years. The massive outflows that are observed from these stars may represent the clearing of material from their neighborhood that reveals the stars. Both the gas chemistry and the properties of the dust near the imbedded O stars appear to be strongly modified by their winds and radiation. Finally, although little is known about the basic questions of what determines the masses of the brightest stars and their Initial Mass Functions, there is some evidence of large scale collapse onto the young O groups, which could provide the means for synchronizing the formation of O sub-groups.

INTRODUCTION

In any discussion about star formation, a number of basic questions arise. 1. What physical circumstances, processes, etc. determine the mass of a star? 2. Studies by Humphreys(1985) and Garmani et al(1982) have shown that the Initial Mass Function for O stars has a substantial difference in its slope between the inner and outer galaxy. What factors determine the IMF? Is it, for example, the metalicity of the gas that forms the stars, and, if so, how? 3. For the massive stars, what governs the (nearly) coeval formation of a number of sub-groups in an association? 4. Do all stars form with disks? 5. What are the effects of the young, newly formed stars on their surroundings?

There are at least two practical difficulties that confront the study of the formation of massive stars. First, the massive stars are relatively rare, and the nearest region forming such stars is in Orion at a distance of 500 pc. The typical best resolutions of telescopes equipped with array cameras in the near and mid-infrared and of the interferometric radio arrays are about one arc second. Thus, we are limited to linear resolutions of the order of 500 AU or worse. Second, O stars arrive on the Main Sequence still heavily imbedded in the clouds in which they form. It is what we observe, and it is to be expected in view of the short thermal collapse times for the massive cores. An important recent study by Wood and Churchwell (1989) showed that about 15 percent of the O stars are found in compact HII regions; that is, they are heavily imbedded. Hence, the O stars are imbedded for about 15 percent of their MS lifetimes. Observers who wish to study high mass star formation are forced to investigate highly obscured objects with angular resolutions best suited for groups of stars rather individual members. Because the young massive objects are obscured, observations are largely restricted to longer wavelengths where the opacity of the surrounding dust and gas is less, that is, to infrared and radio wavelengths.

IMAGES OF YOUNG MASSIVE STELLAR ASSOCIATIONS

Figure 1 is a VLA 6cm wavelength image of the core of the giant molecular cloud W49 (Welch et al, 1987). The radio free-free emission from about 40 separate (mostly) compact HII regions is evident. This region, at about 11 kpc distance, is probably the richest such association in the galaxy. Sizes of the individual regions are as small as .01 pc, and each is probably an entire O sub-group. The extent of the distribution is about 55 pc. Note the remarkable variety of morphologies among the HII regions.

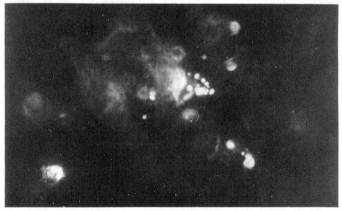

Figure 1. A VLA 6-cm wavelength image of W49.

The findings of Wood and Churchwell (1989), that O stars spend about 15 percent of their MS lives imbedded, argue that the age of the collection of HII regions in Figure 1 is no more than about one half to one million years. The shortness of this time scale raises pointedly the question about the nearly coeval formation of a large association of O sub-groups. Blaauw (1964) has noted the evidence of sequential star formation in the ages of revealed O stars in neighboring sub-groups, and Elmegreen and Lada (1977) have argued that the proximity of some O groups to related oblong molecular clouds is best understood by sequential star formation in which the formation of one sub-group drives ionization waves into the molecular cloud which produce further star formation. While this scheme probably accounts for some of the processes by which fragmentation proceeds, some mechanism must be found for the synchronization of massive star formation in a region as large as the W49 core. If we suppose that the signal which synchronizes the star formation across the core is carried through the gas, a sound speed of 55 km s^{-1} is required to carry the message a distance of 55 pc in a million years. This is large compared to any likely thermal sound speed: neutral gas, ionized plasma, or magnetized plasma. Such speeds are not unusual for runaway O stars, and one might speculate that the synchronization in W49 is due to the passage of such a star. On the other hand, the distribution in Figure 1 is roughly spherical with no obvious linear features as one might expect for such a mechanism.

The million year lifetime is actually long for the ages of individual compact HII regions with dimensions of the order of .05 pc, a typical size for some of the smaller HII regions in W49, for example. If such a region could expand at the plasma sound speed as one would expect in an approximately uniform medium (Spitzer, 1978), its lifetime would be only about 5000 years. To deal with this discrepancy, Van Buren et al (1990) have suggested that most O stars may be moving at several km s^{-1} with respect to their surrounding gas. This supersonic motion will set up a small scale standing shock in front of the star which will keep its small size for as long as the moving star remains in reasonably dense material. This attractive suggestion also explains the cometary appearance of a number of the compact HII regions.

EVIDENCE FOR DISKS

The formation of disks around young stars explains how material of substantial angular momentum is able to collapse onto the stellar core, the formation of planets, and possibly also explains the source of young stellar winds, or at least their collimation (Rodriguez, 1990). For the low mass Pre Main Sequence stars, the T Tauri's and the slightly more massive Ae/Be stars (Herbig, 1960), the principal evidence for the presence of a disk is the overall spectrum, which exhibits the infrared to be expected from a surrounding disk-like structure (Strom et al, 1972).

Strong evidence for disks around O stars is relatively recent.   Two massive stars at relatively different stages of evolution have yielded relevant observational data at unusually high angular resolution.

The remarkable star MWC 349A is a member of a visual binary with a companion which is a BOIII (Cohen et al, 1985). It has a luminosity corresponding to an O9 MS star and an emission line spectrum suggesting circumstellar material. Its stellar wind is slow, 50 km s^{-1}, and massive ($\dot{M} = 10^{-5} M_{\odot}$/yr). A high angular resolution VLA image (White and Becker, 1985) reveals a bipolar morphology to the ionized wind, which suggests the presence of a disk. Leinert (1986) obtained an infrared image using speckle interferometry that showed a .065" elongation in the image consistent with the disk implied by the VLA radio results. Hamman and Simon (1986) found that their high velocity resolution near IR emission spectra are best explained as coming from material in a Keplerian disk. Most intriguing is the recent discovery of maser emission in the 1mm (H32a) recombination line of hydrogen, with two prominent peaks at radial velocities on two sides of the systemic velocity (Martin-Pintado et al,1989; Gordon, preprint).   An important experiment by Planesas et al

(1992) with the Owens Valley Millimeter Wave Interferometer has shown that the two recombination line emission peaks are spatially separated at an angle in agreement with both the VLA image and the speckle image and that the different radial velocities are spatially separated at a distance corresponding to the emitting gas being in Keplerian orbits around the star. The tiny angular separation, .065", could be measured interferometrically because the separate velocity features appear as isolated point sources. Figure 2a from Hamman and Simon (1986) shows the disk morphology implied by the radio map and the near IR spectra. Figure 2b is the double peaked recombination line spectrum mapped by Planesas et al (1992).

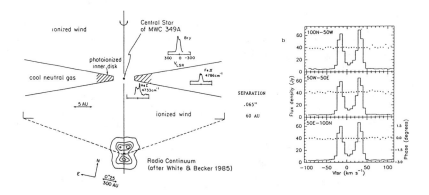

Figure 2. (a) A schematic of the MWC349A disk from Hamman and Simon (1986). (b) The 1-mm wavelength hydrogen recombination line spectrum showing the two spatially separated radial velocity components.

The earlier suggestions of a disk around this star are now borne out by direct imaging. Recent theoretical work by Hollenbach et al (this volume) and Yorke (this volume) shows that material ablated by normal O star winds and ultraviolet radiation from a disk such as that of MWC 349A can attain flows of $10^{-5} M_\odot$/year at velocities of 50 km s^{-1}, both of which are observed for this object. It is also likely that the disk around this star is the remnant of the disk with which the star formed. The association of the star with a B0III companion argues that the pair are at least as old as a few million years. This should prove no difficulty for the suggestion of a primordial disk, since, with the present mass loss rate of $10^{-5} M_\odot$/yr a disk of several tens of solar mass could persist for several million years.

The possibility of a massive disk as a long term source or reservoir for dense material flowing away from an O star suggests another resolution to the problem of the old compact HII regions, discussed above. The massive disk continues to provide dense gas to be ionized close to the star, and if the disk is sufficiently massive initially, the supply may last for a million years or more. From this point of view, MWC349A represents the very late stage of a compact HII region.

The second disk around a massive star for which there is now good direct evidence is the disk surrounding the heavily obscured luminous object IRC2 in the Orion nebula. This young star, whose luminosity may be as large as 10^5 $L_\odot$(Genzel and stutzki, 1989), shows a powerful neutral outflow or wind but only a tiny ionized core. The ionized region may be small because it is trapped by dense gas close to the star. A particularly striking property of IRC2 is that it is the only known young stellar object with strong SiO maser emission (Olofsson et al, 1981). Recently, Plambeck et al(1990) have imaged the various radial velocities of the maser emission with a relative accuracy of .015" and found the striking distribution shown in Figure 3. As in the study of Planesas et al (1992), the fact that each radial velocity component is an isolated point source permitted the unusually high spatial resolution. The spatial velocity pattern is accurately represented by a simple model of maser emission in an expanding, rotating disk. Because of the excellence of the model fit shown in Figure 3, we may regard the disk as spatially imaged in this study. This bears many similarities to the MWC349A disk, but at a younger stage. Even the apparent expansion may be evidence of the ablation which is more clearly seen in the older star. The wind here is also slow and massive and has its maximum velocity, as seen in water masers, normal to the disk. These two examples of disks around O stars appear to be the same phenomena, but at very different stages.

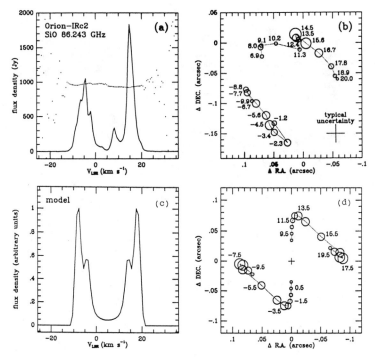

Figure 3.   The Orion SiO Masers.   (a) The observed visibility spectrum. (b) Map of the different radial velocity components. (c) Model visibility spectrum. (d) Model map.

ENVIRONMENTAL EFFECTS

Outflows

The available evidence is that all, or nearly all, newly forming stars show strong outflows (Terebey et al, 1989). Figure 4, taken from the review of Lada(1985), exhibits the run of mechanical luminosity with bolometric luminosity for these outflows. Although the scatter is large, there is a clear trend showing that the more luminous stars have stronger outflows. Lada points out that because of the relatively low speeds and high masses of these outflows, their momenta exceed the radiative momenta of their central stars by several orders of magnitude and are not likely to be driven directly by the stellar radiation. It now seems more likely that the observed outflows are really ambient gas swept up by fast lower density winds which may themselves be driven by stellar radiation pressure or interactions between the star and its disk. Such fast winds are more difficult to detect and have been observed in only one or two instances to date (Lisano et al, 1988).

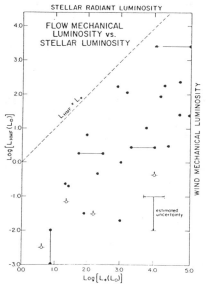

Figure 4. Luminosities of bipolar outflows for both high and low mass stars (Lada, 1985).

The amount of the outflow from a typical O star is consistent with it being just the material that must be pushed away to reveal the star after its having been imbedded in its cloud for about one half to one million years. For the luminous stars in the Ceph A and NGC 7538 regions, for example, the outflow velocities are about 20 km s^{-1} and the mass loss rates are in the range 5×10^{-4} to 1×10^{-3} M$_\odot$/yr. Material moving at 20 km s^{-1} for one million years will traverse 20 pc, a distance needed to uncover an O star from a molecular cloud of that typical size. Also, a mass flow rate of 10^{-3} M$_\odot$/year for a million years yields one thousand M$_\odot$ driven from the star. Note that the core of W49 contains about 40 O stars and has a mass of 40,000 M$_\odot$ (Jaffe et al, 1984). The outflows from the 40 stars should clear away that amount of material in a million years. These numbers are also consistent with the efficiency of star formation being about 1 - 2 percent for the O stars.

Chemistry Effects
The intense radiation fields and the winds from the massive stars do appear to have significant effects on the chemical abundances of molecular gas in their surrounding clouds. Variations in abundances in more quiescent regions are modest, and Figure 5a shows the overlaid maps of NH3 and HC3N obtained by Olano et al (1988) toward the cloud TMC1. The ratio of column depth varies by a factor of about four over the cloud. In contrast to this result, Figure 5b shows maps of a number of different molecules toward the Orion IRC2 region (at one radial velocity) obtained by Plambeck and Wright (in preparation). Here there are differences of as much as two orders of magnitude in the column depths among the different species. These differences are evidently associated with the winds and strong radiation from the imbedded O star. Especially noteworthy is the strong enrichment of HDO (Plambeck and Wright, 1987). Its abundance is about two orders of magnitude greater than would be expected if (a) the solar abundance of oxygen were present in the form of water and (b) the ratio of normal to deuterated water corresponded to the cosmic Helium fraction of 10^{-5}. A large enrichment of the deuterated molecule would not be anticipated at the high gas kinetic temperature in this region, about 250K. One speculation is that the enrichment occurred before the massive stars formed, when the region was cold, and that the ices that formed on the grains are now being driven off into the gas phase by the present winds and radiation.

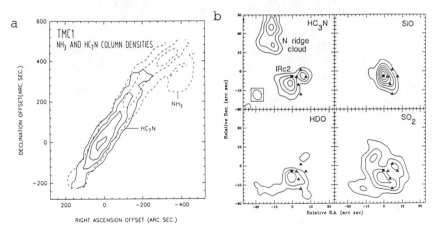

Figure 5. (a) Relative column depths of two molecules in TMC1
(Olano et al, 1988). (b) Various molecular lines toward IRC2
in Orion (Plambeck and Wright).

Effects on the Dust
Emission by interstellar dust, which is observable at wavelengths shorter than 3 mm, provides a valuable tool for the estimation of cloud or clump masses

(Gordon, 1987). Obtaining an accurate estimate of the mass depends critically on knowing the long wavelength emissivity spectrum of the dust, and a reliable theoretical determination of the emissivity spectrum depends on the (unknown) detailed composition of the grains. In general, a steeper spectrum is expected for simple grains as compared with those that might be covered with layers of ices.

One recent study in the Orion region provides fairly convincing evidence of differences in the emissivity spectrum of dust in the neighborhood of the IRC2 source as compared with the dust spectrum in less active regions nearby. Wright et al (Ap. J., in press) have combined Hat Creek interferometer maps at 3mm wavelength with JCMT maps at .35 mm and individual point flux measurements at 1 mm and 20 micrometers to obtain accurate overall spectra of the dust emission in three locations in the Orion region. These spectra have enabled the separation of the dust temperature from the dust emissivity spectrum. Figure 6a shows the .35 mm map and Figure 6b shows the 3mm map, with comparable resolutions of about 7" for both. The dust emissivity spectra are shown in Figure 6b for the three principal clumps. The spectral differences are significant and exhibit a steeper spectral index for the clump near to IRC2. The obvious suggestion is that grains near IRC2 have been modified by having their mantles removed by either the winds or the high temperature, in agreement with the speculation about the HDO enrichment above.

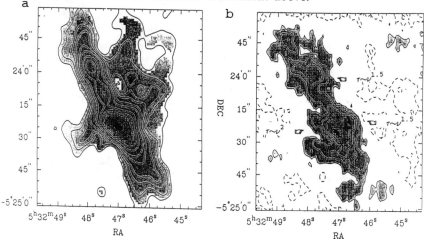

Figure 6. (a) .35 mm map of the Orion ridge. (b) 3mm map of the same region. The dust spectra are shown by the hands.

EVIDENCE FOR PROTOSTELLAR COLLAPSE

Although the collapse of material to form individual stars is at present virtually unobservable because of the limited angular resolution of current telescopes, there has been some recent evidence of collapse of material to form entire stellar associations. The practical difference is that the mass of a single star is too small to produce significant accelerations at any but unobservably small distances, whereas the many thousand solar masses of gas associated with a newly forming O association can produce accelerations that can be distinguished from small random gas motions at significant distances. Keto et al (1987) reported observations of material falling onto a stellar association in G 10.6 . Welch et al (1987) observed inverse P-Cygni profiles toward the core of W49 (see Figure 1) suggesting the remnant of the large collapse that may have resulted in the simultaneous formation of many of the O groups in the core. Similar results were reported by Rudolph et al for W51 (1990) and by Carral and Welch (1992) for G34.2+0.2.

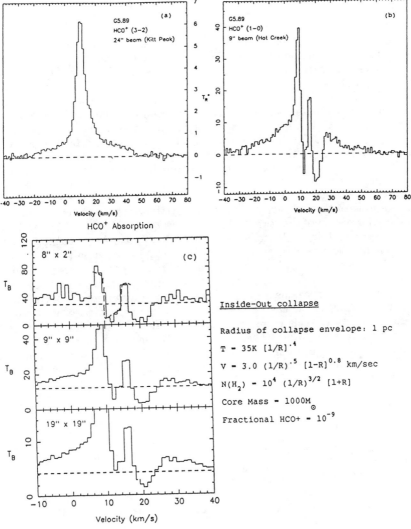

Figure 7. (a) HCO+(3-2) spectrum toward G5.89. (b) HCO+(1-0) spectrum in a smaller beam. (c) HCO+(1-0) spectra at different spatial resolutions. The parameters of the dashed line model fit for the spectrum of an "inside-out" collapse are at the right.

An example of more recent results comes from the study of the isolated compact HII region G5.89 by Forster and Wilner (in preparation). Detailed VLA continuum maps of this region have been obtained by Wood and Churchwell (1989); Forster and Wilner's data is from the BIMA array and the 12m NRAO telescope. Figure 7a shows a spectrum in a 24″ beam in the HCO+(3-2) transition toward this object. Figure 7b is a spectrum in the lower energy HCO+(1-0) transition revealing strong red- shifted absorptions of both background line and the continuum of the HII region. There is no absorption of the blueshifted side of the line. The three panels on the left side of Figure 7c show the absorptions as measured in synthesized beams of three different sizes. There is a significant difference in the way the two absorptions appear at the three resolutions. The strength of the redder absorption system does not change relative to the continuum with beam size, indicating that there is no emission in this system near the source. This places this gas well in the foreground of the source; it is probably not dynamically related to G5.89. On the other hand, the lower velocity system absorption fills in in the larger beams. This means that there is low level emission in this system from the source. Since molecular hydrogen densities of $10^5 cm^{-3}$ or more are required to excite this gas above the 3 degree background, this absorbing gas must be part of the core. It is probably dynamically related to the source.

To summarize, we observe red shifted gas in the core in front of the HII region and blue shifted core gas behind it. The simplest interpretation, as in the cases sited above, is that there is an overall collapse of material onto the neighborhood of the HII region. It is probably the remnant of the collapse that led to the formation of the stars within the HII region. The dotted line in Figure 7c is a theoretical fit to the collapse based on the "inside-out" collapse picture of Shu (1977). The parameters of the model are summarized on the right side of Figure 7c. These results, along with those sited above, provide growing evidence for large scale collapse onto systems of O sub-groups; such coherent collapses could provide the mechanism for synchronizing the formation of O groups.

Whereas, in all the cases sited there is the red shifted absorption that we associate with large scale infall, we see little evidence of the blue shifted gas of the outflows in absorption. The important difference is that the outflows are collimated, and there is only a small probability of the blue wing of one being in front of an HII region; the infall, on the other hand, must be essentially spherically symmetric.

CONCLUSIONS

1. Newly formed O stars are buried in the clouds in which they form during the first 15 percent of their MS lives, and they are distant and can seldom be observed with high linear resolution.
2. Disks which form along with the massive stellar cores may provide the long lasting sources of dense ionized gas which appear as compact HII regions for intervals of up to a million years.
3. Imbedded O stars strongly modify the properties of the gas and dust in their neighborhoods.

4. There is increasing evidence of large scale collapse prior to the formation of O groups; the collapse may synchronize the nearly coeval formation of the O groups.

REFERENCES

Blaauw, A. 1964, Ann. Rev. Astron. Astrophys., 2, 213.
Carral, P., and Welch, W. J. 1992, ApJ, 385, 244.
Cohen, M., Bieging, J. H., Dreher, J. W., and Welch, W. J. ApJ, 292, 249.
Elmegreen, B. G., and Lada, C. J. 1977 ApJ, 214, 725.
Garmani, C. D., Conti, P. S., and Chiosi, C. 1982 ApJ, 263, 777.
Genzel, R. and Stutzki, J. 1989, Ann. Rev. Astron. Astrophys., 27, 41.
Gordon, M. A. 1987, ApJ, 316, 258.
Hamman, F., and Simon, M. 1986, ApJ, 311, 909.
Herbig, G. 1960, ApJS, IV, 337.
Humphreys, R. H. 1984, in IAU Symp. 105, eds. A. Maeder and A. Renzini.
Jaffe, D. T., Becklin, E. E., and Hildebrand, R. H. 1984. ApJ, 279, L51.
Keto, E. R., Ho, P. T. P., and Hashick, A. D. 1987 ApJ, 318, 712.
Lada, C. J. 1985, Ann. Rev. Astron. Astrophys., 23, 267.
Leinart, C. 1986, A&A, 155, L6.
Lisano et al. 1988, ApJ, 328, 763.
Martin-Pintado, J., Bachiller, R., Thum, C., and Walmsley, M. 1989, A&A, 215, L13.
Olano, J., Walmsley, M., and Wilson, T. L. 1988, A&A, 196, 194.
Olofson, H., Hjalmarson, A., and Rydbeck, O. E. H. 1981, A&A, 100, L30.
Plambeck, R. L., and Wright, M. C. H. 1987, ApJ, 317, L101.
Plambeck, R. L., Wright, M. C. H., and Carlsrom, J. E. 1990, ApJ, 348, L65.
Planesas, P., Martin-Pintado, J., and Serabyn, E. 1992, ApJ, 386, L23.
Rodriguez, L. 1990. in The Evolution of the Interstellar Medium, PASP, Conference Series, ed L. Blitz.
Rudolph, A., Welch, W. J., Palmer, P., and B. Dubrulle. 1990, ApJ, 363, 528.
Shu, F. 1977, ApJ, 214, 488.
Spitzer, L. 1978, Physical Processes in the Interstellar Medium,
New York: Wiley.
Strom, S. E., Strom, K. M., Yost, J., Carrasco, L., and Grasdalen, G. 1972, ApJ, 173, 353.
Terebey, S., Vogel, S. N., and Myers, P. C. 1989, ApJ, 340, 472.
Van Buren, D., Mac Low, M., Wood, D. O. S., and Churchwell 1990, ApJ, 353, 570.
Welch, W. J., Dreher, J. W., Jackson, J. M., Terebey, S., and Vogel, S. N. 1987 Science, 238, 1550.
White, R. L., and Becker, R. H. 1985, ApJ, 297, 677.
Wood, D. O. S., and Churchwell, E. 1989, ApJS, 69, 831.

Massive Stars: Their Lives in the Interstellar Medium
ASP Conference Series, Vol. 35, 1993
Joseph P. Cassinelli and Edward B. Churchwell (eds.)

PHOTOEVAPORATION OF DISKS AROUND MASSIVE STARS AND ULTRACOMPACT HII REGIONS

D HOLLENBACH
NASA Ames Research Center

D JOHNSTONE and F SHU
University of California, Berkeley, Ca

ABSTRACT Young massive stars produce sufficient Lyman continuum luminosity ϕ to significantly affect the structure and evolution of the accretion disks surrounding them. In the absence of a stellar wind, a nearly static, photoionized, 10^4 K, disk atmosphere, with a scale height that increases with $r^{3/2}$, forms inside a gravitational binding radius $r_g \approx 10^{15}$ $(M_*/10\ M_\odot)$ cm where M_* is the mass of the central star. For $r \gtrsim r_g$ the diffuse field produced by hydrogen recombinations to the ground state in the atmosphere produces a steadily evaporating disk. The outer region of the disk then has a mass-loss rate of order $3\times10^{-5}\phi_{49}^{1/2}$ $M_\odot$ yr^{-1}, where $\phi_{49} = \phi/(10^{49}$ Lyman continuum photons s^{-1}). In the presence of a stellar wind, we define the wind as strong or weak depending on whether r_w is greater or less than r_g, with r_w being the radius where the ram pressure from the stellar wind balances the thermal pressure of the flowing disk material. A strong stellar wind largely clears away the photoionized atmosphere for $r < r_g$, while the disk evaporates roughly as before for $r > r_w$. Since $r_w \approx 2 \times 10^{16}\phi_{49}^{0.88}v_{w8}^2$ cm typically, where v_{w8} is the wind velocity in units of 1000 km s^{-1}, most O-star winds can be considered strong. The mass-loss rate from the photoevaporating disk for this situation is of order $5\times10^{-5}\phi_{49}^{0.44}v_{w8}$ $M_\odot$ yr^{-1}. In either case, the mass loss has important consequences. The resulting ionized wind from the disk, which blows relatively slowly at 10–50 km s^{-1} and persists for $\sim 10^{5.5}$ yr if the disk mass $M_d \sim 0.3M_*$, may explain the observational characteristics of ultracompact HII regions and their long inferred lifetimes.

INTRODUCTION

Ultracompact HII regions (UCHIIs) are small dense regions of ionized hydrogen first observed by Ryle and Downes (1967). Surveys conducted by Wood and Churchwell (1989a,b) have vastly improved our understanding of these mysterious objects. Using the VLA at 2 and 6 cm, with a beam of 0.4

arcsec, Wood and Churchwell (1989b) found that the peak emission measure (EM) for UCHIIs averages about 10^7–10^8 cm^{-6} pc. Combined with their small sizes, typically less than 10^{17} cm, the high emission-measures require electron densities, $n_e > 10^4$–10^5 cm^{-3}.

The central sources for the ionizing radiation cannot be seen optically, but their properties can be deduced from measurements at radio and far-infrared wavelengths. The first gives the Lyman-continuum luminosity from an estimate for the total amount of ionized gas (assumed to have a temperature $\sim 10^4$ K); the second gives the bolometric luminosity from an estimate for the total amount of reprocessing of starlight by circumstellar dust. Although great uncertainty attaches to any individual object, these two observations yield central sources consistent with B0–O5 main-sequence stars.

Wood and Churchwell (1989a) estimate that 10–20% of all O stars are in the UCHII phase. Given that O stars live approximately $10^{6.5}$ yr, this estimate implies that UCHIIs should exist for $\sim 10^{5.5}$ yr. On the other hand, the high densities and temperatures of these regions compared to the molecular cloud as a whole suggest that UCHIIs are overpressured and should expand at velocities in excess of 10 km s^{-1}, leading to dynamical lifetimes $\lesssim 10^{17}$ cm/10^6 cm s^{-1} $\sim 10^{3.5}$ yr. Thus, a paradox of two orders of magnitude exists concerning the inferred longevity of these regions. Although a number of models have been described in the literature: bow shocks (van Buren *et al.* 1990), ram-pressure confinement by accretion flows (Reid *et al.* 1981), "champagne" flow (Tenorio-Tagle 1979, Yorke *et al.* 1983), no completely satisfactory explanation for the vast number of UCHIIs has emerged.

We propose that many UCHII regions form by the photoionization and photoevaporation of neutral accretion disks that orbit newly-born OB stars. Thus, UCHII regions live for $10^{5.5}$ yr because they are constantly being replenished by a dense circumstellar reservoir – the orbiting disk. Although the high-pressure gas expands away from the disk at speeds of 10–50 km s^{-1}, the reservoir maintains a time-independent density profile and emission measure. The emission measure is dominated by the densest gas found near the origin of the photoevaporated flow off of the disk surface. Our study shows that we may distinguish between two conceptual extremes. When a stellar wind is absent or weak, the high-mass star irradiates the disk, creating a hot ionized atmosphere inside a gravitational radius r_g, defined by where the isothermal sound speed $c_s \approx 12$ km s^{-1} of the HII region equals the escape speed from the system. Most of the mass loss occurs from the flow of photoevaporated disk material which lies just outside of r_g. When a stellar wind is present and strong, the ram pressure from the wind clears much of the atmosphere inside r_g, and most of the disk material photoevaporates beyond a wind radius r_w, defined by where the ram pressure of the wind equals the thermal pressure of the ionized disk material. For a typical massive star, the strong-wind case applies, and the observed physical size of the UCHII region is of order $2r_w \sim 10^{16}$–10^{17}cm, assuming the disk extends to those kinds of distances.

Our proposal that OB stars are born with disks is consistent with observations of masering by hydrogen recombination lines from the Be star, MWC 349A, by Martín-Pintado *et al.* (1989a, 1989b) and Gordon (1992). These authors interpret the double-peaked masering of the H29α, H30α, and H31α lines to originate from a Keplerian disk, at a radius of $r \approx 20$–30 AU. The masering requires a large electron number density near the disk, $n_e > 10^7$ cm^{-3}.

Moreover, continuum observations from 100 μm to 20 cm reveal a power-law slope of $L_\nu \propto \nu^{0.6}$. As we shall discuss below, the continuum measurements can be explained by an optically thick, ionized wind with number density $n_e \propto r^{-2}$. Finally, a slow disk-wind interpretation is consistent with the finding of Altenhoff *et al.* (1981) of a velocity width, $v \approx 50$ km s^{-1}, for the ionized flow associated with this massive star.

DISKS AROUND YOUNG STARS

The current paradigm for star formation begins with the gravitational collapse of a molecular cloud core and the formation of a protostar with an orbiting (accretion) disk (see Adams, this volume). For low-mass stars, considerable observational evidence has accumulated for the existence of these disks, most notably through observations of infrared excesses and spectral signatures indicative of inner-disk kinematics. In addition, blueshifted emission from protostellar winds indicates disk obscuration of the receding portion of the outflow (Edwards *et al.* 1987), and the thermal emission observed at millimeter and submillimeter wavelengths has been interpreted as arising in the outer regions of these dusty disks (*e.g.*, Beckwith *et al.* 1990). Around low-mass stars, the disks are observationally inferred to be of size $r_d \approx 10^2$ AU and mass $M_d \approx 10^{-2}$–10^{-1} $M_\odot$. Recently, Beckwith and Sargent (1992) have analyzed isotropic line emission from the $J = 1$–0 transition of CO and have concluded that around several T Tauri stars, the low-density portions of their disks may extend to at least 400 AU. Hillenbrand *et al.* (1992) have recently modeled the infrared emission from higher-mass Ae and Be stars, and have shown that these objects have larger and more massive disks than low-mass pre-main-sequence stars. We anticipate that these trends will continue into regions of OB star formation.

Theoretical collapse calculations also show that disk formation constitutes a highly likely stage in the creation of a star, due to the angular momentum in the dense molecular cloud undergoing gravitational collapse. In many models the infalling material falls first onto the disk and then accretes through the disk. The accretion disk becomes gravitationally unstable in a global sense when it reaches a mass $\approx 0.3\ M_*$, where M_* is the mass of the star (Shu *et al.* 1990). The instability may cause rapid accretion onto the star; thus, during the collapse of the cloud core, the disk mass may be maintained close to the value $0.3\ M_*$. Once the accretion phase ends and the disk mass falls below the critical value, disk accretion onto the star may rapidly decline and other mechanisms (such as photoevaporation) may disperse the remaining gas and dust.

In summary, it appears likely that spatially thin, neutral disks of size $r_d \gtrsim 10^{15}$–10^{16} cm and mass $M_d \approx 0.3\ M_* \approx 3$–$15\ M_\odot$ exist around young massive stars. These disks should act as a dense source of hydrogen, capable of being photoionized and subsequently evaporated by the UV photon flux from the luminous central star. We contend that this picture constitutes the essence of an UCHII region.

WEAK-WIND MODEL AND RESULTS

Weak stellar wind

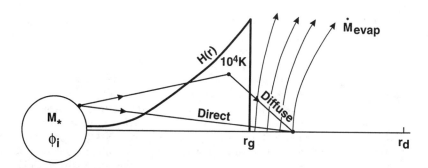

FIGURE 1: Schematic for the weak-wind model. Inside r_g an ionized 10^4 K atmosphere forms with scale height H. Material evaporates beyond r_g.

In the case of a weak wind, photoionization of the neutral hydrogen on the surface of the inner region of the disk should result in the formation of a bound ionized atmosphere with a characteristic temperature of $T \approx 10^4$ K. Beyond the gravitational radius r_g, the ionized hydrogen above the disk can flow at $v \gtrsim 10$ km s^{-1} into the interstellar medium. Figure 1 gives a schematic representation of the situation described above. If we assume ejection of the ionized hydrogen from the disk at the sound speed c_s, the radius out to which the material is bound by the central star equals

$$r_g \approx 10^{15} \frac{M_*}{10 \ M_\odot} \ \text{cm}.$$

Inside this radius the atmosphere can be approximated as hydrostatic and isothermal. The density of ionized hydrogen in the inner region depends on the base density along the disk, $n_0(r)$, and the height above the disk. The scale height reads

$$H(r) = r_g \left(\frac{r}{r_g}\right)^{3/2} .$$

Note that $H(r_g) = r_g$. The number density at any point in the bound atmosphere satisfies the hydrostatic law,

$$n(r, z) = n_0(r) \exp\left[-z^2/H(r)^2\right] .$$

Outside of r_g we assume the evaporated material to flow from the disk at the sound speed c_s; thus $n(r, z) \approx n_0(r)$ yields a reliable estimate near the disk where the flow has little opportunity to expand. We find the mass loss rate from the disk, $\dot{M}_d$, by considering the material which evaporates from beyond r_g,

$$\dot{M}_d = 2m_H c_s \int_{r_g}^{\infty} 2\pi n_0(r) r \, dr,$$

where the first factor of 2 accounts for both faces of the disk.

In order to calculate the base density $n_0(r)$, consideration must be given to the radiation field incident on the disk at r. Very close to the star the direct UV radiation from the central star will maintain a dense atmosphere with $n_0(r) \propto r^{-9/4}$. Farther from the central star the scale height increases (see Fig. 1), and the atmosphere intercepts more of the direct radiation. A diffuse field of ionizing radiation is generated from approximately one-third of the recombinations that go directly to the ground state. This diffuse field dominates the radiation seen at the base of the disk, leading to an atmosphere whose base density drops as $n_0(r) \propto r^{-3/2}$. Two reasons account for the dominance of the diffuse field over the direct UV flux from the star. First, a large volume of material exists above the disk that creates a substantial amount of diffuse radiation. Second, the very dense but spatially thin atmosphere above the disk and close to the star produces a large foreground attenuation of the direct UV radiation field. Modeling of this situation (Hollenbach *et al.* 1992) reveals that the base density in the static region equals

$$n_0(r) = 4 \times 10^7 \phi_{49}^{1/2} M_1^{-3/2} \left(\frac{r_g}{r}\right)^{3/2} \text{ cm}^{-3},$$

where ϕ_{49} is the number of Lyman continuum photons per second from the central star in units of 10^{49} s^{-1} and M_1 is the mass of the star in units of 10 $M_\odot$. The exponent $3/2$ results because it corresponds to the marginal state between an ionization-bounded and a density-bounded HII region.

Beyond the static region the ionized hydrogen is maintained primarily from diffuse photons produced in the atmosphere at r_g. The power-law density profile steepens to $n(r) \propto r^{-\alpha}$, where $\alpha \gtrsim 5/2$. Most of the mass loss occurs, then, in the region just outside of r_g, which results in

$$\dot{M}_d = 3 \times 10^{-5} \phi_{49}^{1/2} M_1^{1/2} \ M_\odot \ \text{yr}^{-1}.$$

STRONG-WIND MODEL AND RESULTS

In the case of a strong wind from the central star, the ram pressure of the stellar wind, $\rho_w v_w^2$, will be too high for a static atmosphere to achieve a full scale height above the disk. Instead, the stellar wind will significantly suppress the height of the ionized layer out to a radius $r_w > r_g$, where the thermal pressure of the ionized hydrogen from the disk balances $\rho_w v_w^2$. The situation is depicted in Figure 2. The radius r_w, beyond which the photoevaporated material freely flows vertically off the disk, is given by

$$r_w = 2 \times 10^{16} \dot{M}_{w-6}^2 v_{w8}^2 \phi_{49}^{-1} \text{ cm},$$

where $\dot{M}_{w-6}$ is the mass loss rate in the stellar wind in units of 10^{-6} $M_{\odot}$ yr^{-1} and v_{w8} is the wind velocity in units of 1000 km s^{-1}. The criterion for a strong wind, $r_w > r_g$, can be written

$$\dot{M}_{w-6} v_{w8} > 0.22 \, \phi_{49}^{1/2} M_1^{1/2}.$$

This criterion is generally met for O and B stars (van Buren 1985).

Strong stellar wind

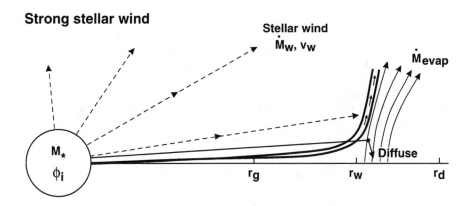

FIGURE 2: Schematic for the strong-wind model. Material evaporates beyond r_g but the dominant flow is from r_w, where the wind ram pressure equals the thermal pressure of the flow.

At r_w the direct and diffuse radiation fields discussed in the proceeding section will create a density of the flowing material of (Hollenbach *et al.* 1992)

$$n_0(r_w) = 4 \times 10^5 \phi_{49}^2 \dot{M}_{w-6}^{-3} v_{w8}^{-3} \text{ cm}^{-3}.$$

The ionized density drops rapidly beyond r_w as the direct field is attenuated by the flowing plasma. The photoevaporation rate from the disk now reads

$$\dot{M}_d \approx 5 \times 10^{-5} \dot{M}_{w-6} v_{w8} \phi_{49}^{-1/2} M_{\odot} \text{ yr}^{-1}.$$

The naive inference that $\dot{M}_d$ and r_w decrease with increasing ϕ holds only for a constant mass-loss rate in the stellar wind. In fact, the stellar-wind mass-loss rate tends to scale with ϕ. For B0–O5 stars, van Buren (1985) and Panagia (1973) show that, with considerable scatter,

$$\dot{M}_{w-6} \approx \phi_{49}^{0.94}.$$

Thus,

$$\dot{M}_d = 5 \times 10^{-5} \phi_{49}^{0.44} v_{w8} M_{\odot} \text{ yr}^{-1},$$

and

$$r_w = 2 \times 10^{16} \phi_{49}^{0.88} v_{w8}^2 \text{ cm.}$$

These results imply that massive and luminous stars have higher photoevaporation rates and larger characteristic radii r_w from which the photoevaporation originates. We emphasize that these approximate analytic solutions hold only when the disk extends to $r_d > r_w$, and when the thickness of the neutral disk is negligible.

APPLICATION TO OBSERVED ULTRACOMPACT HII REGIONS

We now show that photoevaporating disks can explain many of the general features of observed UCHII regions. We begin with their measured sizes. The size of a photoevaporating disk, as seen in the radio continuum, depends on whether the cantral star has a strong or weak wind. When the wind is weak, the emission measure peaks around $r_g \approx 10^{15}$ cm. Such sources would appear unresolved in the Wood and Churchwell survey. When the wind is strong, the emission measure will peak near $r_w > 2 \times 10^{16}$ cm. Nearby sources with this property would correspond to resolved UCHIIs.

For the weak-wind case, the predicted emission-measure averaged over a 0.4 arcsec beam reads

$$EM \approx 10^8 \dot{M}_{w-5}^{4/3} \left(\frac{3\text{cm}}{\lambda} \right)^{0.7} \left(\frac{5\text{kpc}}{D} \right)^2 \text{ cm}^{-6}\text{pc} \qquad \text{for} \qquad \lambda \underset{\sim}{<} 1 \text{ mm,}$$

consistent with the observations of unresolved sources. The radio emission is optically thick for $\lambda \underset{\sim}{>} 1$ mm, and thus the radio luminosity from the UCHII region should vary as

$$L_\nu \propto \nu^{0.6} \qquad \text{for} \qquad \lambda \underset{\sim}{>} 1 \text{ mm.}$$

The strong-wind model predicts an emission measure of

$$EM \approx 1.2 \times 10^9 \phi_{49}^{-0.76} v_{w8}^{-4} \text{ cm}^{-6}\text{pc,}$$

The luminosity will vary as

$$L_\nu \propto \nu^{-0.1} \qquad \text{for} \qquad \lambda \underset{\sim}{<} 6 \text{ mm} \qquad \text{if optically thin,}$$

$$L_\nu \propto \nu^{0.6} \qquad \text{for} \qquad \lambda \underset{\sim}{>} 6 \text{ mm} \qquad \text{if optically thick.}$$

At long wavelengths the source size can get quite large. On the other hand, if the photoevaporation flow is stopped by interaction with ambient gas, L_ν will ultimately attain a ν^2 dependence.

The UCHII region persists as long as the disk survives the mass loss. For both the weak and strong stellar-wind models, the characteristic lifetime of a photoevaporating disk is given by the mass of the disk divided by the rate of evaporation of the disk, $\tau = M_d/\dot{M}_d$. Using the formulae of the previous section, we get

$$\tau \approx 3 \times 10^4 \phi_{49}^{-1/2} M_1^{-1/2} M_d \text{ yr (weak winds),}$$

$$\tau \approx 2 \times 10^4 \phi_{49}^{-0.44} v_{w\,8}^{-1} M_d \text{ yr (strong winds)},$$

if we measure M_d in solar masses. Thus, lifetimes of $10^{5.5}$ years can readily be achieved by 3–15 $M_\odot$ disks.

Consider now a specific application to the source MWC 349A. We claim that this object can be understood as a photoevaporating disk around a star with a relatively weak stellar ($\lesssim 1000$ km s^{-1}) wind. Both the emission lines, due to recombinations in the photoevaporating (disk) wind, and the continuum radiation, from the optically thick, free-free emission of this unresolved wind, are well produced by such a model. The high number density in the wind is also consistent with the photoevaporation model, and the density distribution, $n_e \propto r^{-2}$, is consistent with the disk evaporation becoming a spherical wind at distances large compared to r_g. The velocity of ejection of the evaporating disk material should be of order 12 km s^{-1}, and the pressure gradients in this freely expanding material can increase the velocity by a factor of a few, and make the flow at ~ 50 km s^{-1} quasi-spherical. Thus, MWC 349A may constitute the best-studied individual example of a photoevaporating disk.

SUMMARY

In this paper we have developed the idea that UCHII regions arise as evaporating, photoionized plasmas flowing freely from the surfaces of neutral disks around high-mass stars. The model follows naturally from the current paradigm that *all* stars are born with disks. High-mass stars must photoionize these appendages. If the disks contain a dense reservoir of neutral hydrogen gas in a spatially thin and radially extended layer, the results of our calculations depend only on measurable parameters of the central star, and not at all on the (uncertain) properties of the interior of the disk. In particular, due to the increasing height of the ionized plasma above the disk, a significant percentage of the UV radiation from the central star is intercepted before reaching great distances. This radiation is absorbed and reprocessed, with much of the diffuse ionizing flux irradiating the disk at $r_g \approx 10^{15}$ cm (weak stellar wind), or $r_w \gtrsim 2 \times 10^{16}$ cm (strong stellar wind). The photoevaporation of the disk at nearby radii leads to mass-loss rates on the order $\gtrsim 10^{-5}$ $M_\odot$ yr^{-1}. The emission measures from the models are then $EM \gtrsim 10^8$ cm^{-3} pc averaged over a 0.4 arcsec beam for sources at distances of order 5 kpc. The size of the dominant emission region is of order r_g (weak stellar wind), and r_w (strong stellar wind). Inferred lifetimes for UCHII regions exceed the requisite $10^{5.5}$ yr if disk masses exceed ~ 3 $M_\odot$, as has been suggested by both observational and theoretical studies of star-forming regions.

REFERENCES

Altenhoff, W. J., Strittmatter, P. A., Wendker, H. J. 1981, A&A, 93, 48.
Beckwith, S. V. W. and Sarget, A. I. 1992, ApJ, in press.
Beckwith, S. V. W., Sargent, A. I., Chini, R. S., Güsten, R. 1990, AJ, 99, 924.
Edwards, S. , Cabrit, S., Strom, S. E., Heyer, I., Strom, K. M., Anderson, E. 1987, ApJ, 321, 473.

Gordon, M., 1992, ApJ, in press.

Hillenbrand, L. A., Strom, S. E., Keene, J. Vrba, F. J. 1992, ApJ, in press.

Hollenbach, D., Johnstone, D., Lizano, S., Shu, F. 1992, in preparation.

Martín-Pintado, J., Bachiller, R., Thum, C., Walmsley, M. 1989a, A&A (letters), 215, L13.

Martín-Pintado, J., Thum, C., Bachiller, R. 1989b, A&A (letters), 222, L9.

Panagia, N. 1973, AJ, 78, 929.

Reid, M. J., Haschick, A. D., Burke, B. F., Moran, J. M., Johnston, K. J., Swenson Jr., G. W. 1981, ApJ, 239, 89.

Ryle, M. and Downes, D. 1967, ApJ (letters, 148, L17.

Shu, F., Tremaine, S., Adams, F., and Ruden, S. 1990, ApJ, 358, 495.

Tenorio-Tagle, G. 1979, A&A, 71, 59.

Thum, C., Martín-Pintado, J., Bachiller, R. 1992, A&A, 256, 507.

van Buren, D. 1985, ApJ, 294, 567.

van Buren, D., MacLow, M.-M., Wood, D. O. S., Churchwell, E. 1990, ApJ, 353, 570.

Wood, D. O. S. and Churchwell, E. 1989a ApJ, 340, 265.

Wood, D. O. S. and Churchwell, E. 1989b ApJS, 69, 831.

Yorke, H. W., Tenorio-Tagle, G., Bodenheimer, P. 1983, A&A, 127, 313.

Massive Stars: Their Lives in the Interstellar Medium
ASP Conference Series, Vol. 35, 1993
Joseph P. Cassinelli and Edward B. Churchwell (eds.)

OBSERVATIONS OF NEWLY FORMED MASSIVE STARS

ED CHURCHWELL
Department of Astronomy, University of Wisconsin-Madison
475 N. Charter St., Madison, WI 53706

ABSTRACT Evidence for excitation by stellar clusters and absorption of stellar UV by dust within UC HII regions is discussed. It is shown that a size-density relationship appears to hold for spherical and unresolved UC HII regions, but no such relationship is apparent for cometary and core/halo nebulae. UC HII regions trace the Local spiral arm quite well, but no enhancement is seen in regions of the Sagittarius and Scutum arms inferred from CO emission. Evidence for hot, dense, molecular gas surrounding UC HII regions is discussed along with indications of chemical anomalies and bipolar outflows associated with these regions. Finally, a brief discussion of the cometary nebula G29.96-0.02 is given.

INTRODUCTION

Newly formed, massive stars interact strongly with their environments. The ionization state, atomic and molecular constituents, temperature, density, radiation field, dust properties, relative abundances and properties of the circumnebular molecular gas, and the velocity structure around such stars change dramatically over very short distances (≤ 0.1 pc). The interactions such stars have with their environments have been summarized by Churchwell (1990, 1991). Ultracompact (UC) HII regions are observational manifestations of newly-formed, massive stars that are still embedded in their natal molecular clouds. In this review, I will concentrate on new developments since the Churchwell (1991) review. In particular, in §II, ionization and heating by stellar clusters and absorption by dust of stellar UV radiation are discussed; in §III, a possible correlation between mean electron density and size is examined; in §IV, UC HII regions as tracers of galactic spiral arms is considered; in §V, properties of the circumnebular molecular gas is reviewed; and, in §VI, the bow shocked nebula G29.96-0.02 is discussed.

STAR CLUSTERS AND DUST

Are newly formed, massive stars that ionize UC HII regions accompanied by an associated cluster of lower mass stars? The nature of UC HII regions and their environs depend fundmentally on this queston. For example, the radiation spectrum, ionization structure, and the inferred spectral type of the ionizing star critically depend on whether a single star or a cluster of stars provide the luminosity, ionization, and heating of the gas and dust. Kurtz, Churchwell, and Wood (1992, hereafter KCW) have used infrared (IR) and radio observations, combined with model atmospheres and estimates of the initial mass function (IMF) to determine if stellar clusters are required to understand the energetics of UC HII regions. In the following, we outline their proceedure.

In the Rayleigh-Jeans regime, the free-free flux density of a spherical, homogeneous, optically thin HII region at distance D depends on the stellar ionizing photon flux, N_c^* in photons s^{-1}, as

$$S_v(Jy) = 2.10x10^{-49} \, \xi N_c^* a(v,T_e) \left(\frac{v}{GHz}\right)^{-0.1} \left(\frac{T_e}{K}\right)^{0.45} \left(\frac{D}{kpc}\right)^{-2} \tag{1}$$

where ξ is the fraction of N_c^* that goes into ionization of the gas, as opposed to absorption by dust or escape from a density-bounded nebula. Ionization boundedness is a good approximation for UC HII regions, thus ξN_c^* photons are absorbed by gas while $(1-\xi) \, N_c^*$ photons are absorbed by dust in the HII region.

The far infrared (FIR) flux density, in contrast, is produced by thermal emission from dust heated by all wavelengths of stellar radiation. The dust ultimately reduces all stellar emission to IR wavelengths such that the luminosity at frequency v is

$$L_v = f_v L^* = 4\pi D^2 S_v(\Delta v) \tag{2}$$

where f_v is the fraction of L^* emitted at frequency v in the passband Δv, and S_v is the flux density at v. Using the above relations, the ratio of 100 µm to 2 cm flux densities becomes

$$\frac{S_{100\mu m}}{S_{2cm}} = 3.2x10^{47} \frac{f_{100\mu m}(L^*/L_o)}{\xi N_c^*} \tag{3}$$

where it is assumed that $T_e = 10^4$ K, $v(\text{radio}) = 15$ GHz, $a(v, T_e) = 0.98$, and $\Delta v(100 \text{ µm}) = 9.85x10^{11}$ Hz (the IRAS 100 µm bandpass). UC HII regions typically have their maximum radio flux density near 15 GHz where the optically thin approximation is valid and the maximum in the IR occurs at ~100 µm. Atmospheric models of early type stars shows that $\log(S_{100\mu m}/S_{2cm})$ is a strong function of spectral type, ranging from ~3 for an O4 ZAMS star to ≤ 6 for a B3 ZAMS star. Absorption of UV radiation by dust in the HII region will decrease the radio flux, driving the IR/radio ratio higher. In Figure 1 (taken from KCW), total stellar luminosities (from integrated IR flux densities) is

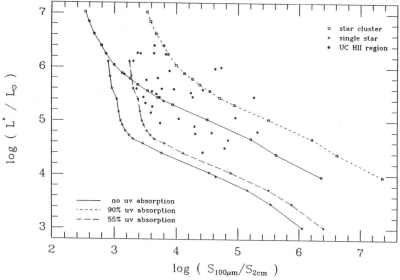

Figure 1 Total luminosity versus the 100µm/2cm flux density ratio.

plotted against $S_{100\mu m}/S_{2cm}$; the lowest solid curve delineated by *s is the predicted relationship for ionization by single stars with spectral types ranging from O4 (upper left) to

B3 (lower right) ZAMS stars using Panagia (1973) model predictions; the dashed curve with the same shape, but shifted to the right is for ionization by single stars, but with 55% of the stellar UV flux absorbed by dust in the HII region. The solid circles are observed values. This figure dramatically illustrates that for a given $S_{100\mu m}/S_{2cm}$ value, the observed luminosities are well in excess of that expected for excitation by a single star. Absorption by dust in the HII region shifts the expected values closer to the observations, but there is a limit to the amount of absorption that can occur before the HII region can no longer be detected at 2 cm. Even with 55% of the UV radiation absorbed by dust, all but two nebulae lie well above the single star curve.

If the ionizing star(s) that produces the UC HII region is (are) accompanied by a cluster of lower mass, non-ionizing stars, then L_* will be greater than that of a single star with the same S_{100vm}/S_{2cm} value. This is shown in Figure 1 by the solid curve with open square symbols. The points on this curve correspond to star clusters with upper mass limits ranging from the equivalent of an O4 (upper left) to a B3 (lower right) ZAMS star. A Miller-Scalo IMF (Miller and Scalo, 1979), the mass-luminosity (M-L) relation from Allen (1973), and the model atmosphere results of Panagia (1973) were used to calculate the curve for star clusters with different upper mass cutoffs. The dashed curve with open squares represents the same stellar clusters as the solid curve, but with 90% of the stellar UV flux absorbed by dust. Clearly, the combination of excitation by star clusters with varying amounts of UV absorption by dust seems to be able to account for the observations quite well. They caution, however, that the calculated curves are sensitive to uncertainties in the stellar models, the IMF, and the M-L relation (all of which are poorly determined for O-stars), while the observations are sensitive to possible diffuse emission detected by IRAS but not by the VLA (likely to cause factors of 2-3 errors but not an order of magnitude error). If taken at face value, Figure 1 implies that ≤ 40% of UC HII regions are likely to be excited by single stars while ≥ 60% require an associated cluster of stars truncated over a range of upper mass limits. Over 95% of all UC HII regions in the sample appear to suffer varying amounts of stellar UC absorption by dust in the HII region. The combination of star cluster plus UV absorption by dust can account for all sources in the sample with no unusual requirement on model parameters.

SIZE-DENSITY CORRELATION
Classically, HII regions are expected to expand to larger sizes and lower densities as they age. The bow shock model (van Buren *et al.* 1990, and Mac Low *et al.* 1991) predicts confinement of the ionized gas by ram pressure as the star moves through the molecular cloud. Hence, the size and density would be governed not by age, but by the balance of the ram pressure against the stellar wind. In this case, one would not expect a size-density-age relationship. Garay *et al.* (1992) and KCW have investigated possible correlations of mean electron density of UC HII regions with their diameters as a function of morphological type. Mac Low *et al.* (1991) showed that bow shocks are consistent with cometary and core/halo morphologies, we therefore, distinguish UC HII regions by their morphological class to investigate a possible size-density correlation. In Figure 2a (taken from KCW), the mean densities of the spherical and unresolved UC HII regions from the Wood and Churchwell (1989) and KCW samples are plotted against their diameters. For these nebulae, mean density is definitely correlated with size. The slope in this log-log plot is -0.65±0.09 and the correlation coefficient is -0.72. However, in Fig. 2b, no correlation of mean density with source diameter is apparent for cometary and core-halo morphologies. It appears that the spherical and unresolved UC HII regions in KCW are consistent with expansion accompanied by decreasing density as they age, but the cometary and core-halo nebulae show no size-density correlation as expected in the bow shock hypothesis.

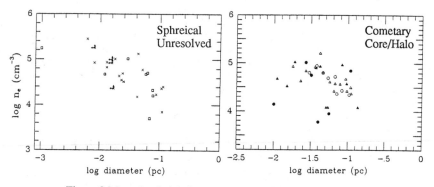

Figure 2 Mean (rms) density versus half-power diameter.

UC HII REGIONS: TRACERS OF SPIRAL ARMS?

In principle, newly-formed, massive stars should be one of the best tracers of spiral arms since they are generally thought to define regions of recent star formation initiated by the passage of a spiral density wave. KCW has plotted the projected positions of UC HII regions in the WC and KCW surveys in the plane of the Galaxy and compared them with the locations of the spiral arms identified by Cohen *et al.* (1980) from CO line emission studies. In Figure 3a, the distribution of UC HII regions in the Galactic plane are plotted as squares with the 4 kpc, Scutum, Sagittarius, Local, and Perseus arms of Cohen *et al.*

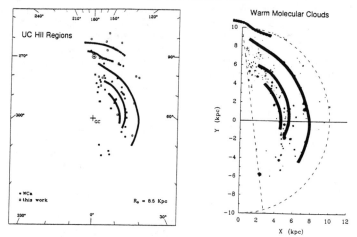

Figure 3 UC HII regions (left) and molecular clouds (right) projected onto the galactic plane. The heavy curves are spiral arms identified by Cohen *et al.* (1980).

(1980) superimposed as thick curves. Fig. 3b shows the same spiral arms superimposed on the molecular clouds of Solomon and Rivolo (1989). The most distinctive feature delineated by the UC HII regions is the Local arm, which agrees well with the Local arm suggested by the CO schematic of Cohen *et al.* (1980). The Perseus, Sagittarius, and Scutum arms are not clearly defined by the distribution of UC HII regions. There are too few nebulae in the region of the 3 kpc expanding arm and the Perseus arm to expect these features to be accurately defined, but this is not the case with the Scutum and Sagittarius arms. These arms also do not appear to be well defined by the molecular clouds of Solomon and Rivolo shown in Fig. 3b. The poor correlation of both UC HII regions and molecular clouds with spiral arms is almost certainly due to inaccurate distances. The distances to both the UC HII regions and molecular clouds are mostly kinematic, which suffer from near/far distance ambiguities inside the solar circle and are probably inaccurate at about the 50% level, particularly within a few degrees of the galactic center and anticenter directions where no enhancement of sources in the suggested region of spiral arms is seen, rather they are almost uniformly distributed in the plane in these directions. The Local arm or "spur", by contrast, is very well defined both by molecular clouds and by UC HII regions. It is noteworthy that the nebulae associated with the Local arm are mostly associated with clusters whose distances have been determined by independent means.

THE CIRCUMNEBULAR MOLECULAR GAS

a) Global Properties

UC HII regions are deeply embedded in molecular clouds. A survey of about 80 nebulae in the lines of NH_3(1,1 and 2,2) and the 1.3 cm maser transition of H_2O has shown that about 70% of the sample had both NH_3 and H_2O emission (Churchwell, Walmsley, and Cesaroni 1990, hereafter CWC). CWC concluded that warm, dense, molecular gas is commonly associated with UC HII regions. Also, the large percentage of sources which have H_2O maser emission at any instant implies that, although a specific maser component may be relatively short-lived, the phenomenon is persistent for most of the UC phase of evolution. Churchwell, Walmsley, and Wood (1992; CWW) observed the 3 mm and 1.3 mm transitions of CH_3CN, CS, and ^{13}CO to measure the temperature, density, total column density, and systemic velocity of molecular gas associated with 11 UC HII regions. This study showed that UC HII regions have large column densities ($> 5 \times 10^{23}$ cm^{-2}) of hot (≥ 100 K), dense ($\geq 10^5$ cm^{-3}) molecular gas lying within a few tenths of a pc of the ionizing star. Due to large optical depths in all three molecules, precise estimates of the abundances, densities, and temperatures of the molecular gas could not be made. More precise values for the density were obtained by Cesaroni et al. (1991) by observing $C^{34}S$ (2-1;3-2; and 5-4) toward 8 selected UC HII regions with resolutions of 25.5", 17.6", and 12", respectively. They found densities of $\sim 10^6$ cm^{-6} in clumps of mass ~ 2000 M_0 and size ~ 0.4 pc. Cesaroni, Walmsley, and Churchwell (1992) observed NH_3 (4,4) and (5,5) inversion transitions toward 16 UC HII regions and combined these data with the (1,1) and (2,2) data from CWC to examine the excitation of NH_3 and to derive kinetic gas temperatures. The (4,4) and (5,5) levels are about 200 K and 300 K, respectively, above ground and are populated only in dense and relatively hot clouds. It was found that in all but two sources, the populations can be fit with a single rotational temperature. The kinetic temperatures found range from 64 K to > 200 K. The hyperfine satellite lines of the (4,4) and (5,5) transitions were detected toward G10.47+0.03A and G31.41+0.31. For these two objects they find (4,4) optical depths > 100, NH_3 column densities $\sim 10^{19}$ cm^{-2}, and kinetic temperatures of order 200 K. The NH_3 (5,5) transition was discovered to mase toward G9.62+0.19; this is the only source known to mase in this transition. The maser has since been confirmed by high resolution VLA observations by Hofner et al. (1992, this volume).

b) Chemistry

Although it is difficult to separate excitation from chemical effects in the warm regions around embedded O stars, NH_3/H_2 appears to be enhanced by an order of magnitude or more near UC HII regions relative to that in cold molecular clouds (Wilson et al. 1983; Mauersberger et al. 1986; Henkel et al. 1987; and Cesaroni et al. 1992). Other Nitrogen bearing molecules such as HC_3N and C_2H_5CN may also be enhanced near embedded O stars (Blake *et al.* 1987). Evaporation of grain mantles may be important in the formation of deuterated molecular species observed toward "hot cores" (Walmsley 1989 and references therein). Abundance anomalies (relative to cool molecular clouds) are not entirely unexpected because the environment around embedded O stars provides several additional processes not generally available in ambient molecular clouds. An enhanced radiation field may modify radiative reaction rates; shock heating and density enhancements both increase reaction rates and make certain endothermic reactions possible; and evaporation of grain mantles due to grain heating may increase mantle constituents such as NH_3, H_2O, CH_3OH, and others. We are a long way from understanding the chemistry near embedded hot stars; much more observational work is needed.

c) Dynamics

The dynamics of the circumstellar gas around embedded O stars is still a matter of some controversy. Tenorio-Tagle, Bodenheimer, and Yorke, motivated by the belief prevalent in the 1970s that O stars form preferentially near the boundaries of molecular clouds, modeled them as "champagne flows" or "blisters" (Tenorio-Tagle 1979; Bodenheimer *et al.* 1979; Yorke *et al.* 1983; and others). Ho, Keto, Haschick, and Zheng have proposed that the molecular gas is gravitationally collapsing toward the central star and spinning up with decreasing distance to the center (Zheng *et al.* 1985; Ho and Haschick 1986; Keto *et al.* 1987, 1988). Hollenbach (1992, this volume) proposes that UC HII regions are the result of flows from photoionization of accretion disks around O stars. Forster *et al.* (1990) proposed that UC HII regions are expanding into surrounding molecular gas with a steep density gradient. Wood and Churchwell (1989); Van Buren *et al.* (1990); and Mac Low *et al.* (1991) have proposed that the cometary and core-halo nebulae are stellar wind supported bow shocks. There is not enough space to explore the strengths and weaknesses of the proposed models here, however, in the next section I will discuss in some detail the arguments in favor of the bow shock model for G29.96-0.02.

d) Bipolar Outflows

One of the most intriguing results regarding the molecular gas associated with UC HII regions is the discovery that at least several have bipolar outflows. Harvey and Forvielle (1988) reported the "most luminous" bipolar outflow ever detected associated with G5.89-0.39. In the meantime, Cesaroni *et al.* (1991) have mapped bipolar outflows associated with G5.89-0.39, G10.62-0.38, and G34.26+0.15 in $C^{34}S$. The case of G5.89-0.39 is particularly interesting because it appears that the flow can be traced back to the ionized shell. In Fig. 4 (left half) a deep 3.6 cm VLA image of the ionized gas is shown with the outer contours enhanced to show that the ionized gas appears to be extended along the NNW-SSE directions. In Fig. 4 (right half), the continuum image is shown superimposed on the $C^{34}S$ bipolar outflow image of Cesaroni *et al.* (1991) at the same angular scale. The central position of the bipolar outflow has been shifted by about 3" (easily within the positional uncertainty of the single dish data) to obtain this overlay. One sees that the extensions of the ionized gas coincide precisely with the peak positions of the bipolar outflow lobes. This strongly suggests that the outflow originates within the HII region, probably at the central star.

Are the bipolar outflows still being driven or are they relics of an earlier accretion phase? Stated another way, are the central stars in UC HII regions still accreting matter and, therefore, premain sequence objects ? What fraction of UC HII regions have associated bipolar outflows? More observations will be required to answer these questions.

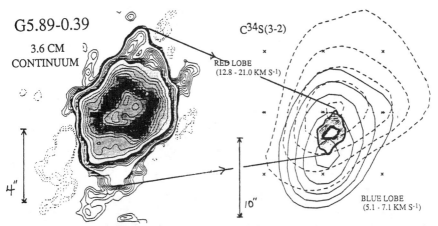

Figure 4 The 3.6 cm f-f emission from G5.89 (left) superimposed on the molecular outflow mapped by Cesaroni *et al.* (1991) in $C^{34}S(3-2)$ emission (right).

A CASE STUDY OF A BOW SHOCKED NEBULA

In this section, observations of the ionized and neutral molecular constituents of the cometary nebula G29.96-0.02 will be discussed and shown to be consistent with a wind supported bow shock model. The properties and kinematics of the ionized gas were studied by Wood and Churchwell (1991) who mapped the H76α line with a spatial resolution of 0.6"x0.5" and a spectral resolution of ~4 km s^{-1}. They found that the distribution of radial velocities in the ionized gas is well ordered. A velocity gradient exists across the nebula ranging from ~80 km s^{-1} on the leading edge of the cometary arc to ~105 km s^{-1} in the "tail"; a shift of ~25 km s^{-1} over an angular distance of ~4" (~0.14 pc). The gradient is about twice as steep on the leading edge of the arc than behind it. They also found that the line widths are significantly broader along the leading edge of the arc than elsewhere in the nebula. The lines are ~40 km s^{-1} wide at the leading edge but only ~25-30 km s^{-1} in the rest of the nebula. It is argued that bulk motions of the ionized gas are responsible for both the velocity gradient and line width distributions. These results are consistent with a flow around a bow shock (Van Buren and Mac Low 1992).

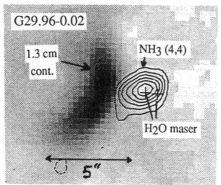

Figure 5 The 1.3 cm continuum and NH3(4,4) emission from G29.96 with a HPBW of 1.2"x1.0". Two H2O masers are seen toward the core of the NH3 clump (HPBW 0.3").

The average properties of the ambient molecular gas near G29.96-0.02 were found by CWC to be: T_k ~136 K, $n(H_2)$ ~ $2x10^5$ cm^{-3}, $N(H_2)$ ~ $2x10^{23}$ cm^{-2}, and $N(NH_3)$ ~ $2x10^{16}$ cm^{-2}. High resolution observations (HPBW ~1") of NH$_3$(4,4) by Cesaroni et al (1993) show a small, hot, dense clump of molecular gas just ahead of the apex of the ionized arc, just where the greatest compression occurs in a bow shock. Fig. 5 shows the free-free emission in grey scale and the NH$_3$(4,4) emission as contours.   The hyperfine structure components indicate that the optical depth in the (4,4) inversion line is >10 and the column density in the (4,4) level alone is ~$2x10^{15}$ cm^{-2}. Perhaps the most intriguing result is the discovery of two H$_2$O masers located almost at the core of the hot, dense, molecular clump (Hofner et al. 1993); they are designated by + symbols in Fig. 5. The H$_2$O profiles shown in Fig. 6 shows that the two

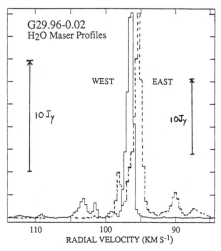

Figure 6 The H$_2$O maser profiles toward G29.96 shown on the same velocity scale but
different flux scales.

masers are almost mirror images of each other; the western H$_2$O maser trails off toward higher positive velocities relative to the main emission line whereas the eastern maser trails off to lower velocities relative to the main line component which has almost the same velocity at both positions. A possible interpretation of the NH$_3$ and H$_2$O results is shown schematically in Fig. 7. Here a newly forming star surrounded by a large, rotating, accretion disk is located at the core of the NH$_3$ clump which is itself probably slowly collapsing toward the star. The western H$_2$O masers are seen along a line that cuts through the disk at a radius of ~$3x10^{16}$ cm where the disk motions are away from the observer. The eastern masers are seen on the opposite side of the disk where it is rotating toward the observer. The rotation speed at ~$3x10^{16}$ cm is inferred to be about 7.5 km s^{-1} if the disk is seen edge-on, which implies a total mass (star plus disk) interior to this radius of ~125 M_0. This model is very preliminary and will have to be subjected to more stringent observational tests before it can be considered viable. However, if correct, it suggests a possible case of induced star formation at the region of highest compression of a bow shock associated with a massive star plunging through a molecular cloud. In this case, the bow shock has probably encountered a preexisting dense clump which may have been pushed into collapse by the encroaching shock.

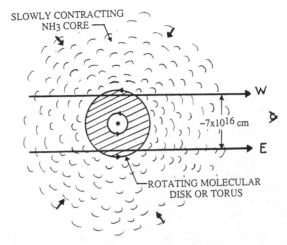

Figure 7 A schematic showing a possible interpretation of the H_2O masers embedded in the dense NH_3 clump. The H_2O masers are formed on opposite sides of a rotating accretion disk around a forming star.

The detailed study of UC HII regions is still in its infancy. There is still much to do observationally and theoretically. In the coming years we should learn more about the nature of the bipolar outflows and the physical properties of the molecular gas, the ionized gas, and the dust associated with these very energetic and active regions. Hopefully, with more high resolution data we will be able to obserationally distinguish between photoevaporating disks, bow shocks, champaign flows, and collapsing spinning-up cores.

REFERENCES
Allen, C. W. 1973, *Astrophysical Quantities,* The Athlone Press, Univ. of London.
Blake, G. A., Sutton, E. C., Masson, C. R., Phillips, T. G. 1987, Ap. J., **315**, 621.
Bodenheimer, P., Tenorio-Tagle, G., Yorke, H. W. 1979, Ast. Ap., **233**, 85.
Cesaroni, R., Walmsley, C. M., Kömpe, C., Churchwell, E. 1991, Ast. Ap., **252**, 278.
Cesaroni, R. Walmsley, C. M., Churchwell, E. 1992, Ast. Ap., **256**, 618.
Cesaroni, R., Hofner, P., Walmsley, C. M., Kurtz, S., Churchwell, E. 1993, Ast. Ap., in preparation.
Churchwell, E. 1990, Ast. & Ap. Rev., **2**, 79-123.
Churchwell, E., Walmsley, C. M., Cesaroni, R. 1990, Ast. Ap. Suppl., **83**, 119.
Churchwell, E. 1991, in *The Physics of Star Formation and Early Stellar Evolution,* eds. C. J. Lada and N. D. Kylafis, Kluwer Acad. Pubs., The Netherlands, pp. 221-268.
Cohen, R. S., Cong, H., Dame, T. M., Thaddeus, P. 1980, Ap. J., **239**, L53.
Forster, J. R., Caswell, J. L., Okumura, S. K., Hasegawa, T., Ishiguro, M. 1990, Ast. Ap., **231**, 473.
Garay, G., Rodriguez, L. F., Moran, J. M., Churchwell, E. 1993, Ap. J., in preparation.
Harvey, P. M., Forveille, T. 1988, Ast. Ap., **197**, L19.
Henkel, C., Wilson, T. L., Mauersberger, R. 1987, Ast. Ap., **182**, 137.
Ho, P. T. P., Haschick, A. D. 1986, Ap. J., **304**, 501.
Hofner, P. , Churchwell, E., Kurtz, S. 1993, Ap. J., in preparation.
Keto, E. R., Ho, P. T. P., Haschick, A. D. 1987, Ap. J., **318**, 712.
Keto, E. R., Ho, P. T. P., Reid, M. J. 1987, Ap. J., **323**, L117.

Keto, E. R., Ho, P. T. P., Haschick, A. D. 1988, Ap. J., **324**, 920.

Kurtz, S., Churchwell, E., Wood, D.O.S. 1992, Ap. J. Suppl., submitted.

Mac Low, M.-M., Van Buren, D., Wood, D. O. S., Churchwell, E. 1991, Ap. J., **369**, 395.

Mauersberger, R., Henkel, C., Wilson, T. L., Walmsley, C. M. 1986, Ast. Ap., **162**, 199.

Miller, G. E., Scalo, J. M. 1979, Ap. J. Suppl., **41**, 513.

Panagia, N. 1973, A. J., **78**, 929.

Solomon, P. M., Rivolo, A. R. 1989, Ap. J., **339**, 919.

Tenorio-Tagle, G. 1979, Ast. Ap., **71**, 59.

Tenorio-Tagle, G., Yorke, H. W., Bodenheimer, P. 1979, Ast. Ap., **80**, 110.

Van Buren, D., Mac Low, M.-M., Wood, D. O. S., Churchwell, E. 1990, Ap. J., **353**, 570.

Walmsley, C. M. 1989, in *Interstellar Dust*, IAU Symp. No. 135, eds. L. J. Allamandola and A. G. G. M. Tielens, Kluwer Acad. Pub., Dordrecht, p. 263.

Wilson, T. L., Mauersberger, R., Walmsley, C. M., Bartla, W. 1983, Ast. Ap., **127**, L19.

Wood, D. O. S., Churchwell, E. 1989, Ap. J. Suppl., **69**, 831.

Wood, D. O. S., Churchwell, E. 1991, Ap. J., **372**, 199.

Yorke, H. W., Tenorio-Tagle, G., Bodenheimer, P. 1983, Ast. Ap., **127**, 313.

Zheng, X. W., Ho, P. T. P., Reid, M. J., Schneps, M. H. 1985, Ap. J., **293**, 522.

Massive Stars: Their Lives in the Interstellar Medium
ASP Conference Series, Vol. 35, 1993
Joseph P. Cassinelli and Edward B. Churchwell (eds.)

THE FORMATION OF MASSIVE STARS – RECENT THEORETICAL RESULTS

HAROLD W. YORKE
Institut für Astronomie und Astrophysik, Universität Würzburg, Am Hubland, D-W 8700 Würzburg, Germany

ABSTRACT   Our theoretical understanding of the formation of massive stars is reviewed.   Emphasis is placed on the numerical problems associated with star formation theory and difficulties due to our poor understanding of basic physical processes. An example of a 2D collapse calculation leading to the formation of an ultracompact HII region via photoionization of the protostellar disk is discussed. It is shown that hydrodynamic processes alone can induce ionized and neutral conical outflows ($v \sim 40$ km s^{-1}) without invoking magnetic fields and without postulating an intrinsic stellar wind from the central source. Numerical models including stellar winds of sufficiently low mechanical luminosity evolve in a similar manner, except for the addition of the outflowing wind confined and focussed along the pole.

INTRODUCTION

Although one of the fundamental astrophysical processes related to the evolution of the universe, the formation of massive stars (here: M $\gtrsim$ 8 M$_\odot$) is still one of the least understood.   In part this is due to the complexity of the processes involved. When account is taken of magnetohydrodynamic effects with clumpiness and motions over a large range of scale sizes, radiation transfer and acceleration effects, chemical evolution and the evolution of dust grains, it rapidly becomes clear that the process is not a 'simple' collapse. Furthermore, massive stars do not form individually but in groups, and the newly formed stars significantly influence their local environment in a complex way (stellar winds, molecular outflows, ionizing radiation, supernova explosions, etc.).

Global aspects of massive star formation such as the formation and destruction of molecular clouds, large scale propagating star formation, the birth and early evolution of OB clusters, the initial mass function, star formation efficiency, and the starburst phenomenon are beyond the scope of this paper. At the risk of being hopelessly incomplete I refer the interested reader to reviews by Blitz (1991, 1992)/Genzel (1991), Elmegreen (1992), Lada (1987)/Larson (1990), Scalo (1986)/Zinnecker et al. (1992), Larson (1988), and Mas-Hesse & Kunth (1991), respectively. In the following I will concentrate on our theoretical understanding of the formation of individual massive stars and attempt to complement the review on massive star formation by Henning (1990) and the reviews in this volume given by Welch (1992), Churchwell (1992) and Hollenbach (1992).

THE BASIC SCENARIO

We are accustomed to interpret observations of star formation in terms of the scenario discussed by Shu *et al.* (1987): (a) formation of cloud fragments or dense molecular cloud cores, (b) collapse of a slowly rotating fragment onto itself and formation of a disk, (c) molecular outflow and/or optical jet fed by disk accretion, and finally (d) disappearance of the outflow and slow dissipation of the disk to leave eventually a star, perhaps with a planetary system. While there are many objects which are observed to be presumably in stages (c) and (d) above (see the reviews on T Tauri stars by Bertout (1989) and Appenzeller & Mundt (1990) and Henning's (1990) discussion of BN stars), we have yet to find a convincing candidate for stage (b), *i.e.* a collapsing fragment without conical outflow. Presumably such an object is to be found deeply embedded inside one of the cold cores of stage (a), such as those studied by molecular line observations of high density tracers (*e.g.* Benson & Myers 1989; Fuller 1990) or by 1 mm dust emission (Mezger *et al.* 1988, 1990).

There are important differences between low mass and high mass star formation.

1. Massive stars have a shorter Kelvin-Helmholtz time scale and thus evolve faster than low mass stars. As a consequence, stars more massive than 3–5 $M_{\odot}$ spend their entire pre-main sequence (PMS) lifetime enshrouded in the dusty remnants of the molecular clump out of which they formed, whereas below this mass they first become optically visible as PMS stars (Yorke 1986).

2. Because of their higher L/M ratio, radiative acceleration can become more important than gravity in the protostellar envelopes of massive (proto)stars, whereas radiation is hydrodynamically unimportant in the low mass case. Wolfire & Cassinelli (1987) argue that the grains must be preprocessed to even allow massive star formation $M \gtrsim 40$ $M_{\odot}$.

3. Whereas optically thick disks are found in around 30% to 50% of young low mass PMS stars (Strom *et al.* 1992), no main sequence stars appear to have such massive, optically thick disks. Excess infrared emission consistent with the existence of optically thin disks – analogous to the structures surrounding β Pictoris and Vega – is found for young stars of all masses. Presumably such disks are the direct descendants of originally massive, optically thick disks. Nondetection of massive disks surrounding massive stars can be interpreted as a correspondingly short time scale for their disk evolution (several 10^5 yr).

4. It has been repeatedly conjectured that massive stars form under special conditions in preferred locations, whereas their low mass counterparts form continuously in dense molecular clouds (*e.g.* Herbig 1962; Mezger & Smith 1977; Larson 1986, to name a few). Direct observational evidence is not conclusive (see *e.g.* discussion by Zinnecker *et al.* 1992). The strongest argument in favor of bimodal star formation is the starburst phenomenon,

where the low M/L ratios are best explained by a relative absence of low mass stars (Larson 1986).

5. Massive star formation is associated with "hot spots" in molecular cloud cores, powerful IR sources $(L \gtrsim 10^3 \, L_\odot)$, (ultra)compact HII regions, H_2 shock emission, outflows, jets, masers. Low mass star formation is less spectacular – associated with weak IR sources $(L \lesssim 10^2 \, L_\odot)$, molecular outlows, T-Tau stars.

Phase 'a': Clump Instability and collapse Initiation

Magnetic support of molecular clumps can be understood in terms of the following simplified argument. The gravitational binding energy of a clump of mass M and radius R is given by $E_G \sim -GM^2/R$ and its magnetic energy by $E_M \sim B^2 M/8\pi\rho$. Assuming spherical compression and flux-freezing we note that $B \propto R^{-2}$. Since $\rho \propto R^{-3}$, we find $E_M \propto R^{-1}$, i.e. $E_G/E_M = $ const., which means that spherical motions alone cannot alter this ratio. Stability is possible when the clump is confined by external pressure and E_M exceeds $|E_G|$ by a factor which depends on the relative distribution of mass, magnetic field and the detailed boundary conditions. This can be translated into a condition for the mass $M < M_M = B^3/(280 \, G^{3/2}\rho^2)$, where the dimensionless number 280 is from Mouschovias & Spitzer (1976).

Ambipolar diffusion is capable of weakening the support of a magnetically subcritical molecular clump and thus allow its slow, quasi-magnetohydrostatic contraction on a time scale of several 10^6 years (Lizano & Shu 1989; Fiedler & Mouschovias 1991). During this contraction the density distribution evolved towards a $\rho \propto r^{-2}$ power law. At sufficiently high density n_{cr} ohmic dissipation becomes important and the magnetic fields decouple from the gas and dust; free fall collapse is possible. Umebayashi & Nakano (1990) estimate the critical density at $n_{cr} \approx 10^{11} \, cm^{-3}$.

Shu et al. (1987) argue that massive stars should form out of magnetized clumps that are supercritical. Instability in a magnetically supported cloud requires two conditions (Mouschovias & Spitzer 1976). First, the field B must be sufficiently weak that gravity dominates, i.e. $M > M_M$. Second, external pressure must be sufficiently large to overcome the internal pressure (with the help of gravity), which can be expressed in the form (Elmegreen 1992) $M > M_P$ with $M_P = 1.37 \, c_S^4 \, G^{-3/2} P_{ext}^{-1/2} [1 - (M_M/M)^{2/3}]^{-3/2}$, where c_S and P_{ext} are the clump's isothermal sound speed and external pressure, respectively. Tomisaka et al. (1989) have extended the instability criterion to the case of a rotating magnetic cloud: $M > M_M$ and $M > M_{crit} = [M_P^2 + (4.8 \, c_S \, j/G)^2]^{1/2}$, where j is the specific angular momentum.

Compression along field lines will modify the ratio E_G/E_M (or M/M_M), however, by decreasing E_M and increasing E_G. In terms of the argument presented above we note that when two equal clumps coalesce, the total magnetic energy increases by a factor of two, whereas the absolute value of the gravitational energy increases by a factor of two *plus* the relative binding energy of the two clumps. Thus, clump-clump collisions and coalescence can lead to instability.

Phase 'b': Numerical Collapse Calculations

A small amount of residual angular momentum in collapsing molecular cores will lead to the formation of a circumstellar disk. Numerical studies of the formation and early evolution of disks surrounding young (proto)stars have concentrated on the low mass case (see *e.g.* the introduction of Bodenheimer *et al.* 1990; Boss 1990). On the one hand, this case pertains directly the early solar nebula and the formation of planets around our Sun. On the other hand there are several difficulties (principally numerical) associated with the intermediate and high mass cases. In Henning's (1990) compilation of published high mass hydrodynamic collapse calculations the most recent paper cited was Yorke (1979) and the second most recent was Yorke & Krügel (1977)! These calculations demonstrated that radiation pressure and relative dust/gas drift can strongly affect the star formation process, and may even be the primary cause of an absolute upper mass limit (see also Wolfire & Cassinelli 1987).

Use of frequency-dependent opacities and self-consistent radiation transport in 1D together with a realistic model for grain evolution has resulted in a basic understanding of the relevant physics, but up to present only the spherically symmetric, non-rotating collapse has been treated in detail – a serious shortcoming. Hydrodynamic codes in 2D and 3D exist that could be adapted to this problem (see Table I). The principal difficulties encountered can be summarized as follows:

1. As discussed above it is not clear what initiates collapse and the initial conditions for numerical simulations are correspondingly uncertain.

2. Radiative acceleration critically influences the evolution of massive protostellar clumps. Radiation transfer in 2D and 3D geometry is difficult to deal with numerically. The exact solution requires determining the distribution of the radiation intensity $I_\nu(r, \Omega; t)$ and dust temperature $T_D(r)$ simultaneously, *i.e.* for a resolution of 10^2 per independent variable (1 frequency, 2 or 3 spatial, and 2 angular variables) 10^{10} (in 2D) or 10^{12} (in 3D) numbers per time step must be calculated. Present-day computers are too slow for such an accurate calculation. Approximate radiative transfer currently used in collapse codes is not sufficiently accurate to deal with high mass ($\gtrsim 20\ M_\odot$) protostellar evolution (Preibisch & Yorke 1992).

3. The numerical problem requires sufficient spatial resolution to resolve the quasi-hydrostatic core $l \lesssim 10^{10}$ cm and to sufficiently resolve shocks over large distance scales $L \gtrsim 10^{15}$ cm. The first condition requires an implicit code (Tscharnuter 1987). The second condition requires sufficiently high resolution in a direction perpendicular to the disk. No present-day code satisfies both conditions.

4. Massive disks provide a reservoir of material with specific angular momentum too large to be directly accreted by a central object. Angular momentum must be transported outwards to allow some of this material to be accreted. Too little is known about angular momentum transport mechanisms, so numerical simulations often resort to a parameterized treatment (so-called "α-disks"). Angular momentum transport by spiral density waves excited by tidal effects can only be dealt with in 3D.

5. Dust properties are poorly known in this environment. The relatively high densities in these disks provide favorable conditions in which dust grains can further coagulate and evolve (Morfill *et al.* 1985; Tielens 1989). This affects both the mechanical dust/gas coupling and the opacity of the disk material (and thus the energetics within the disk and the disk's appearance).

TABLE I   Hydrodynamic codes adaptable to star formation – present state of the art. No present collapse code includes all effects discussed below. However, such a code could be constructed and run on currently available computers in a reasonable amount of time.

PHYSICS	1D	explicit 2D	implicit 2D	3D
hydrodynamics				
spatial resolution	10^6	10^3	10^2	10^2
core resolved?	+	–	o^+	–
multiple components	20	3	×	×
turbulence	o^-	o^-	–	–
convection	o	o^-	o^-	–
magnetic fields	×	o	×	o
ang. mom. transport	×	α–disk	α–disk	α–disk, tidal effects
radiation transfer				
continuum (frequencies)	+ (10^2)	o (grey)	o (grey)	o^- (grey)
lines	o	–	–	×
cont. rad. acceleration	+	o^-	–	–
equation of state				
heating/cooling	o^+	o	–	–
ionization/recombination	o^+	o	–	–
molecule form./destr.	o	–	–	–
dust formation/destr.	o	–	–	–
reliability	95%	80%	80%	70%

notes: + accurate   o approximate   – poor   × not included

6. The disks can be expected to interact with gas outflows. Indeed, such outflows and the related phenomenon of highly collimated jets are presumably a direct consequence of the existence of circumstellar disks.
7. Photoionization of disks will strongly influence the formation and evolution of an ultracompact HII region (*c.f.* Hollenbach 1990, 1992).
8. The role of turbulence and/or magnetic fields is still unclear.

Phase 'c': Outflows and Ultracompact HII Regions
It should be possible to realistically simulate the evolution of outflows and the formation of compact HII regions with present 2D codes, if the evolution during this phase in not totally governed by magnetohydrodynamic effects. If however the production of magnetic fields in a protostellar dynamo and their interactions with a turbulent, partially ionized gas play a dominant role, then we are far from being able to compute this realistically.

In the following we present recent results of such 2D hydrodynamic calculations for which magnetic fields are neglected. In this sense we discuss phase 'b' and early phase 'c' evolution, during which magnetic fields are assumed to have decoupled from the gas and dust, but to have previously (*i.e.* in phase 'a') reduced the total angular momentum of the cloud. The resulting ionized and neutral flow pattern can explain many of the features observed in the ultracompact HII regions surveyed by Wood & Churchwell (1989) – as an alternative to the bow shock model (*e.g.* Van Buren *et al.* 1990, Mac Low *et al.* 1991).

EVOLUTION OF A 10 $M_\odot$ PROTOSTELLAR CLUMP – A CASE STUDY

Recent numerical calculations of the evolution of a 10 $M_\odot$ rotating and collapsing clump (Bodenheimer *et al.* 1992; Yorke *et al.* 1992b) and the subsequent ionization of the protostellar disk (Yorke & Welz 1992; Welz & Yorke 1992) offer a case study of numerical simulations. The cloud is assumed to be originally density peaked with $\rho \propto r^{-2}$ and to rotate as a solid body ($\Omega_0 = 5 \times 10^{-12}$ s^{-1}). The computer code discussed by Yorke *et al.* (1992a) is used, which includes the effects of radiative acceleration (grey flux-limited diffusion) but neglects magnetic fields. The nested grid technique (ENG method) allows us to resolve the central regions to 1.4 AU, whereas the diameter of the spherical clump is 2700 AU. The center is not resolved numerically but is modeled approximately, *i.e.* we prescribe the luminosity L_{core} of the unresolved core as a function of mass M_{core}, angular momentum J_{core} and accretion rate $\dot{M}_{core}$.

The calculations show how a warm, quasi-hydrostatic disk surrounding a central (unresolved) core of only a few $M_\odot$ is formed (see Fig. 1). The disk continues to accrete material from the surrounding molecular clump as it grows in size and mass. The disk is encased in two accretion shock fronts, both of which are several scale heights (≈ 10) above the equatorial plane. These accretion shocks are quite prominent in Fig. 1 (bottom), which shows the detailed density and velocity structure of the disk shortly before the central ionizing source was "turned on" (see below). Negligible velocities indicates that the material here is in hydrostatic equilibrium. The equilibrium disk appears at first glance to be rather "thick". However, the density scale height H_ρ – defined as the height at which $\rho(R, H_\rho)/\rho(R, 0) = 1/e$ in a cylindrical (R, Z) grid – resembles that of a flared "thin" disk with $H_\rho \approx 0.15R$.

At this point 2.7 $M_\odot$ had accreted onto the central core and $\dot{M}_{core} \approx 0$. Directly above the equilibrium disk out to about 100 AU is a well pronounced shock front, the disk's "inner accretion shock". The perpendicular components of the preshock velocities vary from barely supersonic ($v_\perp \approx 3$ km s^{-1} $\approx 2c_S$) at $R = 100$ AU to very supersonic ($v_\perp \gtrsim 30$ km s^{-1} $\approx 10c_S$) at $R = 0$ (c_S is the local isothermal sound speed). Outside $R \approx 125$ AU there is a smooth pressure change rather than a shock front about 10 scale heights above the disk. This configuration is embedded in an outer accretion shock, where the density of the infalling material jumps by more than a factor of 10. This outer accretion shock resembles the bow shock which occurs when a supersonic flow encounters an obstacle (here, the equilibrium disk). Because the gas flow is converging rather than a parallel flow, the "bow shock" is able to wrap itself around the

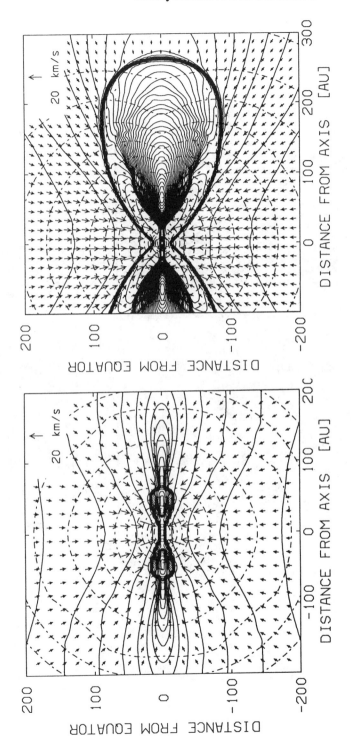

Fig. 1. The density, temperature and velocity structure of a 10 M$_\odot$ rotating molecular clump is shown 2615 (left) and 7074 (right) years after the onset of collapse. Density contours (solid lines) are separated by $\Delta \log \rho = 0.2$, temperature contours (dashed-dotted lines) by $\Delta \log T = 0.05$. Arrows depict the gas velocity at the positions of the tips; the velocity scale is shown in the upper right of each frame. At the times shown 2.45 M$_\odot$ and 2.70 M$_\odot$ have accumulated in the center.

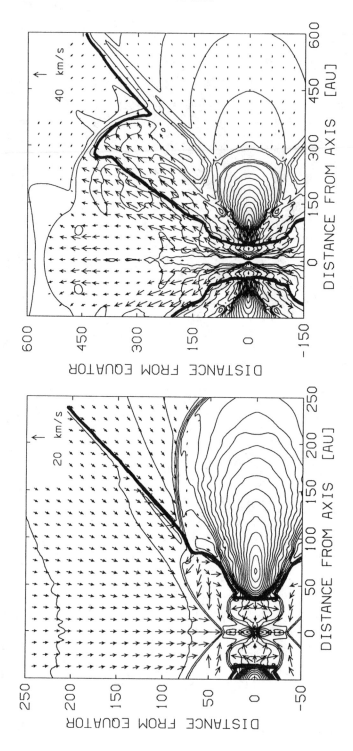

Fig. 2.   The density and velocity structure of the 10 M$_\odot$ case shown in Fig. 1 is displayed at two times (left: $t = 8$ yr and right: $t = 97$ yr) after the ionizing radiation is turned on. 5.7 M$_\odot$ from the inner disk has been added to the core. The location of the ionization "front" ($n_e/n_H = 0.1$) is indicated by the thick line. Here $\Delta \log \rho = 0.4$.

obstacle. Within the spatial resolution possible with this calculation, all shocks appear isothermal.

The initial configuration for the photoionization calculations was taken from the case discussed above. Both the core and the inner portions of the disk ($R \lesssim 30$ AU) were unstable to non-axisymmetric perturbations. Assuming that the partial break up of the material into clumps will subsequently transport angular momentum out of this region via gravitational torques, a central core of 8.4 $M_\odot$ with a stable disk of 1.6 $M_\odot$ should result. After relocating the unstable 5.7 $M_\odot$ of the inner disk into the core we allow the new configuration to relax hydrodynamically and thermally at the modified luminosity $L_{core} = 2600$ $L_\odot$. Then we continue the hydrodynamic calculations assuming a hydrogen-ionizing flux of $S = 10^{46.5}$ s^{-1} and a central stellar wind.

In Fig. 2 the density, velocity and ionization structure is shown for two evolutionary times. The main features of the resulting flow can be summarized:

1. The material close to the symmetry axis within a cone is quickly ionized. The HII region is density bounded in the polar direction, ionization bounded for opening angles $\theta \gtrsim 45°$.

2. The location of the outer accretion shock is shifted outwards as the ionized material expands conically with a velocity $v \approx 40$ km s^{-1}. The inner accretion shock expands initially as a fossil feature in the flow; it later joins the outer expanding shock front.

3. The photoionization of material from the inner portions of the disk close to the equator results in its expansion first towards the symmetry axis, where it encounters a "centrifugal barrier"; it bounces and then moves outwards behind the expanding shock. Note that this also affects material within the radius $R_{esc} \approx 30$ AU, defined as the distance for which the escape velocity equals the local sound velocity $2GM/R_{esc} = c_S^2$.

4. Recombinations in the higher density photoionized disk material results in a closing of the ionization cone to $\theta \approx 30°$. The quasi-steady state outer flow for a conical shell $30° \lesssim \theta \lesssim 40°$ becomes self-shielding in the sense that it recombines and continues to expand as a *neutral* flow $v \approx 30$ km s^{-1}; *i.e.* the HII region is squelched (see Yorke 1986) within this conical shell.

5. We estimate the ionized outflow to be $\dot{M}_{HII} \approx 10^{-7}$ $M_\odot$ yr^{-1}, The neutral outflow was about twice this value.

6. Adding a modest stellar wind ($v_W = 100$ km s^{-1}, $\dot{M}_W = 4 \times 10^{-8}$ $M_\odot$ yr^{-1}) modified the general flow pattern in the following manner. The stellar wind expansion was stopped in the equatorial direction by the ionized disk gas. It was able to expand in the polar direction unimpeded and thus manifested itself as a collimated high velocity component.

CONCLUSIONS

The formation of protostellar disks is a necessary product of collapse; these disks play an important role in the subsequent evolution. Modern computer codes can treat the overall 2D hydrodynamical problem adequately, especially the low mass case, but there are many uncertainties with respect to the detailed physics (initial conditions, grain properties, turbulence, generation of magnetic fields either by the star or within the disk, boundary conditions, H-ionizing radiative flux, stellar winds). Improved approximate methods of radiation transport will soon allow us to calculate the high mass case with the same degree of confidence as the solar nebula. The case study presented here of the collapsing 10 $M_\odot$ clump, which ultimately resulted in a conical, ionized outflow *and* a neutral outflow in a conical shell demonstrates that the photoionization of protostellar disks is a viable model for the explanation of many ultracompact HII regions. Further improvements of the model should include global density gradients, motion within an external medium (in order to relax the symmetry conditions) and an extension to higher mass stars with stronger winds and higher ionizing fluxes.

ACKNOWLEDGEMENTS

Many thanks to Peter Bodenheimer, David Hollenbach, Alexander Welz and Hans Zinnecker for helpful discussions. The author acknowledges support from the Deutsche Forschungsgemeinschaft and the use of computer facilities at the Leibniz Computing Center (LRZ) in Munich and at the Nuclear Research Facility in Jülich (HLRZ).

REFERENCES

Appenzeller, I., Mundt, R. 1990, A&A Rev. 1, 291.

Benson, P. J. & Myers, P.C. 1989, ApJS, 71, 89.

Bertout, C. 1989, ARAA, 27, 351.

Blitz, L. 1991, in The Physics of Star Formation and Early Stellar Evolution, eds. C. J. Lada & N. D. Kylafis, (Dordrecht: Kluwer), p. 3.

Blitz, L. 1992, in Evolution of Interstellar Matter and Dynamics of Galaxies, eds. J. Palous, W. B. Burton, P. O. Lindblad, (Cambridge Univ. Press), p. 158.

Bodenheimer, P., Yorke, H. W., Laughlin, G. 1992, ApJ, in prep.

Bodenheimer, P., Yorke, H. W., Różyczka, M., Tohline, J. E. 1990, ApJ, 355, 651.

Boss, A. 1990, PASP, , 101, 767

Churchwell, E. 1992, this volume.

Elmegreen, B. G. 1992, in Evolution of Interstellar Matter and Dynamics of Galaxies, eds. J. Palous, W. B. Burton, P. O. Lindblad, (Cambridge Univ. Press), p. 178.

Fiedler, R. A., Mouschovias, T. C. 1992, ApJ, 391, 199.

Fuller, G. A. 1990, PhD Thesis, Univ. California at Berkeley.

Genzel, R. 1991, in The Physics of Star Formation and Early Stellar Evolution, eds. C. J. Lada & N. D. Kylafis, (Dordrecht: Kluwer), p. 155.

Henning, T. 1990, *Fund. Cosm. Phys.* 14, 321.

Herbig, G. H. 1962, Adv. Astron. Astrophys. 1, 47.

Hollenbach, D. J. 1990, ASP Conf. Ser. 12, 167.

Hollenbach, D. J. 1992, this volume.

Lada, C. J. 1987, in Star Forming Regions, eds. M. Peimbert & J. Jugaku, (Dordrecht: Reidel), p. 1.

Larson, R. B. 1986, MNRAS, 218, 409.

Larson, R. B. 1988, in Galactic and Extragalactic Star Formation, eds. R. E. Pudritz & M. Fich (Dordrecht: Kluwer), p. 459.

Larson, R. B. 1990, in Physical Processes in Fragmentation and Star Formation, eds. R. Capuzzo-Dolcetta, C. Chiosi, A. Di Fazio, (Dordrecht: Kluwer), p. 389.

Lizano, S., Shu, F. 1989, ApJ, 342, 834.

Mac Low, M.-M., Van Buren, D., Wood, D. O. S., Churchwell, E. 1991, ApJ, 369, 395.

Mas-Hesse, J. M., Kunth, D. 1991, A&AS, 88, 399.

Mezger, P. G., Chini, R., Kreysa, E., Wink, J. E., Salter, C. J. 1988, A&A, 191, 44.

Mezger, P. G., Wink, J. E., Zylka, R. 1990, A&A, 228, 95.

Mezger, P. G., Smith, L. F. 1977, in Star Formation, eds. T. de Jong & A. Maeder, IAU Symp., 75, (Dortrecht: Reidel), p. 133.

Morfill, G. E., Tscharnuter, W., Völk, H. J. 1985, in Protostars and Planets II, eds. D.C. Black & M.S. Mathews, (Tucson: Univ. Arizona), p. 493.

Mouschovias, T. C., Spitzer, L. Jr. 1976, ApJ, 210, 326.

Preibisch, T., Yorke, H. W. 1992, AG Abs. Ser., 7, in press.

Scalo, J. M. 1986, Fund. Cos. Phys., 11, 1.

Shu, F. H., Adams, F. C., Lizano, S. 1987, ARAA, 25, 23.

Strom, S. E., Edwards, S., Skrutskie, M. F. 1992, in Protostars and Planets III, eds. E. H. Levy, J. I. Lunine, M. S. Matthews, (Tucson: Univ. Arizona), in press.

Tielens, A. G. G. M. 1989, in Interstellar Dust, eds. L. J. Allamandola & A. G. G. M. Tielens, (Dortrecht: Kluwer), p. 239.

Tomisaka, K., Ikeuchi, S., Nakamura, T. 1989, ApJ, 341, 220.

Tscharnuter, W. 1987, A&A, 188, 55.

Umebayashi, T., Nakano, T. 1990, MNRAS, 243, 103.

Van Buren, D., Mac Low, M.-M., Wood, D. O. S., Churchwell, E. 1990, ApJ, 353, 570.

Welch, W. J. (1992), this volume.

Welz, A., Yorke, H. W. 1992, AG Abs. Ser., 7, in press.

Wolfire, M. G., Cassinelli, J. P. 1987, ApJ, 319, 850.

Wood, D. O. S., Churchwell, E. 1989, ApJS, 69, 831.

Yorke, H. W. 1979, A&A, 80, 308.

Yorke, H. W. 1986 ARAA, 24, 48.

Yorke, H. W., Bodenheimer, P., Laughlin, G. 1992a, ApJ, in press.

Yorke, H. W., Bodenheimer, P., Laughlin, G. 1992b, in Star Forming Galaxies and their Interstellar Media, EIPC Workshop, June 1992, in press

Yorke, H. W., Krügel, E. 1977, A&A, 54, 183.

Yorke, H. W., Welz, A. 1992, in Star Forming Galaxies and their Interstellar Media, EIPC Workshop, June 1992, in press

Zinnecker, H., McCaughrean, M. J., Wilking, B. A. 1992, in Protostars and Planets III, eds. E. H. Levy, J. I. Lunine, M. S. Matthews, (Tucson: Univ. Arizona), in press.

Massive Stars: Their Lives in the Interstellar Medium
ASP Conference Series, Vol. 35, 1993
Joseph P. Cassinelli and Edward B. Churchwell (eds.)

CIRCUMSTELLAR DISKS ASSOCIATED WITH THE FORMATION OF LOW TO HIGH MASS STARS

FRED C. ADAMS
Physics Department, University of Michigan
Ann Arbor, MI 48109, U.S.A.

ABSTRACT   We review the observational evidence for the existence and properties of circumstellar disks associated with forming stars, from solar-type stars to high mass stars. We discuss the production of gaps or holes, gravitational instabilities, and the implications for massive stars.

1. INTRODUCTION

Over the course of the last decade, a new and reasonably successful paradigm of the star formation process has emerged (Shu, Adams, and Lizano 1987; Lada and Shu 1990). This paradigm is most complete and successful for the case of single stars of low (i.e., solar-type) masses. The extension of the theory to include the formation of high mass stars and the formation of binary companions is currently being studied.

We begin with a quick overview. In the current paradigm, stars form within the *cores* of molecular clouds. These core regions are small subcondensations within the molecular clouds and evolve in a quasi-static manner through the process of ambipolar diffusion. In this process, magnetic field lines slowly diffuse outward and the central regions of the core become increasingly centrally concentrated (see Shu 1983; Lizano and Shu 1989). The magnetic contribution to the pressure support decreases with time until the thermal pressure alone supports the core against its self-gravity. At this point, the core is in an *unstable* quasi-equilibrium state which constitutes the initial conditions for dynamic collapse. When a core begins to collapse, a small hydrostatic object (the protostar) forms at the center of the collapse flow and an accompanying circumstellar disk collects around it. This phase of evolution – *the protostellar phase* – is thus characterized by a central star and disk, surrounded by an infalling envelope of dust and gas.

As a protostar evolves, both its mass and luminosity increase; the protostar eventually develops a strong stellar wind which breaks through the infall at the rotational poles of the system and creates a bi-polar outflow. During much of this *bipolar outflow phase* of evolution, the outflow is well collimated (in angular extent) and infall takes place over most of the solid angle centered on the star. This outflow gradually widens in angular extent and the visual extinction to the central source gradually decreases. The outflow eventually separates the newly formed star/disk system from its parental core and the object enters *the T Tauri Phase* of evolution. The newly revealed star then follows a pre-main-sequence (PMS) track in the H–R diagram and evolves toward the main sequence. Notice, however, that high mass stars evolve to the main sequence during the protostellar phase (Palla and Stahler 1991; Yorke, this volume).

Circumstellar disks play an important role in the process of star formation outlined above. In recent years, compelling observational evidence (cf. the reviews of Appenzeller and Mundt 1989; Bertout 1989; Shu, Adams, and Lizano 1987) has established the presence of disks associated with young stellar objects, although the exact properties of such disks remain controversial. In this review, we discuss the evidence for disks associated with forming stars in various stages of evolution and of varying mass. In particular, we discuss protostellar objects, T Tauri systems, binary systems, Herbig Ae-Be stars, and high mass stars. We also discuss the differences between star/disk systems of high mass and those of low or intermediate mass. Finally, we review our current understanding of gravitational instabilities in star/disk systems; these instabilities play an important role in limiting the disk mass in such systems.

2. OBSERVED PROPERTIES OF CIRCUMSTELLAR DISKS

Many different lines of evidence give rise to our current understanding of circumstellar disk properties. In this section, we discuss several different sources of information, including embedded protostars of low mass, T Tauri systems, binary systems, and Herbig Ae-Be stars. We also discuss the current observational situation for disks associated with O stars.

2.1 Disks Associated with Protostellar Objects

In the protostellar phase of evolution, the central star/disk system is deeply embedded in an infalling envelope of gas and dust. Essentially all of the intrinsic radiation from the star/disk system is absorbed and re-radiated by dust grains in the envelope. As a result, no direct signature can be measured for these disks. However, we can establish the existence of these disks and estimate their properties through indirect methods.

The properties of protostellar disks can be summarized as follows. The physical size of protostellar disks is determined by the angular momentum of the collapse and is likely to be $\sim 10 - 100$ AU for much of the protostellar phase (Adams, Lada, and Shu 1987). We expect the protostellar disk mass M_D to be a significant fraction of the total mass M_{total} during the infall phase because much of the infalling material falls directly onto the disk rather than onto the star (see Shu 1977; Terebey, Shu, and Cassen 1984; Cassen and Moosman 1981). However, as we discuss below

(§4), the disk masses cannot be arbitrarily large because such disks would become gravitationally unstable; disk accretion must thus occur to in order to keep the disks stable. To summarize, in order to simultaneously account for the observed protostellar luminosities, the observed column densities of protostellar envelopes, and the observed final masses of stars, disks must be present in protostellar systems and disk accretion must occur (Adams 1990).

2.2 Disks Associated with T Tauri Stars

For T Tauri disk systems (newly formed stars of roughly solar mass), the spectral energy distributions are directly measurable and we can therefore study the properties of the circumstellar disks. One of the most striking properties of these systems is the large infrared excesses (e.g., Rucinski 1985; Rydgren and Zak 1987) which indicate the presence of circumstellar disks. Here, we will only summarize the basic properties of these disks (see also Bertout, Basri, and Bouvier 1988; Adams, Emerson, and Fuller 1990; Kenyon and Hartmann 1987; Adams, Lada, and Shu 1988). The characteristic radial size is $\sim$100 AU; this size can be deduced from the characteristics of the spectral energy distributions, from studies of spectral lines in outflows (Edwards et al. 1987), or from our own solar system (the mean distance of Neptune from the sun is $\sim$30 AU). The disks in T Tauri systems can be either *passive* and merely reprocess stellar radiation or *active* and possess additional intrinsic luminosity.

For sources with active disks, the intrinsic disk luminosity L_D can be comparable to that of the star, i.e., L_D can be a significant fraction of the total luminosity. In addition, the spectral energy distributions of active disk sources indicate that the effective temperature distribution is often much flatter than the expected form $T_D \sim r^{-3/4}$ for a classical steady Keplerian accretion disk (Lynden-Bell and Pringle 1974); many disks require temperature profiles of the form $T_D \sim r^{-1/2}$, which suggests that *non-local* processes may be at work. The disk mass is estimated to lie in the range

$$0.01 M_\odot \leq M_D \leq 1.0 M_\odot$$

(see, e.g., Weintraub, Sandell, and Duncan 1989; Adams, Emerson, and Fuller 1990; Beckwith et al. 1990). However, these mass estimates depend directly on the assumed dust opacity at submillimeter wavelengths and this quantity remains uncertain at present.

Notice that the disk masses and radii (and hence the angular momenta) for the T Tauri disks are consistent with that expected from the protostellar theory. Notice also that both the protostellar disks and the T Tauri disks must be capable of disk accretion, but no definitive theory of disk accretion exists at this time (see, however, §4).

2.3 Holes in Disks Associated with Binary Systems

The presence of binary companions can greatly alter the configuration of a circumstellar disk. The GW Ori system is the first well studied case of such a binary + disk system (Mathieu, Adams, and Latham 1991, hereafter MAL). The system is observed to be a spectroscopic binary with a period of 242 days. The mass of the primary is estimated to be 2.5 $M_\odot$ and the mass of the secondary lies in the range 0.5 – 1.0 $M_\odot$. The orbit is nearly circular and has a radius of $\sim$1 AU. Since this system also has an extremely large infrared excess indicative of a circumstellar disk, this system is ideal for studying the interaction of binary companions with disks. The spectral energy distribution, as shown in Figure 1, shows a large amount of energy at mid- to far-infrared wavelengths and a very peculiar deficit near 10μm. As shown by the theoretical curves in Figure 1, this sort of spectrum can be understood if there is an annular gap in the disk and optically thin gas resides in the gap. This optically thin gas component produces the 10μm silicate feature observed in emission. By modeling the spectrum and obtaining the best fit, we estimate the inner and outer radii of the gap to be $r_{in} = 0.17$ AU and $r_{out} = 3.3$ AU. Thus, the orbit of the binary companion lies within the estimated gap in the disk. Furthermore, a stellar companion can clear a gap of almost (but not quite) the estimated width of the gap. Although the energetic considerations for this system are complicated and somewhat problematic (see MAL), we interpret this system as a binary pair with an inner disk surrounding the primary and an outer disk surrounding the binary pair.

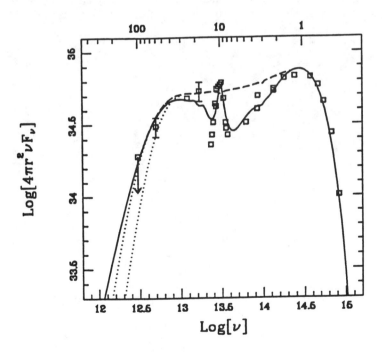

Figure 1. Observed and theoretical spectra of the binary system GW Ori; from Mathieu, Adams, and Latham (1991).

2.4 Central Holes in Disks Associated with Herbig Ae-Be Stars

Recents studies have shown that the disks associated with young stars of intermediate mass (in particular Herbig Ae-Be stars) have disks and that these disks have "holes" in their inner regions near the central star (Lada and Adams 1992; Hillenbrand et al. 1992). Such conclusions can be drawn from modeling of the spectral energy distributions or from examination of the JHK color-color diagrams of these stars (see Figure 2)

We have produced star/disk models of systems with the parameters appropriate for Herbig Ae-Be stars (Lada and Adams 1992). If we consider disk models with no holes or gaps, the observed stars populate a much larger region in the color-color diagram than the theoretical models. However, we find that this discrepancy can be removed provided that the disk models contain central holes. In other words, the inner part of the disk near the central star must be optically thin so that it cannot reprocess stellar radiation and cannot produce intrinsic radiation. Furthermore, we find that the basic physical parameters of these systems always conspire to produce temperatures at the inner disk edge in a narrow range 2000–3000 K. Since this inferred temperature is nearly identical to the destruction temperature for dust grains, we conclude that dust constitutes the primary source of opacity in the disks associated with Herbig Ae-Be stars and that the radiation field of these stars effectively sublimates grains in the inner region of the disks (see also Hillenbrand, this volume).

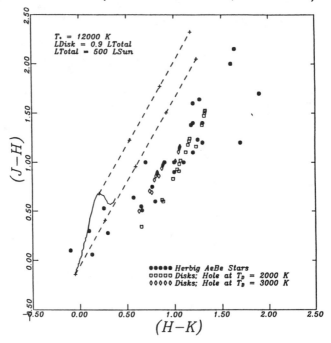

Figure 2. JHK color-color diagram for Herbig Ae-Be stars (solid symbols) and for theoretical star/disk models with central holes (open symbols); from Lada and Adams (1992).

2.5 Disks Associated with High Mass Stars

Since O stars are the brightest and hence the easiest to observe, one might expect that the disks associated with O stars would be the best studied. However, no definitive evidence for disks associated with forming O stars has been published thus far. Although such stars are in fact well observed, the existing observations are consistent with purely spherical dust shells and do not require a disk component for explanation (see, e.g., Chini 1986, 1987; Churchwell, Wolfire, and Wood 1990). In a later state of evolution, early-type stars in OB associations have been studied and found to have near-infrared excesses in their spectra (e.g., Leitherer and Wolf 1984). However, the observed infrared excesses are consistent with that expected for free-free emission and thus the observations (once again) do not require a disk for explanation.

We stress that although the current observations do not provide evidence for disks associated with high mass stars (O stars), these observations do not imply the absence of disks. Since disks are found over a considerable range of lower mass stars and since general physical considerations of angular momentum suggest their presence, disks are expected to be present in these high mass systems (see Hollenbach, this volume).

3. DISKS IN HIGH MASS vs LOW MASS SYSTEMS

In this section we discuss the qualitative differences between systems with high mass and those of intermediate or low mass. During the course of this discussion, we show that the mass scale which divides "high mass" systems from "low mass" systems is approximately $M_* = 7\ M_\odot$. The formation of stars with sufficiently high mass is affected by radiation pressure, in contradistinction to the case of stars with lower masses. We expect that high mass stars form within the same general paradigm that is now fairly well established for low-mass stars (see §1). As many authors have shown previously (e.g., Larson and Starrfield 1971; Kahn 1974; Shu, Adams, and Lizano 1987; Wolfire and Cassinelli 1986, 1987), the radiation pressure from a forming star will be strong enough to affect the infall once the luminosity to mass ratio exceeds a critical value approximately given by

$$\frac{L_*}{M_*} = \frac{6\pi c G}{\kappa_P(T_d)} \approx 700 L_\odot/M_\odot, \qquad (3.1)$$

where $\kappa_P(T_d) \approx 30$ cm^2 g^{-1} is the Planck mean opacity of the dust grains at the dust destruction temperature $T_d \approx 2000$ K. If we use main sequence values, we find that stars more massive than 7 $M_\odot$ have luminosity to mass ratios that exceed the critical value given above. Thus, the formation of stars more massive than about 7 $M_\odot$ (roughly B stars) becomes problematic unless (a) dust grains are somehow depleted in the regions that form massive stars (Wolfire and Cassinelli 1987), or, (b) the infall is highly nonspherical and a substantial fraction of the final mass of the star is accreted through a circumstellar disk (Shu, Adams, and Lizano

1987). Thus, circumstellar disks may be *required* in order to form high
mass stars (stars with $M_* > 7\ M_\odot$).

   Stars with sufficiently large masses thus require disk accretion because of the radiation pressure argument given above. Unfortunately
from a observational perspective, these same stars are sufficiently luminous that the stellar luminosity dominates the disk accretion luminosity,
which is given by

$$L_{disk} = \frac{GM_*\dot{M}}{2R_*},\qquad (3.2)$$

where $\dot{M}$ is the mass accretion rate onto the star from the disk and the
factor of 2 in the denominator arises because half of the energy is stored
in the form of rotational energy. For the stars of interest (with masses
$M_* \sim 5 - 10\ M_\odot$), the Kelvin-Helmholtz timescale,

$$\tau_{KH} = \frac{GM_*^2}{R_*L_*},\qquad (3.3)$$

which determines the rate at which the star itself evolves, is shorter than
the evolutionary timescale of the system. As a result, the stellar structure
is close to the configuration of a star of mass M_* on the zero-age main
sequence; in particular, the stellar radius R_* and stellar luminosity L_*
have nearly their main sequence values. In this case, the mass to radius
ratio M_*/R_* appearing in equation [3.2] is roughly constant and hence

$$L_{disk} \approx 1500 L_\odot \frac{\dot{M}}{10^{-4}M_\odot \text{yr}^{-1}}.\qquad (3.4)$$

If the value of $\dot{M} = 10^{-4}M_\odot\text{yr}^{-1}$ is a plausible estimate, then the stellar
luminosity will be much greater than the disk accretion luminosity for
stars more massive than about 7 $M_\odot$. In order to view equation [3.4]
another way, consider a star with a luminosity of $L_* = 10^5\ L_\odot$ and hence
$M_* = 30\ M_\odot \geq M_D$. In order for the disk accretion luminosity to compete
with the stellar luminosity, the mass accretion rate through the disk must
be $\sim 10^{-2}\ M_\odot\ \text{yr}^{-1}$ and hence the disk mass supply will only last for
$\sim$3000 yr. Notice, however, that the accretion rate in the disk need not
be a constant in time. If disk accretion occurs through episodic outbursts,
then disks associated with these high mass stars could be observable, but
for only a small fraction of the time.

   Although the disk accretion luminosity is expected to be small compared to the stellar luminosity, one might expect reprocessing of stellar
photons by the disk to produce an observable infrared excess (Adams,
Lada, and Shu 1987, 1988; Kenyon and Hartmann 1987). However, just
as in the case of the Herbig Ae-Be stars described above, the intense radiation field of the star is likely to destroy the inner regions of the disk
where most of the reprocessing takes place.

4. GRAVITATIONAL INSTABILITIES IN DISKS

In this section we discuss the stability of circumstellar disks to the growth of self-gravitating modes (see Adams, Ruden, and Shu 1989; hereafter ARS; and Shu, Tremaine, Adams, and Ruden 1990; hereafter STAR). Since the dynamics depends only on the ratio of disk mass to star mass, the results of this section are independent of M_*.

The unperturbed state of the system is chosen to be consistent with the observed disk properties (see §2). The growth of sprial modes is determined by three elements: self-gravity, pressure, and differential rotation. The gravitational forces are determined by the potential of the star and by the disk's surface density distribution $\sigma_0(r)$, which is assumed to be a simple power-law in radius r. The pressure is determined by the distribution of temperature (or sound speed), which is also assumed to be a power-law in radius. The rotation curve $\Omega(r)$ in the disk is then determined self-consistently from the potential of the star, the potential of the disk, and the pressure gradients. Since the potential well of the star dominates that of the disk everywhere except near the disk's outer edge, the rotation curve is nearly Keplerian throughout most of the disk's radial extent. Since observations indicate that "typical" disk sizes are $\sim$100 AU, we consider disk sizes R_D up to $\mathcal{O}(10^4)$ times the radius R_* of the star (only the ratio R_D/R_* enters into the calculations). For the nonlinear SPH simulations discussed at the end of this section, we must adopt a smaller radius ratio of $R_D/R_* = 30$.

4.1 Modes with Azimuthal Wavenumber $m = 1$

Recent work (ARS, STAR) concentrates on modes with azimuthal wave number $m = 1$, since these modes can be global in extent and may also be the most difficult modes to suppress in unstable gaseous disks. Modes with $m = 1$ correspond to elliptic streamlines (i.e., eccentric particle orbits), a special characteristic of Keplerian potentials. Thus, for a disk with an exact Keplerian rotation curve and no interactions between particles, $m = 1$ disturbances correspond to purely kinematic modes of the system; for realistic disks (with pressure), a relatively "small" amount of self-gravity is required to "hold the mode together" and sustain its growth (see ARS and STAR for further discussion).

One unique and important aspect of $m = 1$ modes is that the center of mass of the perturbation in the disk does *not* lie at the geometrical center of the system; hence, the frame of reference centered on the star is *not* an inertial reference frame. The star is actually in orbit about the center of mass (the star is accelerating) and creates an effective forcing potential – the "indirect potential." The interaction of this indirect potential with the outer Lindblad resonance in the disk can be the dominant amplification mechanism for these modes (ARS, STAR).

4.2 Wave Physics and Spiral Instabilities in Gaseous Disks

In the simplest description of spiral instabilities, self-excited disturbances (spiral modes) can grow through the feedback and amplification of spiral density *waves*. In the asymptotic (WKBJ) limit, the dispersion relation for spiral density waves in a gaseous disk has the form

$$(\omega - m\Omega)^2 = k^2 a^2 - 2\pi G \sigma_0 |k| + \kappa^2, \qquad (4.1)$$

where k is the radial wavenumber, κ is the epicyclic freqeuncy, and where ω is the complex eigenvalue of the mode (e.g., Lin and Lau 1979). Since the gravitational term is proportional to $|k|$, this dispersion relation has four branches,

$$k = \pm(k_0 \pm k_1), \qquad (4.2)$$

where

$$k_0 \equiv \frac{\pi G \sigma_0}{a^2}, \quad k_1 \equiv \frac{\pi G \sigma_0}{a^2}[1 - Q^2(1 - \nu^2)]^{1/2}, \quad \text{and} \quad \nu \equiv (\omega - m\Omega)/\kappa. \qquad (4.3)$$

The quantity Q determines the stability of the disk to *axisymmetric* disturbances and is defined by

$$Q \equiv \frac{\kappa a}{\pi G \sigma_0}, \qquad (4.4)$$

where $Q > 1$ implies stability (Toomre 1964). The overall sign of k determines whether the waves are *leading* ($|k| > 0$) or *trailing* ($|k| < 0$); the inner sign determines whether the waves are *short* $[k \propto (k_0 + k_1)]$ or *long* $[k \propto (k_0 - k_1)]$.

The quantity ν is a dimensionless frequency of the spiral density waves. The radius in the disk where $\nu = 0$ (i.e., where $\Re(\omega) = m\Omega$) is known as the *corotation resonance* (CR); the energy and angular momentum of the perturbation (and the action) are positive outside the corotation radius and negative inside. Notice that for $Q > 1$, the wavenumber k_1 becomes imaginary for radii sufficiently close to the CR, i.e., a classical turning point exists for the density waves. The resulting "forbidden" region surrounding the CR is known as the "Q-barrier". Notice also that for long waves, $k \to 0$ at any radius where $|\nu| = 1$. The radius in the disk where $\nu = +1$ is known as the *outer Lindblad resonance* (OLR) and plays an important role in the physics of $m = 1$ modes.

4.3 SLING Amplification and The Four-Wave Cycle

The dominant mechanism for modal amplification arises from the indirect potential, which provides an effective forcing term. The indirect term varies slowly with radius in the disk; since a slowly varying force can only couple to oscillatory disturbances at the disk edges or at the Lindblad resonances, the main coupling occurs at the outer Lindblad resonance for the modes considered here. This amplification mechanism thus differs substantially from the previously studied mechanisms, which utilize the process of *super-reflection* across the corotation resonance. In the analytic treatment, the growth rates are determined under the assumption that *all* of the amplification arises from this coupling of the indirect term to the outer Lindblad resonance in the disk. In other words, the indirect potential acts as an external forcing term and exerts a torque on the disk at the OLR. Since the long-range coupling of the star to the outer disk provides the essential forcing, this new instability mechanism is called SLING: Stimulation by the Long-range Interaction of Newtonian Gravity.

We now present the feedback cycle for $m = 1$ modes in gaseous disks (from STAR):

[A] Begin with the excitation of a long trailing (LT) spiral density wave at the outer Lindblad resonance (OLR) by the indirect term. The LT wave propagates inward (its group velocity is negative) until it encounters the outer edge of the Q-barrier (a classical turning point for the LT wave).

[B] At the Q-barrier, the LT wave refracts into a short trailing (ST) spiral density wave that propagates back outward, through the OLR to the outer disk edge.

[C] The ST wave that propagates to the outer disk edge reflects there to become a short leading (SL) wave. The SL wave then propagates back to the interior, through the OLR, until it encounters the outer edge of the Q-barrier, where it refracts into a long leading (LL) spiral density wave that propagates back toward the OLR.

[D] At the OLR, the LL wave has a classical turning point and reflects into a LT wave. If this reflected LT wave possesses the correct phase relative to the LT wave launched from OLR by the indirect term in step [A] above, then we have constructive reinforcement of the entire wave cycle and the establishment of a *resonant wave cavity*.

Using a WKBJ analysis, we have derived a quantum condition on the basis of the above four-wave cycle. This quantum condition accurately predicts the pattern speeds (i.e., the real part of the eigenfrequencies) for these modes; for strongly growing modes, the analytical and numerical results agree to within $\sim$1 percent (see STAR).

The combined numerical and analytical treatments indicate the dependence of the growth rates (i.e., the imaginary part of the eigenfrequencies) on the parameters of the problem. Most importantly, a finite threshold exists for the SLING Amplification mechanism. When all other properties of the star/disk system are held fixed, this effect corresponds to a threshold in the ratio of disk mass M_D to the total mass $M_* + M_D$. The growth rates are largest for the case of equal masses $M_D = M_*$ and decrease rapidly with decreasing relative disk mass. In the optimal case, $M_D = M_*$, the grow rates can be comparable to the orbital frequency at the outer disk edge, i.e., the modes can grow on nearly a dynamical timescale. On the other hand, the presence of the finite threshold implies a critical value of the relative disk mass, i.e., the maximum value of $M_D/(M_* + M_D)$ that is stable to $m = 1$ disturbances; for the simplest case of a perfectly Keplerian disk and $Q(R_D) = 1$, this critical ratio has the value

$$\frac{M_D}{(M_* + M_D)} = \frac{3}{4\pi}. \tag{4.5}$$

4.4 Nonlinear SPH Simulations of Disk Instabilities

In order to study the nonlinear behavior of the gravitational instabilities which are suggested by the linear analysis, we have performed smooth particle hydrodynamic (SPH) calculations of the behavior of these instabilities in nearly Keplerian disks (Adams and Benz 1992; see also Tomely, Cassen, and Steiman-Cameron 1991 for N-body simulations of similar star/disk systems). The numerical simulations were performed using the smooth particle hydrodynamics code written by W. Benz (see Benz 1990 and Benz et al. 1990 for a description of the code characteristics). All of these simulations use an isothermal equation of state, $p = a^2\rho$, where the sound speed a is a constant. From a physical point of view, this assumption implies that the disk is able to radiate all energy dissipated in shocks or adiabatic compression.

The linear stability analysis (ARS, STAR) suggests that modal growth is relatively insensitive to the inner boundary condition, but rather sensitive to the outer boundary condition (the effects of the outer boundary are quantitfied in Ostriker, Shu, and Adams 1992). In particular, the outgoing ST waves must be able to reflect off of the outer edge. For the SPH simulations, we adopt a "free" outer boundary condition, i.e., the particles on the outer edge are free to wander according to the gravitational and pressure forces exerted on them by the rest of the system. For the initial perturbations of the disk, we start the simulations with 1% amplitude perturbations in density with azimuthal wavenumber $m = 1$.

We find that gravitational instabilities can grow strongly in these systems, i.e., the grow rates are comparable to the dynamical timescale of the outer disk edge. For disks which are not too far from the condition of stability to axisymmetric modes, spiral instabilities with azimuthal wavenumbers $m = 1, 2, 3, 4$ (and higher) are present. As stability is increased (i.e., as the Toomre Q values are raised or the mass ratio M_D/M_* is decreased), the relative strength of the $m = 1$ disturbance increases, although the growth rates of all modes decrease (as expected).

Finally, we find that a spiral arm can often collapse to form a "knot" of bound gas when the equation of state of the disk is isothermal. These collapsed "knots" typically have masses of ~ 0.01 M_D and travel on elliptical orbits. The possibilty that these knots survive to form either giant planets or binary companions is especially interesting. However, at this point, it is necessary to add a note of caution. These knots are formed as a result of the assumption of isothermal evolution of the disk. This assumption limits the local pressure support available. The density enhancements found in our unstable modes are sufficient for pushing small regions over the local Jeans mass and thus triggering the collapse. This collapse would not occur for adiabatic evolution. Hence, the formation of these bound knots of gas can occur in real physical systems only when these systems can efficiently radiate away a large fraction of their energy.

5. DISCUSSION AND SUMMARY

In this talk, we have discussed circumstellar disks associated with young stars ranging in mass from solar-type stars to high mass stars. The most fundamental point to be made is that circumstellar disks are ubiquitous. The combination of observational data and theoretical calculations suggests that circumstellar disks are present during most of the star formation process, beginning with the deeply embedded protostellar phase and ending with the final dispersion of the disk, long after the formation of the star. In addition, disks are found over a wide range of masses. A wealth of observational data implies the existence and properties of disks associated with stars of low to intermediate mass. Although no direct evidence exists for disks associated with the highest mass stars, such disks are expected to be present on theoretical grounds. In addition, the luminosity of these high mass stars is so intense that the disk luminosity (and hence any disk signature) becomes difficult to observe.

We have shown that binary companions can clear gaps in circumstellar disks. For the system that we have studied in some detail (the GW Ori system), the gap size and location required to fit the observed spectral energy distribution are consistent with the observed binary period and inferred binary orbit. We have also shown that the observed locations of Herbig Ae-Be stars in the near-infrared color-color diagram can be understood if the disks are optically thin in their central regions near the star. In all cases considered, the radius at which the disk must become optically thin implies a temperature of $\sim$2000 K, which is near the dust destruction temperature.

Available observations suggest that the disk masses M_D can be a substantial fraction of the stellar masses M_*. This finding is roughly in agreement with theoretical calculations of disk stability. We have argued that gravitational instabilities with azimuthal wavenumber $m = 1$ are likely to be the most unstable in these star/disk systems. Furthermore, the largest disk mass that is stable to these perturbations is estimated to be $M_D/(M_* + M_D) = 3/4\pi$, i.e., the maximum disk mass is about $M_*/3$. In addition, we have shown that nonlinear simulations of disks are in agreement with the linear results and that unstable disks can produce gravitationally bound "knots" of gas. Although these results are preliminary, these "knots" may survive to become either binary companions and/or giant planets in these systems.

ACKNOWLEDGEMENTS

I would like to thank my collaborators Willy Benz, Charlie Lada, Bob Mathieu, Steve Ruden, Frank Shu, and Scott Tremaine. I would also like to thank Paul Ho, Phil Myers, and Anneila Sargent for stimulating discussions related to this paper. This work was supported by NASA Grant No. NAGW–2802 and by funds from the Physics Department at the University of Michigan.

REFERENCES

Adams, F. C., 1990, *ApJ*, **363**, 578

Adams, F. C., and Benz, W. 1992, in IAU Colloquium No. 134,
   Complementary Approaches to Double and Multiple Star Research,
   (Provo: Pub. A.S.P.), in press

Adams, F. C., Emerson, J. E., and Fuller, G. A. 1990, *ApJ*, **357**, 606

Adams, F. C., Lada, C. J., and Shu, F. H. 1987, *ApJ*, **312**, 788

Adams, F. C., Lada, C. J., and Shu, F. H. 1988, *ApJ*, **326**, 865

Adams, F. C., Ruden, S. P., and Shu, F. H. 1989, *ApJ*, **347**, 959 (ARS)

Appenzeller, I., and Mundt, R. 1989, *Astron. Ap. Rev.*, **1**, 291

Beckwith, S.W.V., Sargent, A. I., Chini, R. S., and Gusten, R. 1990,
   AJ, **99**, 924

Benz, W. 1990, in *Numerical Modeling of Nonlinear Stellar Pulsations:
   Problems and Prospects* (Dordrecht: Kluwer Academic Press)

Benz, W., Bowers, R. L., Cameron, A.G.W., and Press, W. H. 1990,
   ApJ, **348**, 647

Bertout, C. 1989, *ARA & A*, **27**, 351

Bertout, C., Basri, G., and Bouvier, J. 1988, *ApJ*, **330**, 350

Cassen, P. and Moosman, A. 1981, *Icarus*, **48**, 353

Chini, R. S. 1986, *A & A*, **167**, 315

Chini, R. S. 1987, *A & A*, **181**, 378

Churchwell, E., Wolfire, M. G., and Wood, D.O.S. 1990, *ApJ*, **354**, 247

Edwards, S., Cabrit, S., Strom, S. E., Heyer, I., Strom, K. M., and
   Anderson, E. 1987, *ApJ*, **321**, 473

Hillenbrand, L. A., Strom, S. E., Vrba, F. J., and Keene, J. 1992,
   ApJ, in press

Kahn, F. D. 1974, *A & A*, **37**, 149

Kenyon, S. J., and Hartmann, L. 1987, *ApJ*, **323**, 714

Lada, C. J., and Shu, F. H. 1990, *Science*, **1111**, 1222

Lada, C. J., and Adams, F. C. 1992, *ApJ*, **393**, 278

Larson, R. B., and Starrfield, S. 1971, *A & A*, **13**, 190

Leitherer, C., and Wolf, B. 1984, *A & A*, **132**, 151

Lin, C. C., and Lau, Y. Y. 1979, *Studies in Applied Math.*, **60**, 97

Lizano, S., and Shu, F. H. 1989, *ApJ*, **342**, 834

Lynden-Bell, D., and Pringle, J. E. 1974, *MNRAS*, **168**, 603

Mathieu, R. D., Adams, F. C., and Latham, D. W. 1991, *AJ*,
   101, 2184 (MAL)

Ostriker, E. C., Shu, F. H., and Adams, F. C. 1992, *ApJ*, in press

Palla, F., and Stahler, S. W. 1991, *ApJ*, **375**, 288

Rucinski, S. M. 1985, *AJ*, **90**, 2321

Rydgren, A. E., and Zak, D. S., 1987, *Pub. ASP*, **99**, 141

Shu, F. H. 1977, *ApJ*, **214**, 488

Shu, F. H. 1983, *ApJ*, **273**, 202

Shu, F. H., Adams, F. C., and Lizano, S. 1987, *ARA & A*, **25**, 23

Shu, F. H., Tremaine, S., Adams, F. C., Ruden, S. P. 1990, *ApJ*,
   358, 495 (STAR)

Terebey, S., Shu, F. H., and Cassen, P. 1984, *ApJ*, **286**, 529

Tomley, L., Cassen, P., and Steiman-Cameron, T. 1991, *ApJ*, **382**, 530

Toomre, A. 1964, *ApJ*, **139**, 1217

Weintraub, D. A., Sandell, G., and Duncan, W. D. 1989, *ApJ*, **340**, L69

Wolfire, M. G., and Cassinelli, J. P. 1986, *ApJ*, **310**, 207

Wolfire, M. G., and Cassinelli, J. P. 1987, *ApJ*, **319**, 850

Massive Stars: Their Lives in the Interstellar Medium
ASP Conference Series, Vol. 35, 1993
Joseph P. Cassinelli and Edward B. Churchwell (eds.)

STAR FORMATION WITH NONTHERMAL MOTIONS

P. C. MYERS and G. A. FULLER
Harvard-Smithsonian Center for Astrophysics, 60 Garden Street,
Cambridge MA   02138

ABSTRACT   The nonthermal component of the velocity dispersion helps determine the density of a massive core before it collapses, and thus helps set the duration of the collapse. We model a self-gravitating core having thermal and nonthermal motions. This "TNT" model is denser than the singular isothermal sphere (SIS), but reduces to the SIS as the nonthermal motions vanish. The TNT model matches observed core and cloud line widths and densities better than does the SIS model. The TNT model core collapses faster than the SIS core at the same temperature, and thus makes a massive star more quickly: at 20 K, a TNT core makes a 10 M_O star in $2\text{-}11 \times 10^5$ yr, while a SIS core takes 20×10^5 yr. The typical core velocity dispersion and pressure are observed to increase with the mass of the most massive associated star. Thus a cluster-forming cloud may form many low-mass stars in its lower-pressure periphery, and a few more massive stars near its higher-pressure center.

INTRODUCTION

In the last decade, much progress has been made in understanding low-mass star formation (e.g., Shu, Adams and Lizano 1987 and references therein). Isolated stars with mass less than about 1 M_O have parent "dense cores" whose thermal motions dominate their nonthermal motions, according to observations of the 1-cm lines of NH_3 in nearby dark clouds. These aspects of low-mass cores match the properties of the isothermal sphere, whose mass infall rate dm/dt during "inside-out" collapse is approximately σ^3/G, where σ is the sound speed and G the gravitational constant (Shu 1977). For these regions, the duration of the infall which produces a star of mass $M*$ is $M*G/\sigma^3$.

Understanding the formation of more massive stars has had four additional difficulties: (1) the lack of a clear link between the properties of the parent cores and the stars they produce (e.g., Silk 1988), (2) the dominance of poorly understood nonthermal motions over thermal motions, (3) uncertainty as to the relative importance of winds and other nonthermal motions in the observed velocity dispersion, and (4) the tendency of massive stars to appear in clusters, suggesting that the formation of massive stars involves interaction with their cluster neighbors (e.g., Larson 1990).

In this report we address these points. We show that formation of massive and low-mass stars can be understood in the same model framework. The key property is the two-component velocity dispersion σ in the parent dense core. The thermal component σ_T is the isothermal sound speed, and the nonthermal component σ_{NT} is closely related to the magnetic Alfvén speed. Low-mass cores have $\sigma_T > \sigma_{NT}$ while massive cores have $\sigma_{NT} > \sigma_T$. Massive cores have greater values of σ, and thus greater values of dm/dt, than do low-mass cores. This difference in mass infall rate leads to formation of more massive stars in more massive cores.

Much of the material in this report is presented in greater detail in Myers and Fuller (1992; hereafter MF92) and Myers and Fuller (1993; hereafter MF93).

OBSERVED PROPERTIES OF CORE LINE WIDTHS

Line width - size relation
The general tendency for the nonthermal motions in a line to increase with the spatial extent of the line is well known. Evidence that the line width and map size of a core are correlated was discussed by Larson (1981), who pointed out its similarity to the better-known tendency for line widths to increase with map size from cloud to cloud in a line. Recently Fuller and Myers (1992; hereafter FM) analyzed six cores without stars in regions of low-mass star formation, each observed in lines of NH_3, CS, and $C^{18}O$. They found that the nonthermal velocity dispersion σ_{NT} increases with radius r in the typical core as $r^{0.7 \pm 0.1}$ for 0.03 pc $\leq r \leq 0.3$ pc.

A similar trend is evident in line width and size data for two regions associated with relatively massive stars: L1688 in the Ophiuchus dark cloud (Myers et al 1978; 8 lines), which contains some 80 young stars, the most massive being the B2 star HD147889 (Wilking and Lada 1989); and L1204 in Cygnus, which contains the S140 H II region, with luminosity $\sim 10^4$ L_O (Blair et al 1978; Ungerechts, Walmsley and Winnewisser 1982; Mundy et al 1987; 5 lines). The best-fit straight line for each region is

$$\log \Delta v_{NT}(L1688) = 0.27 \pm 0.02 + (0.28 \pm 0.03) \log R \qquad (1)$$

$$\log \Delta v_{NT}(L1204) = 0.40 \pm 0.04 + (0.53 \pm 0.10) \log R \qquad (2)$$

In eq. (1-2) the uncertainties are one-sigma in each fit parameter. Thus, in each case $\log \Delta v_{NT}$ is well correlated with $\log R$, but with significantly different slope, 0.3 in L1688, 0.5 in L1204, and 0.7 for the low-mass cores of FM.

Line width - stellar mass relation
MF92 presented a relation in their eq. (21) between L_* and M_*, combining the main-sequence relation $L_* \sim M_*^{3.3}$ for more massive stars with a relation based on the "birthline" for young, low-mass stars presented by Stahler (1983). Figure 1 shows the NH_3 line data compiled by MLF, with the stellar luminosities replaced by stellar mass according to MF92 eq. (21). The best-fit linear relation between $\log \Delta v_{obs}$ and $\log M_*$ is, for $-1.3 < \log M_* < 1.6$,

$$\log \Delta v_{obs} = (-0.26 \pm 0.02) + (0.42 \pm 0.03) \log M_* . \qquad (3)$$

The more massive stars in the observed sample are probably accompanied by lower-mass cluster members in many cases. If so, the mass given in eq. (3) is the stellar mass responsible for most of the observed *IRAS* luminosity, i.e. the mass of the most luminous embedded cluster member.

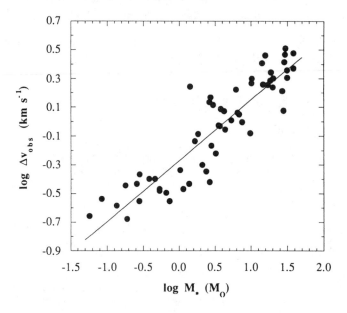

Fig. 1. NH_3 line width in a core whose star has mass M_*.

Role of stellar winds in line broadening

The interaction of stellar winds and outflows with the circumstellar gas in the core increases the velocity dispersion over its value prior to the onset of the wind, and probably also over its value prior to the formation of the star. This increase can be estimated by comparing the typical NH_3 line width of a group of cores associated with stars of similar luminosity, and of a group of neighboring starless cores.

For low-mass cores and stars, eight cores observed by BM have associated stars with L_* within a factor 2 of 1 L_O, and 12 otherwise similar cores have no associated stars. The FWHM line widths for these two groups have mean ± standard error of the mean 0.34 ± 0.02 km s^{-1} for the cores with stars and 0.27 ± 0.01 km s^{-1} for the cores without stars. The quadrature difference in the mean line widths is then 0.21 ± 0.03 km s^{-1}. This difference is statistically significant, but smaller than the FWHM of the velocity distribution of thermal motions, 0.45 km s^{-1}, and smaller than the FWHM of the velocity distribution of nonthermal motions, 0.31 km s^{-1}.

For more massive cores and stars, L1630 and L1641 have 18 cores with associated IRAS sources and 8 starless neighbor cores (Harju, Walmsley, and Wouterlout 1992). The 8 stars with associated cores and with starless neighbor cores have mean luminosity ± standard error 33 ± 6 L_O. Among these, the line widths of cores with stars have mean ± standard error 0.78 ±

0.04 km s^{-1} while the starless neighboring cores have 0.74 ± 0.05 km s^{-1}. The quadrature difference between the groups of cores with and without stars is 0.25 ± 0.19 km s^{-1}, statistically insignificant. This difference is also smaller than the mean thermal FWHM, 0.54 km s^{-1}, and the mean nonthermal FWHM, 0.77 km s^{-1}.

Thus for low-mass and massive cores, line width increases attributable to associated stars are significant in some cases, but are generally smaller than the nonthermal and thermal motions in neighboring starless cores. These starless cores provide the best available estimate of initial conditions for star formation. Consequently the increase in Δv_{NT} with M_* in Fig. 1 reflects a real trend between the velocity dispersion of a prestellar core and the mass of the star which the core will form.

TNT MODEL OF A CORE AND ITS SURROUNDING CLOUD

We assume a spherically symmetric core, whose thermal velocity dispersion $\sigma_T = (kT/m)^{1/2}$ is spatially uniform. Here k is Boltzmann's constant, T the kinetic temperature, and m the mean molecular mass. The nonthermal velocity dispersion σ_{NT} and the thermal velocity dispersion σ_T are derived from the observations as described by FM. We assume as usual that $\sigma_T(m)/\sigma_T(m_{obs})$ $=(m/m_{obs})^{-1/2}$ while $\sigma_{NT}(m)/\sigma_{NT}(m_{obs}) = 1$.

The total one-dimensional velocity dispersion σ is given by $\sigma^2 = \sigma_T^2 + \sigma_{NT}^2$. The velocity dispersion σ_{NT} increases monotonically with radius, while σ_T is constant with radius. Thus σ_{NT} and σ_T are equal at a unique radius, which we denote r_{TNT}. The surface of radius r_{TNT} separates the primarily thermal core interior from the primarily nonthermal core exterior.

We assume that the nonthermal motions exert an isotropic kinetic pressure, related to their velocity dispersion in the same way as for thermal motions. This "microturbulent" assumption probably becomes invalid at small sizes. We assume that these size scales are negligibly small.

We assume hydrostatic equilibrium (HSE). When $\sigma_{NT}(r)$ increases approximately as $r^{1/2}$, as indicated by eq. (2) for L1204, and when σ_T is spatially uniform, the equation of HSE reproduces this behavior for a density law of the form

$$n(r) = \frac{\sigma_T^2}{2\pi m G r_0^2} (x^{-2} + x^{-1}) \qquad (4)$$

where $x \equiv r/r_0$, and where $r_{TNT} \approx r_0/3$. Here the x^{-2} term corresponds to the thermal part of the velocity dispersion and the x^{-1} term corresponds to the nonthermal part of the velocity dispersion. If the nonthermal motions vanish, eq. (4) reduces to the density law for a singular isothermal sphere (SIS). The exact value of r_{TNT}/r_0 depends slightly on the parameter values, as shown in MF92. The velocity dispersion corresponding to eq. (4) in the HSE model is

$$\sigma_{NT}^2(x) = \sigma_T^2 \frac{x}{1+x} \{2 + x[\ln(x_{max}/x) + C]\} \qquad (5)$$

where $x_{max} \equiv r_{max}/r_0$, and where C is a constant, defined in MF92, which is generally close to unity and is independent of r. Eq. (5) shows that for small values of x, $\sigma_{NT} \propto r^{1/2}$, as expected.

Figure 2 (upper) shows σ_{NT} and σ_T, each expressed as the FWHM $\Delta v = (8 \ln 2)^{1/2}\sigma$, and observed values of Δv_{NT} in L1204 (filled circles). The lower plot shows the corresponding density profiles, based on the TNT and purely thermal (SIS) models.

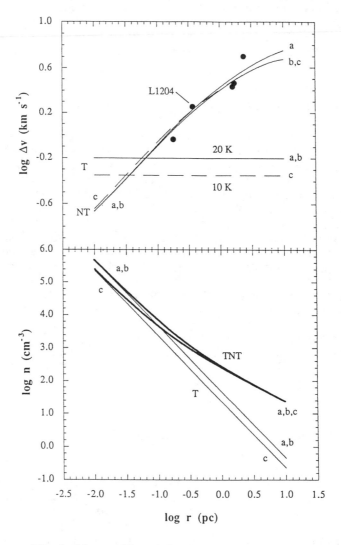

Fig. 2. Line width and density v. radius in TNT and SIS models.

The nonthermal velocity dispersion profiles in Figure 2 fit the observed data points well. Cases a,b, and c are described in MF92. The curves have nearly constant slope, 0.5 on the log-log plot, for r smaller than about 0.3 pc, and then have shallower slope as r approaches the reference radius r_{max}. At larger radii where nonthermal motions dominate, the TNT density profiles vary nearly as r^{-1}, and indicate significantly greater density than do the corresponding thermal profiles. For example, at $r = 0.3$ pc the TNT density is 1.0 to 1.4×10^3 cm^{-3}, in much better accord with estimates based on observations of the J=1-0 line of ^{13}CO (e.g., Bally et al 1987; Loren 1989) than the thermal values of 250 to 500 cm^{-3}.

INFALL OF A TNT CORE

We obtain an approximate solution to the "inside-out" gravitational collapse of a TNT core, by assuming that the free fall of each radial layer is triggered by an "expansion wave" travelling outward at the local sound speed, as was found in the exact solution of the collapse of the SIS (Shu 1977). In contrast to the SIS model, which has a constant sound speed, the TNT sound speed exceeds the thermal speed, and increases with increasing radius. Thus the TNT expansion wave accelerates with increasing radius and increasing time.

The time t_{acc} from the start of infall for all of the mass $M_* = M(<r_{col})$ to collapse to the origin, is the time for the expansion wave to reach r_{col}, plus the time for the gas at r_{col} to fall to the origin. The mass $M(<r_{col})$ is obtained by integrating the density in eq.(4) over radius.

In the thermal limit, the resulting mass accretion rate is

$$\dot{M}(x \ll 1) = \frac{2 \sigma_T^3}{(1 + \pi/4)G}. \tag{6}$$

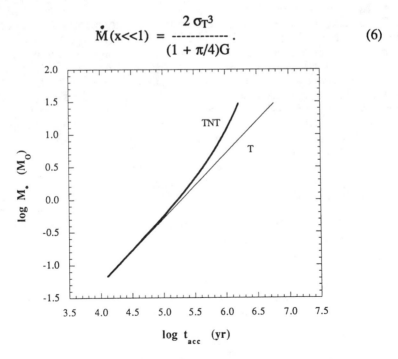

Fig. 3. Accreted mass v. time in TNT and SIS models.

The coefficient of σ_T^3/G, $2/(1 + \pi/4)$, equals 1.12, whereas the corresponding coefficient in the exact SIS solution is 0.975 (Shu 1977). This difference is negligible compared to the difference in magnitude of the mass accretion rates between the relatively large TNT rates and relatively small SIS rates.

Figure 3 shows the accreted mass M_* vs. time from the start of accretion t_{acc}, for the TNT and thermal cases. As the expansion wave radius becomes comparable to r_{TNT} (in 2×10^5 yr), the accreted mass increases at a rate significantly faster than the linear, thermal rate. The time to accrete $10\,M_O$ is 1×10^6 yr, yielding a mean mass accretion rate of $1 \times 10^{-5}\,M_O$ yr^{-1}. In contrast, the accretion rate for a purely thermal low-mass core, at 10 K, is $2 \times 10^{-6}\,M_O$ yr^{-1}, smaller by a factor of 5.

The time to accrete a given stellar mass is shorter in the TNT case than in its thermal counterpart, for two reasons: (a) in the TNT core at radii $r \geq r_{TNT}$ there is significantly more mass than in the thermal core, and (b) at radii $r \geq r_{TNT}$ the expansion wave propagates faster, so it triggers collapse more rapidly in the TNT core than in the thermal core.

CORE-STAR GENETICS

The foregoing results give the time for a "massive" core to make a star of any mass. But low-mass (massive) cores tend to make low-mass (massive) stars. Here we estimate the infall time for the stellar mass that a core with certain values of σ_T and σ_{NT} *tends* to make, rather than the infall time for any stellar mass that it *could* make.

To calculate t_{acc} we use eq.(3) to infer the velocity dispersion in a core of radius r_{obs}, traced by the NH_3 lines, which is typically associated with a star of mass M_*. The nonthermal component of the velocity dispersion is corrected for effects of winds from the associated star as described in MF93. We assume that the resulting σ_{NT} describes the nonthermal motions at r_{obs} in the prestellar core, and that σ_T describes the thermal motions at all r in the prestellar core. We relate $\sigma_{NT}(r \leq r_{col})$ to $\sigma_{NT}(r_{obs})$ according to the following possible cases:

(1) $\sigma_{NT}(r \leq r_{col}) = \sigma_{NT}(r_{obs})$, i.e. the observed value of the nonthermal velocity dispersion, corrected for winds, is uniform for all $r \leq r_{obs}$.

(2) $\sigma_{NT}(r \leq r_{col}) = (r/r_{obs})^{1/2}\,\sigma_{NT}(r_{obs})$, i.e. the nonthermal velocity dispersion is smaller than at r_{obs} according to the relation $\sigma_{NT} \sim r^{1/2}$ characteristic of some star-forming regions, as in eq. (2).

(3) $\sigma_{NT}(r \leq r_{col}) = 0$, i.e. the nonthermal motions are negligible, and the accretion time depends only on the thermal motions.

We consider cases (1) and (2) to bracket the likely range of possibilities, and present the thermal case (3) for comparison.

Figure 4 shows curves of log t_{acc} *vs.* log M_* for these three cases. Cases (1) and (2), which take nonthermal motions into account, have shorter formation times than does the thermal case (3), since greater velocity dispersions imply greater accretion rates. For these two cases, the range of formation times for stars of mass 0.3, 1, 3, 10, and 30 M_O is 1.1-1.5, 3-4, 4-8, 2-11, and 1-12 $\times 10^5$ yr. All of the stars considered form in 1-12 $\times 10^5$ yr, with the greatest range for the most massive stars. Therefore the range of stellar formation *times* is significantly smaller than the range of stellar *masses* formed by molecular clouds.

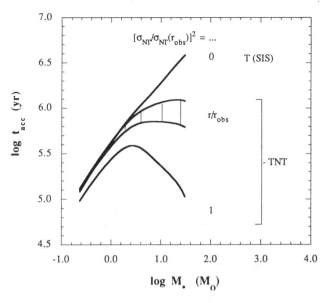

Fig. 4. Accretion time v. stellar mass in TNT and SIS models.

DISTRIBUTION OF STELLAR MASSES IN A CLUSTER

The estimates of gravitational star formation time presented above are for single stars, whereas most massive stars, and perhaps even most low-mass stars, are believed to form in open clusters (Larson 1990; Lada and Lada 1991). Here we use the relations among core velocity dispersion and stellar mass in eq. (5) to estimate the distribution of stellar masses in a cluster-forming cloud. Under simple assumptions, it appears possible to match the form of the initial mass function (IMF) for stars more massive than 2-3 M_O.

The cluster-forming cloud is assumed to consist of a spherically symmetric "cloud" component, whose velocity dispersion varies with radius R as R^q. The cloud is self-gravitating, so its density and pressure vary respectively as R^{-2+2q} and as R^{-2+4q}. Spherical self-gravitating "cores" are embedded in the cloud, and have total mass small compared to the cloud mass. In the star-forming part of a core, the pressure and density are much greater than in the cloud component. At the core radius r_{obs} traced by NH_3 observations, the pressure $P(r_{obs})$ is assumed to be a fixed multiple of the pressure of the cloud component. The variation in cloud pressure across r_{obs} is assumed negligibly small compared to the mean core pressure within r_{obs}. Cores with high pressure have greater velocity dispersion within r_{obs} than do cores with low pressure, and therefore tend to form more massive stars, according to eq. (3). Thus the model cluster-forming cloud has a high-pressure interior, with high-pressure cores, making more massive stars, and a low-pressure exterior, with low-pressure cores, making less massive stars. Stars in a given range of stellar mass tend to form within a particular range of cloud radius and pressure. The tendency for more massive stars to be concentrated toward the centers of young clusters was discussed by Larson (1982).

Figure 5 shows curves of log ξ for two values of q, chosen so that the slope of log ξ *vs.* log M∗ closely matches that of the high-mass part of the IMF. The IMF, shown in filled circles, is taken from Scalo (1986).

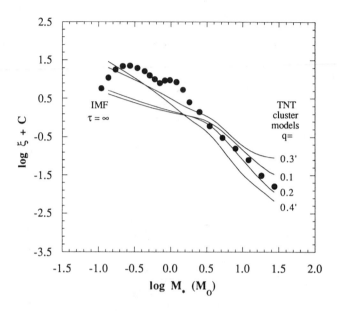

Fig. 5. Stellar mass distributions in TNT models and in IMF.

Figure 5 shows that for each density profile, a range of q exists which closely brackets the slope of the upper part of the IMF, for stars more massive than 2-3 M_O. For constant density, this range is q= 0.1-0.2, while for density which decreases slightly with increasing M∗, this range is q=0.3-0.4. The varying density case appears more realistic, as discussed in MF93. Thus a self-gravitating cluster cloud having cores which follow the observed relations between core velocity dispersion and associated stellar mass can account for the slope of the IMF for stars more massive than 2-3 M_O, provided the cluster cloud velocity dispersion increases with radius as R^q, with q= 0.3-0.4. It is noteworthy that the nearest cluster cloud, L1686 in Ophiuchus, has q = 0.3, as expected in this model. Determination of q for more cluster clouds will be an important test of the model.

ACKNOWLEDGEMENTS

We thank Fred Adams, Peter Barnes, Bruce Elmegreen, Scott Kenyon, Elizabeth Lada, Ned Ladd, Richard Larson, David Leisawitz, Susana Lizano, Mark Reid, Frank Shu, and Steve Stahler for helpful discussions. PCM thanks the organizers of this meeting for their invitation and hospitality. GAF acknowledges the support of a Center for Astrophysics postdoctoral fellowship.

REFERENCES

Bally, J., Langer, W.D., Stark, A.A., and Wilson, R.W. 1987, ApJ, 312, L45.

Benson, P. J., and Myers, P. C. 1989, ApJS, 71, 89 (BM).

Blair, G.N., Evans, N.J., Vanden Bout, P.A., and Peters, W.L. 1978, ApJ, 219, 896.

Fuller, G.A., and Myers, P.C. 1992, ApJ, 384, 523 (FM).

Harju, J., Walmsley, C. M., and Wouterlout, J. G. A. 1992, AAS, in press.

Lada, C.J., and Lada, E. A. 1991, in The Formation and Evolution of Star Clusters, K. Janes, ed.(San Francisco: Astronomical Society of the Pacific), p. 3.

Larson, R.B. 1981, MNRAS 194, 809.

Larson, R.B. 1982, MNRAS, 200, 159.

Larson, R.B. 1990, in Physical Processes in Fragmentation and Star Formation, R. Capuzzo-Dolcetta et al, eds. (Dordrecht: Kluwer), p. 389.

Loren, R.B. 1989, ApJ, 338, 902.

Mundy, L.G., Evans, N.J., Snell, R.L., and Goldsmith, P.F. 1987, ApJ, 318, 392.

Myers, P.C., and Fuller, G.A. 1992, ApJ, in press, September 10 (MF92).

Myers, P.C., and Fuller, G. A. 1993, ApJ, in press, January 10 (MF93)

Myers, P.C., Ho, P.T.P., Schneps, M.H., Chin, G., Pankonin, V., and Winnberg, A. 1978, ApJ, 220, 864.

Scalo, J. M. 1986, Fund. Cosmic Phys.11, 1.

Shu, F.H. 1977, ApJ, 214, 488.

Shu, F.H., Adams, F.C. and Lizano, S. 1987, ARAA, 25, 23.

Silk, J. 1988, in Galactic and Extragalactic Star Formation, R. E. Pudritz and M. Fich, eds. (Dordrecht: Kluwer), p. 503.

Stahler, S.W. 1983, ApJ, 274, 822.

Ungerechts, H., Walmsley, C.M., and Winnewisser, G. 1982, AA, 111, 339.

Wilking, B.A., Lada, C.J., and Young, E.T. 1989, ApJ, 340, 823.

Massive Stars: Their Lives in the Interstellar Medium
ASP Conference Series, Vol. 35, 1993
Joseph P. Cassinelli and Edward B. Churchwell (eds.)

EMBEDDED STELLAR CLUSTERS IN HII REGIONS

MARK McCAUGHREAN
Steward Observatory, University of Arizona, Tucson, AZ 85721
and
Max-Planck-Institut für Astronomie, Königstuhl, D-6900 Heidelberg, Germany

ABSTRACT   This paper discusses some of the motivations, results, and early interpretations of studies of clusters of young low mass stars found near OB stars in their HII regions. Particular emphasis is given to the rôle of high spatial resolution observations and some preliminary results from a large-scale survey for clusters associated with dense molecular cloud cores.

INTRODUCTION

Massive stars dominate the energetics and disruption in star-forming regions via their strong stellar winds and ionising flux. However, massive stars are relatively rare: assuming a Miller-Scalo initial mass function (Miller and Scalo 1979), there should be $\sim$35 stars in the mass range 0.1–3 $M_\odot$ for every star in the range 3–50 $M_\odot$, and roughly 3 times more mass in the low mass stars. For example, in the Trapezium Cluster in Orion, the $\sim$10–20 high mass stars would lead us to expect 350–700 low mass stars in the same region. However, until recently, only 100 or so members of the Trapezium Cluster were known (Herbig and Terndrup 1986). Unfortunately, it is hard to see young low mass stars at optical wavelengths, particularly in those regions associated with high mass star formation, *i.e.*, HII regions. Surveys shortward of 1 μm for low mass stars in HII regions are generally quite incomplete due to dust extinction, nebular 'glare', and the redness of the stars themselves. In the last five years however, infrared cameras sensitive at 1–5 μm have begun to solve these problems, enabling us to reveal the complete stellar population of an HII region. Indeed, infrared imaging of the Trapezium Cluster has now revealed well over 500 members (McCaughrean *et al.*, in preparation).

Figure 1 shows an example of a recently discovered embedded cluster. S 252 C is a radio knot and optical nebulosity in the HII region S 252 (Felli, Habing, and Israël 1977). Optical images of this region show only a few stars, while low-resolution infrared scanning revealed thirteen (Chavarría-K. *et al.* 1989): several hundred are seen in our infrared image. This dense cluster is particularly interesting, as within the same HII region there are two other clusters embedded in the radio knots S 252 A and S 252 E, the optical members of the latter cluster known as the 'Christbaumhäufchen' (Chavarría-K., de Lara, and Hasse 1987). There is also a larger, more diffuse stellar cluster known as NGC 2175 in the HII region (Pişmiş 1970), and a comparative study of the various cluster populations in this region should prove quite illuminating.

Many of the concepts, results, and infrared images that made up the talk this paper results from have recently been discussed and shown elsewhere (*e.g.*, Zinnecker, McCaughrean, and Wilking 1992; McCaughrean, Rayner, and Zinnecker 1992; Zinnecker and McCaughrean 1992). To avoid duplication, this paper will focus on some new material in more detail, material only briefly mentioned at the meeting.

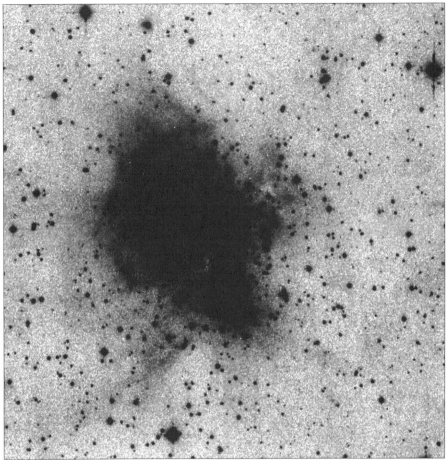

Figure 1. A K' (2.1 μm) mosaic of S 252 C covering ~6×6 arcmin (~3.8×3.8 pc at 2.2 kpc). The image scale is 0.77 arcsec/pixel and the integration time is 2 mins per pixel. The faintest stars visible in this image are at $K' \sim 17^m$. These data were obtained using a 256×256 pixel HgCdTe NICMOS3 array camera on the University of Hawaii 2.2 m telescope on Mauna Kea in October 1990.

MOTIVATIONS

There are two major reasons for studying star clusters at as early an age as possible. Firstly, if we can catch a cluster before 10^7 years or so, the high mass stars should still be there, enabling us to examine the entire IMF at one time. There are also the advantages that the whole cluster should be at more or less the same distance, and should have formed from the same molecular material at roughly the same time. By studying a number of young clusters, we can hope to determine to what degree the IMF has a universal form in the galaxy, across the entire mass spectrum from high to low mass stars.

The second reason is that most stars form in dense clusters, not in isolation, and thus the effects of being born in a very crowded nursery must be studied and incorporated into our models of star formation. Our most detailed observational information about star formation comes, not surprisingly, from the star-forming regions nearest to the Sun, *i.e.,* in the dark clouds of Taurus-Auriga, Ophiuchus, Lupus, Chamaeleon, *etc.* In addition to being quite nearby, the young stars in these regions are fairly well separated, making them readily accessible to a wide range of measurement techniques, even at relatively low sensitivity and spatial resolution.

However, most of the basic material of star formation is in the giant molecular clouds (GMCs), and thus most stars in the galaxy form in GMCs. Within the GMCs, the great majority of stars appear to form in dense clusters as opposed to relative isolation. For example, of the total young stellar populations in the GMCs L 1641 and L 1630 (Orion A and B), greater than 80% and 90% respectively are found in dense clusters (Zinnecker *et al.* 1992; Lada *et al.* 1991). Unfortunately, the nearest GMCs and their massive star-forming regions are far enough away and their stellar members close enough together, that it is hard to obtain the level of detailed information we have for the nearby regions. As a result, more is presently known about the so-called 'isolated' mode of star formation than the apparently dominant 'cluster' mode.

By applying modern techniques to studies of dense clusters, we can begin to examine the effects that a crowded environment might have on star formation. For example, there may be competitive accretion at the protostar stage, and disruption of accretion disks through tidal interactions later on. The formation of binary systems and planets may also be different in a dense cluster, the latter being of particular interest, as the Sun and its planetary system may have formed in a very dense cluster environment.

WHY THE NEAR-INFRARED?

The obvious initial goal in these studies is to find and examine as many members of a given cluster as possible, *i.e.,* a complete sampling of the population. The best wavelength range is the near-infrared (1–$2.5\,\mu$m). As mentioned above, optical observations of embedded clusters in H II regions are hampered by the intense nebulosity, dust extinction, and faintness of the low mass stars relative to the OB stars: all these problems are reduced in the infrared. Also, in the near-infrared we are able to take advantage of large ground-based telescopes, relatively low sky background, and modern imaging detectors: while the very youngest and/or most deeply embedded young stars (*i.e.,* the protostars) may be cool and/or red enough to avoid detection even at $2\,\mu$m, ground-based studies at longer wavelengths are hampered by extremely high thermal backgrounds. Even when the cryogenic space observatories ISO and SIRTF are available, their relatively small apertures will result in somewhat limited spatial resolution.

As a rough rule of thumb, a 10^6 year old $0.1\,M_\odot$ star at $2\,$kpc and embedded behind a visual extinction of 5–10 magnitudes, will have a K magnitude of about 17^m (Zinnecker and McCaughrean 1992). Stars of this brightness are detected at the 3σ level in about 2 minutes on a $2\,$m class telescope, using a high quantum efficiency ($\gtrsim 60\%$) infrared array. Covering a 10×10 arcminute region to this depth through the three near-infrared broad-band filters (J,H,K) takes about 5 hours, using a large-format array (256×256 pixels) with 0.5 arcsecond pixels. This time includes the necessary overheads incurred by obtaining sky frames, standard stars, *etc.,* and gives a realistic estimate of what can be achieved with today's technology. Clearly, we are now able to obtain high-quality infrared imaging and photometry for a large number of clusters embedded in galactic H II regions.

THE IMPORTANCE OF HIGH SPATIAL RESOLUTION

Due to the crowded nature of embedded clusters, high spatial resolution has a key rôle to play in our understanding of them. Fortunately, the seeing is better at near-infrared wavelengths than in the optical, and is frequently 0.4–0.8 arcsec FWHM at the best sites. However, due to the small size of early infrared arrays ($\sim$64×64 pixels), large pixels ($\gtrsim$1 arcsec) were often used to cover a reasonable area of sky. This practice continues today, despite an increase in array size to $\sim$256×256 pixels. When studying point sources in crowded regions, large pixel sizes and low resolution results in a bias towards the more massive stars, which 'drown out' nearby low mass stars. Also, from a technical standpoint, pixels larger than half the seeing disk result in undersampled data, from which it is harder to obtain accurate photometry using the PSF fitting algorithms required in crowded fields. By using small enough pixels to fully sample the seeing, the resulting higher resolution lessens crowding, improves the contrast of faint point sources over the extended nebular background, and improves photometry.

Spatial resolution can be improved even further using adaptive optics techniques, which are also most effective at near-infrared wavelengths, where the isoplanatic patch is relatively large. Clusters embedded in H II regions are prime targets for near-infrared adaptive optics experiments, as there is often (almost by definition) a nearby bright OB star that can be used as a wavefront or tip-tilt sensing star in the optical or infrared. This kind of work has only recently begun, but is very promising.

Figure 2 a shows the BN-KL cluster of young stars and dust clumps, deeply embedded in the molecular cloud OMC-1 behind the Orion Nebula. The spatial resolution in this image is set at about 1.5 arcsec FWHM mainly by the 0.65 arcsec pixel size. Figure 2 b shows the same field imaged using simple adaptive optics, with the bright BN object imaged on one infrared camera to track and compensate for image motion via a fast secondary mirror; the remainder of the stabilised field was imaged onto a second camera with 0.25 arcsec pixels. The resulting resolution is completely pixel limited at 0.5 arcsec FWHM, and probably would have been higher had we chosen to use smaller pixels instead of imaging a larger field-of-view.

Ultimately, we can expect a fully-corrected 4 m-class telescope to yield a diffraction-limited resolution of 0.1 arcsec FWHM at 2.2 μm, *i.e.,* 50 AU at 500 pc, the distance to the nearest clusters associated with H II regions. As this resolution roughly corresponds to the separation at the peak of the PMS binary star frequency distribution (*cf.* Simon *et al.* 1992), such observations would enable us to study the formation of binary systems in crowded clusters and more accurately assess the faint end of the cluster luminosity function. Fully-corrected 8 m-class telescopes will gain us another a factor of two, extending our reach to more distant clusters, and increasing the number of resolvable systems in nearby clusters.

Although adaptive optics systems are not yet delivering this kind of resolution on a routine basis, it *is* now routinely obtained for point sources at optical wavelengths by the HST, despite the spherical aberration. As an example of what might be achieved using ground-based telescopes in the near future, Figure 3 shows a very small section of an *I*-band image taken with the Planetary Camera of the HST. The upside-down 'V' asterism is the same as that seen to the west of the BN object in Figures 2 a and b. The pixel size in this image is 0.043 arcsec: diffraction rings are seen around the stars, and one is clearly resolved into a 0.2 arcsec binary. Preliminary analysis of HST imaging of most of the optical members of the Trapezium Cluster has revealed a binary fraction between 10 and 20% down to 0.1 arcsec, *i.e.,* 50 AU.

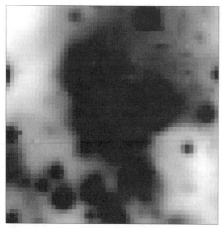 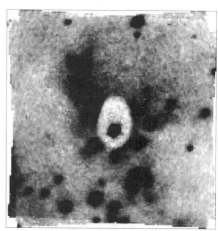

Figure 2 *a,b*. The left-hand figure shows a direct *K* (2.2 μm) image of the BN-KL complex in the Orion Nebula using 0.65 arcsec pixels on the Steward 2.3 m telescope. The resolution is pixel and seeing-limited at 1.5 arcsec FWHM. BN is at the centre, surrounded by a number of lower-mass PMS stars and reflection nebulosities. At this resolution, they are hard to tell apart. The right-hand figure shows a tip-tilt corrected *K* image of the same region with 0.25 arcsec pixels, obtained using a single 1.8 m mirror of the MMT, and the Steward Observatory ACME adaptive optics instrument. A mirror with a small hole scraped in the silvering (seen here as the light-coloured oval) was placed in the beam. Most of the flux from the central BN object passed through the hole to a fast-readout IR camera to sense image motion at 50 Hz, and correct for it by rapid movements of the secondary mirror. A small fraction of the BN flux, and all the flux from the the rest of the field, was reflected off the mirror, and imaged on a second camera using long integration times. The integration time per pixel in this coadded image is 5 mins; the resolution is now entirely pixel-limited at 0.5 arcsec FWHM.

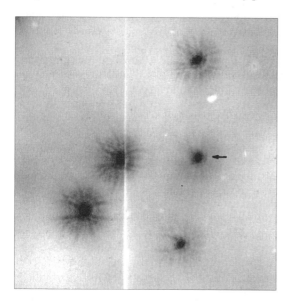

Figure 3. An HST PC *I*-band image of a small group of stars near the BN-KL complex, visible as an upside-down 'V' asterism to the west of BN in Figures 2 *a,b*. The northmost star in the asterism lies on the eastern edge of those images, and the field in the HST image is rotated slightly counter-clockwise relative to them. The pixel size is 0.043 arcsec, and despite the HST spherical aberration, diffraction rings are seen and a 0.2 arcsec binary (marked with an arrow) is clearly resolved. The vertical stripe is the join between adjacent CCDs in the PC.

A SURVEY FOR EMBEDDED CLUSTERS

Most embedded clusters are associated with well-studied regions of massive star formation, and some clusters were already known from low-resolution raster scanning surveys (*e.g.*, Trapezium Cluster, ρ Oph, M 17, NGC 2024), while others have been revealed by infrared imaging (*e.g.*, S 106, NGC 7538, LkHα 101). To widen the sample, we recently started a larger scale survey for stellar clusters embedded in dense molecular cloud cores, partly motivated by the finding of Lada *et al.* (1991) that of the five densest cores in the L 1630 (Orion B) GMC, four are seen to contain young stellar clusters.

A large sample of dense molecular cores was recently published by Plume, Jaffe, and Evans (1992), who used the CS $J = 7 \rightarrow 6$ transition as an indicator of dense gas. As it is impractical to survey large areas of sky in this transition, Plume *et al.* searched for it along the line of sight to all the H_2O masers north of $\delta = -30°$ from the sample of Cesaroni *et al.* (1988) thought to be associated with star-forming regions. H_2O masers provide a good starting point for such a survey, as they themselves must contain at least a small amount of very dense gas; they are seen to lie close to the densest cores in well-studied regions; and they have positions accurate to a few arcseconds. Of the 179 H_2O masers surveyed by Plume *et al.*, 104 were found to show CS $J = 7 \rightarrow 6$ emission, *i.e.*, $\sim 58\%$.

We have been surveying the same sample of 179 H_2O masers in the near-infrared for embedded clusters. We decided to search regions both with and without CS emission, as a wider test of the Lada *et al.* (1991) finding for L 1630. To date, 80 regions have been surveyed at K' (2.1 μm) to a limiting magnitude of $\sim 17^m$ over an area of 4×4 arcmin centred on each maser. As most of the regions surveyed so far have been located in the outer galaxy (*i.e.*, $60° \lesssim l \lesssim 240°$), background field star contamination is relatively small, and most of the regions lie within a few kiloparsecs of the Sun, both factors making cluster detection quite simple.

Of the 80 regions surveyed so far, 55 have some sort of stellar cluster, *i.e.*, ~69%. These clusters range from the very large and dense to the quite small, the latter perhaps more properly thought of as stellar 'groupings' rather than clusters. Figures 4, 5, and 6 show some typical examples of medium, small, and very small clusters, as described in their figure captions. Plume *et al.* found CS emission towards 43 of the regions we have surveyed; of these, 35 have associated stellar clusters, *i.e.*, ~81%. In the 37 regions that have no CS emission, we found 20 clusters, *i.e.*, ~54%. Thus, regions with CS emission are more likely to have an associated stellar cluster, although there is greater than an even chance that *any* region with an H_2O maser will contain a cluster.

Further analysis of our survey images suggests we may seeing some kind of evolutionary sequence of clusters. We have roughly split the sample into older and younger regions: those associated with optical H II regions for example, are classified as older, while those which are optically invisible and perhaps associated with an IRAS source, are classifed as younger. The regions we have surveyed so far are roughly evenly split between older and younger. Almost 60% of the younger regions have clusters, while almost 80% of the older regions do. Perhaps more tellingly, 75% of the clusters seen in the younger regions are small or very small, while over 75% of the clusters in the older regions are large or very large. Thus, older regions are more likely to have clusters, and large ones at that; younger regions are less likely to have clusters, and if they do, they are likely to be quite small. The quite reasonable suggestion then is that more stars form as the parent star-forming region evolves. However, at this early stage, we cannot discount selection effects due to (say) dust extinction, and follow-up multi-wavelength studies will be required to help eliminate such ambiguities.

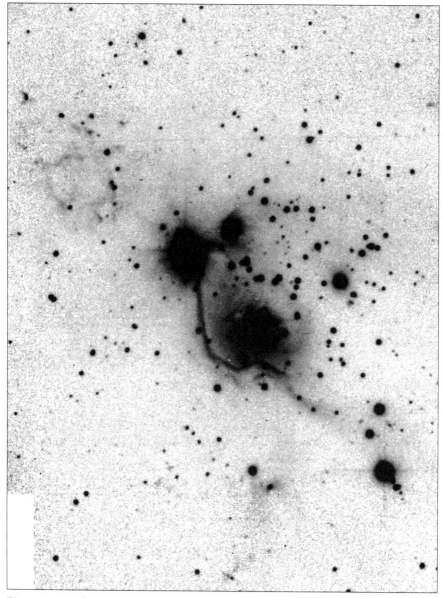

Figure 4. A K' mosaic of BFS 11A/B, a source from the Plume *et al.* (1992) survey with an H_2O maser, CS $J = 7 \rightarrow 6$ emission, and as seen here, a fair sized stellar cluster. The image covers 4.7×6.3 arcmin with an integration time of 4 min per pixel. The region is better known as NGC 7129, a well-studied site of star formation (see *e.g.,* Hartigan and Lada [1985]). In addition to the young stellar cluster to the west of the PMS $\sim$B3 star LkHα 234, considerable nebular structure is seen, some associated with known Herbig-Haro objects (*cf.* Eiroa, Gómez de Castro, and Miranda 1992). The data in this figure and Figures 5 and 6 were obtained using a NICMOS3 array camera at the Steward 2.3 m telescope on Kitt Peak in December 1991.

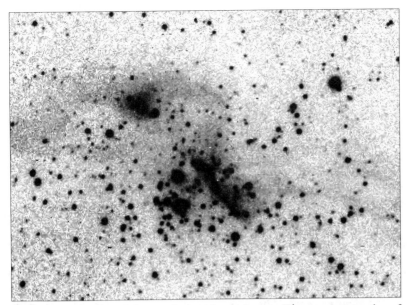

Figure 5. A 3.9×2.8 arcmin section of our K' survey mosaic of IRAS 22314+6033, a source associated with the small H II region S 163. No CS emission was found at this site by Plume *et al.*, although a small stellar cluster is seen here. The integration time was 2 min per pixel.

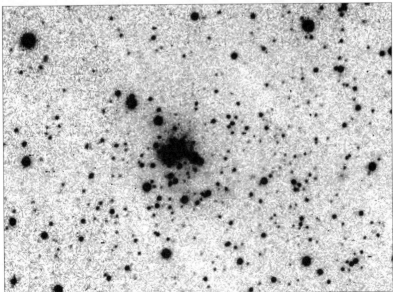

Figure 6. A 3.9×2.8 arcmin section of our K' survey mosaic of IRAS 00117+6412. No CS emission was found here by Plume *et al.* and only a very small stellar 'grouping' and some nebulosity are seen in our image. The integration time was 2 min per pixel. The only optical sign of any activity in this area is a tiny patch of nebulosity on the POSS red plate.

INTERPRETATION

Once we have discovered a cluster and solved the many problems associated with obtaining high quality photometry in these crowded regions, we have to face two kinds of hurdle in any interpretation of the data.

Firstly, there are 'observational' problems to be solved before we can be sure that we have measured the luminosity of a given star. Extinction due to dust must be determined and removed. Any thermal excess must be measured and dealt with: the approach depends on whether the excess is due to reprocessing of stellar energy in a passive disk, or due to additional non-stellar luminosity from an active disk. In the latter case, the very notion of stellar luminosity becomes ill-defined, as the star has probably not yet finished accreting its final mass. Finally, there is the issue of unresolved binaries and the proper division of the flux.

Even if we can define the total luminosity of a star, there is the 'theoretical' problem that we cannot immediately determine its mass, as the mass-luminosity relation is time-dependent for young stars. Young stars are over-luminous for their mass, with extra luminosity released during gravitational contraction and PMS nuclear burning of deuterium. It is only by comparison with detailed models of PMS evolution that we can hope to deduce stellar masses properly. For example, there is the strong temptation to assume that features in the luminosity function must imply corresponding features in the cluster mass function. However, it has recently been shown that deuterium burning in young clusters seriously affects the predicted PMS mass-luminosity relation, and thence the predicted luminosity function. Peaks and turn-overs in the luminosity function can appear, even with an underlying monotonically rising mass function (Zinnecker and McCaughrean 1992).

Thus at present, it is quite hard to interpret infrared imaging photometry of young clusters. However, by incorporating additional data (*e.g.*, thermal-infrared photometry, near-infrared spectroscopy) and more detailed models of PMS evolution, it is possible that we will be able to make considerable advances in the next few years.

CONCLUSIONS

Infrared imaging cameras are now in routine use at the national observatories, and increasingly at private institutions. The *de facto* standard size for astronomical near-infrared arrays in 1992 is 256×256 pixels, with HgCdTe, InSb, and PtSi arrays all available and being used at that size. Although larger format PtSi arrays are available, we still need larger format arrays with high quantum efficiency. Quick, low-sensitivity surveys are possible over large areas of a star-forming region with current arrays, but larger arrays are required to probe the complete stellar population over these areas.

While the observational tools and techniques are maturing, reduction and analysis techniques are lagging. Producing an image of an embedded cluster is one thing, but converting it to an accurate and photometrically calibrated HR diagram for comparison with theoretical models is another. For example, images of the sky are commonly used as a flat-field, despite the night sky airglow being a poor colour analogue to stellar spectra. Tungsten illuminated dome flats are much more effective. Photometric colour equations must be measured and used: modern cameras and array detectors may have quite different colour characteristics to the single-element InSb photometers used to establish the standard star networks. PSF fitting photometry is the tool of choice in dense clusters, but problems arise when it is used on mosaic images with large pixels, due to variations in the composite PSF and the effects of undersampling. There should be a proper assessment of completeness and an accurate determination of measurement accuracy by adding and retrieving fake stars, and by repeated measurements of the same fields. These problems will be familiar to optical CCD users of course; infrared astronomers should be perusing the optical literature for clues!

Finally, improved PMS theory and evolutionary models are required to help fully interpret the observational information on embedded clusters. Hopefully, the early results from infrared imaging surveys will help stimulate these developments. In particular, the effects of a dense cluster environment on star formation must be addressed, as it appears that the great majority of stars form in crowded and often violent regions, as opposed to the isolated, relatively undisturbed dark clouds nearer to the Sun.

ACKNOWLEDGMENTS

All the work presented in this paper comes from various collaborations the author is involved in, and none of it would have been possible without great efforts on the parts of the many collaborators. John Rayner and Hans Zinnecker are the mainstays of the wide-field infrared imaging work, along with Bruce Wilking and Mike Burton. Roger Angel, Don McCarthy, and everyone else in the ACME adaptive optics group have made it possible to start taking diffraction-limited infrared images of embedded clusters. John Stauffer, Charles Prosser, and other members of the HST Trapezium Cluster project, and Italo Mazzitelli are also to be thanked.

REFERENCES

Cesaroni, R., Palagi, F., Felli, M., Catarzi, M., Comoretto, G., DiFranco S., Giovanardi, G., and Palla, F. 1988, *Astr. Ap. Suppl.*, **76**, 445.
Chavarría-K., de Lara, E., and Hasse, I. 1987, *Astr. Ap.*, **171**, 216.
Chavarría-K., Leitherer, C., de Lara, E., Sánchez, O., and Zickgraf, F.-J. 1989, *Astr. Ap.*, **215**, 51.
Eiroa, C., Gómez de Castro, A. I., and Miranda, L. F. 1992, *Astr. Ap. Suppl.*, **92**, 721.
Felli, M., Habing, H. J., and Israël, F. P. 1977, *Astr. Ap.*, **59**, 43.
Hartigan, P. and Lada, C. J. 1985, *Ap. J. Suppl.*, **59**, 383.
Herbig, G. H. and Terndrup, D. M. 1986, *Ap. J.*, **307**, 609.
Lada, E. A., De Poy, D. L., Evans, N. J., II, and Gatley, I. 1991, *Ap. J.*, **371**, 171.
McCaughrean, M. J., Rayner, J. T., and Zinnecker, H. 1991, *Mem. S. A. It.*, **62**, 715.
Miller, G. E. and Scalo, J. M. 1979, *Ap. J. Suppl.*, **41**, 513.
Pişmiş, P. 1970, *Boletín Obs. Tonantzintla Tacubaya*, **5**, 219.
Plume, R., Jaffe, D. T., and Evans, N. J., II 1992, *Ap. J. Suppl.*, **78**, 505.
Simon, M., Chen, W. P., Howell, R. R., Benson, J. A., and Slowik, D. 1992, *Ap. J.*, **384**, 212
Zinnecker, H., McCaughrean, M. J., and Wilking, B. A. 1992, in *Protostars and Planets III*, eds. E. H. Levy, J. I. Lunine, and M. S. Matthews, (University of Arizona: Tucson), in press.
Zinnecker, H. and McCaughrean, M. J. 1992, *Mem. S. A. It.*, **62**, 761.

Massive Stars: Their Lives in the Interstellar Medium
ASP Conference Series, Vol. 35, 1993
Joseph P. Cassinelli and Edward B. Churchwell (eds.)

ANATOMY OF THE GEM OB1 MOLECULAR CLOUD COMPLEX

JOHN M. CARPENTER, RONALD L. SNELL, and F. PETER
SCHLOERB

Five College Radio Astronomy Observatory, University of
Massachusetts, Amherst, MA 01003

ABSTRACT   A 32 deg^2 region toward the molecular cloud nearby
the Gem OB1 association has been mapped in the J=1–0 rotational
transitions of ^{12}CO and ^{13}CO at 50$''$ spacing. The morphology of
this cloud complex is dominated by numerous filaments and arcs of
molecular gas, particularly nearby the known massive star forming
regions, strongly suggesting that the present structure of the Gem OB1
cloud complex has been shaped primarily by past star formation events.

INTRODUCTION

It is well established that our Galaxy is populated by numerous clouds of
molecular gas and that these clouds are the predominant sites of current
star formation. The largest of these clouds can extend over 100 parsecs and
have masses of $\sim 10^6$ $M_\odot$. While small regions within these clouds have been
studied in much detail, little is known about the large scale morphology of
molecular clouds and the physical processes which control their structure and
evolution.

In large part this deficiency can be attributed to the large investment
of telescope time required to map large regions of the sky at high angular
resolution with radio telescopes. At the Five College Radio Astronomy
Observatory (FCRAO), a 15 element receiver array (QUARRY) has been
developed that significantly improves the ability to create large molecular line
maps in a reasonable amount of time. This paper describes the maps that we
have made of the Gem OB1 molecular cloud.

OBSERVATIONS

The Gem OB1 molecular cloud was mapped in the J=1–0 rotation transition
of both ^{12}CO and ^{13}CO with QUARRY on the 14 m telescope at FCRAO.
Boths maps cover a 5.09$°$ x 6.36$°$ region and were sampled every 50$''$ ($\sim$full
beam width spacing), resulting in a total of 166,440 spectra per map. The

spectrometers for each pixel of the array consisted of a 32 channel filter bank
with a resolution of 250 kHz per channel.

OVERVIEW OF THE GEM OB1 MOLECULAR CLOUD

The Gem OB1 molecular cloud is located toward the anti–center region of
the Galaxy at $(l{\sim}190°, b{\sim}0°)$. A number of prominent star forming sites
are located in this cloud complex, including the HII regions S247, S252,
and S254–258. All of these H II regions have independent distance estimates
of ~2.0 kpc, suggesting that they do indeed form a single molecular cloud
complex. Thus the molecular line images presented here cover a 178 × 222 pc
region at 0.49 parsecs resolution and can be used to trace the structure of the
Gem OB1 molecular cloud over two orders of magnitude in size scales.

GLOBAL STRUCTURE OF GEM OB1

A grey–scale image of the observed ^{12}CO peak antenna temperature of
toward Gem OB1 is displayed in Figure 1. The coordinates of the optically
visible HII regions in the Gem OB1 complex are listed in the table below.

HII Region	RA(1950)	DEC (1950)
S247	$6^h 5.5^m$	21° 38′
S252	$6^h 6.5^m$	20° 31′
S254–258	$6^h 10^m$	18° 00′
BFS52	$6^h 12^m$	19° 02′

Not surprisingly, all four of these HII regions are associated with a
concentration of molecular gas. For the Sharpless HII regions, it is also
apparent that either the HII region or the stellar winds from the ionizing
stars has created a cavity in the molecular cloud. Normally only one
hemisphere of these cavities is observed, and rarely are the approaching and
receding sides of the cavities seen. Thus the structure of the molecular cloud
at the time of the cavity formation may well have been highly asymmetric
on the parsec size scale. The molecular gas distant from the HII regions is
significantly more diffuse than the gas nearby star formation sites, although
it, too, possesses many small scale features.

    The most massive concentrations of molecular gas are found along
the arcs around the HII regions. Comparison of the moleculars maps with
IRAS images of Gem OB1 reveal that luminous $(\gtrsim10^{3-5}\,L_\odot)$ point sources
are associated with many of these "knots". We speculate that these regions
represent sites of "triggered" star formation.

Carpenter, Snell, Schloerb

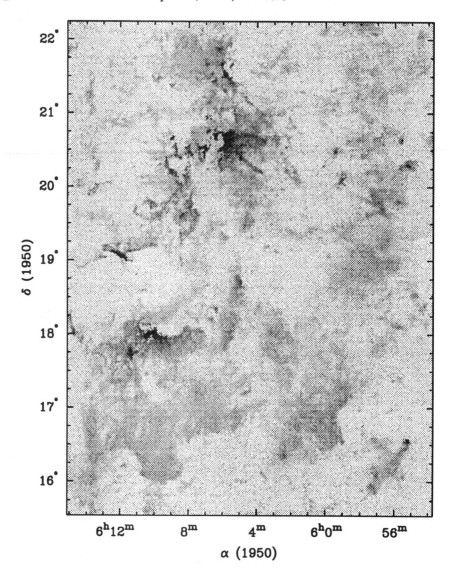

Fig. 1. A grey scale image of the observed peak antenna temperature of ^{12}CO(J=1-0) toward the Gem OB1 molecular cloud complex. The halftones range from 0.25 to 8.0 K.

Massive Stars: Their Lives in the Interstellar Medium
ASP Conference Series, Vol. 35, 1993
Joseph P. Cassinelli and Edward B. Churchwell (eds.)

A MULTI-TRANSITION STUDY OF CS IN REGIONS OF MASSIVE STAR FORMATION.

RENÉ PLUME, DAN JAFFE & NEAL J. EVANS II
University of Texas at Austin, Dept. of Astronomy, RLM 15.308, Austin, Tx 78712.

JÉSUS MARTIN-PINTADO & JÉSUS GOMEZ-GONZALEZ
Centro Astronómico de Yebes, Apartado 148, E-19080, Guadalajara, Spain

ABSTRACT We present the results of a survey for CS J = 7-6, 5-4, 3-2 and 2-1 emission towards a large number star-forming regions containing H_2O masers. Large Velocity Gradient (LVG) excitation calculations have determined the average density of these regions to be $<n> = 4.3 \times 10^5$ cm^{-3}. However, in many sources the data suggest the presence of a small amount of gas of very-high density (n > 10^7 cm^{-3}). We also calculate the CS 5-4 luminosity of the Galaxy to be 7 - 13 $L_\odot$.

INTRODUCTION

Very dense cores (n > 10^5 cm^{-3}) in molecular clouds are sites of massive star formation. Winds and radiation from the newly formed stars can disrupt the surrounding molecular material, but this effect may be moderated by the presence of very dense gas. Therefore, to understand the process of star formation, it is first necessary to understand the properties of dense cores and their effects on the rest of the cloud.

Small beam sizes and limited telescope time rule out a complete survey of the sky. To obtain a relatively unbiased survey of dense gas in star forming regions, we require a pointer to likely locations of dense cores. H_2O masers serve this purpose rather well. The masers must contain at least small amounts of very dense gas (Elitzur *et. al.* 1989) and, in well-studied regions, the H_2O masers lie close to the densest molecular gas. Therefore, using H_2O masers as targets, we have surveyed towards 179 regions of star formation in the J = 7 - 6 transition of CS (Plume, Jaffe and Evans 1992). The high critical density of this transition (n_{crit} ~ 2×10^7 cm^{-3}) means that is not excited in low density gas, and that its emission increases strongly with density.

SELECTION CRITERIA

The H_2O masers were selected from the catalogue of Cesaroni *et. al.* (1988), which is a compilation of all known water masers north of δ = -30°, if their positions were known to an accuracy of better than 8" and if they were not known to be evolved stars. We tested this last assumption for 70 sources by plotting IRAS colour-colour diagrams and checking for sources that fell in the region delineating late-type stars capable of producing maser emission (Figure 1 of Plume *etal* . 1992).

OBSERVATIONS AND RESULTS

The initial CS 7-6 observations were performed at the 10.4m Caltech Submillimeter Observatory and the subsequent CS and $C^{34}S$ 5-4, 3-2 and 2-1 lines were observed at the IRAM

30m telescope. We searched for the CS J = 7-6 transition in 179 sources and for the CS 5-4, 3-2 and 2-1 transitions in 152 sources (chosen from the original 7-6 survey). We also searched for $C^{34}S$ J = 5-4, 3-2, and 2-1 in 50 sources in which we had already observed the main isotope. The detection statistics are given in Table 1.

Table 1
Detection Statistics for CS and $C^{34}S$ Data

Transition	Number of Detections	Percentage	Transition	Number of Detections	Percentage
CS 7-6	104	58%	-	-	-
CS 5-4	117	77%	$C^{34}S$ 5-4	30	60%
CS 3-2	144	95%	$C^{34}S$ 3-2	48	96%
CS 2-1	147	97%	$C^{34}S$ 2-1	35	70%[1]

DISCUSSION

We determined densities and column densities using a 20 x 20 grid of Large Velocity Gradient (LVG) models. The LVG code calculates the line radiation temperatures for a single density and column density. For our LVG grid we used the parameters: T_k = 50K, density (n) range 10^4 to 10^8 cm^{-3}, column density (N) range 10^{11} - 10^{18} cm^{-2}. The observed line temperatures were fit to the LVG grid using the method of least squares. Figure 1 shows the distribution of densities determined from both the CS (solid line) and $C^{34}S$ (dashed line)transitions. The average density is <n> = 4.3×10^5 cm^{-3} and the average column density is <N_{cs}> = 5.2×10^{13} cm^{-2}. Assuming a CS to $C^{34}S$ abndance ratio of 22:1, the LVG fits to the $C^{34}S$ data obtain <n> = 5.4×10^5 cm^{-3} and <N_{cs}> = 2.1×10^{14} cm^{-2}.

Figure 1

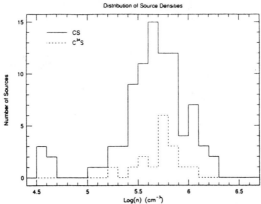

Figure 1 - Distribution of source densities as determined from the LVG code using CS (solid line) data and $C^{34}S$ (dashed line) data. The CS distribution includes 91 sources for which we have observed the J = 7-6, 5-4, 3-2 and, 2-1 transitions. The $C^{34}S$ distribution includes 16 sources for which we have observed the J = 5-4, 3-2 and, 2-1 transitions.

---

[1] The low number of detections of this transition is due to receiver problems.

It is not reasonable to expect that clouds are composed of a single density component (such as that calculated by the LVG code). In fact, we might expect small clumps of very high density gas ($n > 10^7$ cm^{-3}) to exist in these regions. Figure 2 shows the best fit LVG models to the CS line temperatures observed in the source W42. The solid line indicates the line temperatures calculated by the LVG code for Log(n) = 6.0 and Log(N/dV) = 13.6, and is the best fit to all 4 CS transitions. The dashed line is the best fit to the CS J = 5-4, 3-2, and 2-1 transitions only, and is calculated for Log(n) = 5.7 and Log(N/dV) = 13.6. The dotted line fits the CS J = 7-6 and 5-4 data with a curve calculated for Log(n) = 7.3 and Log(N/dV) = 13.3. Obviously, the data is better fit by two density components than by one. We have found 12 sources in which the best fit LVG model to the CS 7-6 and 5-4 data yields densities > 3×10^6 cm^{-3}.

Figure 2

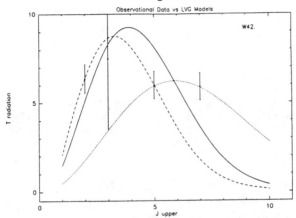

Figure 2 - Best fit LVG model to the observations of W42. The solid line shows the fit to all 4 CS transitions, the dashed line shows the best fit to the CS 5-4, 3-2 and 2-1 data, and the dotted line indicates the best LVG model fit to the CS 7-6 and 5-4 data.

Mapping CS 5-4 in 57 sources has shown that the average area of a CS 5-4 cloud is ≈ 1.1 pc^{-2}. Assuming the CS 7-6 detection rate of 58% also applies to 5-4, and using all the non-stellar H$_2$O masers in the Cesaroni catalogue, we have estimated the CS 5-4 luminosity of the Galaxy to be $\approx$ 7 - 13 L$_\odot$. This luminosity falls in the range of values that we derive for IC 342 (3L$_\odot$)and M82 (28L$_\odot$) from the data of Mauersberger and Henkel (1989) but below the luminosity of NGC 253 (154L$_\odot$).

REFERENCES

Elitzur, M., Hollenbach, D.J., and McKee, C.F. 1989, *Ap. J.*, 346, 1983.
Plume, R., Jaffe, D.T, and Evans, N.J. II 1992, *Ap. J. Supp.*, 78, 505.
Cesaroni, R., *etal* 1988, *A.&A. Supp.*, 76, 445.
Mauersberger, R., and Henkel, C. 1989, *A&A*, 223, 79.

Massive Stars: Their Lives in the Interstellar Medium
ASP Conference Series, Vol. 35, 1993
Joseph P. Cassinelli and Edward B. Churchwell (eds.)

STELLAR WIND ACCELERATION OF ATOMIC GAS
FROM A DISSOLVING MOLECULAR CLOUD?

DAVID LEISAWITZ
Department of Astronomy and Astrophysics, Pennsylvania State
University, 525 Davey Laboratory, University Park, PA 16802

EUGÈNE DE GEUS
Astronomy Program, University of Maryland, College Park, MD 20742

INTRODUCTION AND OBSERVATIONS

To learn about the fate of molecular clouds that produce massive stars and
the formation of molecular clouds from atomic gas we (Leisawitz and de Geus
1991) surveyed the 21 cm line emission from a 162 deg^2 area around NGC 281
($\ell = 123°13$, $b = -6°24$) and several other OB clusters whose neighborhoods
had previously been mapped for CO (J = 1→0) emission (Leisawitz, Bash, and
Thaddeus 1989). NGC 281 contains an O6.5V star and is the most prominent
region of star formation in this part of the sky. Its distance is about 2 kpc.

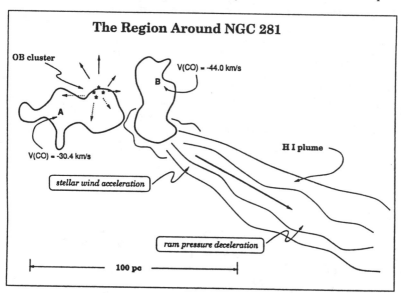

Fig. 1.   Molecular cloud B, at -44.0 km s^{-1}, is the apparent source of an
H I plume. Cloud A, at -30.4 km s^{-1}, is also associated with NGC 281.

The region around NGC 281 is depicted schematically in Fig. 1. A circular
region of radius 0°8 centered on the cluster was surveyed for CO emission.
Contours of CO line intensity delineating molecular clouds A and B are shown

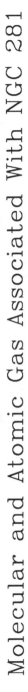

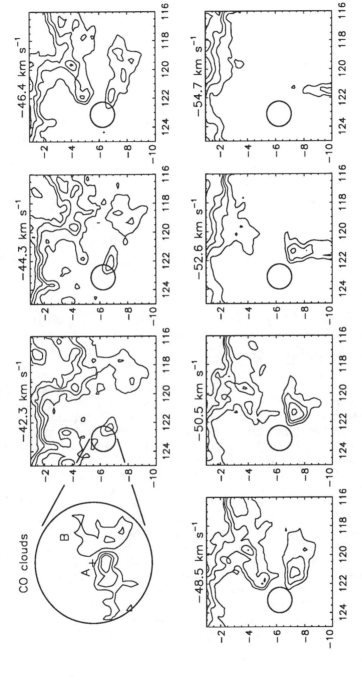

Fig. 2   21-cm emission at velocities close to that of cloud B. Circular inset shows molecular clouds near NGC 281 (+). All plots are in Galactic coordinates and labeled in degrees. Cloud A CO contours are at 3, 6 and 9 K km/s; cloud B contours are at 1 and 3 K km/s; H I contours are at 40, 60, 80, and 100 K km/s.

in the circular inset in Fig. 2, which includes a series of "channel maps" showing the 21 cm line intensity at velocities close to that of cloud B. A 150 pc long H I "plume," evident in the $-46.4\,\mathrm{km\,s^{-1}}$ map, starts at the position of cloud B and ends near $\ell = 118°$, $b = -8°$. Atomic gas near the midpoint of the plume's long axis is concentrated at a more negative radial velocity than gas at the two endpoints; the intensity-weighted velocity is shown as a function of position along the plume axis in Fig. 3a. The $2.4 \times 10^4\ M_\odot$ mass of the plume is distributed as shown in Fig. 3b.

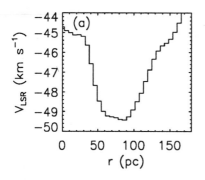 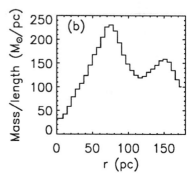

Fig. 3.   21-cm line velocity (a), and H I mass per unit length (b), as a function of projected position along the plume axis. NGC 281 is at r = 0.

PRELIMINARY INTERPRETATION

A model in which stellar wind accelerates freshly photodissociated gas, which eventually sweeps up ambient gas and slows down (see Fig. 1), can explain the observations qualitatively, but not quantitatively. The dynamical time for expansion of the atomic gas is $(r/\Delta V_{LSR})\tan i$, where i is the inclination of the plume to the plane of the sky and $r/\Delta V_{LSR} \simeq 20\,\mathrm{Myr}$. For consistency with the main sequence lifetime of an O6.5 star, i must be $\lesssim 10°$, in which case the observed velocities imply that the atomic gas was accelerated to nearly $30\,\mathrm{km\,s^{-1}}$ and acquired about 10^{50} ergs of kinetic energy. Since the solid angle subtended by cloud B as viewed from NGC 281 is $\ll 4\pi\,\mathrm{sr}$, it is difficult to understand how so much energy could have been absorbed. Indeed, an analytical momentum balance calculation that takes the mass profile (Fig. 3b) into account implies that a cumulative wind energy 1000 times greater than that of an O6 star would be required to accelerate the gas. An alternative model in which the molecular cloud forms as a result of compression of atomic gas already swept from the neighborhood of the cluster has the same deficiency. We conclude that wind from NGC 281 is not responsible for the velocity structure in the H I plume; the acceleration source remains unidentified.

EdG acknowledges financial support from NSF grant AST 89-18912.

REFERENCES

Leisawitz, D., Bash, F., and Thaddeus 1989, *Ap. J. Suppl.*,**70**, 731.
Leisawitz, D. and de Geus, E. 1991, *Ap. J. Suppl.*,**75**, 835.

Massive Stars: Their Lives in the Interstellar Medium
ASP Conference Series, Vol. 35, 1993
Joseph P. Cassinelli and Edward B. Churchwell (eds.)

A MILKY WAY CONCORDANCE

PETER J. BARNES and PHILIP C. MYERS
Harvard-Smithsonian Center for Astrophysics, MS 42, 60 Garden Street,
Cambridge, MA 02138

ABSTRACT   We describe a new graphical concordance for the Milky Way useful for making cross-catalogue comparisons and for identifying otherwise unknown objects. The concordance is interactive and uses the SuperMongo plotting package. It can easily be extended to treat more catalogues than the six currently available.

When conducting surveys of classes of astronomical objects, or in other sorts of studies using large astronomical databases, it is sometimes necessary or convenient to consult previously existing catalogues as an aide to one's research goals.   When several such catalogues need to be consulted, and positional coincidences between program sources and previously known objects are required, one can become bogged down dealing with large volumes of tabular information which are relatively difficult to assimilate.   Marsalkova (1974) produced the first *graphical* concordance for nebulous objects in the galactic plane, enabling much easier comparisons of this kind. The "Ohio overlays" for the POSS prints (Dixon, Gearhart, & Schmidtke 1981) served a similar purpose.   For both of these works, however, the scale of the plots and the selection of data plotted are fixed.   In addition, improvements in the existing catalogues and the appearance of new catalogues have rendered these venerable references somewhat out of date.

During our search for new regions of intermediate-mass star formation based on IRAS PSC data (see the following paper), we were motivated to develop a more up-to-date and interactive graphical resource for determining these cross-catalogue concordances, the consulting of reams of tables being far too labour-intensive.   Thus, to facilitate searches for associations with previously identified objects, we developed a set of plotting routines using SuperMongo (Lupton & Monger 1990). The package consists of a set of macros that are run within SuperMongo. The first macro establishes the user's field of interest (all plots are performed in galactic coordinates).   Using subsequent macros, the user then has the option of plotting all entries of a given catalogue that lie within the chosen field.   The catalogues to plot from can be selected individually, or one can simply plot entries from all available catalogues.

The catalogues' data are assumed to be in tabular ascii files, the macros reading and plotting only positional, size, and labelling information.   At the moment, the package plots entries from the following catalogues: Blitz, Fich, & Stark (1982 – northern galactic HII regions); Rodgers, Campbell, & Whiteoak

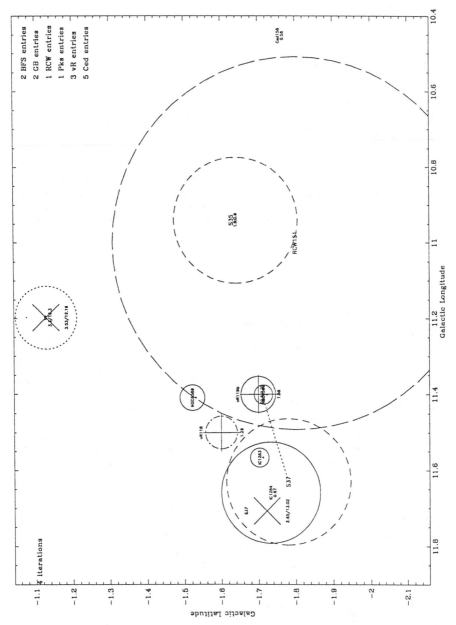

Fig. 1. Sample concordance for a region near galactic longitude 11°. Each entry in the various catalogues is represented as a circle approximating the object's size, where size information is available. Objects are labelled with a name, and distance where available. The objects shown are HII regions (Sharpless/BFS, short-dashed circles; RCW, long-dashed circles; Greenbank, crosses at peak positions; Parkes, dotted circles) and reflection nebulæ (van den Bergh & Racine, dot-dashed circles; Cederblad, solid circles).

(1960 – southern galactic HII regions); Lockman (1989 – northern radio HII regions); Caswell & Haynes (1987 – southern radio HII regions); van den Bergh (1966), Racine (1968), and Cederblad (1946) (all northern reflection nebulae). It would be a simple matter to extend the package to include other catalogues, such as ones of dark clouds, planetary nebulae, or supernova remnants; catalogues of stellar objects; or catalogues of extragalactic objects. For example, selections from the CD-ROM "Selected Astronomical Catalogs" issued by the Astronomical Data Center could easily be accomodated. All that would be necessary would be to obtain a tabular ascii form of the desired catalogue, and generate a macro to plot the appropriate quantities (which could be done using one of the other catalogue macros as a template).

In Figure 1 we display the concordance for a region around S35 and S37, based on the catalogues listed above. Of course any desired region can be plotted at any scale of the user's choosing. The macro package is available from us on request (SuperMongo must be obtained by the user separately). Send email to peterb@cfacx2.harvard.edu to be put on the email information distribution list.

REFERENCES

Blitz, L., Fich, M., & Stark, A.A. 1982, *Ap. J. Suppl.*, **49**, 183
Caswell, J.L., & Haynes, R.F. 1987, *A&A*, **171**, 261
Cederblad, S. 1946, *Lund Astr. Obs. Medd.*, Ser. II, no. 119
Dixon, R.S., Gearhart, M.R., & Schmidtke, P.C. 1981, *Atlas of Sky Overlay Maps*, Ohio State University Radio Observatory, unpublished
Lockman, F.J. 1989, *Ap. J. Suppl.*, **71**, 469
Lupton, R. & Monger, P. 1990, SuperMongo manual, unpublished
Marsalkova, P. 1974, *Ap. Space Sci.*, **27**, 3
Racine, R. 1968, *A.J.*, **73**, 233
Rodgers, A.W., Campbell, C.T., & Whiteoak, J.B. 1960, *M.N.R.A.S.*, **121**, 103
van den Bergh, S. 1966, *A.J.*, **71**, 990

Massive Stars: Their Lives in the Interstellar Medium
ASP Conference Series, Vol. 35, 1993
Joseph P. Cassinelli and Edward B. Churchwell (eds.)

A SEARCH FOR INTERMEDIATE MASS STAR-FORMING REGIONS WITHIN ONE KPC OF THE SUN

PETER J. BARNES and PHILIP C. MYERS
Center for Astrophysics, MS 42, 60 Garden Street, Cambridge, MA 02138

ABSTRACT   We describe an IRAS-based search for nearby intermediate mass star-forming regions (ISFRs).   We have begun observing these regions in several dense molecular gas tracers, in order to characterise the physical conditions in the dense cores associated with these IRAS sources.

BACKGROUND

Physical conditions in dense cores forming low-mass stars contrast markedly with those in cores giving rise to massive stars (Myers 1991*a,b*).   Studies of

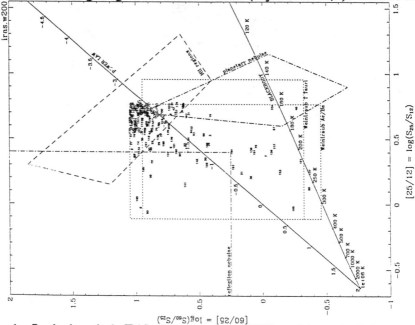

Fig. 1.   Result, shown in the IRAS colour-colour plane, of PSC search (see text).   Each point is labelled by its number in the resulting source list.   Also shown are the Weintraub colours for Herbig Ae/Be stars and T Tauri stars, as well as the rough domains for reflection nebulae, HII regions, and planetary nebulae (after Hughes & MacLeod 1989).   Lines representing loci of blackbodies of various temperatures, and of power-law spectra, are as shown.   Note that our selection method introduces substantial contamination by distant, luminous HII regions which fall within the Ae/Be domain.

ISFRs offer a good opportunity to see how cloud properties change as the mass of the stellar product rises, without the confusion of the more chaotic massive SFRs. (Here intermediate mass $\Leftrightarrow$ Sp ~A9 to ~B3 $\Leftrightarrow$ M ~2–10 $M_\odot$.) We are using the Point Source Catalog to compile an extensive list of nearby ISFRs.

SEARCH TECHNIQUE

We searched the PSC for *all* objects with colours typical of Ae/Be stars (Weintraub 1990) and $S_{60\mu m} \geq 200$ Jy; see Fig. 1. We need to select against non-star-forming contaminants and against those objects >1kpc away that are incidentally included by our procedure. To obtain distances and make associations between each IRAS source and previously catalogued objects we used the concordance described in the previous paper. We identified ~80% and obtained distances for ~50% of the 202 sources in Fig. 1. Of the 124 north of $\delta = -40°$, 15 are at known distances ≤ 1 kpc and 41 more were either unidentified, not clearly associated, or only associated with objects at poorly-known distances (Table I).

TABLE I   Source List

number	α	δ	ID	number	α	δ	ID
104	17ʰ15ᵐ 4.6	–32°24′15″	GL6815S	174	19ʰ22ᵐ45.8	+17°21′35″	
110	17 21 41.4	–38 1 21		175	19 23 0.9	+15 6 28	
116	17 28 14.1	–33 35 59		176	19 23 40.7	+14 56 55	
118	17 38 35.6	–31 16 56		177	19 26 49.5	+17 54 54	GL5381S
119	17 39 34.4	–29 50 21		178	19 27 3.9	+17 50 4	
120	17 39 51.9	–29 48 32	GL 5377	179	19 31 12.8	+19 50 22	
121	17 41 7.4	–31 54 24	GL 5379	180	19 38 19.5	+27 11 33	
123	17 43 18.6	–29 21 25	S 16	181	19 45 53.5	+24 42 48	
125	17 50 29.8	–25 19 21		183	19 49 56.9	+26 13 23	
132	18 14 29.8	–17 23 23		184	19 52 3.0	+27 59 43	
137	18 18 26.7	–13 2 54	MWC922	186	20 2 26.8	+33 30 25	
138	18 19 37.3	–13 31 45	GL 2136	187	20 11 0.9	+33 21 19	
139	18 22 22.3	–12 43 59		188	20 19 46.8	+37 21 34	BC Cyg
141	18 25 0.5	–3 51 49	MWC297	191[*†]	21 0 58.4	+67 58 15	NGC7023
143	18 27 14.5	–12 17 32		194	21 40 43.7	+54 41 22	LDN1084
				195	21 41 21.2	+54 42 30	"
144	18 28 17.3	–10 24 51	IC 1287				
145	18 30 30.4	–8 26 18					
146	18 30 38.0	–8 35 58		197[††]	22 17 41.1	+63 3 41	S140
149	18 31 25.5	–8 20 15		202	23 54 34.1	+65 8 29	GL 5623
151	18 31 46.1	–5 13 23	GL 5508	3[*†]	3 23 39.0	+58 36 33	GL 490
				4[*†‡]	3 25 57.9	+31 5 50	NGC1333
158	18 48 51.1	+0 0 41	L599/604	5[*†‡]	3 26 4.7	+31 11 41	"
159	18 49 27.1	+1 10 2					
160	18 50 44.5	+0 57 7		6[*†]	4 26 57.2	+35 10 1	LkHα101
161	18 51 6.4	+1 46 40	GL 5542	8[†]	5 4 25.8	–3 25 8	NGC1788
165	18 58 18.5	–36 57 2	TY CrA	11	5 35 34.0	+30 39 48	DG 66
				14[††]	5 43 44.1	–0 1 23	NGC2068
166[*†]	18 58 32.9	–37 1 32	R,T CrA	20[*††]	6 38 26.2	+9 32 25	GL 989
167	18 58 43.0	+5 21 27					
168	19 4 22.3	+7 26 55					
171	19 13 17.5	+10 35 55					
173	19 21 22.9	+17 23 6					

* = already mapped by others in NH₃
† = mapped in CO and ¹³CO
‡ = mapped in CS
¶ = mapped in C¹⁸O
§ = mapped in 21cm continuum

DATA COLLECTION

We observed all 56 candidate ISFRs with the FCRAO 14m and NRAO 12m tele-scopes, making small (~4–5´), quick maps in ^{13}CO $J=1\rightarrow0$ around the IRAS po-sition in order to associate a velocity with each source and so derive kinematic distances. Of the 41 unknowns, 14 have d≤1kpc. We are now observing the total of 29 nearby regions (Fig. 2) in C^{18}O $J=1\rightarrow0$ and CS $J=2\rightarrow1$ (resp. 19 & 10 maps so far), and will also eventually map these 29 in NH$_3$ (1,1) and HC$_3$N $J=4\rightarrow3$. The data will be used to obtain physical conditions in each core and should indicate those suitable for further study (*e.g.*, see the next paper). In the future we will extend this work to lists with lower flux cutoffs. In the first list, we have found that ~15% are northern objects satisfying our criteria. If this rate also holds for the fainter sources, we might obtain a list of ~100 objects, of which perhaps half could be new identifications of ISFRs near the Sun.

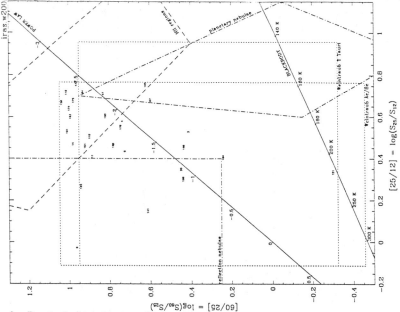

Fig. 2. Result of editing the source list from Fig. 1, using catalogues and our new molecular line observations. Shown are the 29 northern objects that are known to be nearby. Note that the preponderance towards HII regions among these nearby sources is much less pronounced than in the original list of 202 objects (Fig. 1).

REFERENCES

Hughes, V.A. & G.C. MacLeod 1989, *A. J.*, **97**, 786

Myers, P.C. 1991*a*, in *The Formation and Evolution of Star Clusters, A.S.P. Conference Series*, **11**, K.A. Janes, ed.(Chelsea, MI: Astr.Soc.Pac.), p.73

Myers, P.C. 1991*b*, in IAU Symposium 147, *Fragmentation of Molecular Clouds and Star Formation*, E. Falgarone & G. Duvert, eds. (Dordrecht: Kluwer), p. 221

Weintraub, D.A. 1990, *Ap. J. Suppl.*, **74**, 575

Massive Stars: Their Lives in the Interstellar Medium
ASP Conference Series, Vol. 35, 1993
Joseph P. Cassinelli and Edward B. Churchwell (eds.)

LARGE-SCALE MAPPING OF DENSE CORES IN THE COCOON NEBULA, IC 5146

PETER J. BARNES and PHILIP C. MYERS
Harvard-Smithsonian Center for Astrophysics, MS 42, 60 Garden Street, Cambridge, MA 02138

MARK HEYER
FCRAO, 619 Lederle GRC, University of Massachusetts, Amherst, MA 01003

ABSTRACT We report preliminary results of mapping all dense cores in the IC 5146/S125/LDN 1030–1055 complex. We have mapped $C^{18}O$ and HCO^+ $J=1\rightarrow0$ with good sensitivity using the FCRAO 14m (with the new QUARRY focal plane array) and the NRAO 12m telescopes. We are in the process of mapping the same cores in NH_3 and HC_3N at Haystack Observatory.

INTRODUCTION

As a follow-up to the work described in the previous paper, we also wish to survey many cores in a few selected intermediate-mass SFRs. This is because it is important to obtain data on a large sample of cores *without* embedded IRAS sources nearby ("starless" cores), in order to compare the properties of cores both before and after stars have formed. In this way we might hope to ascertain which physical conditions *govern* star formation and which are simply a *consequence* of star formation. The Cocoon Nebula IC 5146 (also known as S125) is an excellent intermediate-mass region to study since it contains an optical cluster whose most massive stars are early-A and late-B at one end of an otherwise dark collection of molecular clouds, L1030–55 (Hallas & Mount 1990). This structure of a prominent cluster at the end of an elongated, quiescent dark cloud with associated IRAS sources is similar to that of the complexes in Orion A, Ophiuchus, and Corona Australis.

PREVIOUS WORK

The L1030–55 dark clouds have been included in several surveys such as that of Benson & Myers (1989), yielding only limited NH_3 maps around some emission peaks. Roger & Irwin (1982) mapped the HII region and HI emission around the cluster area, and constructed a detailed champagne-flow model for the gas motion and distribution next to the molecular material at that end of the

Barnes, Myers, Heyer

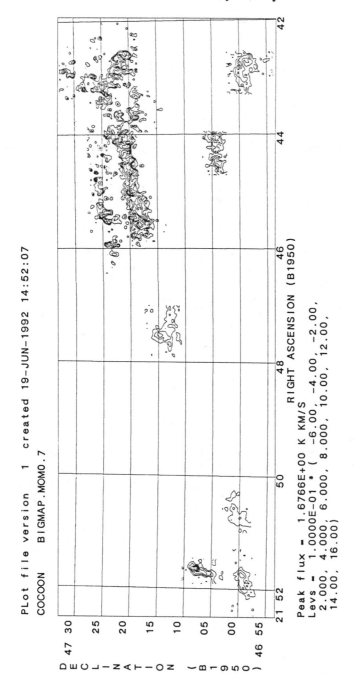

Fig. 1. Map of integrated $C^{18}O$ $J=1\rightarrow0$ emission for the entire complex. Contour levels are as shown, except that the ±0.2 K km s^{-1} levels are not displayed for the long filament to the northwest.

dark complex. McCutcheon, Roger, & Dickman (1982) produced the best published [12]CO and [13]CO maps to date of the vicinity of the optical cluster, which revealed residual, warm molecular material (up to 35 K for [12]CO, 8 K for [13]CO) that was confining the shape of the HII region and directing the general gaseous outflow. These two studies were at resolutions of only ~2–3 arcmin, however. The best published infrared survey to date is that of Elias (1978), who surveyed the entire L1030–55 area for young stellar objects but found only three brighter than a K magnitude of 7. The highest-resolution radio continuum map is that of Israel (1977).

NEW RESULTS

Lada, Lada & Bally (personal communication) and Fukui *et al.* (personal communication) have conducted complete, sensitive surveys in [12]CO, [13]CO and CS of the L1030–55 cloud complex using the Bell Labs 7m and Nagoya 4m telescopes. C. Lada & E. Lada have also recently performed a very sensitive SQIID-IR survey of the clouds, while we have ourselves conducted arcminute-resolution, Nyquist-sampled surveys in $C^{18}O$ and HCO^+ of all the cores revealed in the Lada and Fukui surveys, using the new QUARRY receiver on the FCRAO 14m (Fig. 1). This 3mm receiver is a 15-element focal-plane array which allows very efficient mapping of large areas of sky in various molecular transitions. Using the NRAO 12m, we have also obtained spectra of [13]CO $J=1\rightarrow0$ and $C^{18}O$ $J=2\rightarrow1$ at the peak positions of each core. This will enable us to derive column densities and excitation temperatures for the emitting gas. In addition, we have begun to map the same cores in ammonia at the Haystack 37m, in order to compare the properties of the thermally-dominated gas with the material traced by these other molecules.

Our QUARRY maps have revealed the existence of 20–30 distinct condensations in the dense gas at this resolution (Fig. 1). The distribution of the emission follows very closely that in the earlier surveys, after allowing for the different resolutions. We are now in the process of obtaining the various structural, thermal, and chemical properties of these cores from our observations. We hope to use these data to address some of the questions raised in the introduction. It is already clear, however, from the $C^{18}O$ and HCO^+ observations, that the cores near the HII region are similar in size, brightness, and linewidth to the cores far to the west. This suggests that, for stars as early as B0 (the most luminous star in the cluster), the disruption to the remaining molecular environment is minimal. Preliminary NH_3 results show that the whole region is faint in this line; also, the cluster cores are among the weakest of these cores.

REFERENCES

Benson, P.J. & Myers, P.C. 1989, *Ap. J. Suppl.*, **71**, 89.
Elias, J.H. 1978, *Ap. J.*, **223**, 859.
Hallas, T., & Mount, D. 1990, *Sky & Tel.*, **80**, 368.
Israel, F.P. 1977, *A&A*, **60**, 233.
McCutcheon, W.H., Roger, R.S., & Dickman, R.L. 1982, *Ap. J.*, **256**, 139.
Roger, R.S. & Irwin, J.A. 1982, *Ap. J.*, **256**, 127.

Massive Stars: Their Lives in the Interstellar Medium
ASP Conference Series, Vol. 35, 1993
Joseph P. Cassinelli and Edward B. Churchwell (eds.)

BIPOLAR MOLECULAR OUTFLOW FROM A MASSIVE STAR: HIGH RESOLUTION OBSERVATIONS OF G5.89-0.39

DOUGLAS O. S. WOOD
National Radio Astronomy, P. O. Box O, Socorro, New Mexico, 87106

ABSTRACT Hot ammonia surrounding the ultracompact (UC) HII region G5.89-0.39 imaged at 4" resolution using the Very Large Array (VLA) reveals the origin of the most massive bipolar molecular outflow in the Galaxy. The UC HII region and molecular outflow appear to be produced by a massive star embedded in a dense molecular torus.

INTRODUCTION

The massive star formation region G5.89-0.39 (W28A$_2$) has produced one of the brightest ultracompact HII regions in the Galaxy and the most massive bipolar molecular outflow known. Harvey and Forveille's (1988) ^{12}CO observations found line wings *180 km/s wide* and an outflow mechanical luminosity of ~1500L$_\odot$. Zijlstra *et al.* (1990) mapped the outflow and Cesaroni *et al.* (1992) found a high density C^{34}S outflow oriented slightly west of north. The HII region is a shell (Zijlstra & Pottasch 1988, Wood & Churchwell 1989) and recent 3.6 cm observations by Wood *et al.* (1993) show density enhancements in the shell perpendicular to the outflow with low surface brightness "openings" in the HII emission aligned with the outflow axis.

RECOMBINATION LINE OBSERVATIONS

VLA observations of the H76α recombination line emission from the UC HII region have a remarkable velocity structure, but space limitations prevent me from showing them here. Even at ~1" resolution, the line profile is double peaked at the poles of the outflow. At the equator the lines can be modelled with a single Gaussian. The velocity field is consistent with the ionized gas flowing from the equator of the outflow, over the surface of the UC HII region, converging on the poles. The northern pole appears to be pointing away.

AMMONIA OBSERVATIONS

Imaging the weak ammonia lines is difficult due to the strong continuum emission of the HII region and the presence of both emission and absorption in the molecular lines. Visibility-based continuum subtraction was used to achieve the necessary 1000:1 spectral dynamic range, and a maximum emptiness image reconstruction was used to deconvolve the emission and absorption. This may in part explain the improved quality of these images over those presented by Gómez *et al.* (1991). Fig. 1 shows the NH_3 (3,3) observations. the (2,2) results are similar, but space prevents me from presenting them here. The molecular outflow originates at ~10 km/s where the NH_3 emission surrounds the UC HII region. The gas reaches ~13 km/s toward the northwest, and ~6 km/s in the south. Absorption against the HII region peaks at ~7 km/s.

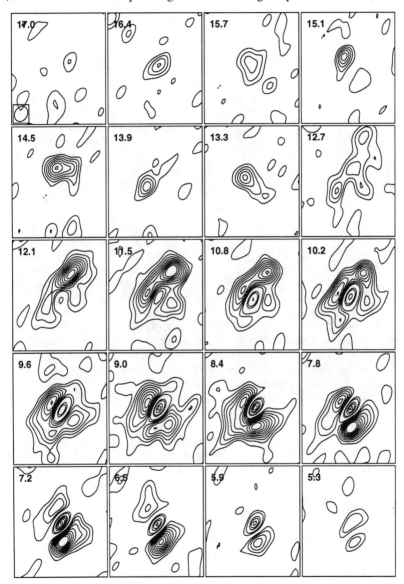

Fig. 1. NH_3 (3,3) channel maps made with the VLA. The 42× 48" fields are centered at $17^h 57^m 26^s8 -24° 03' 56"$ (1950) with contour levels of 20 mJy/beam. The LSR velocity of each image is shown in the upper left of each panel The 4" beam is shown in the lower left of the first panel.

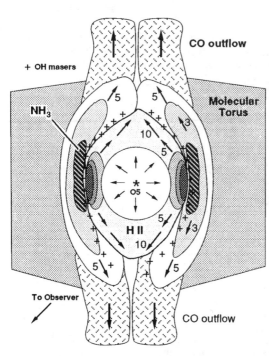

Fig 2. A schematic model for G5.89-0.39.

These observations suggest that an O5 star (or cluster of smaller stars) is surrounded by a dense torus of molecular gas. The inside edge of the torus is ionized by the star, producing the ultracompact H II region. The massive wind of the O5 star blows a bubble in the center of the H II region, compressing the ionized gas onto the surface of a sphere. Gas from the torus is added at the equator where it is ionized and flows over the surface of the sphere toward the poles.

The star shocks and heats the dense molecular gas just outside the H II region in the torus. The molecular gas flows from the equator toward the poles where it reaches a velocity of ±5 km/s before excitation conditions render the high excitation lines invisible. Further along the flow the CO lines are accelerated to ~60 km/s.

The entire torus appears to be rotating about the star with a total enclosed mass of ~200 $M_\odot$. These ammonia lines trace only the hot molecular gas near the star; presumably much more cold molecular gas lies further out in the disk. It is not clear whether the disk is being eroded away or whether accretion of new molecular material will keep the H II region from expanding.

The hot molecular gas traced by the NH_3 lines appears to be simultaneously rotating about the star and outflowing perpendicular to the disk. It establishes the origin of the massive bipolar outflow seen in CO and other molecular lines. Because G5.89-0.39 is excited by a massive star, it provides many observational probes to study the bipolar molecular outflow phenomenon in detail, and is an important link between low and high-mass star formation.

REFERENCES

Harvey, P. M and Forveille, T., 1988, Astr. Ap., 197, L19.
Cesaroni, R., Walmsley, C. M., Kompe, C. and Churchwell, E., 1992, Astr. Ap. (Univ. Wisc. Preprint #401).
Wood, D. O. S., Churchwell, E., Van Buren, D. and Mac Low, M.-M., 1993, Ap. J., in preparation.
Wood, D. O. S. and Churchwell, E., 1989, Ap. J. Suppl., 69, 831.
Zijlstra, A. A. et al., 1989, Astr. Ap., 217, 157.
Zijlstra, A. A., et al., 1990, M.N.R.A.S., 246, 217.
Gómez, Y., Rodríguez, Garay, G, Moran, J. M., 1991, Ap. J., 377, 519.

Massive Stars: Their Lives in the Interstellar Medium
ASP Conference Series, Vol. 35, 1993
Joseph P. Cassinelli and Edward B. Churchwell (eds.)

HIGH-VELOCITY ABSORPTION FEATURES TOWARD EVOLVED
MASSIVE STARS

MARGARET M. HANSON
Department of Astrophysical, Planetary, and Atmospheric Sciences,
University of Colorado, Campus Box 391, Boulder CO 80309

ABSTRACT   We are currently obtaining high resolution optical and
ultraviolet spectra of massive stars in regions of active star formation to
study their effect on the local interstellar medium. We have concentrated
our efforts on two star forming regions in Cygnus and Carina. Here, we
present some preliminary results from that work in progress.

THE CYGNUS REGION

A recent study of the IRAS 60μm/100μm ratio maps (see Fig 1, Shull 1993,
this volume), has revealed a large "peanut shaped" cavity in the southeast
corner of the Cyg X Molecular Cloud complex (Saken et al. 1992). Just to the
north of this cavity, two reasonably well defined shells have been detected. At a
distance of 1.5 to 2.5 kpc, their linear diameters vary from approximately 20 pc
to 40 pc. We are obtaining optical and UV spectra of three stars whose
positions are superimposed on the shells with the hope of detecting the shells
in blue-shifted interstellar lines. In all three stars, blue-shifted absorption lines
have been detected, but it is difficult to know what is being sampled in this gas.
    Toward HD 193514 [O7e, E(B-V)=0.75], moderate velocity (≈ -60 km
s^{-1}) gas has been detected in Fe II, Mg II, Mg I, C II*, Si II and possibly O I
by Phillips et al. (1984) and St.-Louis & Smith (1991) using IUE. We obtained
additional IUE exposures, thus improving the signal-to-noise in the spectra,
and have confirmed their detections of the low ionization species. The higher
ionization species of Al III, Si IV and C IV are more difficult to confirm.
However, the profiles do show broad wings on the blue side of the absorption.
A single low-velocity feature has been detected in Ca II at -20 km s^{-1}. If the
infrared dust shell is associated with HD 193514, it is probably traced in the
lower velocity Ca II feature. The higher velocity features detected with IUE
may be associated with a fast wind-blown bubble closer to the star (Weaver et
al. 1977).
    HD 192660 [B0 Ia, E(B-V)=0.88] sits at the center of a weak but
complete infrared shell. Ca II features were detected at -30 km s^{-1} and $+20$
km s^{-1}, suggestive that the supergiant star is behind the infrared shell which is
expanding at approximately 25 km s^{-1}. The IUE spectra of HD 192660 shows
no convincing high-velocity components. However, the total gas column density
is large enough that any moderate velocity components would be lost in the
wings of the broad zero velocity component.

THE CARINA REGION

Near Eta Carina and around the Carina OB regions, numerous studies have identified multiple high and moderate velocity absorption components with IUE and ground based optical data. The aim of these studies has been to study the large scale motions of the ISM due to the intense star formation (e.g. Cowie et al. 1981; Walborn & Hesser 1982; Nichols-Bohlin & Fesen 1990). The aim of our work on these sightlines is to study indepth a few single sightlines in order to understand better the physical phenomenon responsible for the high velocity features. To this end, we are obtaining additional IUE data on 3 stars that are known to show a single, clean high-velocity feature in them.

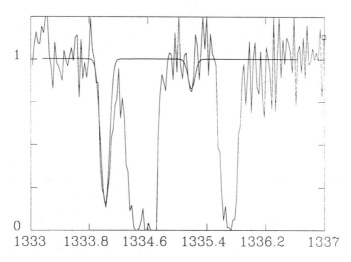

Fig. 1.   The high-velocity component toward one of our program stars in Carina, HD 96946. Inputs for the model profile include: b_{Dop}=6 km s^{-1}, C II Log(N)=15.50 cm^{-2} and C II* Log(N)=13.15 cm^{-2}. The two features lie at velocites of -115 km s^{-1}. Wavelengths are in Angstroms.

HD 96946 shows an exceptionally clean high velocity component at -115 km s^{-1} which appears in over a dozen different lines. Curve of growth analysis of the 2795Å and 2803Å lines of Mg II and the 2599Å and 2382Å lines of Fe II show no hint of saturation. We have completed an abundance analysis of C II, C II*, S II, Si II and Al II assuming the lines are unsaturated. The small abundance of C II* indicates a very low spatial density of electrons, about 0.1 cm^{-3} based on a simple two-level atom model. Because O I is **not** detected, the shell must be fully ionized and the electron density is equal to the hydrogen space density. From the column density of Mg, S and Si, we determine a total hydrogen column density in the expanding shell of approximately N(H)=10^{18} cm^{-2}. The relative abundances of Mg, S, Si, and Fe are very close to solar, indicating nearly complete grain destruction in the shock, though the high abundance of Fe II seems puzzling. At such energies, much if not most of the Fe should exist as Fe III, which is not observable with the IUE (Jenkins, Silk &

Wallerstein 1976).

HD 97152 is a Wolf-Rayet star, spectral type WC7+abs. Toward this star, high velocity features have been detected in 19 lines, including two features in Mg II, two features in Fe II, three features in Si II, two features in Al III and the doublet features in C IV and Si IV. The lower ionization species appear to lie at a slightly lower velocity (approximately -80 km s^{-1}) than the higher ionization species of Al III, Si III, Si IV and C IV (at approximately -110 km s^{-1}). Curve of growth analysis of the Si II, Fe II and Mg II lines indicate these lines to be relatively saturated ($b_{Dop} \approx 6$ km s^{-1}) and possibly slightly depleted from the gas phase, while the higher ionization species show no hint of saturation and must have b_{Dop} of order 20-30 km s^{-1}. It appears likely that the low ionization species and the high ionization species are not spatially related.

WORK IN PROGRESS

The exceptionally clean single components seen toward HD 96946 and HD 97152 have allowed us to determine a number of physical parameters in the expanding shell. We are presently exploring models of wind blown bubbles and circumstellar shells (Hanson & McCray 1992). Because HD 97152 is an evolved helium star, it is possible that one of the two velocity components seen is due to the interaction of the star's wind with a previously expelled circumstellar shell as is now detected around SN1987A.

ACKNOWLEDGEMENTS

This work was made possible through a grant from NASA-*IUE* (contract NSG-5300) at the University of Colorado. M.M.H. acknowledges support from the NASA Graduate Student Researchers Program.

REFERENCES

Cowie, L.L., Hu, E.M., Taylor, W. & York, D.G. 1981, *Ap. J.*, **250**, L25.
Hanson, M.M. & McCray, D. 1992, in progress.
Jenkins, E., Silk, J. & Wallerstein, G. 1976, *Ap. J. Suppl.*, **32**, 681.
Nichols-Bohlin, J. & Fesen, R.A. 1990, *Ap. J.*, **353**, 281.
Phillips, A.P., Welsch, B.Y., & Pettini, M. 1984, *MNRAS*, **206**, 55.
Saken, J.M., Shull, J.M., Garmany, C.D., Nichols-Bohlin, J. & Fesen, R. 1992, *Ap. J.*, **397**, in press.
Shull, J.M. 1993, in *Massive Stars: Their Lives in the Interstellar Medium*, ed. J. Cassinelli & E. Churchwell, (Publ. Astro. Soc. Pac.)
St.-Louis, N. & Smith, L.J. 1991, *A&A*, **252**, 781.
Walborn, N.R. & Hesser, J.E. 1982, *Ap. J.*, **252**, 156.
Weaver, R., McCray, R., Castor, J., Shapiro, P. & Moore, R. 1977, *Ap. J*, **218**, 377.

Massive Stars: Their Lives in the Interstellar Medium
ASP Conference Series, Vol. 35, 1993
Joseph P. Cassinelli and Edward B. Churchwell (eds.)

COMPACT AMMONIA SOURCES TOWARD THE G10.5+0.0
STAR FORMING REGION

GUIDO GARAY
Departamento de Astronomía, Universidad de Chile, Santiago, Chile

LUIS F. RODRIGUEZ
Instituto de Astronomía, UNAM, México, D. F., México

JAMES M. MORAN
Harvard-Smithsonian Center for Astrophysics, Cambridge, MA, USA

ABSTRACT   We present the characteristics of three compact ($\sim$0.3 pc) molecular clouds detected, with the VLA, toward the G10.5+0.0 star forming region. Their physical parameters, kinematics, and association with massive stars are discussed.

Radio continuum observations made with angular resolution of $\sim$ 15" show that G10.5+0.0 consists of three extended ($\sim$ 40") complex sources of ionized gas, spread over a region of $\sim$ 3' in diameter (Garay *et al.* 1992). At resolutions of $\sim$ 1", radio continuum maps show that two of these complexes contain UCHII regions (Wood and Churchwell 1989; Garay *et al.* 1992), indicating that they are sites of recent star formation. Signs of active star formation toward G10.5+0.0 are also indicated by the presence of OH and H_2O maser emission (Genzel and Downes 1977; Braz and Sivagnanam 1987; Churchwell, Walmsley and Cesaroni 1990).

Recently, G10.5+0.0 has been the subject of single dish telescope observations (resolution of $\sim$ 40") in several transitions of NH_3 and $C^{34}S$ (Cesaroni *et al.* 1991; Cesaroni, Walmsley, and Churchwell 1992). Notable characteristics of the observed molecular emission are its strength and broad line profiles, of $\sim$8 km s^{-1} (FWHM). These observations are very important since they show the presence of compact, dynamical structures of dense and hot molecular gas in regions of active star formation. They suffer some shortcomings, however; the low angular resolution does not allow to establish whether there is a close association between the compact molecular cores and the UC HII regions, nor to determine the origin of the broad profiles. In view of these considerations we decided to probe the dense molecular gas toward G10.5+0.0 with higher angular resolution. Our main goals were to determine its structure, physical conditions, kinematics, and location relative to the compact HII regions.

The observations were made with the Very Large Array of the NRAO in the (J,K)= (2,2) and (3,3) inversion transitions of ammonia. In this contribution we briefly report the main results and conclusions of our

VLA study. A detailed description is presented elsewhere (Garay, Moran, and Rodríguez 1992).

We detected three distinct compact molecular cores, spread over a region of ~ 2′ in diameter : NH_3 G10.46+0.03, NH_3 G10.47+0.03, and NH_3 G10.48+0.03 (see Fig. 1). Their ammonia line center velocities are, respectively, 71.3, 66.9, and 65.4 km s^{-1} . The diameter of these cores, assuming they are located at the distance of 5.8 kpc (Churchwell, Walmsley, and Cesaroni 1990), is ~0.3 pc.

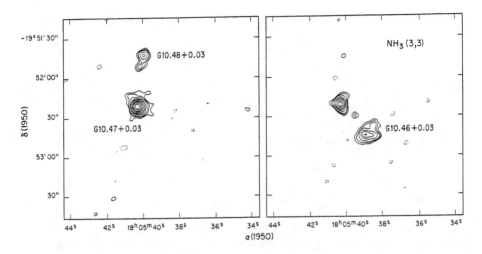

Figure 1.   Line maps of the (3,3) ammonia emission from the G10.5+0.0 region. Right: average of the 63.1 and 65.6 km s^{-1} channel maps. Left: average of the 70.5 and 72.9 km s^{-1} channel maps.

The brightest NH_3 clump is associated with the G10.47+0.03 cluster of UC HII regions (Wood and Churchwell 1989). It shows strong emission in the main and satellite HF components of the (2,2) and (3,3) transitions, broad line profiles, and an halo-core morphology. The core, of ~ 0.08 pc in size, has a peak NH_3 column density of 3×10^{18} cm^{-2} and a peak rotational temperature of ~75 K. It is surrounded by an envelope, of ~ 0.25 pc in diameter, having an NH_3 column density of 4×10^{17} cm^{-2} and an average rotational temperature of ~35 K. Assuming an [H_2/NH_3] abundance ratio of 10^6, we derive H_2 densities of 6×10^5 cm^{-3} for the halo and 1×10^7 cm^{-3} for the core, and a total molecular mass of ~500 M$_\odot$. Position-velocity maps suggest that the halo is slowly rotating with an angular velocity of 9.5±1.1 km s^{-1} pc^{-1}, while the core is undergoing rapid motions (spatially unresolved with the present angular resolution). The broad profiles can be explained assuming that the core gas is slowly falling but rapidly spinning toward the compact HII regions. Alternatively they may reflect outflowing motions driven by the expansion of the hot ionized gas around the newly formed stars.

A second ammonia clump, having an angular size of ~13", is found associated with the UC HII region G10.46+0.03A (Wood and Churchwell

1989). Emission from this cloud was detected in the main HF lines of the (2,2) and (3,3) inversion transitions, having a linewidth of 3.5 km s^{-1}, but not in the satellites. Assuming LTE conditions, we derived a rotational temperature between the (2,2) and (3,3) levels of 48±6 K, a total NH$_3$ column density of 1.3×10^{16} cm^{-2}, and a molecular mass of 28 M$_\odot$. The velocity structure of the ammonia emission suggests that this cloud is probably expanding, with a velocity of ~2 km s^{-1}.

The third ammonia clump, NH$_3$ G10.48+0.03, was detected in the main HF lines of the (2,2) and (3,3) inversion transitions. It has an angular size of ~9". Toward this cloud we did not detect radio continuum sources, brighter than 13 mJy, at 24 GHz. From the (2,2) and (3,3) observations, assuming LTE conditions, we derived a rotational temperature of 47±6 K, a total NH$_3$ column density of 9×10^{15} cm^{-2}, and a molecular mass of 14 M$_\odot$. The velocity structure of the emission suggests that this clump is undergoing a radial flow, with a velocity of ~1.2 km s^{-1}. This structure is consistent with infall generated by the self-gravity of a 20 M$_\odot$ object.

In summary, from spectral line aperture synthesis observations, made with the VLA, we resolved the emission from the dense molecular gas toward G10.5+0.0 (Cesaroni, Walmsley, and Churchwell 1992) into three small clumps with diameters of ~0.3 pc. Ultracompact HII regions are found projected near the peak of two of these clumps, suggesting that the dense and hot molecular emission originates from the inmediate neighbourhood of small regions of ionized gas. These molecular cores are thus useful probes of the physical conditions and dynamics of the interaction zone between the ionized gas and the ambient neutral gas. The third clump does not appear to be associated with a significant source of internal energy and may represent a molecular core undergoing collapse in a stage prior to the formation of a star.

ACKNOWLEDGEMENT

This research has been partially supported by the Chilean FONDECYT under grant 0907/92.

REFERENCES

Braz, M. A., and Sivagnanam, P. 1987, *Astr. Ap.*, **181**, 19.
Cesaroni, R., Walmsley, C. M., Kömpe, C., and Churchwell, E. 1991, *Astr. Ap.*, **252**, 278.
Cesaroni, R., Walmsley, C. M., and Churchwell, E. 1992 *Astr. Ap.*, **256**, 618.
Churchwell, E., Walmsley, C.M., and Cesaroni, R. 1990, *Astr. Ap. Suppl.*, **83**, 119.
Garay, G., Moran, J. M., and Rodríguez, L. F. 1992, *Ap. J.*, submitted
Garay, G., Rodríguez, L. F., Moran, J. M., and Churchwell, E. 1992, *Ap. J.*, submitted
Genzel, R., and Downes, D. 1977, *Astr. Ap. Suppl.*, **30**, 145.
Wood, D. O. S., and Churchwell, E. 1989, *Ap. J. Suppl.*, **69**, 831.

Massive Stars: Their Lives in the Interstellar Medium
ASP Conference Series, Vol. 35, 1993
Joseph P. Cassinelli and Edward B. Churchwell (eds.)

NH₃ AND H₂O MASERS: STAR FORMATION IN G9.62+0.19

PETER HOFNER, STAN KURTZ AND ED CHURCHWELL

Washburn Observatory, University of Wisconsin-Madison
475 North Charter Street, Madison, WI 53706

INTRODUCTION

G9.62+0.19 is a massive star forming region at a distance of 0.7 kpc. High resolution radio continuum observations by Kurtz *et al.* 1992 (*Ap. J. Suppl.*, subm.), show a cluster of three continuum components (Figure 2). According to these authors, the bright, central component is an ultracompact (UC) HII region ionized by a young, massive star of spectral type B1.

Observations of the $NH_3(5,5)$ inversion line with the MPIfR 100 m telescope (FWHM=40″) by Cesaroni *et al.* 1992 (*Astr. Ap.*, **256**, 618), toward G9.62+0.19 reveal a weak thermal component and a strong, narrow component, suggesting maser emission. Here we report VLA observations, obtained on April 19, 1992 (C-configuration, FWHM = 1″.7 × 1″.0) that confirm the $NH_3(5,5)$ maser in G9.62+0.19; this is the only masing metastable transition of para-ammonia known in the interstellar medium.

RESULTS

Ammonia Maser Line

The spectrum of the $NH_3(5,5)$ maser line is shown in Figure 1 and its position is indicated with a triangle in Figure 2. The line parameters and peak positions are summarized in Table 1. Given the peak flux density of 0.88 Jy in the line and an upper limit on the source size of 0″.5 we obtain a lower limit for the brightness temperature of about 10^4 K, thus proving that maser action takes place.

In our maps there is marginal evidence for radio continuum emission at the position of the new $NH_3(5,5)$ maser; this has been confirmed by comparison with continuum maps from Cesaroni *et al.* 1992 (in prep.), and Garay *et al.* 1992 (*Ap. J.*, subm.).

At present, G9.62+0.19 is the only known source with $NH_3(5,5)$ maser emission. Also, it is interesting to note that no maser action has been found in any of the ammonia (1,1), (2,2) or (4,4) inversion line observations toward this source.

Ammonia Thermal Line Observations

The spectrum of the $NH_3(5,5)$ thermal line is shown in Figure 1 and its location is shown in dashed contours in Figure 2. The (5,5) thermal emission extends northward from the region of peak continuum emission. An estimate of the physical parameters in this ammonia clump can be obtained from earlier observations of the (4,4) transition where the hfs components have also been detected (Cesaroni *et al.* 1992, in prep.). We obtain: $\tau_M \approx 18$, $T_{ex} \approx 37\,K$ and $N_{4,4} \approx 6 \times 10^{16}\,cm^{-2}$.

TABLE 1

OBSERVED AMMONIA LINE PARAMETERS

Source	α (1950) (h min sec)	δ (1950) (° ′ ″)	V_{LSR} (km/s)	S_ν (Jy)	ΔV (km/s)
$NH_3(5,5)$ Maser	18 03 15.841	−20 31 52.35	−0.29	0.88	0.4
$NH_3(5,5)$ Thermal	18 03 16.029	−20 31 59.00	4.46	0.15	8.4

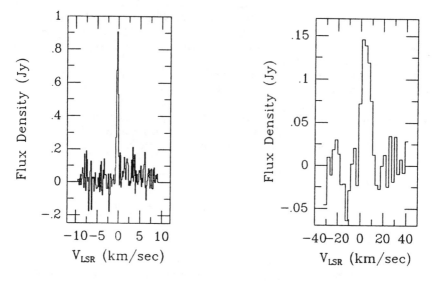

Fig. 1. VLA Spectra of the $NH_3(5,5)$ maser line (left) and thermal line (right). Both spectra have been integrated over the region where line emission is present.

Water Maser Emission

The 1.3 cm water maser emission toward G9.62+0.19 has recently been mapped with the VLA by Hofner *et al.* 1992 (in prep.), (Figure 2). The water masers are conspicuously aligned in a linear structure over a projected length of 0.08 pc, between the bright central UC HII region and the position of the $NH_3(5,5)$ maser, pointing toward the northern continuum component.

This linear structure has been observed by Forster, 1992 (*Astr. Masers*, Springer) who also reports that OH maser spots occur along this line. Furthermore, the map of the thermal $NH_3(5,5)$ line shows that the hot ammonia is located in the same region and elongated in the same direction as the H_2O masers.

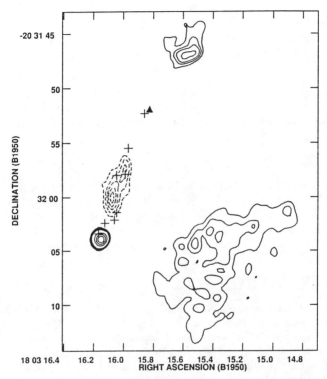

Fig. 2. 3.6 cm continuum image from Kurtz *et al.* 1992 (solid contour lines are 4, 5, 6, 8, 20, 30 mJy/beam). Crosses indicate the position of the water maser features; a triangle shows the location of the $NH_3(5,5)$ maser. Thermal $NH_3(5,5)$ emission is shown in dashed contour levels of (15, 20, 25, 30, 35) mJy/beam.

SUMMARY

• We confirm the discovery of maser action in the (5,5) inversion line of para ammonia. G9.62+0.19 is the first – and at this time the only – known source where this transition shows maser action.

• Hot, dense ammonia appears close to but not coincident with the UC HII region.

• Star forming activity in G9.62+0.19 as traced by water vapor maser emission, highly excited ammonia and young HII regions occurs along a narrow line. This suggests compression along a shock front.

Massive Stars: Their Lives in the Interstellar Medium
ASP Conference Series, Vol. 35, 1993
Joseph P. Cassinelli and Edward B. Churchwell (eds.)

PIG's IN THE TRAPEZIUM

P. R. McCullough (Astronomy Dept., U. California, Berkeley, CA, 94720)
R. Q. Fugate, B. L. Ellerbroek, C. H. Higgins (Phillips Laboratory)
J. M. Spinhirne (Rockwell Power Systems)
J. F. Moroney (Adaptive Optics Associates)
R. A. Cleis (The Optical Sciences Company)

ABSTRACT   The ionized knots in the Trapezium region are known as PIG's, for Partially Ionized Globules. We present an Hα image of the Trapezium made at the Starfire Optical Range of Kirtland AFB in New Mexico.

INTRODUCTION

The PIG's are visible in the optical resonant lines of Hα and [O III] (Laques and Vidal 1979, Vidal 1982) and in the radio continuum (Garay, Moran, and Reid 1987 (GMR), Churchwell, Felli, Wood, and Massi 1987 (CFWM)). They have physical dimensions similar to our solar system, or $\sim 0.25''$ at the distance of the Trapezium. The PIG's are stellar envelopes, photoionized externally by θ^1C Ori.

LASER BEACON ADAPTIVE OPTICS

The 1.5-m SOR telescope uses a closed-loop adaptive-optics system which compensates for seeing in two ways (cf. Fugate et al. 1991). The light from θ^1C Ori is used to detect image wander, which is corrected by a steering mirror operating at 50 Hz bandwidth. The light from a copper-vapor laser backscattered in the atmosphere at a range of 10 km is used to detect wavefront distortion (image blur), which is corrected by a 241-element deformable mirror operating at 75 Hz bandwidth. The laser produces two wavelengths simultaneously, 0.5106 μm and 0.5782 μm, in 50 ns pulses at 5 KHz with a average power of 75 Watts. Because of its finite range, the laser beacon cannot appear in focus to the scientific CCD, which is focused at infinity; in fact, the beacon cannot appear smaller than the angular extent of the primary mirror viewed from the range of the beacon: in this case $\sim 30''$. Interference filters provided enough spectral rejection of the laser beacon as to make it undetectable even without temporally gating the scientific CCD to avoid the laser pulses.

COMPARISON OF Hα WITH RADIO CONTINUUM

It can be shown (cf. Mezger and Henderson 1967, Reynolds 1990) that the ratio of radio flux density to Hα flux for an optically thin ionized medium is

$$\frac{S_{radio}}{I_{H\alpha}} = 3.46 \; \nu_{GHz}^{-0.1} \; T_4^{0.55} \; \frac{mJy}{photons \; cm^{-2} \; s^{-1}},$$

where ν_{GHz} is the observing frequency in GHz, T_4 is the electron temperature of the ionized gas in units of 10^4 K (here we assume $T_4 = 1$). The optically thin

Table I   Comparison of Hα with Radio Continuum

Source[a]	Hα photons (cm^{-2} s^{-1})	2-cm[a] (mJy)	1.3-cm[b] (mJy)	2-cm ratio[c]	1.3-cm ratio[c]
5	2.32	21.8	23.5	0.59	0.52
6	3.60	34.9	33.4	0.57	0.57
7	1.84	8.9	10.3	1.14	0.94
8	0.86	5.5	4.5	0.85	1.01
11	1.72	10.7	13.7	0.88	0.67
13	1.01	16.8	15.8	0.33	0.34
15	2.06	2.3	-	5.00	-

[a]Source numbers and 2-cm data are from CFWM; error = ±0.2 mJy.

[b]1.3-cm data are from GMR; error = ±1.5 mJy.

[c]Hα over Radio ($\times$ 3.46 $\nu_{GHz}^{-0.1}$ $\times$ $10^{0.4\,A_{H\alpha}}$) in dimensionless units.

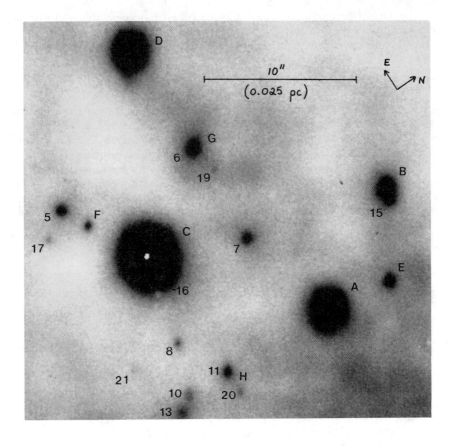

Fig. 1.   Hα image without continuum subtraction showing the star cluster θ^1 Orion (letters) and ionized knots (numbers).

approximation is adequate for $\nu \gtrsim 15$ GHz, except for the object CFWM 15, for which $\tau_{15\ GHz} > 0.56$ (CFWM Table 2).

The Hα fluxes listed in column 2 of Table 1 are based upon a multi-step calibration procedure (Celnik 1983). They refer to the continuum-subtracted Hα flux integrated over a 1.4″ diameter disk, corrected to above the earth's atmosphere. We estimate the accuracy to be $\pm 25\%$. Our Hα fluxes are smaller than those measured photographically by Laques and Vidal (1979), typically by an order of magnitude. Apparently the spatial extent of the PIGs caused the photographic fluxes to be overestimated.

Comparison of the Hα fluxes with the radio data of CFWM and GMR shows the agreement is generally good (a value of 1.0 in columns 5 and 6 of Table 1 would be exact agreement). We assume the interstellar extinction at Hα, $A_{H\alpha} = 0.8$ mag (Cardelli and Clayton 1988). Extinction intrinsic to individual PIGs may account for some of the scatter in columns 5 and 6 of Table 1, because it will reduce the observed Hα flux but not the radio continuum flux density. The poorest agreement in Table 1 (that of CFWM 15) cannot be due to extinction because that would only make the discrepancy worse, but it may be due to the radio emission being optically thick.

Extended (1″ to 4″) Hα emission around the PIGs correlates well with the radio continuum map of Barvainis and Wooten (1987). The total flux in the extended emission, although it has low surface brightness, dominates that of the PIG's cores (within 0.7″ radii).

SUMMARY

An Hα image of the Trapezium made with an adaptive optics system using a laser beacon reveals ionized condensations of the type discovered by Laques and Vidal (1979). The image, although it was made during evening twilight at a zenith angle of 60° from the New Mexico desert, has a spatial resolution (FWHM) of 0.5″ over a 30″ field and sufficient sensitivity to assign optical counterparts to all of the compact objects detected in the radio continuum. For the objects for which we measured the Hα fluxes, the Hα fluxes and the radio continuum flux densities are mutually consistent with the model of the PIGs as externally-ionized by θ^1C Ori.

REFERENCES

Barvainis R. and Wooten, A. 1987,AJ,92,168
Cardelli and Clayton 1988,AJ,95,516
Celnik, W.E. 1983, A&A Supp. Ser., 53,403
Churchwell, E.,Felli, M.,Wood, D.O.S., and Massi, M. 1987 (CFWM), ApJ,321,516
Fugate, R.Q., Fried, D.L., Ameer, G.A., Boeke, B.R., Browne, S.L., Roberts, P.H., Ruane, R.E., Tyler, G.A., and Wopat, L.M. 1991,Nature,353,144
Garay, G., Moran, J.M., and Reid, M.J. 1987 (GMR),ApJ,314,535
Laques, P., and Vidal, J. 1979, A&A,73,97
Mezger, P.G. and Henderson, A.P. 1967, ApJ,147,471
Reynolds, R.J. 1990, in "The Galactic and Extragalactic Background Radiation," S. Bowyer and C. Lienert, eds. p. 162.
Vidal, J. 1982, in "Symposium on the Orion Nebula to Honor Henry Draper," A.E. Glassgold, P.J. Huggins, and E.L. Schucking, eds. p. 176.

Massive Stars: Their Lives in the Interstellar Medium
ASP Conference Series, Vol. 35, 1993
Joseph P. Cassinelli and Edward B. Churchwell (eds.)

FAR-INFRARED OBSERVATIONS AND MODELS OF THE W3-IRS5 PROTOSTELLAR CLOUD

M. F. CAMPBELL, M. B. CAMPBELL, C. N. SABBEY
Colby College, Waterville, ME

P. M. HARVEY, D. F. LESTER, N. J. EVANS II
U. Texas, Austin, TX

H. M. BUTNER
NASA Ames, Moffett Field, CA

P. G. OLDHAM, K. J. RICHARDSON, M. J. GRIFFIN
Queen Mary and Westfield Colleges, University of London

G. H. L. SANDELL
Joint Astronomy Centre, Honolulu, HI

INTRODUCTION

W3-IRS5 is a high luminosity mid- and far-infrared source which has extremely weak radio continuum emission, making it an excellent candidate for a massive protostar. The W3 region has been mapped in far-ir by Werner et al. (1980). We have made new observations of the W3 region in far-ir, submm and mm to take advantage of improved sensitivity and spatial resolution available on the Kuiper Airborne Observatory and at the James Clerk Maxwell Telescope. The IRS5 far-ir profiles and single beam submm and mm photometry are used as data on which to base a radiative transfer model of a centrally heated dusty molecular cloud surrounding a protostar. Near- and mid-ir observations from the literature are also used to constrain the model.

OBSERVATIONS

New 47 and 95 μm maps of W3 were obtained with the Texas 8 bolometer linear array on the KAO in 1987 by repeatedly scanning across the region in azimuth (Lester et al. 1986). A second 95 μm map was made in 1989. The 47 and 95 μm peaks are coincident with each other and with the mid-ir position of IRS5 to within 10" accuracy. The 47 μm map shows a ridge of emission to the NE of IRS5 due to W3A (IRS1), and the 95 μm map shows extended emission N and E of IRS5 which is also due to W3A. There is a secondary peak is due to W3C (IRS4) in each map.

Scanning the Texas array produces very precise profiles which can be used to determine spatial structure for a source model. Scans from the 1987 data are analyzed here; the 1989 data will be included in a fuller analysis.

Scans through IRS5 were made at about 160° position angles (NW-SE) and are asymmetrical due to extended emission from W3A. The SE half of each scan was assumed to be free of extended emission from other sources, and was used to create a source profile of IRS5, which was assumed to be spherical.

Point source profiles (PSP) were obtained by scanning IRC+10420. IRS5 is resolved at a level slightly better than that shown for W3(OH) in Campbell et al. (1989). Flux density calibrations were obtained from IRC+10420: at 47 μm, IRS5 has a peak flux density of 9,800 Jy ±30% in the 20″x 23″ beam; at 95 μm, it has a peak flux density of 15,300 Jy ±30% in the 24″x 35″ beam. These new data are consistent with those of Werner et al.

Submm and mm photometry from JCMT are shown in Table I.

TABLE I   Submillimeter and Millimeter Photometry from JCMT

I(mm)	Beamwidth	Flux Density	Uncertainty
0.45	17.5	112.07	3.43
0.80	14.0	15.14	1.14
0.80	16.0	17.24	1.29
1.10	18.5	6.81	0.46
1.30	19.5	5.45	0.32
2.00	27.2	4.39	0.44

DUSTY MOLECULAR CLOUD MODEL FOR W3-IRS5

We have applied the radiative transfer code of Egan, Leung, and Spagna (1988) to the modeling of IRS5, as was done by Butner et al. (1990) for NGC 2071. Emission of the cloud model is convolved with the KAO far-ir PSPs at 47 and 95 μm to create model scans for comparison with the scan data. At other wavelengths, emission from the model is convolved with Gaussian beams for comparison with single beam photometric measurements. The model's parameters are summarized in Table II.

The shapes of scans across the model (convolved with the beams) compare well to the IRS5 scans except that the model falls below the wings of the 47 μm data. The integrated flux densities along the scan match the data to 10%. Overall, the spectral fit is good (cf Butner et al.)

The model's luminosity is considerably larger than the estimate by Werner et al., probably due to our attributing more of the total emission of W3 to IRS5. We chose 10,000K to represent a high mass protostar approaching the main sequence. A model with a 3000K protostar implies $\alpha = -1.5$ rather than -1 (see Table II). The implied radius of the dust-free cavity is quite large, and the dust temperature at the inner edge of the cloud is well below sublimation point. Our model used dust properties of Mathis, Mezger and Panagia (1983). If we were to fit only the KAO and JCMT data to the models, we could use Draine and Lee (1984) dust. However this dust gives very low predicted near- and mid-ir flux densities .

TABLE II   Source Model

Parameter	Best Value	Comment
Protostar Luminosity	5×10^5	
Protostar Temp	10,000 K	Assumed
Dust Properties	Mathis, Mezger and Panagia (1983)	
Optical Depth	0.5	95 μm, Center to Edge
Density Law	$\alpha = -1$	$n = n_i (r/r_i)^\alpha$
Radius of Cavity	0.08 pc	Dust-free region
Cloud Outer Radius	1.0 pc	Not Critical
Distance	2.4 kpc	Werner et al.

SUMMARY

New long wavelength data are used as the basis of a source model for W3-IRS5. The model is a centrally peaked spherical cloud surrounding a large dust free cavity containing a $10^5 L_\odot$ luminosity protostar.

ACKNOWLEDGEMENTS

This work was supported by NASA Grants NAG 2-67, 2-420, 2-546, and by Colby College. Support was also provided by the Colby Computer Center. L. G. Mundy contributed to the observations on the KAO. A. T. Pickering has provided invaluable assistance with computing.

REFERENCES

Butner, H. M., Evans, N. J. II, Harvey, P. M., Mundy, L. G., Natta, A.,and Randich, M. S. 1990 *Ap. J.*, 364, 164.
Campbell, M. F., Lester, D. F., Harvey, P. M., and Joy, M. 1989 *Ap. J.*, 345, 298.
Draine, B. T., and Lee H. M. 1984, *Ap. J.*, 285, 89.
Egan, M. P., Leung, C. M., and Spagna, G. F. 1988, *Comp. Phys. Comm.*, 48, 271.
Lester, D. F., Harvey, P. M., Joy, M. and Ellis, H. B. 1986 *Ap. J.*, 309, 80.
Mathis, J. S., Mezger, P. G., and Panagia, N. 1983, *Astr. Ap.*, 128, 212.

Massive Stars: Their Lives in the Interstellar Medium
ASP Conference Series, Vol. 35, 1993
Joseph P. Cassinelli and Edward B. Churchwell (eds.)

MOLECULAR LINE OBSERVATIONS OF G34.3+0.2: A CASE OF SPIN-UP COLLAPSE?

M. A. PAHRE, P. T. P. HO, AND M. J. REID
Harvard-Smithsonian Center for Astrophysics, 60 Garden Street, Mail
Stop 78, Cambridge, MA 02138

E. R. KETO AND D. PROCTOR
Institute of Geophysics and Planetary Physics, Lawrence Livermore
National Laboratory, POB 808 L413, Livermore, CA 94550

ABSTRACT   We present low- and high-resolution NH_3 observations
of the molecular material associated with G34.3+0.2 which show a a
differential rotational signature between the envelope and molecular core.
We also detected a blue-shifted emission feature extending 100″ to the
north of the core region, which we suggest to be infall from the backside
of the cloud. We propose that this source shows evidence for spin-up
gravitational collapse without accretion onto the central source.

BACKGROUND

The radio source, G34.3+0.2, has a "cometary"-shaped compact HII region
located at its center (Reid and Ho 1985). Heaton *et al.* (1985) mapped the
extended cloud in $NH_3(1,1)$ and $(2,2)$, but noted no systematic velocity
features across the envelope. A single-dish multi-level study in NH_3 by
Henkel *et al.* (1987) suggested the presence of a hot compact molecular
core. Observations of the core at 3″ in $NH_3(2,2)$ and $(3,3)$ by Garay and
Rodriguez (1990) and at 0.5″ in $NH_3(1,1)$ by Keto *et al.* (1992) show it to be
hot ($T=185\,K$), dense ($n(H_2) \gtrsim 7 \times 10^7\,cm^{-3}$), and optically-thick ($\tau(3,3) \sim 37$).
Other observations in $NH_3(1,1)$ (Keto *et al.* 1987), $(3,3)$ (Heaton *et al.* 1989),
and $HCO^+(1{\to}0)$ (Carral and Welch 1992) have shown the presence of a
systematic velocity gradient across this molecular core consistent with rotation.
    We observed G34.3+0.2 in $NH_3(1,1)$ and $(2,2)$ with the Haystack 37m
telescope at 1.4′ resolution and in $(1,1)$ with the Very Large Array at 4″ in its
D-configuration to determine the relationship between the core and envelope.

RESULTS

Large-Scale Structure
Channel maps of the $NH_3(1,1)$ emission across the main peak are shown in
Figure 1. An asymmetry of the line was detected blue-shifted at $v=55\,km\,s^{-1}$

with respect to the ambient cloud velocity of $v=58.2\,\mathrm{km\,s^{-1}}$ was detected to the north of the center of the cloud.

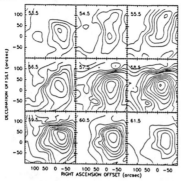

Fig. 1.   Summed channel maps at the Haystack 37m of $NH_3(1,1)$ emission across the main component showing the presence of blue-shifted emission at $v=55\,\mathrm{km\,s^{-1}}$.

Compact Molecular Core and Absorbing Gas

In our VLA observations of $NH_3(1,1)$, we have detected the molecular core at $v=58.1\,\mathrm{km\,s^{-1}}$ to be optically-thick ($\tau(1,1)=8$), hot ($T_{rot}(2,2;1,1)\geq40\,\mathrm{K}$), and spatially-offset to the east of the compact HII region. The absorbing gas at $v=60.8\,\mathrm{km\,s^{-1}}$, however, is optically thin ($\tau(1,1)=0.25$) and cold ($T_{rot}(2,2;1,1)=18\,\mathrm{K}$). We thus concur with Garay and Rodriguez (1990) that the absorbing gas is not directly associated with the compact molecular core, but rather represents gas further out in the extended molecular cloud.

Rotation of the Molecular Cloud

We have detected the signature of rotation of the extended molecular cloud for the first time (Figure 2), measured to be at a rate of $0.7\,\mathrm{km\,s^{-1}\,pc^{-1}}$ about an axis running NE-SW. Our VLA observations also show a rotational rate significantly larger at $14\,\mathrm{km\,s^{-1}\,pc^{-1}}$, with a rotational axis at position angle 60°. We suggest that the variance in rotational rates shows the presence of gravitational spin-up collapse in the molecular clouds surrounding G34.3+0.2. The blue-shifted emission at $v=55\,\mathrm{km\,s^{-1}}$ seen to the north would then represent infall from the backside along the plane of the rotation. It is curious to note that the rotation of the molecular material is in the opposite direction to that of the compact HII region as seen in the hydrogen recombination line observations of Garay *et al.* (1986).

SUMMARY

We have detected a differential rotational signature between the molecular gas at 12″ and 100″ from the core region of G34.3+0.2. Blue-shifted emission detected extending to the north from the core is suggested to be due to infalling motions. We suggest a model for the region in which core gravitational collapse has occured, but that the molecular envelope is still undergoing spin-up collapse. We note that there does not appear to be accretion onto the central source, because the absorbing gas is cold and optically-thin; this is contrasted with the warmer absorbing gas in G10.6-0.4, which has been modelled as an accretion flow (Keto 1990).

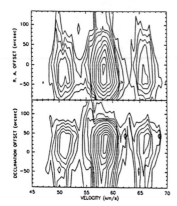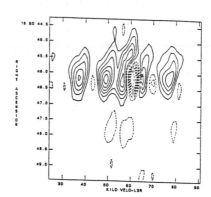

Fig. 2.   (Left) NS and EW position-velocity plots of the peak emission and inner two hyperfines of $NH_3(1,1)$ taken with the Haystack 37m at a velocity resolution of $0.4 \, km \, s^{-1}$. (Right) Position-velocity plot along position angle of -30° of the rotational signature across the compact molecular core as seen with the VLA at 12″ resolution. The contours are at increments of 30 mJy/beam to 0.3 Jy/beam, then 0.36 and 0.42 Jy/beam.

ACKNOWLEDGEMENTS

The authors wish to acknowledge partial support of this work by NSF Grant AST 87-20759. MAP wishes to thank PTPH for his guidance at every stage of this project.

REFERENCES

Carral, P., and Welch, W. J. 1992, *Ap. J.*, **385**, 244.
Garay, G., and Rodriguez, L. F. 1990, *Ap. J.*, **362**, 191.
Garay, G., Rodriguez, L. F., and van Gorkom, J. H. 1986, *Ap. J.*, **309**, 553.
Heaton, B. D., Little, L. T., and Bishop, I. S. 1989, *Astr. Ap.*, **213**, 148.
Heaton, B. D., Matthews, N., Little, L. T., and Dent, W. R. F. 1985,
        M. N. R. A. S., **210**, 23.
Henkel, C., Wilson, T. L., Mauersberger, R. 1987, *Astr. Ap.*, **182**, 137.
Keto, E. R. 1990, *Ap. J.*, **355**, 190.
Keto, E. R., Ho, P. T. P., and Reid, M. J. 1987, *Ap. J. (Letters)*, **323**, L117.
Reid, M. J., and Ho, P. T. P. 1985, *Ap. J. (Letters)*, **288**, L17.

Massive Stars: Their Lives in the Interstellar Medium
ASP Conference Series, Vol. 35, 1993
Joseph P. Cassinelli and Edward B. Churchwell (eds.)

STAR FORMATION IN PHOTOEVAPORATING MOLECULAR CLOUDS

FRANK BERTOLDI
Princeton University Observatory, Princeton, NJ 08544

CHRISTOPHER F. MCKEE
Astronomy Department, University of California, Berkeley, CA 94720

RICHARD I. KLEIN
Lawrence Livermore National Laboratory, Livermore, CA 94450

ABSTRACT   Young massive stars that form in clumpy molecular clouds can, through the dynamical effects of their ionizing radiation, affect the formation of stars in surrounding, photoevaporating molecular clumps.

INTRODUCTION

The idea that star formation is somehow triggered by compressions caused by the sudden exposure of molecular clumps to ionizing radiation from a newly formed massive star has received serious consideration for more than a decade.

Recent observations (Reipurth 1983; Nakano et al. 1989; Sugitani et al. 1989; Duvert et al. 1990; Sugitani et al. 1991) indicate that photoevaporating molecular clumps do frequently show signs of active star formation. This could suggests that the dynamical effects of the ionizing radiation may be directly responsible: they *triggered* star formation in the clumps.

It may be argued, however, that star formation in photoevaporating globules is merely coincidental, i.e., stars were forming *spontaneously* in the clumps anyway, and continue to despite the perturbation. A main theoretical argument that is usually advanced against the effectiveness of induced star formation relates to the interstellar magnetic field: as long as it is a dominant factor in star formation, any fast, compressive effects will leave the gravitational stability of molecular clouds unchanged. However, for a molecular clump that is stabilized by the combined effects of magnetic fields and turbulence, $(M \leq M_\Phi + M_J)$, a compression can enhance the gravity-induced gradual (ambipolar) diffusion of the neutral gas through the field, thereby lowering the magnetic critical mass, $M_\Phi \propto \Phi$ (where Φ is the magnetic flux threading the clump), and it can also enhance the dissipation of hydromagnetic turbulence, thereby lowering the "turbulent" Jeans mass, $M_J \propto \sigma^4/P_0^{1/2}$ (where σ is the clump's gas velocity dispersion and P_0 its surface pressure). Thus compression can speed up the clump's gradual evolution toward gravitational instability.

Considering the photoionization-shock compression of isolated molecular clumps that have diameters small compared to their distance from the ionizing star, we describe several mechanisms that can influence the rate at which stars form in their interiors.

I. SUPERCRITICAL CLUMPS

Molecular Clouds (MCs) are observed to be clumpy, with most of the molecular mass in clumps of much higher than mean density. Active star formation is overwhelmingly concentrated in the most massive $(M \simeq 10^2 - 10^3 M_\odot)$ clumps within a MC (Lada et al.[1991] for Orion B; Blitz [1991] for Rosette MC; Loren [1989] for Ophiuchus), which contain a significant fraction of the total molecular mass in a MC. Those massive clumps are strongly self-gravitating and magnetically supercritical $(M > M_\Phi)$, i.e., magnetic fields without turbulence could not support them against gravitational collapse (Bertoldi & McKee 1992).

Since the supersonic (but sub-Alfvénic) turbulence of the molecular gas dissipates in a time short (or order 5 free-fall times $\simeq 10^6 - 10^7$ yr) compared with the lifetime of the MC ($\sim 5 \times 10^7$ yr), in order for such clumps not to become unstable and turn into stars or disperse, turbulent energy must constantly be replenished by some *internal* source. Bertoldi & McKee (1993) show that a photoionization-regulated balance between the dissipation of hydromagnetic turbulence and the injection of energy by young low-mass stars (McKee 1989) can stabilize the clumps and provide a low-mass star formation rate in their interiors that is consistent with observations.

STABILITY OF SUPERCRITICAL CLUMPS

When a massive star is born in a clumpy molecular cloud, the clumps that become exposed to the ionizing radiation will begin to photoevaporate. The high pressure of the photoevaporation flow off its surface will implode a clump (Klein et al. 1983; Bertoldi 1989a,b). How does a sudden disturbance like an ionization-shock compression affect the gravitational stability of the massive, magnetically supercritical clumps?

- The initial radiation-driven implosion and subsequent rocket acceleration displaces the clump away from the ionizing source and in the course strips off all embedded young low-mass stars that provided for the clump's internal turbulent energy supply. Similarly, all compact cores that were about to form stars are left behind.
- Due to the steady acceleration of the clump, all newly formed low-mass stars that could provide for the turbulent support are also ejected from the clump before they can deposit enough energy to offset the dissipation.
- The mean density of the clump is increased, thereby enhancing the energy dissipation.

Taken together, these effects are likely to destabilize the clump in a dissipation time, thereby inducing a supercritical collapse and possibly fragmentation that could lead to the formation of an association of stars. This could provide for a mechanism of forming massive stars in clusters. Note that most massive stars apparently form in associations and that the canonical picture of low-mass star formation via the slow ambipolar diffusion of the magnetic field in magnetically subcritical clumps is unable to explain the formation of such associations.

II. MAGNETICALLY SUBCRITICAL CLUMPS: MAGNETIC SHOCKS

The small, magnetically subcritical clumps ($M \simeq 1 - 100 \; M_\odot \leq M_\Phi$) can be supported by their magnetic fields alone. For such clumps to become gravitationally unstable, the magnetic flux threading them must be reduced below a critical value. In low-mass star formation, this process occurs gradually through gravitationally induced ambipolar diffusion.

The rate of ambipolar diffusion can be substantially enhanced in shocks, because of the steep gradients in the shock front as well as the higher densities and reduced ionization there; indeed, the entire structure of magnetic shocks (termed C-shocks: Draine 1980) is governed by ambipolar diffusion. Shock-induced ambipolar diffusion can provide a mechanism for rapidly reducing the flux to mass ratio (M_Φ/M) of initially gravitationally stable parts of the clumps, thereby possibly triggering their gravitational collapse.

This mechanism appears particularly suited to demagnetize gas colliding at the symmetry axis of a collimating radiation-driven implosion.

III. SUBCRITICAL CLUMPS: AMBIPOLAR DIFFUSION

After the radiation-driven implosion, when it settles into an equilibrium configuration (while slowly photoevaporating and rocketing away from the ionizing star), the overall gravitational stability of an initially magnetically subcritical clump *increases*, because mass is lost through photoevaporation in such a way that the flux to mass ratio, M_Φ/M, increases.

Now as before the onset of photoevaporation, such a clump can form stars only via the slow, self-gravity driven ambipolar diffusion of the neutral gas through the magnetic field to the point where the clump, or a density condensation within it, becomes magnetically supercritical and collapses.

For the photoevaporating equilibrium clump the ambipolar diffusion rate in its interior can be significantly enhanced and a central core can in some cases become magnetically supercritical and collapse (Bertoldi 1989a). However, only rarely can star formation occur before such a clump evaporates, is rocketed outside the H II region, or becomes too transparent to the strong FUV radiation of the ionizing star. Star formation through the slow gravity-driven ambipolar diffusion is probably not very efficient in photoevaporating globules (Bertoldi, McKee & Klein 1990).

CONCLUSIONS

We discuss how a newly formed massive star in a clumpy molecular cloud can, through the dynamical effects of its ionizing radiation, affect the formation of stars in surrounding photoevaporating molecular clumps. Three basic mechanisms were briefly outlined:

- The stability of massive, initially magnetically supercritical clumps can be severely disturbed because such clumps are stabilized through a fragile equilibrium between dissipation of hydromagnetic turbulence and turbulence-creating low-mass star formation. The collapse of a magnetically supercritical clump may lead to the formation of an association of massive stars, and possibly even the self-propagation of massive star formation.
- During the radiation-driven implosion, a small fraction of the highly compressed molecular gas can be demagnetized through the rapid ambipolar diffusion in magnetic shocks. Low-mass stars may form in the focal points of such implosions.
- Although the gravity-driven ambipolar diffusion in an initially subcritical, photoevaporating clump is enhanced because of it's highly compressed state, star formation is unlikely because of the competing evaporative mass loss.

REFERENCES

Bertoldi, F. 1989a, Ph.D. Thesis, University of California at Berkeley.
Bertoldi, F. 1989b, *Ap. J.*, **346**, 735.
Bertoldi, F., & McKee, C. F. 1990, *Ap. J.*, **354**, 529.
Bertoldi, F., McKee, C. F., & Klein, R. I. 1990, in *Fragmentation of Molecular Clouds and Star Formation*, ed. E. Falgarone (Dordrecht: Kluwer), 391.
Bertoldi, F., & McKee, C. F. 1992, *Ap. J.*, **395**, 000.
Bertoldi, F., & McKee, C. F. 1993, in preparation.
Blitz, L. 1991, in *Physics of Star Formation*, ed. C.Lada & N.Kylafis (Kluwer).
Draine, B. T. 1980, *Ap. J.*, **241**, 1021.
Duvert, G., Cernicharo, J., Bachiller, R., Gómez-González, J. 1990, *AA*, **233**, 190.
Klein, R.I., Sandford, M.T., II, & Whitaker, R.W. 1983, *Ap. J.*, **271**, L69.
Lada, E., DePoy, D.L., Evans, N.J., & Gatley, I. 1991, *Ap. J.*, **371**, 171.
Loren, R. B. 1989, *Ap. J.*, **338**, 902.
McKee, C. F. 1989, *Ap. J.*, **345**, 782.
Nakano,M., Tomita,Y., Ohtani,H., Ogura,K., Sofue,Y. 1989, *PASJ*, **41**, 1073.
Reipurth, B. 1983, *AA*, **117**, 183.
Sugitani, K., Fukui, Y., Mizuno, A., & Ohashi, N. 1989, *Ap. J.*, **342**, L87.
Sugitani, K., Fukui, Y., Ogura, K. 1991, *Ap. J. Supp.*, **77**, 59.

Massive Stars: Their Lives in the Interstellar Medium
ASP Conference Series, Vol. 35, 1993
Joseph P. Cassinelli and Edward B. Churchwell (eds.)

AN INVESTIGATION OF PROTOSTELLAR DISC DISPERSAL BY LUMINOUS YOUNG STELLAR OBJECTS

JOHN M. PORTER and JANET E. DREW
Department of Astrophysics, University of Oxford, Keble Road,
Oxford OX1 3RH, U.K.

ABSTRACT   There is evidence that stars of A and later spectral type continue to be encircled by dust discs long after joining the main sequence (the so-called Vega-excess stars). In contrast, OB stars may disperse their natal discs while still heavily obscured. We have begun to investigate whether radiation pressure has a rôle to play in achieving this dispersion by developing a generalization of the CAK model that can describe mass loss from Keplerian orbits.

INTRODUCTION

There is now an emerging consensus that late in the obscured stage of a young star's evolution, it is encircled by a disc-like structure. A particularly striking example is afforded by the S106 nebula, an ionised outflow with a markedly bipolar morphology, that is centred on a late-O star still embedded in a nearly edge-on highly-obscuring lane of gas and dust (Solf & Carsenty 1982, Felli *et al.*1984). The O star may be clearing a hole for itself in its neighbourhood, of a mass loss rate of the order of 10^{-6} $M_\odot yr^{-1}$. It is unlikely that the outflow responsible for such emission is simply a stellar wind – the expansion velocities are too low. We have begun to look at the part radiation pressure might play in this flow. Other potential means of dispersing remnant discs include ram-pressure stripping (due to the mechanical effect of the wind from the hot star on the CSM) and MHD effects.

Here, we consider the form that must be taken by an equation of motion for a radiation-driven transsonic outflow from the inner ionised edge of a passive Keplerian disc.

THE DISC MODEL

The basis for the model used here is that for radiatively-driven winds developed by Castor (1974) and Castor, Abbot and Klein (1975) (CAK). Friend and Abbot (1986), extended this model to include the finite disc correction which takes into account the non-zero solid angle subtended by the star as the gas is accelerated. Also, a rotational term was added to the equation of motion. However, the rotational velocities considered were sub-Keplerian. The model presented here does not use the finite disc correction as it was felt it could be left out in gaining an initial grasp of the characteristics of a disc 'erosion' model.

The standard CAK equation of motion must include the pressure acting toward the star, due to the material of the disc. To allow for this, a decaying exponential term was introduced to the pressure gradient to simulate this, with amplitude A and scale length Δr. The amplitude A is a measure of the disc back pressure in the rotational plane and Δr is a characteristic length over which the flow rides up over the disc, exterior to the starting radius. The introduction of this term renders a 2–D theory superfluous, as the dimension normal to the plane of the disc is then suppressed. The requirement that $\partial v/\partial r \geq 0$, imposes a minimum value on A for a given Δr. A restriction is placed on the detailed geometry of the inner edge of the disc in the case of stable outflow . Larger Δr allows smaller A. For the particular model treated here with $\Delta r \sim 5H$, where H is the scal e height for an α-disc model (Shakura & Sunyaev, 1972), $A \geq 10^3 \mathrm{cm\ s^{-2}}$. There are problems in defining an exact value for A as only a lower limit is obtained. This lower limit is used, however, because then a maximum value for the mass loss rate is obtained. For convenience, the model assumes that the inner portion of the disc has already been dispersed. We have investigated the specific case of the inner edge at $5R_*$ and stop our calculations at $50R_*$.

RESULTS

Two cases have been considered. The first was a $60M_\odot$ early-O star and the second was a $20M_\odot$ late-O, early-B star. The disc's α was set to a very low value to simulate a passive disc. The temperature of the ionised inner edge of the disc was taken to be 10^4K.

The mass loss rate is proportional to the solid angle presented to the star by the disc edge and also as $\dot{M} \propto r_{crit}^{-2(1-\alpha)/\alpha}$ with α being the CAK force multiplier parameter expressing the mixture of optically thick and thin lines (in the limit of only optically thin lines, $\alpha = 0$). The maximum mass loss rate for a $60M_\odot$ star is $\sim 10^{-8} M_\odot\ \mathrm{yr^{-1}}$. The mass loss rate is proportional to $A^{-(1-\alpha/\alpha)}$, and so decreases as A increases.

Disc lifetimes were calculated from the mass loss rate profiles. For a $60M_\odot, 1.2 \times 10^6 L_\odot$ star, the disc would be dispersed in around 250yrs, and around 7×10^3yrs for a $20M_\odot, 3 \times 10^4 L_\odot$ star.

COMMENTS

Our simple treatment provides some useful insights. First of all, the mass loss rates achieved clearly fall well below estimates based on observation of BN-type objects (for a 20 $M_\odot$ star, the discrepancy is about 3 orders of magnitude). Correspondingly, the dispersal timescale for a disc as massive as those believed to be associated with the lower-mass T Tauri stars would be too long. A small part of the discrepancy may be made up by using a custom made force multiplier. Also, a simplification of our model is that mass is driven only from the inner edge of the disc. While the angular diameter of the star as viewed from the inner portion of the disc remains significant, more of the disc than its innermost edge can be ionised and driven off. A pseudo 3-D hydrodynamic model which will take this and also ram pressure into account is being considered.

Nonetheless, the hope that radiation pressure can be the dominant dispersal mechanism seems a slender one. If the high opacity offered by dust can be exploited (i.e. dust is not destroyed in the region of interest), the situation might be transformed. What of ram-pressure stripping? Assuming similar stellar wind characteristics to those observed in field OB stars, ram pressure can be an effective erosive agent for the relatively low-mass disc ($\sim 10^{-6}$ $M_\odot$) we have treated here but, in the case of a disc a thousand or more times as massive, the wind will flow over the disc surface relatively unnoticed. A significant ram pressure contribution is bound to be accompanied by the formation of bow shock structures which should have testable observational consequences. The other obvious option is the centrifugal MHD mechanism (based on the Blandford & Payne 1982 model) of the sort favoured for the highly focussed bipolar outflows seen in association with some low mass young stellar objects. These models require a highly ordered magnetic field topology.

While the estimated mass loss rates are low, the calculated wind speeds are satisfactory, being of the order of 4–500km s^{-1}.

REFERENCES

Blandford, R.D. and Payne, D.G., 1982, *M.N.R.A.S.*,**199**, 883
Castor, J.I. 1974, *M.N.R.A.S.*,**169**, 279
Castor, J.I., Abbott, D.C. and Klein, R.I., 1975, *Ap. J.*,**195**, 157
Felli, M., Staude, H.J., Reddman, Th., Massi, M., Erioa, C., Hefele, H., Neckel, Th. and Panagia, N., 1984, *Astr. Ap.*,**135**, 261
Friend, D.B. and Abbott, D.C. 1986, *Ap. J.*,**311**, 701
Poe, C.H., and Friend, D.B, 1986, *Ap. J.*,**311**, 317
Shakura, N.I. and Sunyaev, R.A., 1973, *Astr. Ap.*,**24**, 337
Solf, J. and Carsenty, U., 1982, *Astr. Ap.*,**113**, 142

Massive Stars: Their Lives in the Interstellar Medium
ASP Conference Series, Vol. 35, 1993
Joseph P. Cassinelli and Edward B. Churchwell (eds.)

MOLECULAR LINE EMISSION MODELS OF HERBIG-HARO
OBJECTS. HCO+ EMISSION

MARK WOLFIRE
Harvard-Smithsonian Center for Astrophysics, Mail Stop 78, 60 Garden
St., Cambridge MA, 02138

ARIEH KÖNIGL
Department of Astronomy & Astrophysics and Enrico Fermi Institute,
University of Chicago, 5460 S. Ellis Ave., Chicago, IL 60637

ABSTRACT   We present time-dependent models of the chemistry and
temperature of molecular gas clumps that are exposed to the radiation
from propagating stellar-jet shocks. The X-ray, and UV radiation from
the shock initiates ion chemistry and enhances the abundances of several
ionized molecular species and also heats the gas in the clump. The warm
gas collisionally excites these molecules and produces measurable line
emission. We apply these results to the interpretation of the HCO+
emission that has been detected in several HH objects. This picture
provides a natural explanation of the fact that the line intensity typically
peaks ahead of the associated shock as well as of the reported low
line-center velocities and narrow linewidths.

INTRODUCTION

Several rotational transitions of HCO+ have been detected in association with
HH objects: Rudolph & Welch (1988) mapped the $J = 1 - 0$ transition in HH
7-11, Davis, Dent, & Bell-Burnell (1990) observed $J = 3 - 2$ emission in HH
1-2, and Rudolph & Welch (1992) mapped the $J = 1 - 0$ transition in HH 34S.
All the objects detected so far exhibit the following characteristics: (1) The
HCO+ emission is not coincident with the optical emission but generally
appears farther from the source of the outflow than does the optical: as far as
$\sim 10^{17}$ cm in the cases of HH 1-2 and HH 34S. This strongly suggests that the
HCO+ is *not* due to collisional excitation *behind a shock* (or even in the
magnetic precursor of a shock). (2) The observed line center velocities are close
to the ambient cloud velocities, indicating that the emission originates in
stationary material. (3) The observed line widths are quite narrow ($\lesssim 0.5$ km
s^{-1} in HH 7-11), indicating that the material is quiescent.

MODEL CALCULATIONS

In this work we extend the steady-state calculations of Wolfire & Königl (1991) to follow the time-dependent chemical and thermal behavior of a layer of molecular gas that is exposed to the radiation field of a fast shock. The layer is represented by a one-dimensional grid that starts at the face of the shock and extends into the preshock gas. The radiation field from the shock is stepped along the grid as the gas chemistry and temperature are advanced in time.

The density distribution is intended to mimic a two-phase medium consisting of an interclump gas, n_{ic}, and a dense molecular clump, n_{cl}, that initially lies a distance D_0 ahead of the advancing shock. The shock propagation speed v_{bs} is generally higher than the effective (normal) speed v_{eff}, because the shock is curved. The parameters v_{eff} and n_{ic} determine the characteristics of the radiation field produced by the shock.

Results

We choose parameters applicable to the HH 1-2 system giving $v_{bs} = 350$ km s^{-1}, $v_{eff} = 200$ km s^{-1}, and $D_0 = 2 \times 10^{17}$ cm and consider the case $n_{cl} = 10^4$ cm^{-3} and $n_{ic} = 10^2$ cm^{-3}. Figure 1 shows the fractional abundances of several ions as a function of optical depth in the preshock gas after a travel time $t \approx 40$ yrs. Also shown is the steady-state abundance of HCO$^+$ calculated for the same parameters. Note that a strong enhancement in the HCO$^+$ abundance results when the time-dependent nature of the gas chemistry is taken into account. The high abundance is a result of two processes: (1) the partial ionization of the illuminated gas by the shock-produced radiation (producing mainly H$^+$, H$_3^+$ and C$^+$), and (2) the slow dissociation of H$_2$ and CO. These two conditions drive the ion chemistry towards the production of HCO$^+$. For $A_V \lesssim 0.1$ the gas is heated mainly by photoionization of H$_2$, H, and He, while at $A_V > 0.1$ the dominant heating process is photoelectric ejection of electrons from dust grains. The resulting gas temperature is ~ 600 K at the clump edge which then decreases to ~ 15 K at $A_V \sim 1$.

Figure 2 presents the calculated HCO$^+$ line intensities as a function of time. We find that the HCO$^+$ $J = 3 - 2$ emission comes predominantly from the clump and that collisions with electrons are extremely important in exciting the HCO$^+$ molecule. The interclump density is too low to appreciably excite the $J = 3$ level. Molecular hydrogen 1-0 S(1) line emission is also produced and arises from a thin layer of gas close to the shock in which the gas temperature is greater than ~ 2000 K.

For the case $n_{ic} = 10^3$ cm^{-3} and $n_{cl} = 10^5$ cm^{-3}, we find that the peak HCO$^+$ $J = 3 - 2$ line intensity increases by a factor of ~ 25 from that of Fig. 2. The higher emission is due partly to enhanced collisional excitation in the denser medium and partly to the stronger X-ray flux that penetrates deeper into the clump.

Comparing the observations of Davis et al to our model calculations we conclude that the HCO$^+$ and H$_2$ line intensities emitted from a shock-irradiated clump can in principle account for the observations if the clump densities lie in the range 10^4 cm$^{-3} \lesssim n_{cl} \lesssim 10^5$ cm^{-3} and if the interclump densities satisfy 10^2 cm$^{-3} \lesssim n_{ic} \lesssim 10^3$ cm^{-3}. We also note that the emitting material in the clumps must cover a sufficiently large area of the observing beam or else the flux would be diluted. These results place constraints on the volume filling factor of clumps, the clump covering factor, and the characteristic clump size.

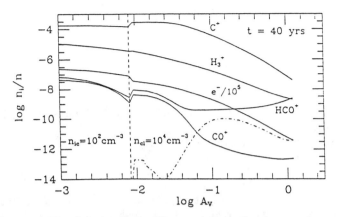

Fig. 1.   The fractional abundances in a molecular gas layer that is irradiated by a fast bowshock. Abundances are plotted as a function of the optical depth at a time $t \approx 40$ yr after the shock was turned on. The shock irradiates a layer of density $n_{cl} = 10^4$ cm^{-3} and propagates with a speed $v_{bs} = 350$ km s^{-1} in a medium of density $n_{ic} = 10^2$ cm^{-3}. The radiation field corresponds to a shock speed $v_{eff} = 200$ km s^{-1}. The vertical dashed line marks the upstream boundary of the layer and the steady-state abundance of HCO$^+$ is indicated (*dash-dotted*).

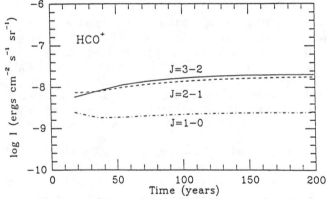

Fig. 2.   Line intensities emitted along the normal to the layer are plotted as a function of time. Model parameters are the same as in Fig. 1. HCO$^+$ $J = 3 - 2$ (*solid*), $J = 2 - 1$ (*dashed*), and $J = 1 - 0$ (*dash-dotted*).

REFERENCES

Davis, C.J., Dent, W.R.F., & Bell-Bernell, S.J. 1990, MNRAS, 244, 173
Rudolph, A., & Welch, W. J. 1988, ApJ, 326, L31
————————————————. 1992 submitted to ApJ
Wolfire, M.G., & Königl 1991, ApJ, 383, 205

Massive Stars: Their Lives in the Interstellar Medium
ASP Conference Series, Vol. 35, 1993
Joseph P. Cassinelli and Edward B. Churchwell (eds.)

EMBEDDED STAR CLUSTERS ASSOCIATED WITH LUMINOUS
FAR-INFRARED SOURCES

RONALD SNELL, JOHN CARPENTER, F.PETER SCHLOERB,
AND MIKE SKRUTSKIE
Department of Physics and Astronomy, University of Massachusetts,
Amherst, MA 01003

ABSTRACT   Twenty bright *IRAS* point sources in the 2nd and 3rd
quadrants of the Galaxy have been imaged at the J, H, and K infrared
bands and mapped in CS. These images reveal rich clusters of stars,
and based on their infrared colors and absolute magnitudes are thought
to be primarily young Ae/Be and T-Tauri stars.

INTRODUCTION

The development of sensitive infrared arrays has provided an opportunity to
study the earliest stages of stellar formation and explore the stellar content
of molecular clouds. This potential has been well recognized and many
researchers have obtained infrared images of prominent star forming regions
that reveal previously obscured clusters of stars. Since the formation of rich
clusters of stars may be the dominate mode of star formation for stars of any
mass, the study of the evolution of these clusters is essential to understand
the evolution of molecular clouds and the star formation history of the
Galaxy.

We report the results of near-infrared and CS observations of 20
bright *IRAS* point sources in the 2nd and 3rd quadrants of the Galaxy. The
sources were selected from a list of sources with 100μm flux densities greater
than 500 Jy that Snell *et al.* (1988) and Snell, Dickman, and Huang (1990)
surveyed for molecular outflows. Carpenter, Snell, and Schloerb (1991) have
studied the molecular line, radio continuum, and FIR emission from these
20 sources and concluded that they were associated with a molecular clouds
of mass 10^3 to 10^4 M$_\odot$, had a FIR luminosity of 10^3 to 10^5 L$_\odot$, and that
typically a single ZAMS star of spectral type B2-O9 could account for the
observed radio continuum emission and most of the FIR emission.

OBSERVATIONS

The near-infrared observations were obtained in Fall 1989 and 1990 with the
Wyoming Infrared Observatory 2.3 m telescope using a 64x64 HgCdTe array
detector. The image scale was $2''$ per pixel giving a field of view of $2.1'$ x

2.1$'$. A 3x3 frame mosaic at J, H, and K bands were made centered on each *IRAS* source and toward two reference fields; the total field size is ~4$'$ x 4$'$. For stars brighter than 15.5^m , 14.5^m , and 13.7^m at J, H, and K bands, respectively, the photometric accuracy was better than 0.2 .

We have also obtained maps of the J=2-1 transition of CS using the 14 m telescope of the Five College Radio Astronomy. Data were obtained spaced by 30$''$ covering the same 4$'$ x 4$'$ region as the infrared images. Figure 1 shows the J, H, and K band images and the CS map of the *IRAS* source 06056+2131.

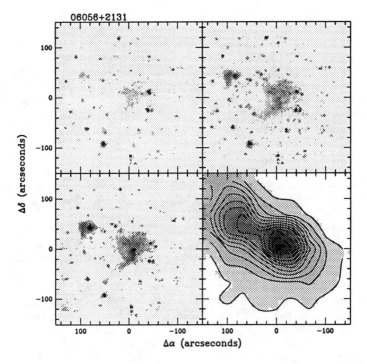

Fig. 1. J band (upper left), H band (upper right), K band (lower left), and CS (lower right) image of the *IRAS* source 06056+2131. The two bright K band sources surrounded by nebulosity correspond to the two CS peaks. Eighty–five clusters members have been identified in the H band image.

RESULTS

Nineteen of the 20 *IRAS* sources contain a significant number of stars in excess of that observed in the reference fields. Undoubtedly, each of these *IRAS* sources are in fact well defined clusters of stars; the number of observed cluster members vary from 17 to 91 stars to a limiting H magnitude of 15.5^m. Thus, one can conclude that massive stars do not form singly but instead as parts of rich clusters. The size of the clusters, defined as the diameter that encloses one-half of the cluster members, are typically about 1 pc. Thus, if the clusters are spherical, they have stellar densities of between 16 and 431 stars pc^{-3}.

Many of the stars in these clusters have large (J-H) and (H-K) colors. These stars have colors and absolute magnitudes consistent with those of known T Tauri and Ae/Be stars. We believe that these clusters are very young, and consist primarily of pre-main sequence stars.

The fitted slopes of the luminosity functions vary from 0.18 to 0.94. χ^2-tests of these observed distributions indicate that some differences in the luminosity functions are statistically significant. Using the empirical relation derived between mass and infrared absolute magnitude, the slopes of the luminosity functions permit mass functions with slopes ranging from -0.8 to -4.3. The average slope of the luminosity functions would imply a mass function slope of -1.7. However, due to effects such as variable extinction from cluster to cluster and within one cluster, these slopes must be taken with a certain skepticism. The most massive stars in each cluster have masses of about 10 solar masses, and probably are the source of the ionizing photons that produce the compact HII regions detected in these regions by Carpenter, Snell, and Schloerb (1990).

The CS observations revealed dense cores in all 20 of the sources. The mass of the cores range from 8 to 270 solar masses. The CS cores exhibit a variety of morphologies, however the most massive tend to be more centrally concentrated and are associated with optically invisible clusters, while the low mass cores are usually more fragmented and associated more often with optically visible clusters.

Estimates of the mass of the young clusters have been estimated using the relationship found between infrared absolute magnitude and mass for young stars. The star formation efficiencies within the clouds mapped by Carpenter, Snell and Schloerb (1990) have been estimated to be between 1 and 16%, with an average of about 5%. However, the efficiency at which dense core material is converted into stars (10 to 86%) is much higher, thus once cores form, the gas is efficiently converted into stars. This high efficiency of converting core material into stars suggests that many of these clusters may remain bound after the core gas is dispersed.

Finally, it is tempting to speculate that the clusters studied here represent an evolutionary sequence. Clusters that are optically invisible, have centrally concentrated, massive cores, and contain extremely red stars, may represent the youngest clusters in our sample, while the optically visible clusters with fragmented cores and relatively few red stars are the oldest clusters. Such an evolutionary sequence is also supported by the dynamical ages of the outflows detected in many of these regions (Snell, Dickman, and Huang 1990).

REFERENCES

Carpenter, J. M., Snell, R. L., & Schloerb, F.P. 1990, *ApJ*, **362**, 147.
Snell, R., Dickman, R., & Huang, Y.-L. 1990, *ApJ*, **352**, 139.
Snell, R., Huang, Y.-L., Dickman, R., & Claussen, M. 1988, *ApJ*, **325**, 833.

Massive Stars: Their Lives in the Interstellar Medium
ASP Conference Series, Vol. 35, 1993
Joseph P. Cassinelli and Edward B. Churchwell (eds.)

IDENTIFICATION OF NEW CANDIDATE HERBIG Ae/Be STARS IN EXTREMELY YOUNG GALACTIC CLUSTERS: M8, M16, M17, M42, NGC 2264

LYNNE A. HILLENBRAND and STEPHEN E. STROM
Five College Astronomy Dept., University of Massachusetts, Amherst
MA 01003

K. MICHAEL MERRILL and IAN GATLEY
Kitt Peak National Observatory, Tuscon AZ 85726

ABSTRACT   The youngest known optically visible pre-main
sequence stars are the Herbig Ae/Be stars. These objects appear to
represent a phase in the pre-main sequence lifetime of intermediate
mass stars during which they are surrounded by circumstellar accretion
disks. The census of Herbig Ae/Be stars amounts to only some 70
candidate objects scattered throughout the plane of the galaxy. No
systematic search, which would enable the identification of a complete
sample in a single young cluster, has yet been undertaken. Such an
unbiased sample is crucial for establishing the properties of both
the stars and the circumstellar accretion disks which are thought to
surround them. *Observational determination of the full range of disk
parameters (e. g. size, mass, lifetime, accretion rate) is a necessary
precursor to understanding the process of disk accretion and, in turn,
of high mass star formation.* We examine five young open clusters
in order to search for disks around intermediate and high mass stars
which are still approaching or which have just reached the main
sequence. Recent infrared studies of these clusters reveal that the
fraction of early type stars surrounded by disks is only about 10% in
clusters of ages $\tau \lesssim 3 \times 10^6$ yr. This suggests that the survival times
of disks around $M \gtrsim 5$ M$_\odot$ stars in young open clusters are shorter
than a few 10^5 years, and are thus much shorter than the pre-main
sequence lifetime of their central stars.

CHARACTERISTICS OF DISKS SURROUNDING HERBIG Ae/Be STARS

In order to identify new candidate Herbig Ae/Be stars, we search for objects
whose near-infrared spectral energy distributions (SEDs) are morphologically
similar to those of the known Herbig Ae/Be stars. Figure 1 shows the
reddening-corrected $0.36\,\mu m - 100\,\mu m$ SEDs of one Herbig Ae/Be star and,
for comparison, one T-Tauri star. The large infrared excesses can be modeled

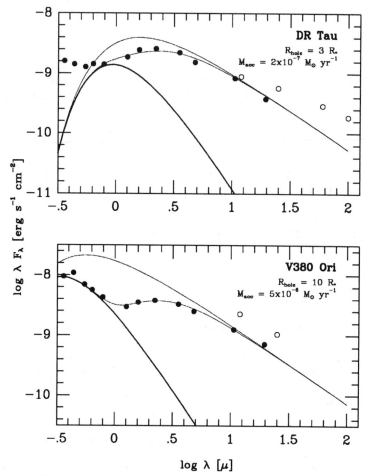

Figure 1. *Spectral energy distributions.* for a T Tauri star (DR Tau) and a Herbig Ae/Be star (V380 Ori). Solid circles are ground-based observations; open circles represent large beam IRAS points. The solid line is the blackbody SED of an appropriate standard main sequence star. Superposed are the spectral energy distributions predicted for (1) a flat, optically thick accretion disk which extends inward to the stellar surface (top dotted curve), and (2) a flat, optically thick accretion disk which extends inward to only $3R_*$ (DR Tau; lower dotted curve) and $10R_*$ (V380 Ori; lower dotted curve); interior to these distances, the disk is assumed to be optically thin. These inner "holes" produce inflections in the near-infrared spectral energy distributions as a direct result of the absence of warm emitting material near the star.

by a central young star surrounded by a dusty, optically thick circumstellar accretion disk with a characteristic infrared spectral slope $\lambda F_\lambda \sim \lambda^{-4/3}$. Our "best fit" disk models include the effects of both disk accretion and inner disk optically thin regions or "holes."

For the sample of Herbig Ae/Be stars studied by Hillenbrand *et al.* (1992), the ranges of disk characteristics are as follows: (1) masses for the disks have upper limits ranging from $0.01 < M_{disk}/M_\odot < 6$; (2) outer disk radii have lower limits ranging from $15 < R_{disk}/AU < 175$; (3) inner optically thin disk regions have sizes $3 < R_{hole}/R_* < 25$; (4) disk mass accretion rates, as diagnosed from the magnitude of the infrared excess, range from $6 \times 10^{-7} < \dot{M}_{acc} < 8 \times 10^{-5} M_\odot$ yr^{-1} ; and (5) disk survival times as optically thick structures are less than 1 Myr.

With the positive identification of disks around young intermediate and high mass stars, the presence of circumstellar accretion disks has now been inferred around pre-main sequence stars of all masses ranging from $0.1 - 20$ M$_\odot$. The disks surrounding Herbig Ae/Be stars appear to be more massive, larger in size, more rapidly accreting, and have shorter lifetimes (*calculated* as M_*/M_{acc}) than the disks surrounding lower mass T Tauri stars. Our next step is to place *empirical* limits on the disk lifetimes around stars of different mass. This requires identifying the fraction of stars surrounded by optically thick accretion disks as a function of mass and age. Study of a complete, unbiased sample of young stars is necessary, and is most comprehensivly done through imaging surveys of extremely young clusters. Only in this way can the full range of disk properties as a function of stellar mass and age be ascertained.

THE (J-H) − (H-K) COLOR-COLOR DIAGRAM

The identification of young intermediate and high mass stars surrounded by accretion disks (analogs of Herbig Ae/Be stars), and those which are no longer surrounded by disks, can be effected from near-infrared photometry. We employ the (J-H) − (H-K) color-color diagram as a diagnostic in order to statistically study the disk populations of extremely young open clusters. All Herbig Ae/Be stars and many classical T Tauri stars lie outside the region of this diagram populated by normal dwarf and giant stars. Figure 2 shows the location of a sample of Herbig Ae/Be stars and, for comparison, the Taurus sample of classical T Tauri stars in the (J-H) − (H-K) color-color diagram.

In order for an object to lie to the right of the reddened main sequence in the (J-H) − (H-K) diagram, a high M_{acc} accretion disk seems to be required. Additionally, as is seen in the SEDs of Herbig Ae/Be stars, the presence of an inner disk hole moves objects even further away from the standard color-color relationships (also see Lada & Adams, 1992). All Herbig Ae/Be stars and some T Tauri stars can be distinguished from normal stars in the (J-H) − (H-K) plane − even if they are heavily reddened. Thus, locating stars in the (J-H) − (H-K) diagram can provide the means for rapid classification of optically obscured, luminous pre-main sequence stars with accretion disks and inner disk holes discovered, for example, in the course of infrared imaging surveys of star-forming regions.

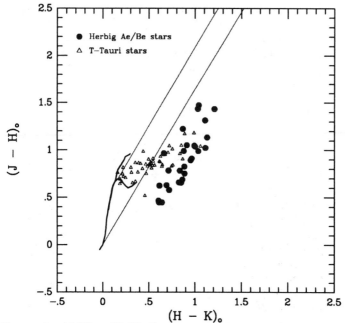

Figure 2. *(J-H) – (H-K) diagram.* of (filled circles) Herbig Ae/Be stars with SEDs similar to that illustrated in Figure 1 and (triangles) classical T Tauri stars in Taurus-Auriga. The solid lines are the standard relations for unreddened luminosity class V and class III stars (Bessell & Brett, 1988); the dotted lines indicate the effects of normal reddening from the main sequence. Note that the Herbig Ae/Be stars lie well to the right of the domain occupied by normal reddened stars, and reflect the presence of both an inner disk "hole" and disk accretion. Most of the T Tauri stars lie within the region occupied by normal reddened stars. The exceptions are those like DR Tau (see Figure 1) which are believed to have unusually high accretion rates.

Classical Be stars, which are found to populate young open clusters, also lie slightly to the right of the reddened main sequence in the (J-H) – (H-K) diagram. However, their small (H-K) excesses relative to their (J-H) colors arise due to free-free emission in a *gaseous* circumstellar disk or shell.

INFRARED SURVEYS OF YOUNG GALACTIC CLUSTERS

We are in the process of diagnosing the fraction of intermediate and high mass stars with disks as a function of stellar mass and age. To search for candidate Herbig Ae/Be systems and establish the disk lifetimes, we have undertaken a study of the young clusters M8, M16, M17, M42, and NGC 2264. Intermediate mass pre-main sequence populations appear in these clusters with ages of only 1×10^5 years. Thus they are prime arenas in which to search for the presence of circumstellar disks associated with the formation of high mass stars - even if the disk lifetimes are extremely short.

As a first step in this study, in order to identify candidate stars surrounded by disks, we have obtained JHK photometry using the SQIID camera and the OTTO InSb detector at the KPNO 1.3m telescope.

SQIID Studies of M8, M16, and M17 (d = 1.5 - 3.2 kpc)
The estimated completeness limit in a 180 second SQIID exposure is to magnitude 15 at J (15σ dectection) which, assuming the flux at J to be photospheric in origin, corresponds roughly to a spectral type of B8 and a mass of 5 $M_\odot$ at the cluster distances. The sample is unbiased and includes all detected objects within a central $15' \times 15'$ field for M8 and M16, and within a central $5'.5 \times 5'.5$ field for M17.

The (J-H) – (H-K) diagrams of these young clusters (see Figures 3,4,5) show the presence of a significant number of objects lying to the right of the region populated by normal reddened stars. We suggest that these objects are either (1) candidate Herbig Ae/Be stars - intermediate mass pre-main sequence stars surrounded by accretion disks or (2) candidate classical Be stars - early B stars of unknown evolutionary state surrounded by gaseous circumstellar disks.

Lower limits to the fraction of objects considered to be candidate Herbig Ae/Be stars or candidate classical Be stars are as follows:

M8 : 5% in a cluster whose age estimates range from $2 - 5$ Myr .
M16 : 13% in a cluster whose age estimates range from $3 - 10$ Myr .
M17 : 13% in a cluster whose age estimate is 1 Myr .

The J – (J-H) diagrams of these clusters (see Figures 3,4,5) show that the fraction of objects which are possibly surrounded by disks increases with decreasing J luminosity. Assuming that the flux at J is dominated by photospheric emission, stars of lower J luminosity should have lower mass. Hence, it appears that the disks may survive longer around stars of lower mass. If so, then the disks surrounding high mass stars must evolve on shorter timescales than those surrounding low mass stars. This result is consistent with the finding in Hillenbrand $et\,al.$ (1992) that disk accretion rates are higher for disks around stars of higher mass ($\dot{M}_{acc} \sim M_*^{2.2}$), and that disk accretion times ($M_*/\dot{M}_{acc}$) are shorter for disks around stars of higher mass.

The J – (J-H) diagrams also seem to indicate the presence of a large number of background objects in our samples. If we assume that all stars with $(J - H) > 1.6$ are in fact background sources, then the fraction of candidate Be and Herbig Ae/Be stars rises to 13% in M8 and to 30% in M16.

OTTO Studies of M42 and NGC 2264 (d = 460 pc; 800 pc)
The sample is optically selected and includes only proper motion cluster members brighter than B = 14. The estimated range in spectral type is from early B to middle G in both clusters.

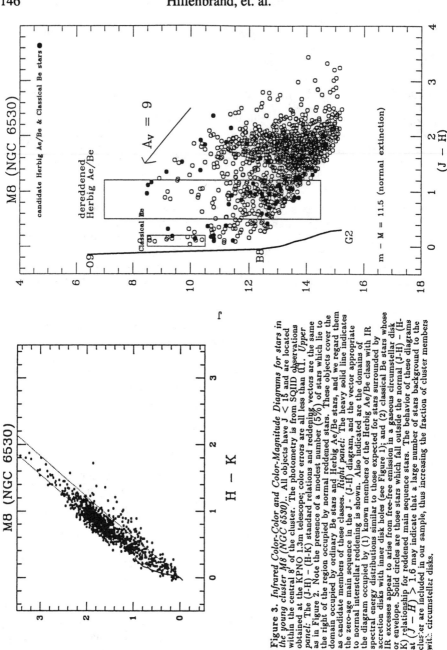

Figure 3. *Infrared Color-Color and Color-Magnitude Diagrams for stars in the young cluster M8 (NGC 6530).* All objects have $J < 15$ and are located within the central $8'$ of the cluster. The photometry is from SQIID observations obtained at the KPNO 1.3m telescope; color errors are all less than 0.1. *Upper panel.* The $(J-H) - (H-K)$ standard relations and reddening vectors are the same as in Figure 2. Note the presence of a modest number (5%) of stars which lie to the right of the region occupied by normal reddened stars. These objects cover the domain occupied by ordinary Be stars and Herbig Ae/Be stars, and we regard them as candidate members of these classes. *Right panel:* The heavy solid line indicates the zero-age main sequence in the $J - (J-H)$ diagram, and the vector appropriate to normal interstellar reddening is shown. Also indicated are the domains of the diagram occupied by the Herbig Ae/Be class with IR spectral energy distributions similar to those expected for stars surrounded by accretion disks with inner disk holes (see Figure 1); and (2) classical Be stars whose IR excesses appear to arise from free-free emission in a gaseous circumstellar disk or envelope. Solid circles are those stars which fall outside the normal $(J-H) - (H-K)$ relationship for reddened main sequence stars. The behavior of these diagrams at $(J - H) > 1.6$ may indicate that a large number of stars background to the cluster are included in our sample, thus increasing the fraction of cluster members with circumstellar disks.

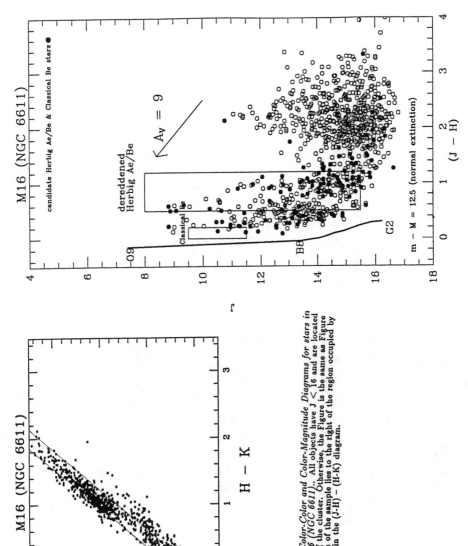

Figure 4. *Infrared Color-Color and Color-Magnitude Diagrams for stars in the young cluster M16 (NGC 6611).* All objects have J < 16 and are located within the central 8' of the cluster. Otherwise, the Figure is the same as Figure 3. Approximately 13% of the sample lies to the right of the region occupied by normal reddened stars in the (J-H) - (H-K) diagram.

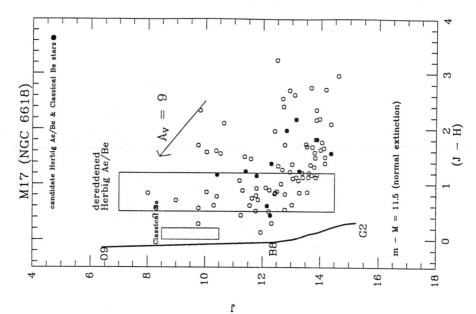

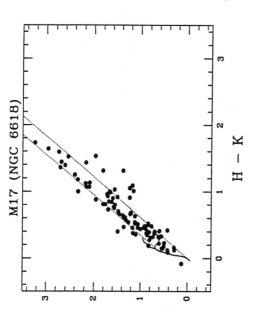

Figure 5. *Infrared Color-Color and Color-Magnitude Diagrams for stars in the young cluster M17 (NGC 6618)..* All objects have J < 14 and are located within the central 3′ of the cluster. Otherwise, the Figure is the same as Figure 3. Approximately 13% of the sample lies to the right of the region occupied by normal reddened stars in the (J-H) − (H-K) diagram.

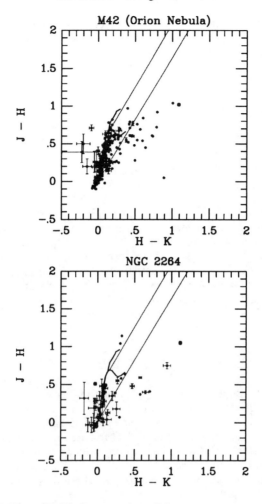

Figure 6. *(J-H) – (H-K) diagram for all known proper motion members.*
with B < 14 located *(upper panel)* within the central 45′ of the Orion Nebula
and *(lower panel)* within the central 20′ of NGC 2264 . These observations were
obtained with the OTTO single channel InSb photometer mounted on the 1.3m
telescope at KPNO; error bars are indicated. In Orion, of a sample of 225 stars, 30
fall in the domain occupied by classical Be or Herbig Ae/Be stars. In NGC 2264,
of a sample of 65 stars, 10 fall in the domain occupied by classical Be or Herbig
Ae/Be stars. We note that our survey sample in these regions includes stars of
spectral types ranging from early B though middle G.

The fraction of objects considered to be candidate Herbig Ae/Be stars or
candidate classical Be stars (see Figure 6) is as follows:

<u>M42</u>: 13% in a cluster whose age estimates range from $3-5$ Myr .
<u>NGC 2264</u>: 15% in a cluster whose age estimates range from $3-20$ Myr .

Consequences for Disk Lifetimes and Disk Accretion Rates

Evidence is rapidly accumulating that stars of all masses seem to be surrounded by dusty circumstellar disks at "birth." If so, then our empirical results indicate that most of the disks initially surrounding intermediate and high mass stars have disappeared by ages of less than 10^6 yr. Strict limits on disk survival times, however, await the construction of proper HR diagrams for these young clusters. We suggest that because the disk lifetimes for early type stars appear to be short, the accretion rates through the disks ($\dot{M}_{acc} = \varepsilon M_{disk}/\tau_{disk}$) must be high.

Sizeable accretion rates are not only suggested, but are required in the simple argument that for a star of mass 5 $M_\odot$ to be built during its pre-main sequence lifetime of 5×10^5 yr, it must accumulate material at the time-averaged rate of 1×10^{-5} $M_\odot$ yr^{-1}. Such high accretion rates imply short disk lifetimes, consistent with our observations. We note that the required steady-state accretion rates are similar to those derived by Hillenbrand *et al.* (1992) from the magnitude of the infrared excesses in Herbig Ae/Be stars. Finally, if the disk survival time is comparable to the pre-main sequence lifetime of the star, these accretion rates imply that a significantly large percentage of the material which makes up the final mass of the star passes through its disk during the pre-main sequence phase.

CONCLUSIONS AND FUTURE DIRECTIONS

Because some of the Be and Herbig Ae/Be candidates we have identified will, in fact, turn out to be classical Be stars, and will not show the large excess infrared emission expected from analogs of Herbig Ae/Be stars, we conclude that *by the age of just a few* 10^5 *years, less than* 10% *of pre-main sequence O and B stars are surrounded by optically thick accretion disks.* If these stars were all initially surrounded by disks of mass comparable to that of the current star, then the accretion rates for these disks must be on the order of 10^{-5} $M_\odot$ yr^{-1}.

For the SQIID surveyed regions, CCD imaging in combination with our IR imaging allows the objects to be properly dereddened and placed in an $M_V - (B - V)_0$ observational HR diagram. Using theoretical evolutionary tracks, we can then estimate the number of stars surrounded by disks as a function of stellar mass and age. We are currently carrying out such investigations.

REFERENCES

Bessell, M.S., & Brett, J.M. 1988 *Pub. A.S.P.*, 330, 350.
Hillenbrand, L.A., Strom, S.E., Vrba, F.J., & Keene, J. 1992 *Ap. J.*, in press.
Lada, C.J., & Adams, F.C. 1992 *Ap. J.*, in press.

Massive Stars: Their Lives in the Interstellar Medium
ASP Conference Series, Vol. 35, 1993
Joseph P. Cassinelli and Edward B. Churchwell (eds.)

FORMATION OF MASSIVE STARS AT THE EDGE OF OUR GALAXY

EUGENE J. DE GEUS
Department of Astronomy, University of Maryland, College Park, MD 20742

ALEXANDER RUDOLPH
NASA Ames, Moffett Field, CA 94035

ABSTRACT   We present preliminary results of a high-resolution radio-continuum and molecular line study of high-mass star forming regions in the outer parts of our Galaxy.

1. INTRODUCTION

Extensive studies have made it clear that all star formation in our Galaxy takes place in molecular clouds, with the most massive (O and B-type) stars being formed in Giant Molecular Clouds (GMCs), which have typical masses $\sim 2 \times 10^5 \; M_\odot$, sizes $\sim$50 pc, and average densities of ~ 100 cm^{-3} (Blitz 1978; Sanders, Solomon, and Scoville 1984). Most of the observational work on GMCs and massive star formation has concentrated on regions in the solar neighborhood and in the inner Galaxy. However, to obtain a complete picture of molecular clouds and star formation in the Milky Way, and to be able to use our locally gathered knowledge for interpretation of observations in external galaxies, we need to understand cloud- and star formation in a physical and chemical environment which is different from that in the inner parts of the disk. The ideal locations for such a study are the massive-star forming clouds at the edge of our own Galaxy. Their galactocentric radius is large enough for the physical environment to be significantly different, *e.g.* molecular clouds are more sparsely distributed (Wouterloot *et al.* 1990), the diffuse galactic interstellar radiation field is weaker, the metallicity is lower (Shaver *et al.* 1983, Fich and Silkey 1991), and the cosmic-ray flux is down (Bloemen *et al.* 1984), and yet these objects are near enough to resolve structures as small as 0.25 pc diameter cores.

The goal of this study is to establish the presence of high-mass star formation at large galactocentric radii, to determine the properties of the star forming molecular clouds down to the scales of the star forming cores and compare these to nearby molecular clouds (*e.g.* Orion), and to use the HII regions to obtain measurements of the metallicity out to much larger galactocentric radii than possible so far.

2. SOURCE SELECTION AND OBSERVATIONS

We used the Wouterloot and Brand (1989) catalog of CO observations toward IRAS point sources in the second and third Galactic quadrants, which they selected based on color criteria indicating the presence of H_2O masers and dense molecular cloud "cores" (Wouterloot and Walmsley 1986) and hence star forming regions ($\Rightarrow \sim 1300$ objects). Distances quoted in that catalog (and used in our study) are based on: $R_0 = 8.5$ kpc, and $\Theta_0 = 220$ km s^{-1}. We selected sources which have $L_{FIR} > 10^4 L_\odot$ (indicating the presence of stars with spectral type B1 or earlier), galactocentric radii $R_{G.C.} > 15$ kpc, and $|l - 180°| > 15°$ (because of large distance uncertainty close to the direction of the anti-center). This results in 31 sources, of which 23 are easily observable with existing mm-interferometers (i.e. they are in northern hemisphere).

Preliminary results on the comparison between VLA 6-cm continuum date and BIMA interferometer observations of CO and CS (J = 2→1) are presented in the next sextion. For the interpretation of the VLA data we used the model by Mezger and Henderson (1967) with an electron temperature of 10^4 K, which is likely to be appropriate for these objects at large $R_{G.C.}$ (Shaver et al. 1983, Fich and Silkey 1991).

3. DISCUSSION

S127 is an optical HII region with $R_{G.C.} = 15$ kpc and $R_{hel} = 11.5$ kpc. The IRAS far infrared luminosity of this source is 10^5 $L_\odot$, which indicates presence of the equivalent of an O7 main sequence star. The two strongest peaks in the 6-cm observations indicate Ly-α photon production rates of 1.5 and 5 $\times 10^{48}$ s^{-1}, which indicates the presence of one O9 V and one O7.5 V star (cf. Felli and Harten 1981 from Westerbork observations). The radio continuum and far-IR luminosities are therefore in good agreement.

Figure 1a shows the Kitt Peak 12-m large scale CO map of the molecular cloud associated with this HII region. From the single dish observations we derive a CO luminosity of $3.3\,10^4$ K km s^{-1} pc^2. The CO-to-H$_2$ conversion factor (X) has been suggested to be a factor of 4 higher in the outer Galaxy than locally (Digel et al. 1991), which would imply a molecular mass for S 127 of $6\,10^4$ $M_\odot$, typical for a GMC. Figure 1b shows the HII region with an overlay of the BIMA interferometer CO map. The mm-interferometer data have been combined with the Kitt Peak 12-m data in order to obtain the zero-spacing information normally missing from interferometer maps. The necessity of this procedure is clear from the fact that the flux seen in the interferometer maps only was about 30% of the total. The two main peaks in the continuum are evidently associated with peaks in the CO distribution. A third peak in CO is devoid of continuum radiation, and might be a good candidate to look for evidence of new signs of star formation. The interferometer observations show clump radii between 0.12 (beam size) and 0.75 pc. We are currently undertaking a full analysis of the clump properties (sizes, linewidths, CO luminosities).

IRAM observations of the CS molecule show a very weak structure associated with the CO/6-cm complex at $(\alpha, \delta) = (21^h\ 27^m\ 08^s, +54°\ 24'\ 30'')$. The CS(3→2) line was not detected suggesting a small column density for CS. This, combined with the presence of the strong continuum source, might indicate that the embedded young star(s) are in the process of destroying (dissociating, ionizing, blowing away) the dense molecular gas. The complete absence of CS emission associated with the southern continuum source might indicate that this process has proceeded further in this part of the cloud.

WB 380 (#380 in Wouterloot and Brand 1989), at a distance of 10.7 kpc from the Sun ($R_{G.C.} = 17$ kpc) has a far-IR luminosity of $1.1\,10^5$ $L_\odot$, which indicates the presence of an O7 V star. The source does not have a known optical counterpart, which may be due to extinction in the foreground and local to the source. The continuum source is unresolved in right ascension, and marginally resolved in declination. From this, we can derive an upper limit to the source diameter of 0.2 pc (compact or ultracompact) and a lower limit to the Ly-α photon production rate of $3.2\,10^{48}$ s^{-1}, which is equivalent with the presence of an O8 V star or earlier. Given the uncertainty in the size of the 6-cm source, the continuum and far-IR luminosities are in good agreement. The continuum source is well correlated in position with peaks in the interferometer maps of CO and CS. This suggests that this object is in a very early stage of its evolution. The high-resolution interferometer observations show the presence of sub-structure ("clumps") with radii of 0.8 pc down to about 0.1 pc. A full analysis of the clump properties is being carried out.

EdG acknowledges financial support from NSF grant AST-8918912 and a research grant funded by the Margaret Cullinan Wray Charitable Lead Annuity Trust.

REFERENCES

Blitz, L., 1978. Ph.D. Dissertation Columbia University.
Bloemen, J.B.G.M., et al., 1984. *Astr. Ap*, **135**, 12.
Digel, S., Bally, J., and Thaddeus, P., 1990. *Ap. J.*, **357**, L29.
Felli, M., and Harten, R.H., 1981. *Astr. Ap*, **100**, 28.
Fich, M., and Silkey, M., 1991. *Ap. J.*, **366**, 107.
Lada, E.A., 1990. Ph.D. Dissertation University of Texas at Austin.
Mezger, P.G., and Henderson, A.P., 1967. *Ap. J.*, **147**, 471.
Sanders, D.B., Solomon, P.M., and Scoville, N.Z., 1984. *Ap. J.*, **276**, 182.
Shaver, P., McGee, R.X., Newton, L.M., Danks, A.C.,and Pottasch, S.R., 1983. *M.N.R.A.S.*, **204**, 53.
Wouterloot, J.G.A., Brand, J., Burton, W.B., and Kwee, K.K., 1990. *Astr. Ap*, **230**, 21.
Wouterloot, J.G.A., and Brand, J., 1989. *Astr. Ap. Suppl.*, **80**, 149.
Wouterloot, J.G.A., and Walmsley, C.M. 1986. *Astr. Ap*, **168**, 237.

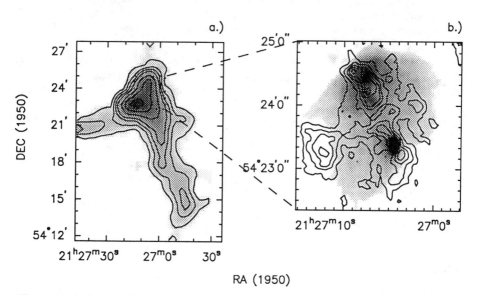

Figure 1. a). Integrated intensity of the CO line between -98 and -92 km/s towards S 127. Levels are 3,5,7,9,13,17,21 Kkm/s. The GMC shows a very striking, highly elongated morphology. b.) High-resolution image of the S 127 region. The grey scale shows the 6-cm continuum radiation, with a peak flux in the map of 21 mJy/beam. The contour overlay shows the BIMA interferometer CO observations, with the contour levels: 10,20,...,60 Jy/beam. The interferometer data have been combined with Kitt Peak 12-m observations, in order to obtain the correct zero spacing flux.

Massive Stars: Their Lives in the Interstellar Medium
ASP Conference Series, Vol. 35, 1993
Joseph P. Cassinelli and Edward B. Churchwell (eds.)

CHEMICAL DIFFERENTIATION BETWEEN THE ORION HOT CORE AND COMPACT RIDGE

P. CASELLI[1,2], T. I. HASEGAWA[1], E. HERBST[1]

1. Department of Physics,The Ohio State University,174 W. 18th Ave.,Columbus,OH 43210

2. Dipartimento di Astronomia,Universita' di Bologna,Via Zamboni 33, 40126 Bologna, Italy

ABSTRACT   We have constructed a dynamical chemical model of massive star formation regions, in which gas and grains are considered. We find that differences in thermal history and gravitational collapse of the cloud can account for the observed chemical differences between the Orion Hot Core and Compact Ridge, two clumps of OMC-1 very close to the luminous protostar IRc2.

INTRODUCTION

The recent gas-grain models of interstellar chemistry in dense and quiescent clouds (Hasegawa, Herbst & Leung 1992, Hasegawa & Herbst 1992) reproduce observations of CO and H_2O ices in the Taurus region reasonably well. They are also successful in reproducing observed gas phase abundances in TMC-1 (see Herbst & Leung 1990). Here we focus on star forming regions, in which higher temperatures and densities can seriously alter the chemistry compared with that of quiescent clouds. In particular, we try to reproduce the well-known chemical differentiation between two clumps of OMC-1, very close to the luminous protostar IRc2: i) the *Hot Core* ($n > 10^7$ cm^{-3}; $T \sim 200$ K), characterized by high abundances of complex N-bearing molecules, like vinyl cyanide (CH_2CHCN) and ethyl cyanide (CH_3CH_2CN), and ii) the *Compact Ridge* ($n > 10^6$ cm^{-3}; $T \sim 100$ K), characterized by high abundances of complex O-bearing molecules, like methyl formate ($HCOOCH_3$), and dimethyl ether (CH_3OCH_3) (Blake et al. 1987). Several attempts have been made to model these regions (e.g. Brown, Charnley & Millar 1988; Millar, Herbst & Charnley 1991; Charnley, Tielens & Millar 1992), all considering a sudden evaporation of the molecules formed on grain surfaces due to the heating of the new-born protostar. However, different initial abundances of key molecules (NH_3, CH_3OH; Charnley, Tielens & Millar 1992), or injection of methanol (Millar, Herbst & Charnley 1991), were needed to reproduce the different chemical evolution of the Hot Core and the Compact Ridge. We have recently found that differences in thermal history and gravitational collapse can account for the chemical differences in the two clumps, at least for the more complex molecules.

MODEL

Schematically, the chemical scenario on grains is the following: gas phase neutral atoms and molecules 1) accrete onto surfaces of classical grains, 2) migrate from potential wells to adjacent wells on grain surfaces, 3) encounter reaction partners, 4) react to form more complex molecules. Cosmic ray-induced desorption and thermal evaporation return surface molecules to the gas phase. There are 255 surface reactions in our chemical network involving 146 neutral surface species; some examples are $C_nH_m + H \longrightarrow C_nH_{m+1}$, $C + C \longrightarrow C_2$, $C + O \longrightarrow CO$. The surface reaction network is combined with our gas phase reaction network that consists of 2918 reactions involving 304 gas phase species. We have introduced free-fall collapse to the previous pseudo-time dependent chemical models, and we solve the rate equations from a set of initial abundances.

For the Hot Core we start with a dense ($n = 2 \times 10^5$ cm^{-3}), warm ($T_{gas} = T_{dust} = 40$ K) cloud, close to a massive protostar. All the hydrogen is in molecular form and the easily ionized heavy atoms are found in their ionized form. We follow the gravitational collapse of the cloud (or protostellar circumstellar shell) for 1×10^5 yr, until the density reaches 10^7 cm^{-3}. At this point we assume a further and sudden heating of the cloud (due, for example, to the increased closeness to the heating source that, in the meanwhile, has accreted mass and increased its luminosity). The grain mantles are evaporated and we follow the gas phase chemistry at $T = 200$ K, for 10^6 yr (Fig.1a).

The initial density ($n = 2 \times 10^4$ cm^{-3}) and temperature ($T_{gas} = T_{dust} = 20$ K) of the Compact Ridge were chosen considering this region as a more external shell of the same collapsing cloud (see, for example, Scoville & Kwan 1976). Here the hydrogen is in atomic form. The gravitational collapse lasts 3.4×10^5 yr, until $n = 10^6$ cm^{-3}; afterwards the sudden increase of temperature ($T = 100$ K), likely due to the interaction with the outflow driven by IRc2 (Blake et al., 1987; Genzel & Stutzki 1989), causes the complete desorption of the molecules. The gas chemistry is then developed for 10^6 yr (Fig.1b).

RESULTS AND CONCLUSIONS

Good agreement with observations, at least for the complex molecules, and the very abundant CO and H_2O, is obtained for both regions (Tab.1). In particular, we succeed in producing high abundances of N-bearing molecules in the Hot Core, whereas the O-bearing molecules are just at the detectability limit (Fig.1a). The best agreement is obtained 10^5 yr after the evaporation of the grain mantles. For the Compact Ridge we obtain good agreement with the observed abundances of HCOOCH$_3$ and CH$_3$OCH$_3$, 10^4 yr after the evaporation. The complex N-bearing molecules are less abundant than in the Hot Core.

Differential thermal evaporation during the gravitational collapse seems to be an important point in producing the strong chemical differentiation observed in the Orion Hot Core and Compact Ridge. In the Hot Core we produce smaller abundances of methanol (the basic molecule for the formation of CH$_3$OCH$_3$ and HCOOCH$_3$) than in the Compact Ridge, by simply increasing the thermal evaporation of surface molecules. In particular, this drastically reduces the rate of the surface process   CO --> HOC --> HOCH --> HOCH$_2$ --> CH$_3$OH, because of the volatility of CO compared with other molecules. Several discrepancies in both sources exists for some lighter molecules, like NH$_3$ and HCN. However, especially for the simplest and more abundant species, there are often uncertainties associated with the observations, given the clumpy nature of the Orion Hot Core and Compact Ridge; the more extended emission of lighter species could "mask" any contribution from the more compact sources (see, for example, Blake et al. 1987). In any case, large N-bearing or O-bearing molecules do not appear to have an extended counterpart; therefore they stand out in producing the different spectral signatures of the two regions discussed above. High resolution observations around IRc2 are needed to spatially resolve the Hot Core and Compact Ridge from the surrounding environment.

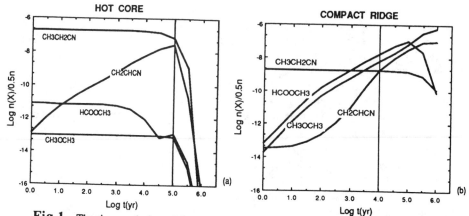

Fig.1 The time evolution of the fractional abundances of gas phase complex molecules after the evaporation of the grain mantles (a) in the Hot Core, and (b) the Compact Ridge.

Tab.1

FRACTIONAL ABUNDANCES

	COMPACT RIDGE			HOT CORE		
	OBSERVED	Ref.	THEORY (10^4 yr)	OBSERVED	Ref.	THEORY (10^5 yr)
CO	5.0(-5)	1*	1.5(-6)	1.2(-4)	1	1.4(-4)
CS	--		6.0(-10)	<5.5(-10)	1	7.9(-13)
H₂O	1.0(-4)	16	2.8(-4)	1.0(-5)	10	4.2(-5)
HCN	--		8.7(-7)	3.0(-7)	1	1.3(-10)
HNC	--		4.0(-7)	1.0(-9)	6	4.8(-9)
NH₃	2.0(-8)	1*	1.2(-5)	1.0(-5)	1	2.4(-5)
C₂H₂	--		5.9(-9)	1-10(-7)	4	1.7(-8)
H₂CO	4.0(-8)	13	3.0(-5)	2.6(-8)	1	1.2(-8)
CH₄	--		7.7(-5)	0.1-3(-6)	12	4.1(-7)
CH₂CO	6.7(-10)	1	4.6(-10)	--		2.1(-8)
HCOOH	5.0(-10)	1	2.5(-10)	--		9.4(-9)
HC₃N	--		6.9(-9)	1.6(-9)	1	2.8(-8)
CH₃OH	1-10(-7)	14	3.2(-5)	--		1.7(-10)
CH₂CHCN	--		1.7(-9)	1.8(-9)	1	2.6(-8)
HC₅N	2.3(-11)	11	7.3(-11)	--		5.1(-10)
CH₃OCH₃	1.0(-8)	1	6.0(-9)	--		7.8(-14)
C₂H₅OH	<5.0(-10)	1	1.2(-11)	--		6.6(-13)
CH₃CH₂CN	--		1.7(-9)	9.8(-9)	1	6.3(-8)
HCOOCH₃	8.7(-9)	1	1.7(-8)	--		1.2(-13)

* For the Extended Ridge. Reliable observations are unavailable for the Compact Ridge.

REFERENCES

1. Blake,G.A., Sutton,E.C., Masson,C.R., Phillips,T.G. 1987, ApJ, 315, 621
2. Brown,P.D., Charnley,S.B., Millar,T.J., 1988, MNRAS, 231, 409
3. Charnley,S.B., Tielens,A.G.G.M., Millar,T.J. 1992, ApJL, (in press)
4. Evans II,N.J., Lacy,J.H., Carr,J.S. 1991, Ap.J., 383, 674
5. Genzel,R., Stutzki,J. 1989, ARAA, 27, 41
6. Goldsmith,P.F., Irvine,W.M., Hjalmarson,A., Ellder,J. 1986, ApJ, 310, 383
7. Hasegawa,T.I., Herbst,E., Leung,C.M. 1992, ApJS, (in press)
8. Hasegawa,T.I., Herbst,E. 1992, MNRAS, (submitted)
9. Herbst,E., Leung,C.M. 1990, ApJ, 233, 177
10. Jacq,T.,Jewell,P.R.,Henkel,C.,Walmsley,C.M.,Baudry,A. 1988, A&A,199,L5
11. Johansson,L.E.B., Andersson,C., Ellder,J., Friberg,P., Hjalmarson,A., Hoglund,B., Irvine,W.M., Olofsson,H., Rydbeck,G. 1984, A&A, 130, 227
12. Lacy,J.H., Carr,J.S., Evans II,N.J., Baas,F., Achtermann,J.M., Arens,J.F. 1991, ApJ, 376, 556
13. Magnum,J.G., Wootten,A., Loren,R.B., Wadiak,E.J. 1990, ApJ, 348, 542
14. Menten,K.M., Walmsley,C.M., Henkel,C., Wilson,T.L. 1988, A&A, 198, 253
15. Millar,T.J., Herbst,E., Charnley,S.B. 1991, ApJ, 369, 147
16. Moore,E.L., Langer,W.D., Huguenin,G.R. 1986, ApJ, 306, 682
17. Scoville,N.Z.,Kwan,J. 1976,ApJ,206,718

SECTION 2

POST-MAIN-SEQUENCE MASSIVE STARS

Massive Stars: Their Lives in the Interstellar Medium
ASP Conference Series, Vol. 35, 1993
Joseph P. Cassinelli and Edward B. Churchwell (eds.)

MASSIVE STAR EVOLUTION

NORBERT LANGER
Universitäts-Sternwarte Göttingen, Geismarlandstr. 11, W-3400 Göttingen, F.R.G.

ABSTRACT   In this review we intend to describe the state of the art of massive single star models by considering both, their success and failures. We find that even the main sequence phase of massive stars is not yet satisfactory understood. The discussion of the post main sequence evolution is divided into two parts, where uncertainites in mixing processes or in mass loss processes are more relevant, respectively. We summarize the evidence for molecular weight gradients preventing convective mixing in massive stars, and discuss the relevance of convective overshooting. The origin and evolution of Wolf-Rayet stars, and the importance of mass dependent mass loss for their structure and final evolutionary phases is discussed. We briefly report on perspectives for supernovae from massive stars with emphasis on non-standard progenitors as blue supergiants and Wolf-Rayet stars.

1. INTRODUCTION

It is well known that massive stars (i.e. $M_{ZAMS} \gtrsim 10\,M_\odot$) have a large impact on their environment, which is due to three components: the radiation field, the stellar wind, and the final supernova explosion. In order to correctly predict the time dependent characteristics of these components, it is necessary to consider the theory of structure and evolution of massive stars. The following examples eluminate the connection of stellar interior physics and ISM.

First, consider the radiation driven winds of massive main sequence stars. Very recently, the radiative opacity coefficient κ has been revised (cf. Rogers and Iglesias 1992), which greatly affects the internal structure of the most massive, radiation dominated stars. The radii of these stars on the main sequence are found to be much larger, the effective temperatures much smaller compared to previous models (cf. e.g. Schaller et al. 1992). Since according to the radiation driven wind theory (Castor et al. 1975) mass loss rate and wind velocity depend sensitively on radii and surface temperatures of luminous hot stars, they are certainly subject to large changes, too. This means that predictions of energy and mass input into the ISM from massive main sequence stars depends on the opacity coefficient used in stellar models. In this case, the stellar radiation field is of course also changed as a consequence of smaller surface temperatures.

The second example concerns the effect of internal mixing processes on main sequence mass loss. Additional mixing due to convective overshooting or rotation is widely discussed for luminous main sequence stars (cf. section 2). The subsequent change of the internal structure leads to a larger stellar luminosity, which will then increase the mass loss rate due to the radiation driven wind. However, there is even a second factor which increases the mass input into the ISM, namely a longer main sequence lifetime.

The third example deals with the winds of Wolf-Rayet (WR) stars. Many authors have adopted a constant mass loss rate for WR models in evolutionary calculations (e.g. Prantzos et al. 1986). In order not to run out of mass for the least massive objects, that constant rate was relatively small and consequently yields rather large final masses for moderately massive and very massive initial masses (cf. Maeder and Meynet 1987). According to our current understanding of core collapse (cf. Woosley and Weaver 1986) such large final masses would result in a black hole rather than in a supernova explosion. If mass dependent WR mass loss (Langer 1989) is invoked, on the other side, small final masses may be the result independent of the initial stellar mass (cf. Langer 1989, Maeder 1990, Schaller et al. 1992) and a final supernova explosion is very likely (Woosley et al. 1992).

These examples show, that in order to quantify the effects of massive stars on the ISM, one needs to study the structure and evolution of massive stars. They show also, that massive star evolution is not a closed theory but rather an open field of research with large changes in the recent past and many remaining fundamental problems. We will try to outline in this work what major progress could be obtained during the last years, but also which basic problems were recently brought up by the observations. We divide our discussion into four parts, i.e. massive main sequence stars (section 2), post main sequence evolution for stars below (section 3) and above (section 4) an initial mass of roughly $35 \, M_\odot$, an on supernovae (section 5).

2. MAIN SEQUENCE STARS

Here, we define a star as being on the main sequence by the condition of central hydrogen burning, i.e. the central temperature should be well below $10^8 \, K$. For massive stars, in contrast to low and intermediate mass stars, we are already in trouble when we want to single out stars of which spectral types correspond to this definition. Clearly, O and early B dwarfs are massive main sequence stars. O type supergiants are probably also main sequence stars, though the term 'supergiant' usually corresponds to post main sequence stars. Anyway, the early B type supergiants are problematic. The reason is, that observations give no indication of the right (cool) borderline of the main sequence band in the HR diagram, i.e. there is no 'terminal age main sequence' (TAMS). This "TAMS problem" is striking in the observed HR diagrams of Blaha and Humphreys (1989), Fitzpatrick and Garmany (1990), and Garmany and Stencel (1992). Taken at face, this means that hydrogen- and helium burning stages overlap in the upper HR diagram. This is in contradiction to all stellar evolution calculations, which predict a pronounced gap between main sequence and post main sequence stages (cf. e.g. Langer 1991, Schaller et al. 1992).

One might think of the TAMS problem as being due to problems of correctly placing observed hot stars into the HR diagram; however, first of all, the observers don't agree on this (cf. Fitzpatrick and Garmany 1990), and moreover, two more fundamental problems cast doubt on current massive star models, i.e. the "mass problem" and the "helium problem". The mass problem means, that masses of O and early B stars derived from observations — either spectroscopic masses or masses derived with the radiation driven wind theory — are on average much smaller ($\sim 50\%$) compared to masses derived from stellar evolution models (cf. Herrero et al. 1992, and references therein). In other words, one of the most fundamental properties of stellar models, the L/M-ratio, apparently disagrees with observations. Finally, the helium problem stands for significant observed helium enrichment in OB stars (see again Herrero et al. 1992), which is not predicted by standard stellar evolution models.

Altogether, the TAMS, helium, and mass problem indicate that massive main sequence stars are far from being well understood. This is as much more true as the variations of weakly restricted parameters in massive star models, which concerns the mass loss rate, convective overshooting or semiconvection, cannot solve the above problems (cf. discussion in Herrero et al. 1992).

In the following we will assume that the standard evolution models of massive main sequence stars (e.g. Langer 1991, Schaller et al. 1992) are more or less correct for at least a fraction of all massive stars, on which we concentrate hereafter. To identify the physical reasons for the TAMS, helium, and mass problems and to investigate its effects for main sequence and post main sequence evolution of massive stars may be a major challenge to stellar model builders in the near future. First attempts to cope with these problems are discussed in a separate paper in this volume.

3. POST MAIN SEQUENCE EVOLUTION FOR $M_{ZAMS} \lesssim 35\,M_\odot$

The reason for dividing the discussion of the post-MS evolution of massive stars into two parts is the different role of mass loss for the two mass regions considered here. For stars initially more massive than a certain critical mass, the post-MS evolution is dominated by mass loss, i.e. the theoretical evolution depends predominantly on what we assume for mass loss in the RSG, LBV, or WR stages. For stars below this critical mass, uncertainties in the theory of convection are more relevant than uncertainties in the mass loss rates. The value of the critical ZAMS mass — which is no sharp defined number anyway — depends greatly on metallicity due to the fact that radiation driven mass loss decreases with lower metal content (Kudritzki et al. 1987). The value of $35\,M_\odot$ mentioned in the heading above should thus be considered only as a rough estimate.

In this section we will consider only stars, which do not evolve into LBV or WR stars. Consequently, we have to cope only with blue or red supergiant phases (BSGs, RSGs). The intermediate stage ("yellow supergiants") is in general thermally unstable and is consequently rather short lived (cf. Tuchman and Wheeler 1990; Langer 1991), a circumstance which this time is rather well supported by the observations (cf. Fitzpatrick and Garmany 1990, and Langer 1991).

As mentioned above, the radiation driven mass loss decreases with decreasing metallicity. Furthermore, opacity effects (cf. section 1) are reduced at lower metallicity. Thus, it appears instructive to focus on the Large or Small Magellanic Cloud (LMC, SMC) in order to investigate effects of various assumtions on the convection theory, since the metallicity of these galaxies is small (roughly $Z_\odot/4$ and $Z_\odot/10$ for LMC and SMC, respectively). Moreover, the distance to the stars of the LMC and SMC is reasonably well known, and it is almost the same for the stars of each galaxy. Due to their proximity we have a rather complete knowledge of their massive star content (Blaha and Humphreys 1989, Fitzpatrick and Garmany 1990), and it is very fortunate that SN 1987A occurred in the LMC.

The two main problems in convection theory, which have been studied in the literature in connection with massive stars, and which bring considerable uncertainty into post-MS models within the considered mass range (cf. Langer 1992) are semiconvection and convective overshooting. Semiconvection, in this paper, shall designate the instability which developes in chemically inhomogeneous superadiabatic regions of a star (cf. Kato 1966, Langer et al. 1983), while convective overshooting describes the expansion of convectively unstable regions into the radiatively stable regime (cf. Zahn 1991).

The problem of semiconvection is closely related to the question, whether the Schwarzschild or the Ledoux criterion for convection should be used in stellar evolution calculations: If semiconvection would be a very rapid process, one could use the Schwarzschild criterion for convection also in chemically inhomogeneous regions (which is actually done in most stellar evolution calculations; e.g. Chiosi and Maeder 1986, Schaller et al. 1992), while for infinitly slow semiconvection this process could be ignored completely, but the Ledoux criterion for convection should be used. (Note that both criteria are identical in chemically homogeneous regions.) The truth is certainly somewhere in between these two extremes. Langer et al. (1983) have estimated the local time scale of semiconvection, and found it to depend on the (destabilizing) superadiabaticity and the (stabilizing) molecular weight gradient, i.e. there may be no unique answer, but the efficiency of semiconvection depends on the local conditions inside the star.

However, in recent years, more and more evidence accumulates that — for the mass range under consideration — semiconvection is no "fast" process, i.e. that molecular weight barriers are not quickly wiped out when superadiabaticity develops: First, it was found by Woosley (1988) and confirmed by Woosley et al. (1988), Weiss (1989), Langer et al. (1989), Langer (1991a), and Arnett (1991), that the assumption of "slow" semiconvection yields stellar models at $M \simeq 20\,M_\odot$ and with LMC composition, that explode as blue supergiants in a post-RSG phase. Alternative ways to construct a BSG supernova progenitor exist (cf. Arnett et al. 1989), but the explanation in terms of semiconvection theory appears to be most natural.

This is as much more so as independent arguments in favour of "slow" semiconvection exist. Langer (1991) has calculated a grid of masses in the range considered here and with LMC composition, and found that the semiconvection model which yields good agreement with the SN 1987A progenitor results in a supergiant distribution in the theoretical HR-diagram which is quite consistent with the observed one, including such intricate features as the Fitzpatrick and Garmany (1990) "ledge". Furthermore, Stothers and Chin (1992) found the supergiant distribution in the young open cluster NGC 330 in the SMC to be consistent with stellar models using the Ledoux criterion, but inconsistent with the Schwarzschild criterion for convection. In the latter case, no RSGs are predicted but 15 of 24 supergiants are observed to be RSGs.

Finally, it is a characteristic of models computed with "slow" semiconvection, that they ignite helium burning as RSGs, where (at least for galactic and LMC metallicity, while probably not for SMC and lower Z; cf. Arnett 1991, Stothers and Chin 1992) nitrogen and helium is mixed into the whole envelope by convective dredge-up. Consequently, the models in the subsequent BSG stage are N-enriched. This agrees with the suggestion of Walborn (1988) that most BSGs are nitrogen enriched. Lennon et al. (1992) find N-enrichment in many galactic BSGs, and conclude that the enrichment is stronger for more luminous stars — an effect predicted by the models of Langer (1991).

Let us mention at this place, that independent evidence in favour of "slow" semiconvection comes from more massive stars: the occurrence of a subgroup of WR stars which display nitrogen and carbon emission lines simultaneously, called WN/WC stars (Conti and Massey 1989; Massey and Grove 1989), is a natural product of "slow" semiconvection (Langer 1991b) but remains unexplained otherwise.

Finally, "slow" semiconvection is also observed to occur in nature and in experiments, which is summarized and discussed in a theoretical framework by Spruit (1992).

Convective overshooting, on the other side, has been considered as necessary in order to explain the HR-diagram of massive stars — in particular the width of the main sequence band — by several groups (cf. e.g. Doom 1985, Chiosi and Maeder 1986, Schaller et al. 1992). However, due to the new opacity data provided by the Livermore group (cf. Rogers and Iglesias 1992) convective overshooting appears no longer to be required (Stothers and Chin 1991). Moreover, model sequences with "slow" semiconvection and overshooting do not yield any BSG phase for LMC composition (Langer 1991) and have thus to be disregarded. Note that also the supergiant distribution of the models of Schaller et al. (1992), which assume a moderate overshooting, is rather unsatisfactory: For solar metallicity, almost no BSGs are predicted, while at low metallicity almost no RSGs are predicted.

In summary, we believe that convective overshooting may be much less efficient in massive stars than assumed by many stellar theorists in the recent past. In any case, the width of the main sequence band should not be considered as a good test for convective overshooting, since we have seen that it is sensitive to other things as well, as e.g. internal opacities (see above), or rotationally induced mixing (cf. Langer, this volume).

4. POST MAIN SEQUENCE EVOLUTION FOR $M_{ZAMS} \gtrsim 35 M_\odot$

As mentioned at the beginning of the previous section, the evolution of the most massive stars is dominated by mass loss, while uncertainties in mixing processes are of minor importance, at least according to the standard evolutionary scenarios. Mass loss on the main sequence is non-negligible (cf. Schaller et al. 1992, for example), and at the Humphreys-Davidson (HD) limit in the HR diagram (Humphreys and Davidson 1979) large mass loss is invoked in order to prevent the stellar models from entering the empty upper right corner of the HR diagram, i.e. to cross the HD-limit. This phase of rapid mass loss, for which the mass loss rate itself can be predicted with some confidence (see Langer 1989a) though the underlying physics is rather unknown (cf. Davidson et al. 1989), and which is assumed to be related with the Luminous Blue Variables (LBVs) (cf. Schaller et al. 1992), leads the star into the Wolf-Rayet phase.

It has been emphasized before, that the recent improvements of the interior opacities have a large impact especially on the evolution of the most massive stars. It appears possible, that — at least at high metallicity — the cool edge of the main sequence band and the HD limit meet in the HR diagram (see again Schaller et al. 1992). Consequently, the LBV phase and the ensueing WR formation may, in this case, occur during core hydrogen burning, which can give rise to a long-lasting WNL phase.

All this indicates, that the way a very massive star looses its hydrogen envelope is a quite complex story, of which many details remain to be investigated. However, once the hydrogen envelope is lost — and we know this happens from the sizable population of hydrogenless WR stars — the internal structure of the remaining compact star is completely independent of its previous evolution (Langer 1989b). The reason is that the internal structure is almost independent of the chemical composition due to the importance of radiation pressure and the dominance of electron scattering for the opacity coefficient. This explains the existence of a narrow mass-luminosity relation for WR stars found by Maeder (1983), and similar relations for radii or surface temperatures (Langer 1989b). It is also a major motivation to assume a mass dependence of the mass loss rate of hydrogenless WR stars of the form $\dot{M}_{WR} \sim M_{WR}^{\alpha}$ (Langer 1989). Note that $\alpha \gtrsim 1$ leads to a convergence of the final masses of WR stars for largely different initial masses, and several observational clues, especially the existence of many WR stars with luminosities well below $\sim 10^5 L_{\odot}$ (Schmutz et al. 1989), indicate that $\alpha \gtrsim 2$ (Langer 1989). Consequently, final masses of the order $5 M_{\odot}$ are found for a wide range of initial masses (cf. Maeder 1990, Schaller et al. 1992, Woosley et al. 1992).

Let us finally mention that predictions for the evolution of stars in the mass range $30 M_{\odot} \lesssim M_{ZAMS} \lesssim 40 M_{\odot}$ (at $Z_{\odot}$) are particularly uncertain, since they are largely affected by uncertainties involved in mixing processes and in the various mass loss rates. E.g. we still don't know whether these stars evolve through an LBV phase in general, and which consequences that would have, or whether the corresponding very luminous RSGs finally evolve into WR stars or not. However, note that according to the current ideas about the mass loss rates in the various evolutionary stages, it is the $30 - 40 M_{\odot}$ stars which end up with the largest final masses and which are therefore the most likely candidates for black hole formation (cf. Woosley et al. 1992).

5. SUPERNOVAE

It is widely accepted that the progenitors of Type II supernovae (SNe) are in general red supergiants which experience iron core collapse at the end of their thermonuclear evolution (cf. Woosley and Weaver 1986). This picture is greatly supported by the neutrino signal detecded from SN 1987A (see Arnett et al. 1989, and references therin). However, in this case it was a blue supergiant which exploded, and it is still a matter of debate in how far BSG explosions are a "normal" and thus potentially frequent event in the universe, especially in low metallicity galaxies. Langer (1991) has estimated that — assuming that nothing special happened during the evolution of the SN 1987A progenitor — about 40% of all supernovae due to massive single stars may occur in BSGs in LMC-like galaxies. Though this number is not completely model-independent, many stellar evolution calculations indicate a trend in the sense that BSG explosions are more likely at lower metallicity: At $Z = Z_\odot$ even models computed with "slow" semiconvection (cf. section 3) end up as RSGs (cf. Woosley et al. 1988, Langer et al. 1989), while even models where the Schwarzschild criterion has been invoked explode as BSG at sufficiently low metallicity (Brunish and Truran 1982a,b; Hillebrandt et al. 1987). "Slow" semiconvection leads also to a final BSG structure for stellar models with a significant helium enrichment in the envelope (cf. Langer 1992). Note that moderate convective core overshooting, on the other side, only leads to RSG pre-supernovae even at metallicities as low as $Z_\odot/20$ (cf. Schaller et al. 1992).

Also Wolf-Rayet stars are considered as supernova progenitors (cf. Langer 1991c, and references therein). The most likely WR supernova progenitors are the low mass WC stars which are predicted due to mass dependent WR mass loss (cf. previous section). It has been investigated in detail by Woosley et al. (1992), to what extent and for which reasons these low mass descendants of very massive stars remember their massive star origin. Woosley et al. find more oxygen and silicon and less helium to be ejected in potential explosions of He- or WR stars of a given final mass for larger initial masses. The reason is that, though the internal structure of hydrogenless core helium burning WR stars is independent of their initial mass (see previous section) they still keep a "chemical memory" of their earlier evolution which affects structure and evolution during their final evolutionary phases (see Woosley et al. 1992, for more details).

Due to the lack of hydrogen, explosions of low mass WC stars would be classified as Type I SNe, and the question whether Type Ib or Ic SNe are the observed counterparts is still a matter of debate (cf. Branch et al. 1991). The synthetic light curve of an exploding $4.24\,M_\odot$ WC star reveals similarities to observed SN Ib light curves (Langer and Woosley 1991, Woosley et al. 1992), but only synthetic spectra will allow a conclusive discrimination between the various Type Ib/c progenitor models.

6. CONCLUDING REMARKS

In the previous sections we intended to show that the theory of massive star evolution is far from being in a final stage. Even massive main sequence stars are not yet well understood: several fundamental discrepancies with observations may require a substantial helium enrichment in a large fraction of them (cf. Langer, this volume).

Our understanding of the post-main sequence evolution of moderately massive stars is hampered by uncertainties in current theories of convective and semiconvective mixing. A detailed comparison with the supergiant distribution in the Magellanic Clouds appears to strongly favour "slow" semiconvection (cf.

section 3) — which also yields BSG progenitor models for SN 1987A — and inefficient convective overshooting.

The evolution of the most massive stars is largely determined by mass loss, while convincing physical explanations for the origin of the LBV and WR mass loss remain to be found. Mass convergence in the evolution of the most massive stars due to mass dependent WR mass loss makes the very massive stars to potential supernova progenitors, while the most likely black hole formation candidates (at $Z \simeq Z_\odot$) may be stars with initial masses of $\sim 35\,M_\odot$.

Our picture of the evolution of massive stars has dramatically changed over the last decade (cf. de Loore 1980), which indicates that much progress could be achieved in this field. However, the large number of basic problem mentioned above indicates that still much work has to be done before we can say that we do understand massive stars.

ACKNOWLEDGEMENT The author is very grateful to the organizers of this conference, and especially to Prof. J. Cassinelli, for their generous support.

REFERENCES

Arnett W.D., Bahcall J.N., Kirshner R.P., Woosley S.E., 1989, ARA &A 27, 629

Arnett W.D., 1991, ApJ 383, 295

Blaha C., Humphreys R.M., 1989, A.J. 89, 1598

Branch D., Nomoto K., Fillipenko A.V., 1991, Comments on Ap. XV, 221

Brunish W.M., Truran J.W., 1982a, ApJ 256, 247

Brunish W.M., Truran J.W., 1982b, ApJ Suppl. 49, 447

Castor J.I., Abbott D.C., Klein R.I., 1975, ApJ 195, 157

Chiosi C., Maeder A., 1986, ARA &A 24, 329

Conti P.S., Massey P., 1989, ApJ 337, 251

Davidson K., Moffat A.F.J., Lamers H.J.G.L.M. (eds.), 1989, *Physics of Luminous Blue Variables*, Proc. IAU-Colloq. 113, Kluver

Doom C., 1985, A&A 142, 143

Fitzpatrick E.L., Garmany C.D., 1990, ApJ , 363, 119

Garmany C.D., Stencel R.E., 1992, A&A Suppl. 94, 211

Herrero A., Kudritzki R.P., Vilches J.M., Kunze D., Butler K., Haser S., 1992, A&A , in press

Hillebrandt W., Höflich P., Truran J. W., Weiss A., 1987, Nature 327, 597

Humphreys R.M., Davidson K, 1979, ApJ 232, 409

Kato S., 1966, P.A.S.J. 18, 374

Kudritzki R.P., Pauldrach A., Puls J., 1987, A&A 173, 293

Langer N., 1989, A&A 220, 135

Langer N., 1989a, Rev. Modern Astron. 2, 306

Langer N., 1989b, A&A 210, 93

Langer N., 1991, A&A , 252, 669

Langer N., 1991a, A&A 243, 155

Langer N., 1991b, A&A 248, 531

Langer N., 1991c, in: *Supernovae*, S.E. Woosley ed., Springer, p. 549

Langer N., 1992, in *Inside the Stars*, proc. IAU-Colloq. 137, W.W.Weiss ed., in press

Langer N., Sugimoto D., Fricke K.J., 1983, A&A 126, 207

Langer N., El Eid M.F., Baraffe I., 1989, A&A Letter 224, L17

Langer N., Woosley S.E., 1991, in: IAU Symp. 143 on *Wolf-Rayet stars and Interrelations with Other Massive Stars in Galaxies*, eds. K.A. van der Hucht, B. Hidayat, Kluver, p. 566

Lennon D.J., Dufton P.L., Fitzsimmous A., 1992, A&A , submitted

de Loore C., 1980, Space Sci. Rev. 26, 113

Maeder A., 1983, A&A 120, 113

Maeder A., 1990, A&A Suppl. 84, 139

Massey P., Grove K., 1989, ApJ 344, 870

Prantzos N., Doom C., Arnould M., de Loore C., 1986, ApJ 304, 695

Rogers F.J., Iglesias C.A., 1992, ApJ Suppl. 79, 507

Schaller G., Schaerer D., Meynet G., Maeder A., 1992, A &AS , in press

Schmutz W., Hamann W.-R., Wessolowski K., 1989, A&A 210, 236

Spruit H.C., 1992, A&A , 253, 131

Stothers R.B., Chin C.-w., 1991, ApJ 381, L67

Stothers R.B., Chin C.-w., 1992, ApJ 390, L33

Tuchman Y., Wheeler J.C., 1990, ApJ 363, 255

Walborn N., 1988, in *Atmospheric Diagnostics of Stellar Evolution*, IAU-Colloq. 108, K. Nomoto, ed., p. 70

Weiss A., 1989, ApJ 339, 365

Woosley S.E., 1988, ApJ 330, 218

Woosley S.E., Weaver T.A., 1986, ARA &A 24, 205

Woosley S.E., Pinto P.A., Weaver T.A., 1988a, Proc. Astron. Soc. Austr. 7, 355

Woosley S.E., Langer N., Weaver T.A., 1992, ApJ , submitted

Zahn J.P., 1991, A&A 252, 179

Massive Stars: Their Lives in the Interstellar Medium
ASP Conference Series, Vol. 35, 1993
Joseph P. Cassinelli and Edward B. Churchwell (eds.)

THE MASSIVE STAR CONTENT OF THE GALAXY AND MAGELLANIC CLOUDS: METHODS AND MADNESS

PHILIP MASSEY
Kitt Peak National Observatory, National Optical Astronomy Observatories, P.O. Box 26732, Tucson, AZ 85726

ABSTRACT   The techniques needed to investigate present-day mass functions and stellar content of OB associations in the Magellanic Clouds and the Galaxy are discussed. Our findings so far have revealed stars of similar high mass in the LMC, SMC, and Galactic OB associations, suggesting that radiation pressure on the collapsing proto-stellar material is not the limiting factor in how massive a star can form. The IMF slopes of the regions studied in the Magellanic Clouds are similar, and somewhat steeper than the two regions studied as part of this project in the Milky Way. What these differences are telling us is not yet clear.

INTRODUCTION

For the past several years my collaborators and I have been studying the stellar content of OB associations in the Magellanic Clouds; recently we have extended this work to young Galactic clusters and associations. In this paper I will summarized our work to date, with emphasis on how we are going about this task, and why this "madness" is necessary.

What We Learn From This
The motivation for this work is three-fold:

- **Direct determination of the initial mass function:** By placing stars on the theoretical (M_{BOL}, $\log T_{eff}$) H-R diagram (HRD), and overlaying theoretical evolutionary tracks, one can simply count up the number of stars in each mass bin. This gives a direct measure of the present-day mass function (PDMF). To whatever extent stellar birth has been coeval, this then provides a direct measure of the initial mass function (IMF).

Is the IMF the same in regions of differing metallicity? Does the slope of the IMF depend upon other environmental factors? Garmany, Conti and Chiosi (1982) claim to have found a galactocentric gradient in the slope of the IMF, with proportionally more higher mass stars found in towards the Galactic center. If true, is this the result or the cause of the Galactic metallicity gradient? Our preliminary work on the Magellanic Cloud OB associations was summarized at the 1989 Boulder-Munich Workshop on Hot Stars by Massey (1990), suggesting that the slope of the IMF and the upper-mass cutoff appears to be constant over a factor of 5 in metallicity (LMC to SMC) and possibly over a factor of 20 in metallicity (Galactic to SMC).

- **Stellar evolution at the upper-end of the HRD:** Models of massive star evolution are highly dependent upon how convective over-shooting and mass-loss are handled. By studying the stars presumed to all be born at the same time, one then obtains an instantaneous snap-shot of stellar evolution. Humphreys, Nichols, and Massey (1985) used cataloged data on OB associations in the Milky Way and M33 to conclude that Wolf-Rayet stars must come from stars whose initial masses are greater than 30-50 $\mathcal{M}_\odot$, and that red supergiants (RSG) and Wolf-Rayet stars do not come from the same mass progenitors. Studies such as these can answer the evolutionary connection between massive O stars and their evolved descendents: red supergiants, Wolf-Rayet stars, and Hubble-Sandage variables (aka "luminous blue variables", or LBVs).

- **Hα as a measure of Star Formation Rate (SFR):** For galaxies beyond the Sculptor and M81 groups, the Hα luminosity is one of the few indicators of current star formation that we can measure (Kennicutt 1983). However, if some HII regions are UV-"leaky" (density bounded rather than radiation bounded), one may confuse other environmental factors with a difference in star-formation rate. For HII regions excited by OB associations in the Clouds or the Milky Way, we can actually compare the measured Hα fluxes to that expected from the ionizing flux from the stars we find.

  In M31 there are two large OB associations—OB78 (NGC206) and OB48— that appear to have similar massive star populations judged from the numbers of Wolf-Rayet stars found; nevertheless, OB78 sits in an HI bubble and has very little Hα emission, while OB48 sits in one of the strongest HII regions of M31 (Massey, Armandroff and Conti 1986). However from studies to date of the Magellanic Clouds, we find excellent (factor of 2) agreement between the Hα fluxes and that expected based upon the stellar census (Massey *et al.* 1989a; Massey, Parker, and Garmany 1989b).

Why This Is Hard

Stars that begin their lives with initial masses $\geq 10\mathcal{M}_\odot$ are not only very luminous ($L \geq 10^4 L_\odot$) but are also very hot ($T_{eff} \geq 25,000°$ K). These high temperatures result in two complications which perhaps have not been sufficiently appreciated.

1. **Degeneracy of colors:** The optical colors [$(U - B)_o$ and $(B - V)_o$] are virtually indistinguishable for main-sequence stars of even $100\mathcal{M}_\odot$ and $30\mathcal{M}_\odot$ (Massey 1985). These stars are so hot that the optical continuum distributions are far out on the tail of the Rayleigh-Jeans distribution. This is nicely illustrated by a cartoon presented by Conti (1986), and reproduced here in Fig. I. These three creatures are all very different, but have identical (optical) tails. Even with the UV capabilities of *IUE* and now *HST* we cannot distinguish among these critters using the continuum energy distribution alone, all issues of reddening aside.

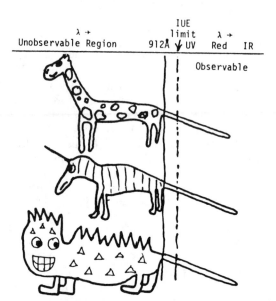

FIGURE I   This figure from Conti (1986) illustrates the difficulty in using optical spectral energy distributions (UBV) to characterize hot stars. All the "action" is below the Lyman limit.

2. **The visually brightest stars are not necessarily the most massive:** Since the bolometric correction (BC) is a steep function of the effective temperature (BC$\approx 6 \times \log(T_{eff})$, for $T_{eff} \geq 10,000°$ K), we find ourselves with the unpleasant fact that within a cluster or association the *visually* brightest stars are not necessarily the most massive. This is illustrated in Fig. II, where we show an H-R diagram for stars in the SMC. At $V = 13$ we see that we are in fact picking up a handful of slightly evolved 60-85 $\mathcal{M}_\odot$ stars, but missing most of the unevolved stars of lower mass. However, the sample at $V = 13$ and brighter is dominated not by this small handful of very massive stars, but rather by $15 - 40\mathcal{M}_\odot$ evolved supergiants. Thus comparison of luminosity functions will in fact be telling you more about ages and stellar evolution than about anything having to do with the mass function.

So How Do We Go About This?

Historically there have been two approaches to constructing an IMF. The first of these is to take photometry (and perhaps spectroscopy) from the literature. The advantage of this method is that you can do it indoors during the day. The disadvantage is that it does not tell you what you want to know. As we have seen in the previous section, an inappropriate magnitude-limited sample of stars can be dominated by lower-mass (evolved) stars.

For the Magellanic Clouds, the existing samples of photometry simply do

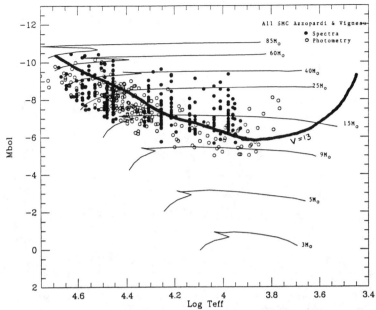

FIGURE II   H-R diagram for stars in the SMC. The thick line corresponds to $V = 13.0$.

not go deep enough to construct a meaningful IMF, even if spectra were available for all these stars. This is illustrated by the histogram of number of stars in the Azzopardi and Vigneau (1982) catalogue of SMC members shown in Fig. 2 of Massey (1990), and reproduced here in Fig. III. One can see that in fact $V = 13$ is a reasonable estimate of completeness of this catalogue, and hence referring to Fig. II again, the data are not complete at *any* mass range.

For the Galaxy, the problem of reddening renders the problem nearly intractable without full photometric (UBV) data. This can be seen from inspecting the $B - V$ color-magnitude diagram of stars in the field of Cyg OB2. We see

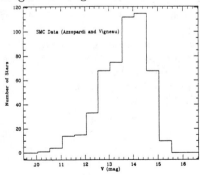

FIGURE III   Number of stars in half-magnitude bins listed in the catalogue of Azzopardi and Vigneau (1982).

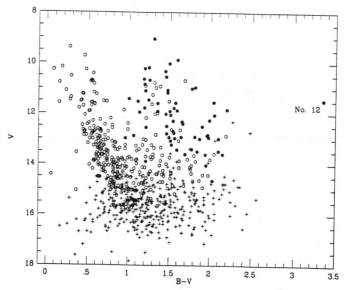

FIGURE IV   The color-magnitude diagram of CygOB2 (Massey and Thompson 1991) appears to contain a nice, tight "main-sequence". However, it is the filled circles which turn out to be the association members, while the apparent main-sequence is nothing more than foreground stars.

in Fig. IV what appears to be a nice, strong main-sequence extending up from $B - V = 1.1$ and $V = 16$ to $B - V = 0.3$ and $V = 10$. However, it is the filled circles in Fig. IV which are actually the association members, as shown from the spectroscopy of Massey and Thompson (1991).

The alternative approach is to simply start from scratch. Using modern CCDs one can obtain UBV images of an OB association or cluster. Digital photometry then reveals members by the use of a reddening-free color index

$$Q = (U - B) - 0.72 \times (B - V).$$

Spectra are needed for the hotter stars ($Q < -0.7$, say) in order to accurately place them on the H-R diagram. The spectral type determines the effective temperature, which is used to place the star on the abscissa of the H-R diagram. However, the critical need for spectra (in order to determine the effective temperature accurately) is actually due to the fact that the bolometric correction is a strong function of the effective temperature. Thus a good determination of the effective temperature is needed if the star is to be placed correctly on the *ordinate* of the HRD as well. In addition, the spectral type determines the color excess $E(B - V)$ and hence the correction for interstellar absorption; it also provides a spectroscopic parallax if the distance needs to be determined. In the case of the cooler stars ($Q > -0.7$) the photometry alone serves adequately for placing the star in the HRD (see Table IX of Massey, Parker, and Garmany 1989b).

The advantage of this method is that it has some hope of yielding useful answers. Note that I did not say "the right answer"—one is still at the mercy of the calibration of spectral type to effective temperature, effective temperature to bolometric corrections, and the conversion of location in the HRD to masses via evolutionary tracks. Worse still is the assumption that star formation has been coeval. Given what we will see later on in this paper, we know that *some* stars of lower mass have formed first. Thus to whatever extent we have constructed a present-day mass function (PDMF), its interpretation as an IMF is somewhat suspect. However, by starting from scratch we haven't *introduced* any of these problems—we are at least starting from a better point observationally.

The disadvantage of this method is the "madness" part of the title: it requires lots of bits, lots of work, and lots of telescope time. We began plugging away at OB associations in the Magellanic Clouds by obtaining UBV images in October 1985. Each subsequent year we have been using the Tololo 4-m telescope to obtain spectra. Although we "multiplex" by rotating the spectrograph so we average 2-3 interesting stars per observation, it has been slow going, and we are still an observing run away from having obtained complete spectroscopic data on the OB associations we intend to study.

As for Galactic OB associations, their large angular extent and the reddening problem (which can be both huge and differential within a cluster or association) has made these studies ironically easier to do in the Magellanic Clouds than in our own Galaxy. It has only been recent technological advances that has made this practical. The photometry of twelve northern Galactic OB associations has now been obtained, thanks to the advent of the Tektronix 2048^2 CCDs on the Kitt Peak 0.9m telescope (providing a 20 arcmin × 20 arcmin field of view). This UBV imaging consisted of 138 frames (46 fields of 12 OB associations in 3 colors), and resulted in the photometry of 16,069 stars, of which 1300 are sufficiently bright ($V < 16$) and blue ($Q < -0.75$) to warrant spectra. (I note in passing that the photometry of this many stars on so many very large CCD frames would have been completely impractical with the computers and software available 3 or 4 years ago; the introduction of digital photometry algorithms in IRAF, and the proliferation of fast, desk-top workstations, made this the work of a few weeks rather than years.) Still, one is left with the question of how to obtain spectra of 1000+ fairly faint stars in one lifetime. Fortunately Kitt Peak's multi-fiber feed HYDRA has recently become available on the Mayall 4m telescope (Barden et al. 1992). This instrument consists of 97 blue and 97 red fibers which feed a high-thoughput, bench-mounted spectrograph. Although my one effort at spectroscopy with this was pretty much lost to clouds, enough data was obtained to be able to illustrate this method on NGC 6611.

NGC6611: A Quintessential Example
NGC6611 is also known as M16, The Eagle Nebula, and is the core of Ser OB1. Hillenbrand *et al.* (this volumne) have a poster on this interesting cluster, combining IR data with the observations described herein.

Humphreys (1978) lists 8 stars with spectra in NGC6611, and 21 altogether in Ser OB1. My CCD photometry yielded UBV colors of 900 stars, of which roughly 100 were sufficiently interesting to require spectra. With HYDRA, I was able to obtain 120 spectra good enough for classification of 73 stars.

The color-magnitude plot is shown in Fig. V. Members established by spectra are shown by filled circles; compared to Fig. IV we see that the overall redden-

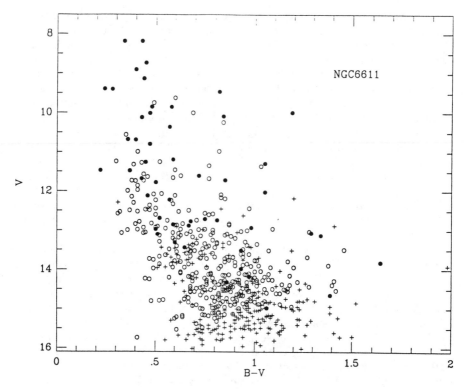

FIGURE V   The $B - V$ color-magnitude diagram of NGC6611.

ing is lower (the "main-sequence" in this plot is actually the main-sequence!), but that there is strong differential reddening within the cluster—some unevolved, hot stars are found with $B - V > 1.0$.

We show in Fig. VI what we obtain from a single exposure with this instrument: at the left is the raw exposure, and to the right is the montage of spectra we classified from this one effort. A few individual spectra are shown in Fig. VII.

The H-R diagram for the association is shown in Fig. VIII. We see immediately a few interesting things: (a) there are some very massive stars present in NGC6611, and (b) there is at least one evolved star of $15\mathcal{M}_\odot$. This in itself demonstrates that star formation has not been strictly coeval. However, if we simply count up the number of stars per mass bin we can construct the PDMF. This plot is also shown in Fig. VIII. The slope we find here is $\Gamma = -1.3 \pm 0.1$, in Scalo's (1986) notation.

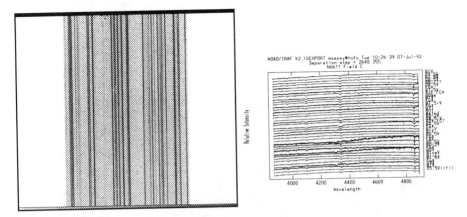

FIGURE VI   A single exposure with Hydra yields a plethora of spectra.

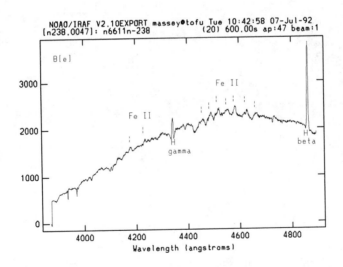

FIGURE VII   A spectrum from the preceding figure.

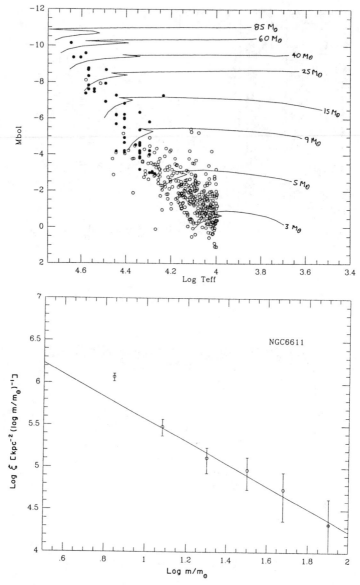

FIGURE VIII   The H-R diagram and Mass Function of NGC6611. Filled circles in the HRD denote stars with spectra; open circles have been placed in this figure from photometry alone. These figures are preliminary. The slope of the mass function corresponds to $\Gamma = -1.3$.

RESULTS TO DATE

In this section I summarize what we have found for regions in the Magellanic Clouds and Milky Way to date.

TABLE I   IMF Slopes for Regions in the LMC, SMC, and Milky Way

Region	Γ	Reference
LMC:		
LH117/118	-1.8 ± 0.1	Massey et al. (1989a)
LH58	-1.9 ± 0.1	Garmany et al., in prep.
LH9/10	-1.4 ± 0.2	Parker et al. (1992)
SMC:		
NGC346	-1.8 ± 0.2	Massey et al. (1989b)
MW:		
CygOB2	-1.0 ± 0.1	Massey and Thompson (1991)
N6611	-1.3 ± 0.1	Hillenbrand et al., in prep.

In all these cases we see H-R diagrams that contain massive stars up to $85\mathcal{M}_\odot$, and a few RSG's and B supergiants of lower mass ($15\mathcal{M}_\odot$) suggesting that some lower mass stars formed first.

For our studies of NGC346 in the SMC, and LH117/118 in the LMC, we find that the Hα fluxes of HII regions agree with what you expect from the stars present.

What can we conclude from these data so far? For one thing, the "traditional" Milky Way IMF slope of $\Gamma = -1.6$ (Garmany, Conti, Chiosi 1982) or -2.4 (Humphreys and McElroy 1984) is considerably steeper than what we have found for the two OB associations studied in the Milky Way so far; a "classic" Salpeter (1955) IMF slope, of $\Gamma = -1.3$, is more appropiate to what we are finding.

Why are the *nouvelle tradition* values different from those we have found? We don't know yet if these two Galactic OB associations are unusually rich in massive stars compared to their lower-mass members, but further studies are in progress. However, we note that both Garmany, Conti, and Chiosi (1982) and Humphreys and McElroy (1984) started from the same (literature) database, and got very different answers for the IMF slope. In part this is because Humphreys and McElroy "corrected" the lower-mass stars for incompleteness by extending the slope of the luminosity function from brighter to fainter stars. A little reflection reveals that this correction is circular, and that the relative number of low and high mass stars will be fixed coeresponding to the relative number of medium and high mass stars (see extensive discussion in Massey 1985). Garmany, Conti and Chiosi, on the other hand, freely mixed OB association members and field stars in their sample; given the need to correct for age in one case, but not in the other, it is a little unclear how to properly deal with this ensemble.

It does appear that these two Galactic regions have flatter IMF slopes than the regions studied so far in the Magellanic Clouds. We emphasize again that while these IMF slopes may not be "right", the intercomparison from one region to another should be valid, since the assumptions have remained the same.

We do not see any difference in the upper mass cut-off (m_u) between the

Milky Way, LMC, and SMC, despite the factor of 20 difference in metallicity. Since simple models predict a dependence of $m_u \propto Z^{-1/2}$ (see discussion in Shields and Tinsley 1976) we conclude that radiation pressure on the collapsing material is probably not the limiting factor in determining the largest mass star that can form.

Finally we note that the Garmany, Conti and Chiosi (1982)'s *gradient* in the IMF slope within the Galaxy from $\Gamma = -1.3$ towards the Galactic center to $\Gamma = -2.3$ towards the anti-center is at least too simple a picture, as CygOB2 is at a Galactic longitude of 80°, while that of NGC6611 is 15°.

ACKNOWLEDGMENTS

I have grateful to my various collaborators on these projects, and in particular to S. Storm and L. Hillenbrand for continuing to remind me what it is all about.

REFERENCES

Azzopardi, M., and Vigneau, J. 1982, *A&AS*, 50, 291.

Barden, S., Armandroff, T., Massey, P., Groves, L., Rudeen, A., Vaughn, D., and Muller, G. 1992, in Fibre Optics in Astronomy II, edited by P. Gray (A.S.P., San Francisco), in press.

Conti, P. S. 1986, in Luminous Stars and Associations in Galaxies, IAU Symp. 116, edited by C. W. H. de Loore, A. J. Willis, and P. Laskarides (Reidel, Dordrecht), p. 199.

Garmany, C. D., Conti, P. S., and Chiosi, C. 1982 *ApJ*, 263, 777.

Humphreys, R. M. 1978, *ApJS*, 38, 309.

Humphreys, R. M., and McElroy, D. B. 1984, *ApJ*, 284, 565.

Humphreys, R. M., Nichols, M., and Massey, P. 1985, *AJ*, 90, 101.

Kennicutt, R. C., Jr. 1983, *ApJ*, 272, 54.

Massey, P. 1985, *PASP*, 97, 5.

Massey, P. 1990, in Properties of Hot Luminous Stars, edited by C. D. Garmany (A.S.P., San Francisco), p. 30.

Massey, P., Armandroff, T. E., and Conti, P. S. 1986, *AJ*, 92, 1303.

Massey, P., Garmany, C. D., Silkey, M., and DeGioia-Eastwood 1989a, *AJ*, 97, 107.

Massey, P., Parker, J. W., and Garmany, C. D. 1989b, *AJ*, 98, 1305.

Massey, P., and Thompson, A. B. 1991, *AJ*, 101, 1408.

Parker, J. W., Garmany, C. D., Massey, P., and Walborn, N. R. 1992, *AJ*, 103, 1205.

Salpeter, E. E. 1955, *ApJ*, 121, 161.

Scalo, J. M. 1986, Fundam. Cosmic Phys. 11, 1.

Shields, G. A., and Tinsley, B. M. 1976, *ApJ*, 203, 66.

Massive Stars: Their Lives in the Interstellar Medium
ASP Conference Series, Vol. 35, 1993
Joseph P. Cassinelli and Edward B. Churchwell (eds.)

THE MOST LUMINOUS STARS - EJECTIONS, ERUPTIONS, AND EXPLOSIONS

ROBERTA M. HUMPHREYS
University of Minnesota, Department of Astronomy,
116 Church St., Minneapolis, MN 55455

ABSTRACT   Most of the stars near the top of the HR diagram show evidence of instability. Some such as the LBV's and the most luminous cool hypergiants can eject matter with much higher mass loss rates than commonly observed. This matter is later seen as circumstellar shells and nebulae and is evidence for more violent events which we call eruptions and explosions.

INTRODUCTION - THE UPPER HR DIAGRAM

The most luminous stars of all types in Local Group galaxies define the upper boundary of the HR diagram (Humphreys and Davidson 1979, 1984).

The most luminous hot stars reveal an upper envelope of declining luminosity with decreasing temperature which for the cooler stars (<8000-10,000K) becomes an upper boundary of essentially constant luminosity.

The boundary for the hot stars is temperature dependent and suggests it is therefore mass dependent in contrast to the nearly temperature-independent upper limit to the luminosities of the cool hypergiants.

In addition to the 'normal' high mass losing OB supergiants and red supergiants there are many stars near this upper boundary that show evidence of much greater instability. These stars appear to be in a unique stage of their evolution where they have reached a limit to their stability. They provide the empirical evidence that the observed luminosity boundary is due to the instability of these evolved massive stars. The most likely cause of the instability in the hot stars is radiation pressure, an opacity-dependent Eddington limit as proposed by several investigators (Humphreys and Davidson 1984, Appenzeller 1986, Lamers 1986, Davidson 1987, Lamers and Fitzpatrick 1988). The unstable cool hypergiants could be those stars just slipping under the Eddington limit; however, they could also be unstable due to turbulence in their atmospheres as suggested by de Jager (1980, 1984) independent of the Eddington limit for hot stars.

THE LUMINOUS BLUE VARIALBES

LBV's are evolved, very luminous, hot and unstable stars which suffer irregular ejections of mass. Some famous members of this group are: η Car, P Cyg, and S Dor. The LBV's also include the Hubble-Sandage variables in M31 and M33.

Their characteristics include:

1. An outburst or ejection phase at visual maximum with an enhanced mass outflow of 10^{-4} to 10^{-5} M$_\odot$/yr, a slowly expanding (100-200 km/s), cool (8000-9000 K) and dense (N ~ 10^{11}cm^{-3}) envelope or false photosphere.

2. At visual minimum or the quiescent stage the star resembles a high temperature supergiant ($\geq$25000K) with emission lines of hydrogen, helium and permitted and forbidden Fe II and with a lower mass loss rate.

3. During these variations in visual light, the bolometric luminosity remains the same. The visual variations are caused by the shift in the star's energy distribution.

4. Many of the LBV's are surrounded by visible (and IR) evidence of their past eruptions including circumstellar nebulae, visible rings and shells.

5. The ejecta and atmospheres of the LBV's are hydrogen and helium rich showing that the LBV's are evolved, post-hydrogen burning stars.

Typical mass loss rates are 10^{-6} to 10^{-7} during quiescence and 10^{-4} to 10^{-5} during maximum light. Lamers (1989) suggested a time-averaged normal mass loss rate of ~ 10^{-5}M$_\odot$/yr assuming the LBV spends half the time in each phase. But there is increasing evidence that LBV's pass through a more violent stage during which the mass loss rate is much higher.

η Car is the most extreme LBV known. It's 'great eruption' (or explosion) about 150 years ago exceeded all other LBV's by orders of magnitude in ejected mass and energy. During its 1840's outburst, η Car lost 2-3M$_\odot$ in a single event. Its luminosity actually increased (M$_v$ ~ -1 mag and M$_v$ < -14mag!) (Davidson, 1989). During the explosion its average luminosity was 10^7 L$_\odot$ and between 1840-1860 it radiated 10^{49}ergs in excess energy.

P Cygni during the 1600's is another example of a star that had a violent eruption. We have no measurement of its mass loss rate during its eruption, but its extended Hα and NII emission measured by Leitherer and Zickgraf (1987) corresponds to a continuous mass loss of 4×10^{-4}M$_\odot$/yr. Others such as AG Car, He 3-519 and R127 (Stahl 1987) have circumstellar gaseous shells or rings. The mean rate of mass loss from these LBV's is $>2\times10^{-4}$ M$_\odot$/yr (see Hajian et al., this volume for recent work on this star).

The total mass lost during the LBV stage is uncertain because we do not know the duration of the LBV phase or how often these more violent eruptions occur. I (Humphreys 1991) have estimated the LBV lifetime at about 25000yrs based on the numbers of LBV's and WR stars in the LMC. Then assuming a time-averaged mass loss rate of $>2\times10^{-4}M_\odot/yr$, an LBV can lose $>5M_\odot$ during its lifetime! This is close to the mass a 50-100 $M_\odot$ star must shed after core hydrogen burning to become a WR star.

Figure 1 shows a schematic HR diagram with the location of all of the well-studied LBV's at both maximum and minimum light plus the cool hypergiants near the upper luminosity boundary. At maximum light the LBV's all occur at essentially the same temperature, near 8000K. This is due to the opaque wind of the LBV. Davidson (1987) showed with his opaque wind model that the temperature in the LBV envelope cannot fall below ~7500K even at very high mass loss rates. At minimum light the LBV's all appear to lie on an inclined strip which Wolf (1988) has suggested is an instability strip. This is a very intriguing possibility; however, the LBV's may divide into two groups; those above ~40-50$M_\odot$ initial mass and above the luminosity cutoff ($M_{Bol} \cong -9.5mag$) in the HR diagram and those below it. The less luminous LBV's include R71 and R110 in the LMC and the newly discovered R40 in the SMC (Szeifert et al 1992); the first LBV in the SMC. These stars are cooler, show smaller light variations and lower mass loss rates than the more luminous LBV's. The instability in the luminous, hotter LBV's can be understood as due to radiation pressure at the Eddington limit, but what is the cause of the instability for the less luminous LBV's which are presumably post red supergiant stars?

Very recently Allen et al. (1990) and Krabbe et al (1991) have described a cluster of He I emission stars in the galactic center whose properties-He I emission, high mass loss ~$10^{-4}M_\odot/yr$ and temperatures ($<35000°K$) are like those of the Of/WN9-LBV stars. The only difficulty is that all 17 stars must be of this type and the radiation field requires that there be no stars hotter than O8 main sequence. This implies that a very restricted burst of star formation occurred $2-3\times10^6$ years ago.

Thus LBV's could be very important contributors to the energetics of starburst galaxies. Imagine what a cluster of η Car-like stars all going off at about the same time would be like!

THE COOL HYPERGIANTS (SPECIAL TYPES A TO M)

The very luminous F, G, K and M stars in our galaxy and the LMC define the upper luminosity boundary for the cool stars (see Figure 1). All of these stars show evidence of instability including light and spectral variability, high mass loss plus circumstellar dust around many.

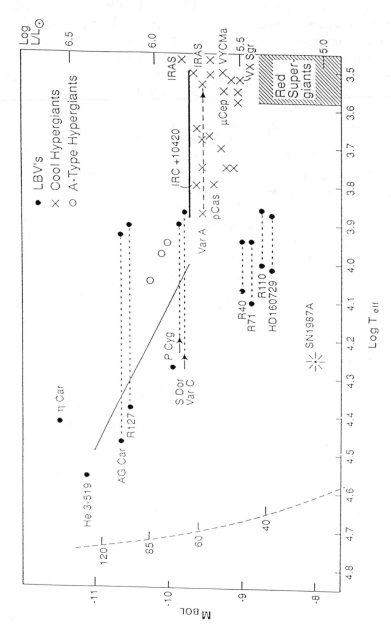

Figure 1 - A schematic HR diagram shows the upper luminosity boundary (solid line) for the true hot and cool stars with the location of the LBV's and the cool hypergiants discussed in the text.

In this review I will just describe two exceptional stars that belong to this group:

Variable A in M33 is one of the original Hubble-Sandage variables (Humphreys et al 1988). In 1950 it was one of the visibly brightest stars in M33 with an F-type spectrum. It then rapidly declined in brightness by 3.5mag becoming faint and red. It is still faint and red and today has the spectrum of an M supergiant not an emission-line hot star like other H-S variables. It also has a large infrared excess and is today as bright at 10μ as it was at its visual maximum in 1950. Its bolometric luminosity has remained constant and its decline in brightness was due to the shift in its energy distribution. Variable A is very luminous ($M_{Bol} \approx -9.5$mag) with a very high mass loss rate (2×10^{-4} $M_\odot$/yr) estimated from its infrared excess. Its present spectrum is probably produced in an expanded false photosphere and it is shedding its mass in a dense, low velocity wind. Variable A at its maximum light was at a very critical point on the HR diagram, near the turnover in the upper luminosity boundary. Thus it is tempting to speculate that it is an evolved supergiant of ~40 $M_\odot$ encountering the Eddington limit near the point where the opacities reach a maximum (~8000°K) and has ejected an optically thick expanding, cool shell.

In contrast, IRC +10420 may be a candidate for a post red supergiant. IRC +10420 has the largest IR observed among the cool hypergiants (Humphreys et al. 1973). Its optical spectrum is that of a very high luminosity F supergiant and it is also one of the warmest known OH/IR sources (Giguere et al. 1976). Lewis et al. (1986) have reported that the 1665MHz OH feature is weakening while the 1612MHz feature is growing. This could be due to expansion of the dust/OH shell.

Our new multi-wavelength observations (Jones et al., this volume) support earlier suggestions by Humphreys that IRC +10420 is a post-red supergiant OH/IR star blowing off its cocoon of gas and dust. A recent high resolution optical spectrum slows a double peaked Hα profile. The absorption dip has the same velocity as the absorption lines in the F supergiant spectrum and therefore arises in the stellar photosphere. The two symmetrically displaced emission components are most likely produced in an equatorial disk where the circumstellar dust is concentrated and where the OH emission also originates. The photometric history of IRC +10420 suggests that the circumstellar dust is thinning. IRC +10420 brightened by about a magnitude in the optical from 1930 to 1970 (Gottlieb and Liller 1978). Recent optical photometry has remained essentially constant although the near-infrared continuum shows it has faded by a factor of two since the early 1970's. This suggests the continued clearing of the dust from the hotter, inner (disk) regions of the dust shell.

Both the optical and IR (8.7μ) imagery show a clear N/S extension of the image due to circumstellar ejecta. The infrared and visual polarimetry at two different epochs show significant charges in the geometry of the dust around the star. The interstellar component is significant about 6% in the visual indicating substantial ($A_v \sim$ 6 mag) interstellar reddening. This is very close to the IS reddening (~6.5 mag) implied by the optical colors and we conclude that there is very little circumstellar reddening if any in the optical. Thus we are viewing the F supergiant photosphere directly and the disk where most of the dust probably exists is not blocking the line of sight to the star.

Our interpretation of IRC +10420 as an evolved massive star depends on its luminosity and therefore its distance. The high IS reddening implies a distance greater than 4 kpc in its direction and its large positive velocity (+75 km/s LSR) indicates a kinematic distance of near 6 kpc. With a distance range of 4 to 6 kpc its bolometric luminosity is -9.2 to -10.0 mag. It is thus far above the regime of the proto-planetary nebulae and even the tip of the AGB.

We suggest that IRC +10420 is a true supergiant traversing the HR diagram to the left, analogous to a post AGB star on its way to becoming a planetary nebula. IRC +10420 is an excellent candidate for a former red supergiant evolving to the Wolf-Rayet stage.

The red supergiants are discussed later in this volume by Knapp. However, I want to point out that half of the most luminous M supergiants ($M_{Bol} \leq$ -9 mag) are known OH/IR stars. This is a very high mass losing stage for these evolved stars (few x 10^{-4} $M_\odot$/yr) and like the LBV's they play a significant role in the evolution of stars near 40 $M_\odot$.

Three stars in Figure 1 deserve special mention. These are the three A type hypergiants, the visually brightest stars in their respective galaxies - the Milky Way, the LMC, and M33. They have bolometric luminocities between -10 and -10.5 mag, and must be approaching the limit to their stability. Massey and Thompson (1991) suggest that Cyg OB2 #12 may be an incipient LBV largely because of light variations up to ~.5 mag and small spectral variations. Depending on how one wants to define (or draw) the upper luminosity boundary for the hotter stars, one can say that these A-type hypergiants are either just above it or that they define its cool limit.

FINAL REMARK

There are many very luminous, highly unstable stars all along the upper luminosity boundary. Two stages are especially important, the LBV and the OH/IR stage for the cool stars. In these stages, their mass loss may exceed 10^{-4} $M_\odot$/yr and the stars may lose a few solar masses in a relatively short time, significantly influencing their eventual fate.

REFERENCES

Allen, D.A., Hyland, A.R., and Hilber, D.J., 1990, M.N.R.A.S, 224, 706.

Appenzeller, I., 1986, IAU Symposium #116, Luminous Stars and Associations in Galaxies, p. 139.

Davidson, K., 1987, Ap. J. 317, 760.

Davidson, K., 1989, IAU Coloquim #113, Physics of Luminous Blue Variables, p. 101.

de Jager, C., 1980, The Brightest Stars (Reidel: Dordrecht).

de Jager, C., 1984, A&A, 138, 246.

Giguere, P.T., Woolf, N.J., and Webber, J.K., 1976, Ap. J. (Letters), 207, L195.

Gottlieb, E.W., and Liller, W., 1978, Ap. J., 225, 488.

Humphreys, R.M., 1991, IAU Symposium #143, Wolf-Rayet Stars and Interrelations with other Massive Stars in Galaxies, p. 485.

Humphreys, R.M., and Davidson, K., 1979, Ap. J., 232, 409.

Humphreys, R.M., and Davidson, K., 1984, Science, 223, 243.

Humphreys, R.M., Strecker, D.W., Murdock, T.L., and Low, F.J., 1973, Ap. J. (Letters), 179, L49.

Krabbe, A., Genzel, R., Drapatz, S., and Rotacuic, V., 1991, Ap. J., 382, L19.

Lamers, H.J.G.L.M., 1986, IAU Symposium #116, Luminous Stars and Associations in Galaxies, p. 157.

Lamers, H.J.G.L.M., and Fitzpatrick, E., 1988, Ap. J., 324, 279.

Leitherer, C., and Zickgraf, F-J., 1987, A&A, 174, 103.

Lewis, B.M., Terzian, Y., and Eder, J., 1986, Ap. J. (Letters), 302, L23.

Low, F.J., 1973, Ap. J. (Letters), 179, L49.

Massey, P., and Thompson, A.B., 1991, A.J., 101, 1408.

Stahl, O., 1987, A&A, 182, 229.

Szeifert, Th., Stahl, O., Wolf, B., and Zickgraf, F-J., 1992, preprint.

Wolf, B., 1988, A&A, 217, 87.

Massive Stars: Their Lives in the Interstellar Medium
ASP Conference Series, Vol. 35, 1993
Joseph P. Cassinelli and Edward B. Churchwell (eds.)

THE VARIETY OF MASSIVE STARS OF THE UPPER
HERTZSPRUNG-RUSSELL DIAGRAM: A GUIDE FOR THE
PERPLEXED

STEVEN N. SHORE
GHRS Science Team/Computer Sciences Corporation
Goddard Space Flight Center, Code 681, Greenbelt, MD 20771 USA
and DEMIRM, Observatoire de Meudon, 92190 Meudon Principal
Cedex, FRANCE

ABSTRACT This paper deals with some of the philosophical
and methodological issues raised by the observations of upper
Hertzsprung-Russell diagram (UHRD) stars. These stars present an
interesting challenge to many of the standard approaches to inferring
the details of stellar evolution by the observation of a morphologically
selected population. It is argued that their spectra are determined
largely by the circumstellar environment and not uniquely by the star
itself. A schematic model is presented that attempts to unify some
of the disparate classes of UHRD stars and that link the Of/WN,
B[e], and LBV classes.

INTRODUCTION

The upper Hertzsprung-Russell diagram (UHRD) stars are a highly time-variable
group, many of whose intrinsic properties are not yet well known. They are selected by their visual brightness in external galaxies and, within the Galaxy,
by their unusual spectral properties. The stars I will discuss here are generally the strongest emission line stars, usually spectral type O and B, and are
often associated with circumstellar nebulae. Stellar evolution calculations have
been extremely difficult for these objects and stretch the limits of the physics
of stellar interiors. For one thing, most of the most luminous stars are at or
near the Eddington limit. Because they are so luminous, radiative opacities
are extremely important, and small uncertainties can magnify into large effects
in the final state of the stars. Even the simplest sequences require very high
time resolution in order to properly account for the effects of stellar winds, and
secular and pulsational instabilities.

Observationally, the Galactic UHRD stars are rare, and their intrinsic properties have often to be determined by indirect means. Their luminosities are
not well known, and their masses and stellar winds are often difficult to study.
Although few are known to be in binaries, many are members of compact groups
that make observations difficult even with very high angular resolution (*cf.*
Malumuth and Heap 1992). We have merely to recall the history of work on
R136a in 30 Dor to realize how precarious are many of our inferences about the

intrinsic properties of these stars (see Malumuth and Heap 1992). They are also quite varied in their spectral peculiarities, and they really strain the classifier's ingenuity to find some common threads in their combined properties.

THE INDISTINCTNESS OF THE TAXA

There are a number of different classes that occupy the niche of the upper HR diagram. With the exceptions of the WR and Of stars, the rest of these cohabit in the region between the O supergiants and the Humphreys-Davidson limit (Humphreys and Davidson 1979).

The Luminous Blue Variables (LBVs) have been reviewed by Humphreys (1989; 1992, *these proceedings*) and Wolf (1992). These stars come under a variety of names in the literature, which makes the nominalistic act of grouping them under a single rubric especially significant (they have also been called S Dor variables, Hubble-Sandage variables, or P Cyg stars). But the LBV class is only associated with a star once it varies, an obvious but important fact. Fingering a candidate before an eruption is hard, irrespective of which direction the star is moving in across the HR diagram. They show extremely strong, relatively low excitation, emission spectra in the optical while in their optical minimum phase, but rise rapidly with an A or later supergiant spectrum at the outset of an episiode, at which time they fade in the UV. Their motion across the HR diagram is always initially toward higher visual magnitude, but they may go in either direction in $B - V$.

The Of/WN, or *slash*, stars are defined by strong but narrow He II (like the Of stars) but strong optical nitrogen lines (Bohannen 1989). They often show only weak variability. Their primary characteristics are that they show Of characteristics, that is N III-V and He II emission, weak narrow P Cyg components on the Balmer lines, and a strong He I spectrum. The He II $\lambda4686\text{Å}$ line is broader than a normal Of star and the He I spectrum is much stronger. The prototype is HD 269227 (Bohannan and Walborn 1989) and the *slash* stars correspond to the Class A stars described by Shore and Sanduleak (1984).

The B[e] stars are defined by their hybrid characteristics, much like the sub-classes of Seyfert galaxies. Selected optically, they display emission lines with broad wings surmounted by sharp cores and a wide range of ionization. They show Fe II and [Fe II] lines that are narrow and quite strong, broad wings on all of the Balmer lines, and relatively low velocity P Cyg absorption within the wings (Zickgraf 1992). Some of these stars look like R50 in the SMC or S 22/LMC. Others, like HD 38489 = S 134/LMC, look more like *slash* stars. They display strong iron peak absorption in the UV, which is sufficiently optically thick that the forbidden and permitted emission appears in the optical. However, there is a range of characteristics in the class. For instance, the strangest of them all is S 18/SMC (Shore, Sanduleak, and Allen 1987, Zickgraf *et al.* 1989). Like other B[e] stars, S 18/SMC does not show large-amplitude photometric variations, but it does display a very strong emission line spectrum in the UV that varies over a timescale of a few years.

All of these stars share a single common characteristic: They are all extremely luminous, greater than $10^5 L_\odot$. Except for the WR and Of stars, none of these classes is really distinct; there is plenty of evidence that the stars can change from one to the other in time. The *slash* stars have shown up later as LBVs, and this is also true for the B[e] stars. The hypergiants can make the swing to the left and turn into B supergiants, like R143. And finally, even the most luminous M stars can simply be a way station for the LBVs at the extrema of their range. Perhaps this is the reason why the upper HR diagram is so morphologically confusing. None of the stars is content to stay where it is placed taxonomically. The key element is time. If we take a snapshot of this luminous group, we will obtain a highly time variable picture because we will be surveying a relatively small number of objects and because the amount of time spent in each stage is quite similar, and cyclic without being periodic.

VARIABILITY: THE IMPORTANCE OF THE BOLOMETRIC CORRECTION
The LBVs are discovered by their optical variations. Over a period of years, these stars increase by several magnitudes in V, and eventually decline over perhaps decades. They have very hot spectra when they enter an episode which gradually become redder as the outburst proceeds. I won't dwell on these details because Humphreys has already summarized them at this conference. But the nature of the outburst is important for the clues that it provides about the ejection mechanism, and also about how one of these stars returns matter to the ISM.

When at minimum light, the LBVs emit most of their radiation in the FUV, shortward of Lyα. At optical maximum, the peak of the emission shifts to the visible and then toward the infrared. Although there are few bolometric observations of any known LBV throughout its cycle, there is a strong case for both R127 in the LMC and AG Car in the Galaxy having evolved at constant total luminosity. The optical and UV are essentially in antiphase with one another (*e.g.* Stahl *et al.* 1983; Stahl and Wolf 1986).

Now what does this bolometric constancy tell us? In novae, which are essentially LBVs in fast forward, the UV peak generally occurs days to weeks *after* optical maximum and the rate of decline is slower (Fig. 1). A comprehensive multiwavelength view of the initial fireball phase has been caught in only one outburst, Nova Cygni 1992. The optical rise was accompanied by a very rapid, roughly 2 day, drop in the UV between $\lambda 1200$ and $\lambda 3300$Å. The UV rise began almost immediately thereafter, and the optical faded rapidly as the UV became progressively more optically thin. However, all of the change was occurring in the *iron curtain*, the very large number of iron peak singly and doubly ionized transitions that blanket the ultraviolet continuum. The fact that the two wavelength regions were in nearly precise antiphase means that the central source must have been completely covered, and this is the lesson that we learn from novae for the interpretation of the LBVs. The fact that the source evolves

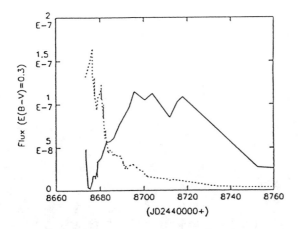

Figure 1: Initial outburst light curves for Nova Cyg 1992. Solid: $\lambda\lambda$1200-2000Å; dashed: visual (IAU Circulars). Note that the UV rise, as the line blanketing decreases, tracks the visual decline until the UV turns completely optically thin. E(B-V)=0.3 is assumed.

at nearly constant bolometric luminosity means that *the stellar wind, when an outburst occurs, completely covers the star and must be optically thick*. It also means that any continuum contributions must be from absorption and not from electron scattering opacities.

AG Car, as the prototypical Galactic LBV, is an excellent example of the simple blanketing effect of the iron curtain. Its optical and UV light curves are essentially in antiphase (although the case is weak because of poor UV coverage) (Shore 1992). This fact makes its circumstellar environment more of a puzzle. It is now well known that AG Car shows an optical jetlike nebulosity, which is seen only in the continuum and is consequently ascribed to a dusty ejection from the central source. This structure is even more remarkable in that it shows no line emission at all, in fact it is completely invisible on narrow band Hα images (Nota *et al.* (1991). Clearly, somewhere in a scale between a few hundred AU and about 1 pc something happens to the outflow or else the "jet" is a completely independent event from the main mass ejection. Some recent work on HR Car shows that its variations in the UV and optical are much like those of AG Car (Fig. 2). The $\lambda\lambda$1200- 2000Å spectrum varies in antiphase with V. It is also surrounded by a nonspherical nebula, one that may display cylindrical symmetry.

Now the total column density that is needed to produce the UV spectrum is about 10^{23}cm^{-2}. This translates into a mass loss rate of $\dot{M} \approx 10^{-5}$M$_\odot$yr^{-1}. This assumes that the wind velocity is of order a few thousand km s^{-1} and that the stellar radius is of order 100 R$_\odot$. This is the same order of magnitude seen in the most active stars. like P Cyg. Once the optical depth in the wind reaches this value, the ionization drop causes the surrounding high density emission line

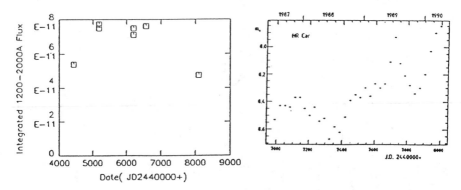

Figure 2: Recent ultraviolet (left) and optical light curves (right. Kutsmékers and Van Dorn 1991) for HR Car (He3-407). Note that even with the relatively sparse data set, the antiphase behavior of the two wavelength regimes is clear.

region to recombine and shut off, any surrounding nebulosity will also recombine. But this will take some time, so it is possible that one can have a surrounding nebula, with only a modest ionization parameter, and a very cool star with strong emission lines superimposed.

An additional point about the bolometric constancy is that any of the UV transitions that are responsible for the blanketing will also produce strong emission lines in the optical. One of the S Dor characteristics (when that star was the best available prototype and before the definition of the LBV class) is the strong optical Fe II and [Fe II] emission. These lines are formed by trapping and fluorescence. Few of the absorption lines are seen in the optical because of the weakness of the transitions. So when we look at the optical spectrum, the strong emission and absence of photospheric lines bespeak the combined effects of veiling by an electron scattering continuum (see Leitherer *et al.* 1990) and a very optically thick ultraviolet iron curtain. The stronger the forbidden line spectrum, the more distended the wind and the stronger the absorption by the Fe II. So even without a proper survey of emission line stars in the ultraviolet, the optical spectrum can in this case clue us into what is happening below $\lambda 2000\text{Å}$.

CIRCUMSTELLAR DUST

At optical minimum, an LBV is usually a late O supergiant. It is able to form a compact H II region from the circumstellar surroundings, especially from matter ejected during an earlier outburst. What is more important, the star will be quite efficient at heating the dust because of its strong UV flux between 1200 and 3000Å when the star returns to the left after an eruption. It will be imbedded in an extensive new environment, one that can easily be ionized when

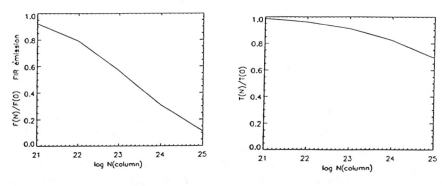

Figure 3: a. Predicted FIR emission from a simple model assuming Galactic-type dust absorption and a 30000 K, $\log g = 3.5$ star as a function of blanketing column density. b. Variation in the expected dust temperature as a function of column density assuming a λ^{-2} emissivity. Normalized to zero column density.

the star reaches surface temperatures in excess of 25000K and one that likely contains the dust formed in the outer ejecta during the previous episode.

Dust appears to be associated with many of the Galactic and Magellanic Cloud LBV candidates, with a typical $L_{IR}/L_{bol} \approx 0.01$ (Shore et al. 1990). Several of these stars have been found to show near IR colors consistent with the presence of dust (see also McGregor et al. 1988a). If the ejection velocity of the wind is about 1000 km s^{-1}, then this means that the material will reach about 1 pc in about 1000 years. Let's for the moment assume that this is the typical timescale between eruptions. Then the heating of the dust results from the contemporaneous spectrum of the underlying star, which for the typical LBV means that most of the flux is coming out at or shortward of 3000Å. Now taking the luminosity to be the Eddington limit for any of these stars, the typical dust temperature will be of order 50 K, and the optical depth of order 0.01 results in a shell luminosity in the FIR of a few percent of the bolometric luminosity.

This dust will have only a small opacity at the 2175Å feature because of the low overall optical depth, and this may help explain why even those stars that show the dust do not show anything that looks like a bump. Alternatively, among the LMC stars, there is considerable evidence for dusty environments. The newly discovered LBV, R143, has a large color excess, much larger than allowed for a field star in the LMC (Parker et al. 1992). About 10% of the stars in the LMC emission line sample (Shore and Sanduleak 1984) have large color excesses and a few even show the 2175Å feature (and in these cases it is not due to the iron curtain absorption).

In order to illustrate the effect, consider this trial calculation. Take an O supergiant with an effective temperature of 30000 K, $\log g = 3.5$, and solar abundances (using the Kurucz grid). Over the surface of this object, place a cool iron curtain with $T_{ex} = 5000$K, $\log n_e = 8$ and $v_{turb} = 20$ km s^{-1} (in LTE, the increase to 7500 K, for instance, merely decreases the equivalent column density by about a factor of 10). Then calculate the heating produced on dust using a Galactic extinction curve as the iron curtain optical depth is increased. The heating is given by the integral over the UV, $\Gamma \sim \int F_\lambda Q_{abs}(\lambda)d\lambda$ and the cooling is $\Lambda \sim \int B_\lambda(T_d)Q_{em}(\lambda)d\lambda$. Here F_λ is the stellar flux, Q_{abs} is the absorption efficiency, $B_\lambda(T_d)$ is the Planck function for dust temperature T_d, and Q_{em} is the emission efficiency. The results for the far infrared emission and the dust temperature are shown in Fig. 3. Note that if the column density reaches about 10^{25}cm^{-2} (for solar abundances) the FIR emission drops by about an order of magnitude. The most important effect is shortward of 2000Å. So the bolometric correction variation of the central star has the additional effect that the increase in the UV also increases the emission of the surrounding medium.

As these models show, the temperature and brightness of the circumstellar material track the UV input to the dust, and consequently the far infrared and visual luminosities should vary in antiphase. There aren't any FIR light curves available for any of the LBVs in the Local Group, but this would be a very valuable clue to what is happening in the circumstellar environment. It is interesting to note, however, that the IRAS observations that detected R127 were taken during the stage when the central star was on the decline in the UV and that this means the dust was experiencing a radiation field that was much harder than the UV indicated at the time of the outburst.

ENVIRONMENTS: CIRCUMSTELLAR NEBULAE

Several Galactic LBVs show circumstellar nebulae. The best known is AG Car (*cf.* Thackeray 1950, Nota *et al.* 1992). This star shows a ring nebula with a non-symmetric extension that is frequently called a jet (Nota and Paresce 1989). The most remarkable thing about the nebula of this LBV is that it is extremely dusty. Optical observations show that the nebula has a very different spectrum than the jet, which is invisible in Hα. The elliptical ring is actually a complete shell, as seen from detailed long slit spectroscopy.

HR Car is also surrounded by a compact nonsymmetric nebula (Kutsmékers and Van Dorn 1991). Since the distance is not as well known, it is harder to infer the properties of this shell and to date there has not been a study comparable to the attention received by its more famous cousin. But there can be no doubt that the HR Car nebula is going to be similar to that observed around AG Car.

He3-519 is also surrounded by a resolved nebula that looks a lot like the AG Car ring (Davidson, *these proceedings*). Again, it has a relatively small value of the thickness, $\Delta R/R$ is low. HD 316289 (= He3-1482) does not display a resolved optical emission nebula, but it does appear to be resolved at cm wavelengths (Shore *et al.* 1990). This is a star that should be studied at high angular resolution – it might show something interesting.

I won't say anything more for the moment about η Car, other than to point out that it is still the best-known LBV nebula and the best studied (see Walborn *et al.* 1991). Yet it is so large and complex that the dynamics of the ejecta are still difficult to sort out. Hillier and Allen (1992) have made considerable progress in this direction. They show that much of the emission spectrum is actually scattered light from the core, which looks a lot like a B[e] or LBV star. So we have a dense, dusty environment veiling the central star, and the spectrum is viewed indirectly through environmental reprocessing. The detailed HST imaging has not yet been followed up with detailed UV line mapping (Hester *et al.* (1991).

There are a few shells detected in the LMC, and these are extremely important for their similarities to the properties of the Galactic stars. The R127 nebula is most striking because it bears many strong similarities to AG Car. First discovered by Stahl (1987), this nebulosity is well resolved from the ground even with conventional techniques. There is some structure seen in recent imaging by Clampin *et al.* (1992, preprint) and it appears that again there is a shell with a radius of order 3 pc. This star was detected by IRAS, but of course it was unresolved. R143 was noticed to have a circumstellar nebula by Feast (1961) and the recent work by Parker *et al.* (1992) finds that many of the properties of this star are similar to Galactic LBVs. Finally, S 12/LMC, R4, S 134/LMC, R50 (=S 65/SMC), S 22/LMC, R66, R71, R82, and R126 have all been found to be imbedded within dust (either from groundbased or IRAS measurements) (see Stahl *et al.* 1985). These are all excellent examples of what I have been discussing, early-type stars that aren't very reddened but are surrounded by cold dust.

The LBVs thus provide an unusual view of their circumstellar environments. McGregor *et al.* (1988a, 1988b) find that CO overtone emission is detected in many of these stars. Certainly this cannot be too close to the photosphere because of the photodissociation of CO. It would seem more reasonable to place the CO at large distance, excited by the UV (like the 2μm Na I emission that is fluoresced by the 3000Å continuum) but otherwise safely situated. The CO requires of order 10^{18}cm^{-2} in order to become self-shielding so there must be a considerable volume of cold material located at relatively large distance from the star. In fact, the dust observed around many of these stars certainly supports this.

PLANETARY NEBULAE AS GUIDES TO THE LBVs

It seems more than a coincidence that the environments around highly evolved stars almost always display axial symmetry. If it is not the result of rotation at the time of the initial mass loss, then some rotation or orbital motion may be involved in the one outflow that formed the current shells.

In a planetary nebula ejection, angular momentum appears to be important even if the precise connection doesn't seem obvious. The best case is NGC 7027, one of the youngest known planetaries. In the optical the nebula is a mess. In

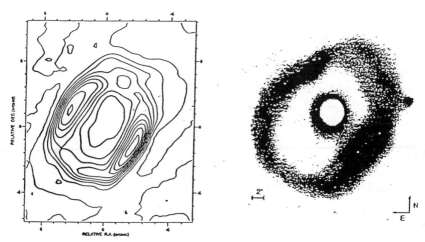

Figure 4: Left: NGC 7027, 2cm (Masson 1989); right: AG Car, optical (Nota *et al.* 1992).

the radio and IR it is clearly a shell structure, with a hollowed-out inner region
of ionized material. For comparison, Fig. 4 shows the AG Car nebula compared
with the radio shell of NGC 7027. The cylindrical symmetry is evident as well
in infrared images of the planetary (Gezari *et al.* 1992, *in preparation*), and the
dust lies just outside the H II region. The covering factor for the nebula is still
quite high since the central star is completely obscured by the H II region and
the neutral dust-containing old wind. But it is well known that some planetaries
have extensive cold material lying beyond the edge of the ionized gas, and this
is likely the key to what's happening with the B[e] stars.

Zickgraf *et al.* (1985) mention something like this in their study of R126
(= S127/LMC) (Shore and Sanduleak 1984). However, they attribute the cold
matter to the active ejection of a circumstellar disk by a rotationally unstable
hypergiant. They invoke an *excretion* disk to explain the high density, low ionization spectral features. The striking similarity of the Galactic star He3-395
and R126 argues against such a fine tuned model (Shore *et al.* 1990). Instead,
the idea of placing the colder material in a dense outer shell near the star could
be retained by simply making it the relic of some earlier stage of the mass ejection by the LBV.

TOWARD A UNIFYING MODEL

When viewed with the structure of planetary nebulae in mind, the LBVs come
easily into the general picture of highly evolved stars. There is a large range of
masses here, from the planetary nebulae with a few solar masses through the

LBVs with more than $20M_\odot$ to the most massive of the Galactic ones, like η Car. But a Platonic instinct we all seem to possess deems such analogies interesting if not compelling!

If the circumstellar environment is so important, we should be able to see manifestations of it in the spectra of the LBVs. After all, these stars make several excursions across the HR diagram in their lifetimes, and they will accumulate quite a bit of mass in their vicinity. Each eruption leaves about 0.1 $M_\odot$ of optically thin gas around the star. The accumulated masses ($e.g.$ HR Car, AG Car, R127) of about 1 $M_\odot$ at a few parsecs means that the mean density in the surroundings will be about the same as in a typical H II region. Thus during those stages when the central star is an O star, the circumstellar environment will be completely ionized. Emission lines from various ions like [N II] will be seen against the stellar photosphere. Now for the density to freeze in the ionization requires that locally, $n_e < 10^3 \mathrm{cm}^{-3}$. This ensures that the recombination time is comparable to the time between outbursts.

Ultraviolet observations of AG Car show that the nebula has a very low excitation and that only O I λ1300 and Fe II λ2600 are seen in emission with a resolution of about 300 (IUE low resolution) (Viotti $et\ al.$ 1988, Altner and Shore 1990). Nothing is known directly in the UV about the emission from HR Car or P Cyg. The optical nebulae of these stars don't appear to be very high excitation, and the excitation of the AG Car nebula is also apparently rather low (Mitra and Dufour 1990).

Assume the scenario that the LBVs are a stage in the progression from comparatively normal O stars through the Of class to the Of/WN stars. The $slash$ stars show the mixed characteristics associated with transitional forms (not unlike archaeopteryx), and the current thinking says that they are transiting between the two long timescale evolutionary states. What we see in R127, however, is that a $slash$ star changed its characteristics from OB supergiant to A supergiant in just one decade. We also see that R143 (Parker $et\ al.$, these proceedings) seems to be on its way back from one of these events (it is now a B star). It is quite clear that the appearance of a star at the time of a spectral census is not definitive in this part of the HR diagram. This leads to the greatest confusion when trying to unravel the evolutionary state of stars in the upper part of the HR diagram.

We seek, in Aristotelian language, to separate what is essential, or $esse$, from what is accidental, or $per\ accidens$. These stars have very high luminosities. This is $esse$. They are very advanced in their evolution. Whenever these stars are found in the LMC, and often in the Galaxy, they are members of compact groups (Lortet and Trestor 1991). Again, these properties are $esse$. The stars seem to always occur in the upper left (hot, luminous) part of the HR diagram. However, this is where the catalog of the essential properties seems to end. All the rest is accident. There is a taxonomic method in evolutionary biology called $clines$ which eschews lines of descent in favor of morphological distance. The $slash$ stars are an example of a cline, a grouping that links the characteristics of the Of stars with the WN sequence. However, R127, by marching across the HR diagram shows that this particular cline isn't secular. It's more like a mood swing of the star.

At what point does a star become a *slash* star? First, it must be hot enough to show the [N II] and N III] lines in the optical. Then it must also show some evidence for He II. The spectrum thus requires that the star, for some portion of its life, must be sufficiently luminous to produce a visible nebula and sufficiently hot to ionize a large volume. It must be one of the stars that has been picked up on an emission line survey, and this is partly a contrast problem.

Must a star be an Of/WN to become an LBV? No. One of the most recently discovered LBVs was classified as a B[e] star. S 22/LMC (Shore 1990, 1992). Another was a late-type hypergiant, R143 (Parker *et al.* 1992). When does a *slash* star become an LBV? Once it varies. This is difficult to assess because the database on variability is still quite poor. Another serious problem is that we don't know the cycle of the variation. Are the B[e] stars really a stage that the LBVs go through and get stuck in for some time? Do the Of/WN stars emerge from the B[e] stars without the full swing to the red?

How does this scenario relate to the observables, the spectrum and energy distribution of a UHRD star? In the earliest stage, the star is surrounded by the initial environment created by the massive wind of the pre-LBV stage. When the star first undergoes the instability that leads to the LBV, it does so within a massive circumstellar environment. The wind slamming into this environment creates a very bright eruption; perhaps this is Davidson's (1989) *Plinean* phase. During this stage, the mechanical efficiency of energy conversion can be very high, and if the dissipation takes place rapidly, the luminosity may actually become quite super-Eddington. It depends on the rate at which the wind kinetic energy is radiated and this is not an equilibrium or hydrostatic process. As the star clears out its environment, there is a slowly moving wind that is ionized and excited by the central star as it swings back and forth along lines in the HR diagram of constant bolometric luminosity. The expanding environment becomes progressively more optically thin and more detached, eventually ending up as a clear star like AG Car where only a few percent of the starlight is absorbed by the circumstellar gas.

The emission lines arising from the inner wind region are the fast hot permitted transitions and the low density probes. The low ionization stuff lies outside in the compacted dense shell from which the infrared emission also arises. Any molecules that have been formed during the coldest phase of the LBV will also be excited by the nonionizing UV.

To summarize the scenario, these are the proposed stages: (a) Initial outburst leads to a massive red supergant wind laden with dust that forms the equivalent of an AGB-like envelope; (b) First recovery leading to a compact H II region; (c) Next eruption leads to a Plinean eruption and initial breakout of the fast wind from the dense circumstellar environment; (d) Cold nebula phase with expanding H II region and invisible central star or B supergiant for central star: (e) Recovery leading to luminous Of/WN star; (f)Successive eruptions lead to the B[e] stage or transitions between Of/WN and B[e]; (g) Progressively cleared environment forms LBV surrounded by a dense but compressed shell and cold circumstellar dust with a massive H II emission region: (h) WR star finally emerges surrounded by the last portions of the shell now swept up by the very fast massive WR wind.

A PARTING WORD

There is no clear evidence for a *secular* progression through the various stages from *slash* through LBV to WR. In fact, the evidence is far stronger that the massive stars get confused at the end of their lives, and that the B[e], *slash*, and classical LBV phases are all intermingled. The large mass loss that domiante all of these phases strips the star down to the mass convergence that was discussed at this meeting by Langer and Castor. The life of UHRD stars is thus a lot like the description of war (or baseball): long periods of boredom punctuated by moments of stark terror.

ACKNOWLEDGMENTS

I want to thank Kris Davidson and Joe Cassinelli for their kind invitation and many discussions, and especially to thank Kris for suggesting the specific topic of this talk. I wish to thank Bruce Altner, Mark Clampin, Laurent Drissen, Jay Gallagher, Dan Gezari, Sally Heap, John Hillier, Ivan Hubeny, Roberta Humphreys, Henny Lamers, Claus Leitherer, Antonella Nota, Joel Parker, Carmelle Robert, George Sonneborn, Otmar Stahl, Michele Thornley, Nolan Walborn, Bill Waller, Bernard Wolf, and Franz-Josef Zickgraf for discussions and communicating results (sometimes in advance of publication). This work has been supported by NASA.

REFERENCES

Altner, B. and Shore, S. N. 1991, *BAAS*, **22**, 1258.

Bohannan, B. 1990, in *Properties of Hot Luminous Stars* (ed. C. Garmany) (San Fransisco: ASP Conf. Proc.), p. 39.

Bohannan. B. and Walborn, N. R. 1989, *PASP*, **101**, 520.

Clampin, M., Nota, A., Golimowskim D. A., and Leitherer, C. 1992, in *New Aspects of Magellanic Cloud Research*, preprint.

Davidson, K. 1989, in *Physics of Luminous Blue Variables - IAU Coll. 113* (eds. K. Davidson, H. Lamers, and A. Moffat) (Dordrecht: Kluwer).

Feast, M. W. 1961, *MNRAS*, **122**, 1.

Hester, J. J. *et al.* 1991, *AJ*, **102**, 654.

Hillier, D. J. and Allen, D. A. 1992, *A&A, in press*.

Humphreys, R. M. 1989, in *Physics of Luminous Blue Variables - IAU Coll.*

113 (eds. K. Davidson, H. Lamers, and A. Moffat) (Dordrecht: Kluwer), p. 3.

Humphreys, R. M. and Davidson, K. 1979, *ApJ*, **232**, 409.

Hutsmkers, D. and Van Dorn, E. 1991, *A&A*, **248**, 141.

Leitherer, C., Appenzeller, I., Klare, G., Lamers, H. J. G. L. M., Stahl, O., Waters, L. B. F. M., and Wolf, B. 1985, *A&A*, **153**, 168.

Lortet, M.-C. and Testor, G. 1991, *A&AS*, **89**, 185.

Malumuth, E. and Heap, S. R. 1992, *preprint*.

McGregor, P. J., Hyland, A. R., and Hillier, D. J. 1988*a*, *ApJ*, **324**, 1071.

McGregor, P. J., Hillier, D. J., and Hyland, A. R. 1988*b*, *ApJ*, **334**, 1071.

Masson, C. R. 1989, *ApJ*, **304**, 623.

Mitra, M. P. and Dufour, R. J. 1990, *MNRAS*, **242**, 98.

Parker, J. W., Clayton, G. C., Winge, C., and Conti, P. S. 1992, *preprint*.

Shore, S. N. 1992, in *Nonisotropic and Variable Mass Outflows from Stars* (eds. L. Drissen, C. Leitherer, and A. Nota) (San Francisco: ASP Press) p. 342.

Shore, S. N., and Sanduleak, N. 1984, *ApJS*, **55**, 1.

Shore, S. N., Brown, D. N., Bopp, B., Robinson, C., Sanduleak, N., and Feldman, P. 1990, *ApJS*, **73**, 461.

Shore, S. N., Sanduleak, N., and Allen, D. A. 1987, *A&A*, **176**, 59.

Stahl, O. 1987, *A&A*, **182**, 229.

Stahl, O. and Wolf, B. 1986, *A&A*, **158**, 371.

Stahl, O. and Wolf, B. 1986, *A&A*, **154**, 243.

Stahl, O., Wolf, B., de Groot, M., and Leitherer, C. 1985, *A&AS*, **61**, 237.

Stahl, O., Wolf, B., Leitherer, C., and Zickgraf, F.-J. 1985, *A&A*, **131**, L5.

Thackeray, A. D. 1950, *MNRAS*, **110**, 526.

Viotti, R., Cassatella, A., Pons, D., and Thé, P. S. 1988, *A&A*, **190**, 333.

Walborn, N. R. 1977, *ApJ*, **215**, 53.

Walborn, N. R. 1982, *ApJ*, **256**, 452.

Walborn, N. R., Evans, I. N., Fitzpatrick, E. L. and Phillips, M. M. 1991, in *IAU Symp. 143: Wolf-Rayet Stars and Interrelations with Other Massive Stars in Galaxies* (eds. K. A. van der Hucht and B. Hidayat) (Dordrecht: Kluwer), p. 505.

Wolf, B. 1992, in *Nonisotropic and Variable Mass Outflows from Stars* (eds. L. Drissen, C. Leitherer, and A. Nota) (San Francisco: ASP Conf. Series), p. 327.

Wolf, B. and Stahl, O. 1986, *A&A*, **164**, 435.

Zickgraf, F.-J. 1992, in *Nonisotropic and Variable Mass Outflows from Stars* (eds. L. Drissen, C. Leitherer, and A. Nota) (San Francisco: ASP Conf. Series), p. 75.

Zickgraf, F.-J., Wolf, B., Stahl, O., and Humphreys, R. M. 1989, *A&A*, **220**, 206.

Zickgraf, F.-J., Wolf, B., Stahl, O., Leitherer, C., and Appenzeller, I. 1986, *A&A*, **163**, 119.

Zickgraf, F.-J., Wolf, B., Stahl, O., Leitherer, C., and Klare, G. 1986, *A&A*, **143**, 421.

Massive Stars: Their Lives in the Interstellar Medium
ASP Conference Series, Vol. 35, 1993
Joseph P. Cassinelli and Edward B. Churchwell (eds.)

MASS LOSS FROM COOL SUPERGIANT STARS

G. R. KNAPP AND MICHAEL WOODHAMS
Department of Astrophysical Sciences, Princeton University, Princeton NJ 08544

1. INTRODUCTION

This contribution contains a discussion of the circumstellar gas and dust shells produced by mass loss from cool ($T_{eff} \leq$ 3000 K), high luminosity ($L_* > 10^5$ $L_\odot$), high mass ($M_* > 10$ $M_\odot$) supergiant stars.

Supergiant stars have very characteristic spectra; their lines, especially the hydrogen lines, are sharp and narrow, indicating a very extended chromosphere. However, a variety of observational evidence suggests that there are at least two quite distinct groups of stars with 'supergiant' spectra, one group comprising the massive, high - luminosity stars which are the subject of this conference and the other stars of much lower mass ($\sim$ 1 $M_\odot$) and luminosity (at most a few thousand $L_\odot$). The latter group contains the so-called 'high-latitude' supergiants, which are the focus of much current interest because of their possible evolutionary status as the precursors of planetary nebulae. Section 2 attempts a deconstruction (partly historical) of the term 'supergiant' so as to focus on the selection and properties of the stars to be discussed in the present paper. A more extensive account is given by Keenan (1963) and some interesting early correspondence on the subject is summarized by Schlesinger (1911).

Section 3 discusses millimeter wavelength molecular line observations of some few supergiants and the derivation of their mass loss rates. Section 4 is a brief diversion on an example of a cool mass losing supergiant, VY CMa, and presents some new data on the evolutionary state of this famous star. Section 5 discusses the infrared excess emission from circumstellar dust and uses it to derive mass loss rates for a larger sample of supergiants, those from the compilation of Jura and Kleinmann (1990). These data are used to estimate the total rate of mass return to the interstellar medium by massive supergiants.

2. SUPERGIANT STARS

The spectral classification scheme used in the Henry Draper Catalogue (Cannon and Pickering 1918-1924) was at once recognized as a sequence of stellar temperatures. Maury and Pickering (1897) noted that the spectra of stars of the same surface temperature show significant differences due to different line shapes, which range from broad to narrow, and made the first attempt at a two dimensional classification. Maury's stars of subtype 'c' have unusually sharp and/or narrow lines and are found for all spectral types. Some of them are variables, some not. Herzsprung (1909) showed that many of the variable c stars are of very high luminosity. Payne's (1930) monograph on the spectra of high luminosity stars contains a list of the then-known c type stars of spectral types B0-K5. The stars are rare; they comprise about 0.5% of stars of all spectral types. Wilson's (1941) study of the non-variable c stars showed that their mean luminosities are a few $\times 10^3$ $L_\odot$, similar to the luminosities of stars of 1 $M_\odot$ on the red giant and asymptotic giant branch stars. Bidelman (1951) and Münch (1951) drew similar conclusions from a reclassification of the stellar spectra on the MK system.

There are thus at least two groups of stars of spectral type later that OB which have supergiant-like spectra. There are stars of luminosity a few thousand $L_\odot$ and masses ~ 1 $M_\odot$ which numerically dominate at high galactic latitudes, and the stars of much higher mass and luminosity which dominate at low latitudes (Humphreys 1978; Chiosi and Maeder 1986). The high latitude supergiants may be post-AGB stars evolving towards the planetary nebula stage, where some of the remnant circumstellar envelope produces a false photosphere of apparently low surface gravity. Thus very similar spectral characteristics can apparently be produced by quite different evolutionary paths. There are several other examples of 'parallel evolution' which involve the massive stars. A second is the existence of very hot low mass carbon rich nuclei of planetary nebulae, whose spectra resemble those of the much more massive Wolf-Rayet stars enough to be placed in the same class (these stars are also found at high galactic latitudes while the massive WR stars are found at lower latitudes). A third potential red herring in the study of massive supergiants is the existence of some low mass evolved stars whose circumstellar material has high outflow speeds. If cool giant stars lose mass by radiation pressure on dust grains, the equation of motion of the circumstellar wind is

$$v\frac{dv}{dr} \; = \; \frac{1}{r^2}\left(\frac{\chi_e L_\star}{4\pi c} \; - \; GM_\star\right) \tag{1}$$

where χ_e is the wavelength averaged opacity of the envelope (cf. Jura 1984). If dust condenses at a distance r_o above the stellar photosphere, determined by the condensation temperature of the dust, then $r_o \sim L_\star^{1/2}$ and is the location where the final acceleration of the wind begins. The terminal velocity is then

$$V_o \; \sim \; \chi_e^{\frac{1}{2}} L_\star^{\frac{1}{4}} \tag{2}$$

i.e. more luminous stars have higher outflow speeds, in agreement with data on OH/IR stars (Jura 1984). However, as a star evolves away from the AGB, it becomes hotter and bluer. Since the opacity of dust is wavelength dependent ($Q_\lambda \sim \lambda^{-1}$) the effective opacity of the remnant circumstellar envelope increases as the star evolves and, as equation (2) shows, the circumstellar material can be accelerated to high velocities. Consider an example of an object in each class. VY CMa (see section 4) has an M5 supergiant spectrum. Its membership of the τ CMa cluster gives fairly secure values for the distance, mass and luminosity (5×10^5 $L_\odot$), and its molecular wind has the relatively high speed of 40 km s^{-1}. The Egg Nebula, CRL2688, also has a supergiant spectrum (F8Ia, Crampton, Cowley and Humphreys 1975) and some of the circumstellar material has a quite high velocity (100 km s^{-1}, Young et al. 1992). However, it is unlikely to be a high mass or high luminosity star; it is an isolated object, and its molecular shell has several velocity components, with most of the material moving at the more modest, and more typical, speed of 18 km s^{-1}. In this paper, then, we discuss only supergiants in clusters and associations so that the distances are reasonably well determined, and consider only cool stars with luminosities $> 10^5$ $L_\odot$.

3. MOLECULAR LINE OBSERVATIONS OF MASS LOSS

Table 1. CO Observations and Mass Loss Rates of Supergiant Stars

Star	Spectrum	Distance (pc)	L_{bol} ($L_\odot$)	V_o (km s^{-1})	$\dot{M}$ ($M_\odot$ yr^{-1})
α Ori	M2Ia $-$ Iab	450	5.1×10^5	15	5×10^{-6}
VY CMa	M5eIab	1700	6.3×10^5	40	3×10^{-4}
VX Sgr	M4Ia	1900	2.2×10^5	30	2×10^{-5}
NML Cyg	M6III	2000	5.0×10^5	32	4×10^{-4}
μ Cep	M2Ia	900	5.0×10^5	10	4×10^{-7}
PZ Cas	M2Ib	2000	1.3×10^5	30	5×10^{-5}
S Per	M4Ia	2300	2.0×10^5	20	1×10^{-5}
IRC + 10420	G8Ia	3400	5.0×10^5	52	3×10^{-4}

The mass shed from cool high-luminosity stars is in the form of molecules and dust, and the resulting extended circumstellar envelopes can be studied with observations of emission in the infrared continuum and in a wide variety of molecular lines. These observations give reasonably well-determined values of the mass loss rate, the gas to dust ratio, and the terminal outflow velocity of the wind for a large number of cool evolved stars, including both carbon and oxygen rich AGB stars, young planetary nebulae, protoplanetary nebulae and massive cool supergiants. Thermal emission in the CO rotational lines is a particularly useful probe, and has to date been detected in eight massive supergiants. The results are listed in Table 1. The mass loss rates and outflow speeds are calculated from CO data obtained by Zuckerman and Dyck (1986), LeBorgne and Mauron (1989), Knapp et al. (1992), and Woodhams (this conference), and from HI data for α Ori by Bowers and Knapp (1987). The distances to the stars are those of the OB associations of which they are members (cf.

Humphreys 1978) except for that of IRC+10420, which is taken to be at its near kinematic distance. Betelgeuse is assumed to be at the distance of the Orion star forming complex, while NML Cygni's membership of the Cyg OB2 association is discussed by Morris and Jura (1983).

A glance at Table 1 shows that the wind outflow speeds are often quite high compared with those measured for AGB stars (typically, $V_o = 10 - 15 \, \mathrm{km \, s^{-1}}$), in agreement with the expected velocity - luminosity relationship given in equation (2). However, this is not always the case - α Ori and μ Cep have quite modest wind speeds. It is possible that radiation pressure is not the only process determining the final outflow speed of the winds, or that the circumstellar material is in a flattened rather than spherical configuration, so that projection effects are important.

Table 1 also shows that some of the mass loss rates are very high; that of NML Cyg, indeed, is the highest observed for any cool evolved star. For several of the other stars, though, the mass loss rates are considerably lower. The data in Table 1 show that the rate of mass return to the ISM from cool supergiants is at least $4 \times 10^{-11} \, M_\odot \, \mathrm{pc^{-2} \, yr^{-1}}$.

4. VY CANIS MAJORIS

This section contains a brief description of the nature of VY CMa and a discussion of its mass loss. It is a bipolar nebula on the edge of a molecular cloud which is part of the NGC2362 OB association. Although the central star is not visible, the reflection nebulosity gives a spectral type of M5Ia and an effective temperature of 2700 K (Herbig 1970). The distance to the association is 1500 pc (Lada and Reid 1978), giving a bolometric luminosity of $5 \times 10^5 \, L_\odot$.

Since both pre- and post-main sequence stars are often observed to be surrounded by dense circumstellar clouds with axisymmetric, or bipolar, structure, (LkHα198 and OH231.8+4.2 are examples in each category - another example of objects with similar properties produced by different evolutionary paths) it is not obvious whether VY CMa is a pre- or post- main sequence star; its irregular variability and association with dense interstellar matter argues in favor of the former interpretation. However, there are several arguments which suggest that VY CMa is in fact a post main sequence star. Wallerstein (1978) has pointed out that the irregular variability can be attributed to structure in very dense circumstellar material, whose motion allows variable amounts of starlight to escape. Further, there is now a fair amount of evidence that VY CMa is an intrinsic long period variable star. While the near infrared emission (2 μm) shows no detectable variations (Harvey et al. 1974) monitoring observations of maser line emission from H_2O (Cox and Parker 1979; Gomez-Balboa and Lépine 1986), SiO (Martinez et al. 1988) and OH (Harvey et al. 1974) and, more recently, of the millimeter wavelength continuum emission (Knapp, Sandell and Robson 1992) show that the object is a quite well behaved long period variable star with a period of about 870^d. The period - density relationship for fundamental pulsators gives a mass of about 50 $M_\odot$, in reasonable agreement with the turnoff mass of NGC2362. Molecular line observations also show that the star is losing mass at a high rate, 2.5×10^{-4} $M_{\odot}$ yr^{-1}, at a speed of about 40 km s^{-1} (cf. Zuckerman and Dyck 1986). The ratio of the stellar wind momentum flux to that of the starlight, $\dot{M} V_o c/L_{bol}$, is about 1 for VY CMa. To all intents and purposes, then, VY CMa is behaving like a cool asymptotic giant branch star, save that it is much more luminous; it is thus very likely to be a high mass post main sequence star. VY CMa has other things in common with AGB stars; its gas to dust ratio is about 100:1 by mass (Rowan Robinson and Harris 1983; Knapp et al. 1992), and the dust properties (composition and structure) are also similar to those of grains around other evolved stars. Submillimeter continuum observations with the JCMT telescope also give some idea of the extent of the dust shell; the spectrum between 350 μm and 2 mm has a power law slope $S_{\nu} \sim \nu^{2.5}$. Since emission at this wavelength is at the Rayleigh-Jeans limit throughout the envelope and since the dust emissivity depends on frequency as $Q_{\nu} \sim \nu$, this spectral shape shows that the JCMT observations resolve the envelope and that its extent is $\geq 5 \times 10^{17}$ cm. The timescale for mass loss is then > 4000 years.

5. <u>INFRARED EXCESSES AND MASS LOSS RATES</u>

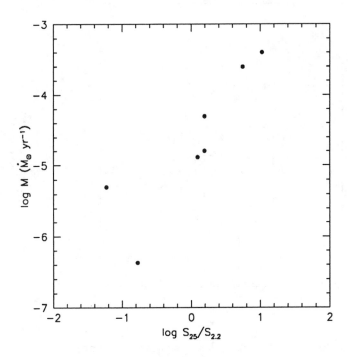

Fig. 1.   Mass loss rate measured by CO versus infrared excess for seven cool massive supergiants (see Table 1).

Jura and Kleinmann (1990) have compiled lists of cool supergiants from
the IRC catalog (Neugebauer and Leighton 1969) and of a local sample
within 2.5 kpc. Not all of these stars have been observed in the CO
line, but since the gas to dust ratio appears to be reasonably constant in
circumstellar envelopes, we can use the observed long wavelength infrared
excesses to estimate mass loss rates. We measure the infrared excess using
the flux density ratio $S_{25}/S_{2.2}$, where S_{25} and $S_{2.2}$ are the $25\mu m$ and $2.2\mu m$
flux densities measured by IRAS and the IRC surveys. This recipe is
useful only for cool stars, since we use $S_{2.2}$ to measure the photospheric flux
and $S_{25}/S_{2.2}$ the fraction of photospheric light absorbed and re-radiated
by circumstellar dust. We use the 25 μm flux density because (1) the 12
μm flux density is affected by emission or absorption in the strong 10.2
μm silicate feature, which, though it depends on the mass loss rate, does
so non-linearly and (2) because the $60\mu m$ and 100 μm flux densities can
be contaminated by strong emission from cool Galactic dust, and many of
the high mass supergiants are at low Galactic latitudes. IRAS measured
flux densities for all of the stars in Table 1 except for NML Cyg; we use
the 25 μm flux density observed by Hagen et al. (1975).

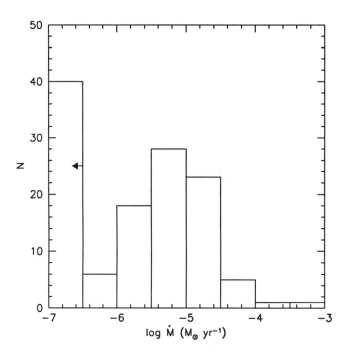

Fig. 2.   Distribution of mass loss rates for supergiant M stars
from Jura and Kleinmann (1990). The leftmost bin contains upper limits,
i.e. stars for which $\dot{M} < 3 \times 10^{-7}\ M_{\odot}\ yr^{-1}$.

Figure 1 shows $\dot{M}$ versus $S_{25}/S_{2.2}$ for all of the stars in Table 1 except IRC+10420 (which is of considerably earlier spectral type). There is a rough proportionality but a lot of scatter. At least part of this is due to a significant contribution to the 25 μm emission from the photospheres of stars with low mass loss rates. The data in Figure 1 give:

$$\dot{M} = 4 \times 10^{-5} \left(\frac{S_{25}}{S_{2.2}} - C \right) \tag{3}$$

where we assume that the circumstellar dust does not radiate significantly at 2.2 μm. C is the $25\mu/2.2\mu$ ratio for the photosphere, and is 0.025 for a 3000 K black body.

Figure 2 shows the resulting distribution of mass loss rates for the M supergiant stars compiled by Jura and Kleinmann (1990). We find no relationship between the mass loss rates and the stellar spectral types. Figure 2 shows that the mass loss rates for cool supergiants range over a factor of 10^3, and suggests that roughly equal total amounts of mass are returned in each decade of mass loss rate. The very few stars losing the largest amount of mass (in this case NML Cyg and VY CMa, each with $\dot{M} > 10^{-4}$ $M_\odot$ yr^{-1}) contribute a significant fraction of the total mass lost.

The total mass loss rate for the stars within 2500 pc is then 6×10^{-11} $M_\odot$ pc^{-2} yr^{-1}; about half of this amount is due to NML Cyg and VY CMa. This number is very uncertain; it is about twice the value found by Jura and Kleinmann (1990). It represents about 10% of the total rate of mass return for cool giants and about 20 % of the amount returned by oxygen-rich stars (cf. Knapp and Wilcots 1987). Cool supergiants thus are an important, but not dominant, source of mass return to the ISM.

ACKNOWLEDGEMENTS

We would like to thank the A.S.P. and the Madison astronomical community for inviting us to this meeting, and M. Jura and R. Humphreys for valuable discussions. We are very grateful to the staffs of the Caltech Submillimeter Observatory and the James Clerk Maxwell Telescope, where much of the data described here was obtained. In particular we thank our collaborators K. Young, T.G. Phillips, G. Sandell and E.I. Robson. This research is partly supported by N.S.F. grant AST90-14689 to Princeton University.

REFERENCES

Bidelman, W.P. 1951, ApJ, 113, 304.
Bowers, P.F. and Knapp, G.R. 1987, ApJ, 315, 305.
Cannon, A.J. and Pickering, E.C. 1918-1924, Harvard Ann., 91-99.
Chiosi, C. and Maeder, A., 1986, ARAA, 24, 329.
Cox, G,G. and Parker, E.A. 1979, MNRAS, 186, 197.

Crampton, D., Cowley, A.P. and Humphreys, R.M. 1975, ApJ(Letters), 198, L135.

Gomez-Balboa, A.M. and Lépine, J.R.D. 1986, A&A, 159, 166.

Hagen, W., Simon, T. and Dyck, H.M. 1975, ApJ(Letters), 201, L81.

Harvey, P.M., Bechis, K.P., Wilson, W.J. and Ball, J.A. 1974, ApJS, 27, 331.

Herbig, G.H. 1970, ApJ, 162, 557.

Herzsprung, E. 1909, AN, 179, 373.

Humphreys, R.M. 1978, ApJS, 38, 309.

Jura, M. 1984, ApJ, 282, 200.

Jura, M. and Kleinmann, S.G. 1990, ApJS, 73, 769.

Keenan, P.C. 1963, in 'Basic Astronomical Data', ed. K. Aa. Strand, University of Chicago Press, p78.

Knapp, G.R., Sandell, G. and Robson, E.I. 1992, submitted to ApJ, .

Knapp, G.R., Young, K., Gammie. C.F. and Phillips, T.G. 1992, in preparation.

Knapp, G.R. and Wilcots, E.M. 1987, in Late Stages of Stellar Evolution, ed. S. Kwok and S.R. Pottasch, D. Reidel Co., p171.

Lada, C.J. and Reid, M.J. 1978, ApJ, 219, 95.

LeBorgne, J.F. and Mauron, N. 1989, A&A, 210, 198.

Martinez, A., Bujarrabal, V. and Alcolea, J. 1988, ApJS, 74, 273.

Maury, A.C. and Pickering, E.C. 1897, Harvard Ann., 28, part 1.

Morris, M. and Jura, M. 1983, ApJ, 267, 179.

Münch, L. 1951, ApJ, 113, 304.

Neugebauer, G. and Leighton, R.B. 1969, Two Micron Sky Survey (NASA SP-3047).

Payne, C. H. 1930, in 'The Stars of High Luminosity', Harvard Observatory Monographs No. 3, McGraw-Hill Book Company, NY.

Rowan-Robinson, M., and Harris, S. 1983, MNRAS, 202, 767.

Schlesinger, F. 1911, ApJ, 33, 260.

Wallerstein, G.H. 1978, Observatory, 98, 224.

Wilson, R. 1941, ApJ, 93, 212.

Zuckerman, B. and Dyck, H.M. 1986, ApJ, 304, 394.

Massive Stars: Their Lives in the Interstellar Medium
ASP Conference Series, Vol. 35, 1993
Joseph P. Cassinelli and Edward B. Churchwell (eds.)

MASSIVE RUNAWAY STARS

ADRIAAN BLAAUW
Kapteyn Laboratory, P.O.Box 800, 9700 AV Groningen, Netherlands

ABSTRACT   We review the properties of runaway OB stars with emphasis on the most massive ones, the O-types, and the two scenarios to explain this phenomenon: binary-supernova release and cluster-dynamical ejection. Evidence based on duplicity and kinematical properties so far seemed to leave room for the actual occurrence of both. However, if rotational velocity and Helium abundance are also taken into account, the binary-supernova scenario seems preferable for the O-type runaways. We point out that therefore great caution must be exercised in using these as reference stars in high-precision quantitative spectroscopy, and we present some considerations with regard to the role of the runaways in externally induced star formation.

1. INTRODUCTION   O and B-type runaway stars are a subset of the O and B stars, characterized by space velocities up to 200 $km\,s^{-1}$ and almost complete absence of duplicity and multiplicity. (For principal studies see Blaauw 1961 and Gies and Bolton 1986, Gies 1987.) Their frequency as a function of spectral type decreases steeply from about 20% among the O-types to 2.5% among B0-B0.5, and still lower among B1-B5. As we shall see below, the most massive runaways are also characterized by generally high rotational velocities and high Helium abundance. The runaway OB stars have been the subject of numerous studies in which the emphasis was especially on the problem of their origin. A very useful review of work done up to 1986 has been given by Gies and Bolton (1986).

The present paper deals mainly with the most massive runaways, those of O-type. In section 2 we describe in some detail some nearby examples, in section 3 we review their principal properties, and in section 4 we discuss the question of the origin. In section 5 some implications of the runaway phenomenon for externally induced star formation will be briefly discussed.

An important side-aspect of the study of the runaway O-type stars is the evaluation of their role in current high-resolution spectroscopic analyses that aim at the prediction of intrinsic properties such as luminosity, masses and chemical abundance; this subject, too, is briefly dealt with in Section 4.

2. EXAMPLES   In the present context I shall rely preferably on the limited number of (the most "bona fide") massive runaways for which it either has been well established, or may be reasonably assumed, that they originate from an identified OB association or cluster, the "parent". In this way we avoid the

207

confusion that sometimes has entered studies that classify as runaways certain
stars simply because their measured space velocity exceeds an adopted limit
or because a high velocity-at-birth is inferred from a large distance from the
galactic plane. (Particularly studies based on samples with high projected
velocities inferred from proper motions tend to be misleading, unless great care
is exercised in recognizing the effects of systematic and accidental errors of the
proper motions of the distant stars). Even then a certain ambiguity remains.
For the sake of clarity we describe in some detail and by means of Figures 1 to
4 some examples of O-type runaways with their parents; they are arranged in
the order of increasing distance. Data on these objects together with those for
other runaways are summarized in Table 1.

Figure 1: The association subgroup Upper Sco (distance about 160 pc)
with the runaway ζ Oph (O9.5V). Its runaway nature follows unambiguously
from the accurately (FK5) determined relative proper motion shown here,
and the distances of star and parent. The relative space velocity between star
and parent is 40 km s^{-1} and the kinematic age, counted from the centre of U
Sco, 1.0 million years. The configuration allows an accurate determination
of the star's luminosity (see section 3.5 and Figure 7). The high speed of
the star with respect to the ISM is also apparent from the bow shock effect
(communication by Van Buren at this conference).

Figure 2: The association Per OB2 (distance about 350 pc)
with the runaway ξ Per (O7.5III). Its runaway nature follows mainly from
the relative radial velocity of +36 km s^{-1} with respect to the mean velocity
of the remaining stars of Per OB2. The dashed arrow represents the proper
motion relative to the mean proper motion of Per OB2, it amounts to 0.″005
per year (corresponding to about 8 km s^{-1}) so that departure from the central
region of the association would imply a time scale of about 1.4 million years.
ξ Per is typical for those cases where the relation to the parent leaves little
doubt, but the ejection age has a high percentage uncertainty. The solid arrow
represents the very accurately (FK5) determined proper motion of 0.″010
after elimination of the standard solar motion. The difference between the two
arrows represents the relatively high projected relative motion of Per OB2 with
respect to the local standard of rest (about 10 km s^{-1}; sharing the kinematics
of the Gould Belt).

Figure 3: The association Ori OB1 (distance about 460 pc)
with the runaways μ Col (O9.5V), AE Aur (O9.5V), and 53 Ari (B2V). Their
runaway nature follows from the well determined proper motions and radial
velocities (see Blaauw 1989) and the distances of the stars (460 pc, 610 pc, and
310 pc, respectively) and the parent. The space velocities relative to Ori OB1
are: 117, 140 and 40 km s^{-1} and the times elapsed since ejection, counted from
the centre of Ori OB1, 2.3, 2.4. and 7.3 million years, respectively. The unique
nature of the pair μ Col + AE Aur with their opposite runaway motions and
nearly equal masses, speeds and kinematic ages (Blaauw and Morgan 1954)
has made it a candidate for the dynamical ejection scenario (Gies and Bolton
1986).

Figure 4: The runaway α Cam (O9.5Ia) and its presumed parent,
the young open cluster NGC 1502, both at distance about 1 kpc. The arrows
show the directions of the proper motions corrected for standard solar motion.
This is very well (FK5) determined for α Cam: 0.″00126 per year. For NGC
1502 it is much less accurate and based on analysis of the relatively few

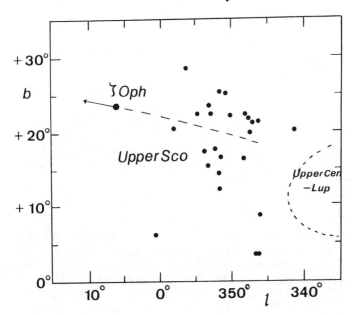

Fig. 1   Runaway ζ Oph with parent Upper Sco; see text

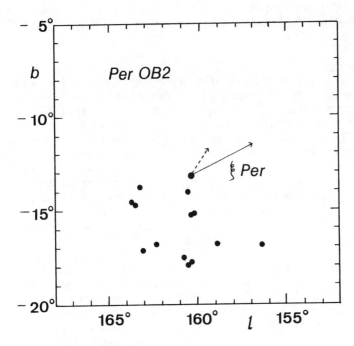

Fig. 2   Runaway ξ Per with parent Per OB2; see text

meridian observations in Washington Catalogues W(3)50, W(4)50 and W(5)50, combined with the GC positions at epoch around the year 1900, for three relatively bright members of NGC 1502: GC4898, GC4931 and GC4932. The radius of the dashed circle corresponds to the probable error of this proper motion. An alternative possibility as a parent for α Cam would seem to be the association Aur OB1 located around $(l,b) = (174°,+2°)$ at distance about 1300 pc (beyond this Figure). Aur OB1 would require the unlikely high kinematic age of about 10 million years (if we ignore galactic gravitational deceleration). The (uncertain) relative motion of NGC 1502 with respect to α Cam suggests ejection from NGC 1502 some 2 million years ago with relative velocity of 48 km s^{-1}. The high speed of α Cam shows also in the bow shock effect in the ISM (De Vries 1985, Van Buren at this conference).

3. PROPERTIES
3.1 The incidence of duplicity
Gies and Bolton (1986) conducted an extensive programme of radial velocity observations for 36 proposed runaway stars, intended mainly for testing carefully the (lack of) binary content that had been found as one of the striking properties of the runaways, and at the same time improving data on their velocity distribution. A follow-up paper by Gies (1987) extended the analysis to include a broader sample of O stars. Their results confirmed the early finding that visual duplicity is absent and spectroscopic duplicity with semi-amplitudes exceeding the limit of detectability - which depending on the star's spectral characteristics for the majority lies between 5 and 25 km s^{-1} - is very rare, of the order of 10 to 20 per cent only. This, in strong contrast to the high incidence (30 to 40 per cent) of duplicity or multiplicity of the O stars belonging to the OB associations.

3.2 The velocity contrast
Radial velocities with respect to the local standard of rest at the location of the star for stars in Gies and Bolton's program lie between 20 and 60 km s^{-1} with an average value about 35 km s^{-1}. Assuming isotropic velocity distribution, this would correspond to peculiar space velocities of 40 to 120 km s^{-1} with an average about 70 km s^{-1}; these figures confirm earlier estimates. It is the sharp contrast between these high velocities and the low relative velocities of the OB stars at their birth places, the OB associations, of the order of 5 km s^{-1} (Blaauw 1991) that has always so strongly pointed to a mechanism for the cause of the runaway velocities different from that determining the velocity distribution at birth for massive stars in general.

3.3 The spectral type (mass) - velocity dependance
Whereas for most of the O-type runaway stars space velocities do not exceed 80 km s^{-1}, velocities up to nearly 200 km s^{-1} are not exceptional among the early B types. An anticorrelation between mass and velocity seems indicated: see Figure 5, adapted from Gies and Bolton (1986). The filled circles were taken from data in Blaauw (1961) and the open circles represent peculiar radial velocities of the sample of Gies and Bolton (1986) multiplied here by a statistical factor of two as it will apply in the case of isotropic velocity distribution.

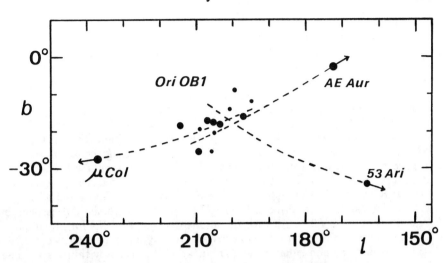

Fig. 3 Runaways μ Col, AE Aur and 53 Ari with parent Ori OB1; see text

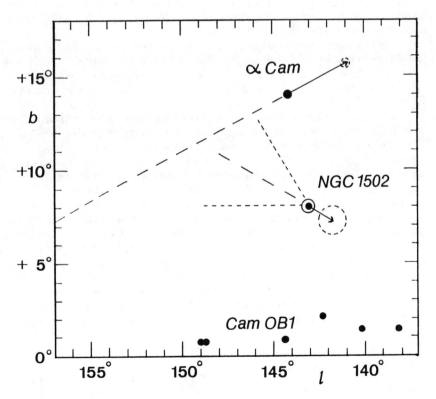

Fig. 4 Runaway α Cam with parent NGC 1502; see text

3.4 Rotational velocity and Helium abundance

It turns out to be essential to also include in this review the rotational
velocity V.sin$i$ and the Helium abundance ϵ (by number relative to Hydrogen),
although determinations of the latter still are scarce. Moreover, because for our
statistics we want to be sure that a star is either a "bona fide" runaway, or a
genuine low-velocity association member, the sample from which we can draw
is limited: I have restricted it to OB associations and related runaways located
within about 2 kpc in the "local arm".

Table 1 lists eight O-type runaways thus selected together with their
"parent" association or cluster. Five of them occur among the examples
in Figures 1 to 4. μ Col has been included although no accurate Helium
abundance is available, because of its special relation to AE Aur (see section
4). For HD 66811 (ζ Puppis) the parent was found to be associated with the
Vela Molecular Ridge and most likely to be identified with the OB association
Vela R2 (not with Vela OB2 with the implied canonical value for M_v about
- 6); see Blaauw and Sahu (1992). For the purpose of comparison, we use 16
stars known to be non-runaway O stars, nearly all of them within 2 kpc. The
following are their HD numbers and parents: 34656 in Aur OB1; 36486 (δ Ori)
and 37742 (ζ Ori A) in Ori OB1; 46150, 46223, and 46966 in Mon OB2; 47839
(15 Mon) in Mon OB1; 164794 (9 Sgr) in Sgr OB1; 191612 in Cyg OB3; 192639
and 193514 in Cyg OB1; 207198 and 209975 (19 Cep) in Cep OB2; 210809 in
Cep OB1; 214680 (10 Lac) in Lac OB1; 217086 in Cep OB3. Basic lists that
served for these selections were the sections "Cluster and Association Stars"
and "Runaway Stars" of Table 1 of Gies (1987).

Helium abundance measures (by numbers) were taken from Herrero et al
(1992), Bohannan et al (1990), Voels et al (1989), and Kudritzki and Hummer
(1990). The rotational velocity data were mostly taken from the compilations
in these same tables of Gies, who however refers for these data to Conti and
Ebbets (1977).

Whereas this final selection with only seven "complete" runaway O stars
is very limited, we shall see that it suffices for pointing to certain important
conclusions with regard to the nature of the O-type runaways. Obviously, when
more extensive observational data will have become available, particularly
kinematic data based on Hipparcos and spectroscopy for rotational velocities
and Helium abundances, in combination with more identifications of the parent
associations, strengthening of these conclusions will be of great interest.

In Figure 6 we show the (projected) rotational velocities V.sin$i$ and the
Helium abundance ϵ for the runaways of Table 1 (dots) and the nonrunaways
mentioned before (circles). Notwithstanding the limited number of runaways, a
striking difference in the patterns for the two groups is apparent: the majority
of the nonrunaways lie within a limited range of both quantities: 0 to 150
km s^{-1} for V.sin$i$ and 0.04 to 0.13 for the Helium abundance, whereas only one
of the runaways lies within this area and the remaining ones scatter up to 350
km s^{-1} in V.sin$i$ and 0.20 in ϵ. We return to this interesting diagram in section
4.

3.5 Blue straggler nature of some runaways

Finally, we show an interesting feature in Figure 7: the position in the HR
diagram for the three O-type runaways ζ Oph, AE Aur, and μ Col, all of
them of type O9.5V, and the B2V-type runaway 53 Ari. For each of these

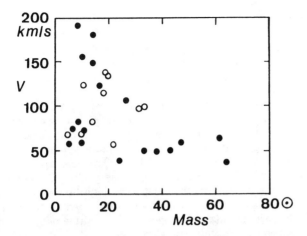

Fig. 5 The velocity-mass relation, adapted from Gies and Bolton (1986, Fig.16). Dots: space velocities according to Blaauw (1961). Open circles: radial velocities from Gies and Bolton, multiplied here with a statistical factor 2 although this leads to some bias in favour of high velocities.

Table 1 O-type runaways with known V.sin$i$ and He abundance

Name Parent	HD HIC FK5	MK	M_v	S^* km s^{-1}	T_{kin} Myrs	V.sin$i$ km s^{-1}	ϵ
ζ Oph Upp-Sco	149757 81377 622	O9.5Vnn	−4.1	40r	1.0	350	0.16
ξ Per Per OB2	024912 18614 148	O7.5III(n,f)	−5.3	36r	2::	200	0.18
AE Aur Ori OB1	034178 24575	O9.5V	−3.9	117r	2.3	27	0.095
μ Col Ori OB1	038666 27204	O9.5V	−3.8	140r	2.4	97	?
ζ Pup Vela R2?	066811 39429 306	O4I(n)f	−7.2:	> 70	1.5	208	0.20
68 Cyg Cyg OB7	203064 105186 396	O7.5IIIn(f)	−5.5	50:	?	274	0.12
λ Cep Cep OB2	210839 109556	O6I(n)fp	−6.3	64:	?	214	0.17
α Cam NGC1502?	030614 22783 178	O9.5Ia	−6.9	48:r	2::	85	0.18

* Space velocities marked r are relative to the parent association or cluster

it was possible to determine with high accuracy the luminosity, in a way entirely independent of the star's luminosity classification. This determination makes use of the fact that, once the parent association of a runaway and the parent's distance are known - as is the case here: Upper Sco and Ori OB1 - the observed radial velocity and proper motion of the runaway uniquely define the distance of the runaway, or, more explicitly, the difference between the distance moduli of parent and runaway. (Essentially: there is only one distance that fits both proper motion and radial velocity in such a way that the past path of the runaway leads back to the parent.) The observational error in the luminosity of the star relative to the zero-age main sequence does not exceed 0.15 magnitudes for the cases considered here. The zero-age main sequence is also shown together with, as an illustration, the positions of the stellar population of Upper Sco and four intermediate and high luminosity O9-B0 stars of Ori OB1. Note that, within their observational errors, the four runaways are located on the zero-age main sequence, and the three of type O9.5V in the region that is typical for blue stragglers.

4. ORIGIN

4.1 The two scenarios

Until recent years, the most generally accepted hypothesis was that of the binary-supernova scenario (Blaauw 1961), postulating the high-velocity release of the - initially - least massive component of an O or B-type binary as a consequence of the supernova explosion of the primary component. In more recent years, an alternative scenario proposed in its original form by Poveda et al. (1967), has received increasing attention: ejection of the star from a young, compact cluster as a consequence of dynamical interaction between the stars within the cluster. We shall refer to the binary-supernova scenario as BSS, and to the cluster ejection scenario as CES.

A major improvement of the BSS was the recognition, since around the year 1980, of the role of mass transfer from the initially most massive star to the (also massive) secondary, implying that at the moment of the supernova explosion of the primary the system did not necessarily become unbound. (See, for instance, scenarios described by Van den Heuvel (1983, 1985) and work in this field by others referred to in these papers). Hence, the runaways would be expected to remain binaries. This seemed to be in contradiction with the observed near-absence of spectroscopic doubles mentioned before (but note that for primary masses reduced to pulsar size the semi-amplitude of the observed initial secondary of course would become quite small). A further complication arose from the introduction of the kick velocity imparted to the nucleus of the primary at the moment of the explosion; it was estimated to be of the order of 100 km s^{-1} and justified on the basis of both the observed high space velocities of pulsars and the likelihood of lack of complete spherical symmetry of the explosion mechanism. This kick would partly or fully compensate the effect of the mass transfer in that it would contribute to the release of the remaining nucleus from the runaway secondary. An important field of studies developed, dealing with the occurrence of accreted massive secondaries with a compact companion (X-ray binaries, Wolf-Rayet stars).

The CES also underwent important improvements. Early numerical experiments using only single stars - sometimes aiming at explaining the

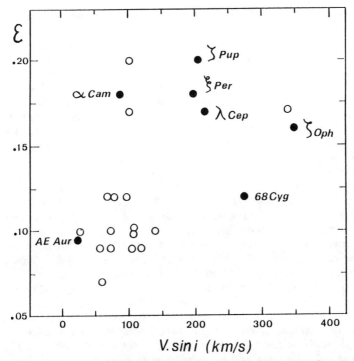

Fig. 6 Projected rotational velocities and He abundances for the runaways of Table 1 (filled circles) and the sample of nonrunaways mentioned in the text (open circles)

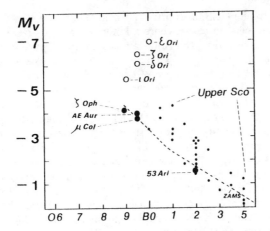

Fig. 7 Position in the HR diagram for the runaways ζ Oph, AE Aur, μ Col and 53 Ari for which accurate absolute magnitudes could be determined independent of luminosity classification (see text), showing blue straggler character. Also shown: the zero age main sequence, stars in Upper Sco and nonrunaways of Ori OB1.

origin of (wide) binaries as a result of close triple encounters - showed that also the runaway stars might thus find a (partial) explanation. Considerable improvement was introduced by the notion of encounters of binaries supposedly belonging to the original cluster population: interaction between these binaries and/or triple encounters, increasing with the gradual concentration of the binaries toward the cluster centre, appeared to be able to contribute to the explanation of the runaway phenomenon; see in particular the work of Leonard and Duncan (1988, 1990). In such cases the runaway would have been originally a member of a close double, and the result of the rather intricate interaction might even be the formation of a new close binary and the ejection of two runaways.

Quite useful reviews of the work along the lines of the two scenarios have been given in recent papers, of which we mention in particular those by Gies and Bolton (1986), Leonard and Duncan (1988) and Conlon et al (1990).

As far as the O-type runaways are concerned - the subject of the present paper - the evidence in favour of either the BSS or the CES until now did not appear entirely conclusive, but note that, for instance, the occurrence of a pulsar-binary system like PSR 1259-63 consisting of a pulsar and a - mass-accreted - component of at least 12 solar masses, with separation of at least 7 a.u. and eccentricity at least 0.95(!) (Johnston et al, 1991) virtually unavoidably leads to the conclusion that the BSS is a realistic one that may lead as well to a runaway surviving binary, as to the release of both a runaway massive star and a neutron star. The overall significance of the BSS can probably be fully evaluated only once detailed calculations have been performed for a wide grid of initial conditions (primary mass, mass ratio, separation, eccentricity) and the subsequent evolution, including full or partial mass transfer. Work now in progress along these lines reported by the Amsterdam school (van den Heuvel and collaborators) may lead to the desired clarification on this point.

With regard to the CES, also pros and contras can be raised. A difficulty remains that even in certain cases where a kinematical age of a few million years only must be assigned to the runaway, the compact parent cluster can not be identified on the past path. This question is, of course related to what happened during this short lapse of time to the parent cluster. One way to clarify this might be a more complete description than was given so far, of the velocity distribution within the cluster after the shake-up of the binary interaction. On the other hand, there is at least one interesting case pointing to the cluster ejection process: the pair of runaways AE Aur and μ Col described in Figure 3. The remarkable "twin nature" of this pair has been an enigma since its discovery (Blaauw and Morgan, 1954): the two stars move in nearly opposite directions from the central region of Ori OB1, with identical spectral type and luminosity and approximately equal speed and kinematic age. For this case, it was suggested by Gies and Bolton (1986) that the parent cluster may be identified with the multiple system ι Ori containing two O9III components. We return to this pair in section 4.3.

4.2 Figures 6 and 7: evidence for the binary-supernova scenario
Confrontation of theory with observation so far was based mainly on the kinds of data described in sections 3.1 to 3.3: incidence of duplicity, velocity distribution, and mass-velocity relation, but not yet on the O-star data

presented in Figures 6 and 7: the high rotational velocities and Helium abundance of the O-type runaways, and the "blue straggler" nature of some of them. For this - admittedly small - sample, those data would seem to fit much better the BSS than the CES, for the following reasons.

It is hard to see that the CES can discriminate between the ejected and the non-ejected stars in such a way, that the former would be given high rotational velocity and Helium abundance and the latter would not. On the other hand, in the BSS both Helium enrichment and fast rotation will have been a natural consequence of the mass transfer to the secondary component prior to the supernova event. There are, in fact, two striking features in Figure 6: a) the fact that almost all of the runaways avoid the low rotation-low Helium abundance area, and b) the fact that a small fraction of the non-runaways do share with the runaways "their" domain. The former may be taken as an indication how very strongly the process of mass transfer prior to the supernova event did determine the kinematics and the composition of the (outer layers of) the initial secondary.

Property (b) might be explained in two ways. Calculations of the BSS have shown that not in all cases does BSS lead to a high velocity of release of the initial secondary; the release velocity is virtually identical to the orbital velocity just before the explosion of the (reduced) primary, and therefore depends on the initial elements of the binary: it does not always exceed the limit beyond which we classify the star as a runaway. Moreover, a proper choice of the size and direction of the kick velocity may have reduced the runaway velocity of the initial secondary. As a second explanation, the high Helium abundance might perhaps be a natural consequence of a rapidly rotating star's individual evolution as it has been pointed out by Maeder (1987) - hence independent of a companion star's influence. However, whereas this may contribute to explaining part of the pattern of Figure 6, it seems unable to account for most of it; for in that case one would not expect the runaways with their kinematic ages of a few million years to avoid so emphatically the "non-runaway domain".

The "blue straggler" nature shown in Figure 7 would also seem to find a natural explanation in the BSS: the evolution of these stars has been retarded in comparison to stars that started out with their current mass; the early rate of evolution was that of a star of considerably lower mass.

4.3 The pair AE Aur - μ Col

At this point it is of interest to return to the pair AE Aur and μ Col (Figure 3). We note that AE Aur is the only runaway located in the "nonrunaway" domain, and this strengthens the suspicion that here, indeed, the CES may have been at work. In that case we would expect μ Col also to lie here. Its rotational velocity, 97 km s^{-1} does agree. A measurement of the Helium abundance therefore would be of great interest.

4.4 Runaway O stars and high-precision quantitative spectroscopy

High-precision quantitative spectrophotometry in combination with comprehensive modelling of stellar atmospheres recently has opened a possibility for predicting intrinsic stellar parameters like mass, radius and luminosity in a way independent of "outside information" (like the star's parallax). This progress has been possible mainly by the inclusion of stellar wind observations which provide information on the star's gravitational field over a certain range of distance from the stellar centre. Work in this field was reviewed by Kudritzki and Hummer (1990). More recent work is mentioned below. We wish to draw attention to an important bearing of the runaway phenomenon on this work.

For the comparison of theoretical prediction with observation, these papers appear to rely to a considerable extend on O-type runaway stars. For instance, in studying stars of type O9.5 Voels et al (1989) chose as representatives of a large range of luminosities the stars AE Aur, δ Ori, ζ Ori and α Cam. We note that the two extremes in this sequence are among the runaways of Table 1. For a range of luminosities of the hottest stars, Bohannan et al (1990) and Kudritzki et al (1991) use as the most luminous star the runaway ζ Pup; and recent work by Herrero et al (1991) makes use of the runaways 68 Cyg, AE Aur, ξ Per and ζ Oph. It was pointed out in the preceding paragraph, that these stars probably have reached their present status through the BSS, i.e. they have acquired considerable (processed) mass and rotational momentum from a (past) massive and more evolved companion. Since in that case their observed properties have not been reached through single star evolution, clearly great caution is required when using them as reference stars in this high precision spectrophotometric work.

5. IMPLICATIONS FOR EXTERNALLY INDUCED STAR FORMATION

If we assume the runaway stars to pass through late stages of evolution identical to those of "regular" massive stars - and there is no reason not to do so - then we must expect the majority of them to explode as supernovae far outside the parent association. If this happens inside, or in the immediate neighbourhood of, an other molecular cloud, then this may perhaps induce "external" star formation (or prevent current star formation?) in that cloud. Two aspects may be studied: a) given a molecular cloud subject to supernovae produced "internally", to what degree does this "external" agency compete with the internal one?; b) does this process play a role in stochastic propagation of star formation?

In the present context, I shall limit myself to aspect (a) only and briefly summarize a provisional estimate. The rate of supernova explosions, H, affecting a given cloud "M" and due to this external cause depends on the inducement-sensitive volume V of cloud M; on the rate of production, P per Myr, of runaways per cloud in the surrounding clouds; and on the number density, N per cubic kpc, of these clouds: $H = V.N.P$ per Myr. The product V.N might be called the effective volume filling factor of cloud M. For the solar neighbourhood, adopting $V = 0.00010$ corresponding to the effective radius of a spherical cloud of 29 pc, and $N = 10$ corresponding to a mean separation of about 500 pc (Blitz 1991) and $P = 1$, we have $H = 0.0010$, or 0.050 over a lifetime of 50 million years for cloud M. This is two to three orders of magnitude lower than the number of supernovae produced internally in a star forming (standard) cloud and hence not of competitive nature. (Taking into account the flattening of the local ISM layer H becomes even lower.) For the external agency to become competitive, N would have to be raised by two orders of magnitude at least, implying an effective volume filling factor of about one tenth and a mean cloud separation of about 100pc. The interesting question then arises whether or not such conditions do occur elsewhere, for instance in the Large Magellanic Cloud.

REFERENCES

Blaauw, A.: 1961, Bull. Astr. Inst. Netherlands **15**, 265
Blaauw, A.: 1989, Soviet Astron. **29**, 417
Blaauw, A., and Morgan, W.W.: 1954, Ap.J. **119**, 625
Blaauw, A.: 1991, in *The Physics of Star Formation and Early Stellar Evolution*, ed. C.J.Lada and N.D.Kylafis, NATO ASI Series C, Vol.342, p 125
Blaauw, A. and Sahu, M.: 1992 (in preparation)
Blitz, L.: 1991, in *The Physics of Star Formation and Early Stellar Evolution*, ed. C.J.Lada and N.D. Kylafis, NATO ASI Series C, Vol.342, p 3
Bohannan, B., Voels, S.A., Hummer, D.G., Abbott, D.C.: 1990, Ap.J. **365**, 729
Conlon, E.S., Dufton, P.L., Keenan, E.P., Leonard, P.J.T.: 1990, Astron.Astroph. **236**, 357
Conti, P.S. and Ebbets, D.: 1977, Ap.J. **213**, 438
De Vries, C.P.: 1985, Astron.Astroph. **150**, L15
Gies, D.R., and Bolton, C.T.: 1986, Ap.J.Suppl. **61**, 419
Gies, D.R.: 1987, Ap.J.Suppl. **64**, 545
Herrero, A., Kudritzki, R.P., Vilchez, J.M., Kunze, D., Butler, K., Haser, S.: 1992, Astron.Astroph. **261**, 209
Johnston, S., Manchester, R.N., Lyne, A.G., Bailes, M., Kaspi, V.M., Guojun, Q., D'Amico, N.: 1991, (in press)
Kudritzki, R.P., and Hummer, D.G.: 1990, Ann. Rev. Astron. Astroph., **28**, 303
Leonard, P.J.T. and Duncan, M.J.: 1988, Astron. J., **96**, 222
Leonard, P.J.T. and Duncan, M.J.: 1990, Astron. J. **99**, 608
Maeder, A.: 1987, Astron.Astroph. **178**, 159
Poveda, A., Ruiz, J., Allen, C.: 1967, Bol. Obs. Tonanzintla y Tacubaya 4, 860
Van den Heuvel, E.J.P.: 1983, in *Accretion-Driven Stellar X-ray Sources*, ed. W.H.G.Lewin and E.J.P. van den Heuvel (Cambridge), p 303
Van den Heuvel, E.J.P.: 1985, in *Birth and Evolution of Massive Stars and Stellar Groups*, ed. W.Boland and H. van Woerden, Astroph. and Space Sc. Library Vol. 120, p 107
Voels, S.A., Bohannan, B., Abbott, D.C., Hummer, D.G.: 1989, Ap.J. **340**, 1073

Massive Stars: Their Lives in the Interstellar Medium
ASP Conference Series, Vol. 35, 1993
Joseph P. Cassinelli and Edward B. Churchwell (eds.)

THE AGE SPREAD AND INITIAL MASS FUNCTION OF THE OPEN CLUSTER NGC 6531

DOUGLAS FORBES
Dept. of Physics, Grenfell College, Memorial University of Newfoundland

ABSTRACT   A photometric study of NGC 6531 shows neither evidence of an age spread among cluster stars, nor a relative deficiency of low-mass stars, as seen by Herbst and Miller (1982) in the similar cluster NGC 3293.

INTRODUCTION

Since the seminal work of van den Bergh and Sher (1960) there has been condsiderable interest in the notion that many open clusters appear to have relatively few low mass stars. This apparent deficiency, even in clusters too young for dynamical evolution to be significant, has suggested there is no universal IMF (Scalo 1978), and that star formation may be bimodal (Larson 1981; Miller and Scalo 1978). One study lending weight to these ideas is that of NGC 3293 (Herbst and Miller 1982). This cluster appears to show an age spread of some 10-20 Myr between the contraction age of the lower-mass stars and the nuclear age of the high-mass stars. In addition, the IMF derived for NGC 3293 is not as steep as that for field stars, indicating a marked deficiency of low-mass stars.

In an effort to see if other clusters of similar age show an age spread and IMF similar to those in NGC 3293, a photometric UBV study of several open clusters has been initiated. Presented here are the results for one cluster, NGC 6531.

AGE SPREAD

Figure 1b, the extinction-free colour-magnitude diagram, shows several features worthy of note. There is a very clear gap at $0.00 \leq (B-V)_o \leq 0.20$, with the stars blueward of this lying on the ZAMS, and stars redward of the gap lying well above the ZAMS. Based on a comparison with Figure 1a the gap does not seem to be the result of the selection of possible cluster

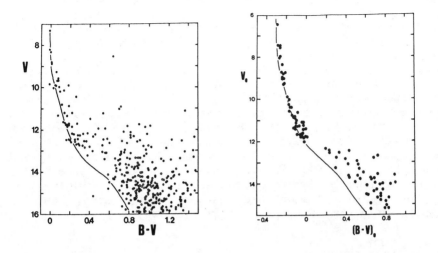

Fig. 1.   Observed (a) and extinction-free (b) colour-magnitude diagrams for NGC 6531.

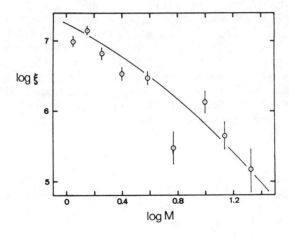

Fig. 2.   Initial mass function (IMF) for NGC 6531.
Masses are in solar units and the logarithm of the mass
function ξ (log M) is plotted.   Solid line is the field star
IMF (Miller and Scalo 1979) normalized to match the high-mass
end of the cluster IMF.   Error bars reflect ± 1σ Poisson
statistics.   Note the absence of any deficiency of low-mass
stars, when compared to the IMF for NGC 3293 (Herbst and Miller
1982; their Fig. 6).

members.   Such gaps are seen in colour-magnitude diagrams of
other clusters, and are thought (Ulrich 1971) to result when
gravitational contraction toward the ZAMS is temporarily halted
by the "burnoff" of ^{3}He.   The gap provides a very clear
indication of the turn-on point at Mv = +1.35 ± 0.2 (est.
error). This corresponds to a mass of 2.2 ± 0.2 $M_\odot$ and yields
an estimate of the cluster contraction age of (8 ± 5) Myr (Iben
1965; Ezer and Cameron 1967).   The nuclear age of the cluster
has been derived from a comparison of the bright end of the
cluster main sequence with the evolutionary models of Maeder
and Meynet (1988), giving an age estimate of (8 ± 2) Myr.
**There appears to be no evidence of any significant age spread
in NGC 6531.** For NGC 6531 to show an age spread of some 20 Myr
years (as seen by Herbst and Miller in NGC 3293) would require
a main sequence turnon point at Mv = +3, more than 1.5
magnitudes fainter than is actually observed.

INITIAL MASS FUNCTION

Figure 2 shows the inital mass function for NGC 6531, derived
from the cluster luminosity function (based on star counts) and
the mass-luminosity relation given by Miller and Scalo (1979).
The common logarithm of the mass function ξ (log M), as
defined by Scalo (1978), is plotted against log M.   Strictly
speaking, Figure 5 is the *observed* mass function of NGC 6531,
but given the youth of the cluster this should be a good
approximation to the *initial* mass function.   The solid line in
the figure is the field star IMF derived by Miller and Scalo
(1979), normalized to match the high-mass end of the cluster
IMF.   **There does not appear to be any significant difference
between the IMF of NGC 6531 and that of field stars, and hence
no deficiency of low-mass stars.**

REFERENCES

Ezer, D. and Cameron, A.G.W. 1967, Can. J. Phys. 45, 3429
Herbst, W. and Miller, D.P. 1982, AJ 87, 1478
Iben, I. 1965, ApJ 141, 513
Larson, R.B. 1982, MNRAS 200, 159
Maeder, A. and Meynet, G. A&AS, 76, 411
Miller, G.E. and Scalo, J.M. 1978, PASP 90, 506
Miller, G.E. and Scalo, J.M. 1979, ApJS 41, 513
Scalo, J.M. 1978, in *Protostars and Planets*, IAU Coll. 52, ed.
    T. Gehrels (Univ. Arizona), p. 265
Ulrich, R. 1971, ApJ 168, 57
van den Bergh, S. and Sher, D. 1960, Pub. David Dunlap Obs. 2,
    203

Massive Stars: Their Lives in the Interstellar Medium
ASP Conference Series, Vol. 35, 1993
Joseph P. Cassinelli and Edward B. Churchwell (eds.)

A STUDY OF ORBITAL CIRCULARIZATION IN A-TYPE BINARIES

LYNN D. MATTHEWS AND ROBERT D. MATHIEU
Astronomy Department, University of Wisconsin, Madison, WI 53706

We have made a study of the orbital period-eccentricity distribution of a sample of radiative-envelope binaries with A-type primary stars selected from the eighth Batten catalogue of spectroscopic orbits (1989). The goal was to assess empirically whether tidal circularization is less efficient in binaries with radiative envelopes than in systems in which at least one component is convective, as argued by some theories (e.g., Zahn 1977).

Figure 1 is a plot of the period-eccentricity relationship for our sample of A-type binaries. We found that all binaries with $P \leq 3^d$ have circular or nearly-circular orbits. Between 3 and 10^d, there is a mixed distribution of circular and eccentric orbits, the maximum eccentricity increasing with increasing period, while beyond $P = 10^d$, all orbits are eccentric. We have taken 10^d to be the cutoff period (P_{cut}) for the sample. Such a cutoff period is significantly longer than predicted by theories such as Zahn (1977) and is comparable to cutoff periods found among samples of older, solar-mass, main-sequence binaries with convective components, namely those in M67 and in the field (e.g., Mathieu *et al.* 1992). Because the longest-period circularized orbits in our sample are double-lined, the secondary components are known to be early-type, main-sequence stars. Hence their circularization cannot be attributed to post-main-sequence evolution or to the presence of convective-envelope secondaries.

We suggest two possible explanations for the similarities between the circularization cutoff periods of radiative- and convective-envelope binaries. First, tidal circularization occurs during the main sequence at a rate which is not strongly dependent upon the stellar envelope. This is consistent with the circularization theory of Tassoul (1988). Second, tidal circularization is accomplished prior to the main sequence, when the stars have larger radii and possibly convective envelopes. This possibility was suggested by Zahn and Bouchet (1989) to account for circularization in convective-envelope binaries, and a similar argument may plausibly be made for binaries possessing radiative envelopes. The above two scenarios are not necessarily exclusive.

A more detailed discussion of the results which preceed can be found in Matthews and Mathieu (1992). Since the time of that analysis, improved orbital elements have become available for nine additional A-type binaries (Stockton and Fekel 1992; Margoni *et al.* 1992). Among these are HD 159560 at a period of 38.0^d and an eccentricity of 0.03 and HD 27628 with $P = 2.1^d$ and $e = 0.13$ (Margoni *et al.* 1992). These two cases are intriguing in light of the issues

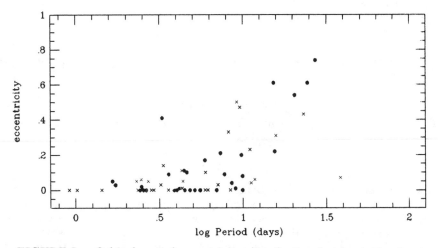

FIGURE I   Orbital period-eccentricity distribution for a sample of spectroscopic binaries with A-type primaries. Single-lined binaries are shown as crosses; double-lined binaries as filled circles.

discussed above. However, because both systems are single-lined, the nature of their secondaries is unknown; thus circularization by an evolved secondary component in the past cannot be ruled out. In addition, since HD 27628 is a memeber of the Hyades cluster, dynamical influences may be responsible for its eccentricity. The other recently-observed A-type binaries have periods and eccentricities consistent with the trends observed in the Batten sample.

REFERENCES

Batten, A.H., Fletcher, J.M., and MacCarthy, D.G. 1989, *Eighth Catalogue of the Orbital Elements of Spectroscopic Binaries* XVII, (Victoria: Dominion Astrophysical Observatory).

Margoni, R., Munari, U., and Stagni, R. 1992, *A&AS*, **93**, 545.

Mathieu, R.D., Duquennoy, A., Latham, D.W., Mayor, M., Mazeh, T., and Mermilliod, J.-C. 1992, in *Binaries as Tracers of Stellar Formation*, ed. A. Duquennoy and M. Mayor, (Cambridge Press).

Matthews, L.D. and Mathieu, R.D. 1992, in *Complementary Approaches to Double and Multiple Star Reasearch, IAU Colloquium 135*, ed. W. Hartkopf and H. McAlister, (ASP: San Francisco), in press.

Stockton, R.A. and Fekel, F.C. 1992, *MNRAS*, **256**, 575.

Tassoul, J.-L. 1988, *ApJ*, **360**, L47.

Zahn, J.-P. 1977, *A&A*, **57**, 383.

Zahn, J.-P. and Bouchet, L. 1989, *A&A*, **223**, 112.

Massive Stars: Their Lives in the Interstellar Medium
ASP Conference Series, Vol. 35, 1993
Joseph P. Cassinelli and Edward B. Churchwell (eds.)

THE EFFECTS OF OVERSHOOT, MASS LOSS, AND ROTATION ON THE EVOLUTION OF A 15 $M_\odot$ STAR

JAMIE M. HOWARD
Department of Astronomy, Yale University, New Haven, CT 06511

ABSTRACT   Non–rotating models have been run for a 15 $M_\odot$ star with no mass loss or convective overshoot, varying amounts of overshoot, and mass loss at a rate given by an empirical parameterization. The results of these calculations agree well with similar models in the literature. Rotating models have been run for the case of no overshoot or mass loss. Preliminary results suggest that the inclusion of rotation may help explain some of the discrepancies between theory and observation.

THE YALE ROTATING EVOLUTION CODE

The models in this study have been calculated using the Yale Rotating Evolution Code (YREC),a Henyey code which was originally designed by Prather (1976) and later modified by numerous researchers at Yale (see Seidel *et. al*, 1987, and Pinsonneault *et. al*, 1989.) Recent modifications to YREC include up-to-date nuclear reaction rates supplied by Bahcall, surface boundary condtions from the most recent Kurucz atmosphere models (Kurucz, 1992), OPAL opacities (Inglesias & Rogers, 1991; Rogers & Inglesias, 1992), and a weighted mean reaction rate scheme for energy generation in convective cores.

CONVECTIVE OVERSHOOT

The extent of convective overshoot is treated as a free parameter in YREC and specified in the form of a fraction of the pressure scale height H_p at the edge of the convective zone. Separate overshoot parameters can be specified for convective cores, intermediate zones, and envelopes. The overshoot region is mixed together with the associated convective zone, and the temperature gradient is assumed to remain non- adiabatic.

The most convincing results from a number of researchers indicate that only a small amount of convective overshoot is necessary to produce reasonable agreement with observations. For the sake of comparison, however, I have generated models with small (0.1 H_p), moderate (0.3 H_p), and large (0.7 H_p) amounts of convective overshoot at the edges of all convective zones. The results can be seen in figure 1a. Convective overshoot widens the main sequence and extends the core H–burning lifetime by 6%, 17%, and 34%. Even small amounts of overshoot suppress the formation of intermediate convective zones. This suggests that convective overshoot may inhibit or eliminate semiconvection.

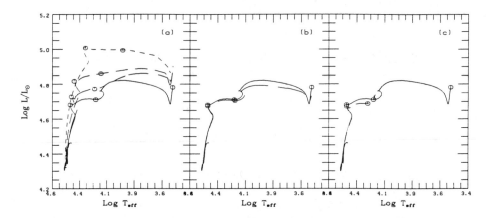

Fig 1.   15 M$_\odot$ evolutionary tracks: solid track has no overshoot,
masssloss, or rotation. (a) overshoot 0.1 (long dash), 0.3 (medium dash),
0.7 (short dash); (b) massloss (long dash); (c) rotation (medium dash).

MASS LOSS

The mass loss rate for YREC is specified in a parameterized form and can
depend on luminosity, radius, mass, surface gravity, and Reimer's η, as well as
a constant term. Mass is removed from the envelope and outermost shells of
the model at the specified rate for each timestep. Mass was removed from the
15 M$_\odot$ model using a parameterization determined by Waldron (1984) for stars
all across the HR diagram.

As can be seen from figure 1b, mass loss has minimal effects during the
main sequence phase, resulting in almost no increase in the main sequence
lifetime. It is in the red supergiant phase that the mass loss rate becomes truly
significant and the effects of mass loss become important. Extension of the
mass-losing track to core He exhaustion is necessary to determine just how
important. In the current models the star has lost roughly 10% of its original
mass by the time it reaches the Hayashi track.

ROTATION

The treatment of rotation incorporated in YREC is conceptually the same
as that used by Endal & Sofia (1976,1978) but the technical details are
different,as YREC was rewritten using Prather's (1976) code as a starting
point. YREC differs from all non-rotating codes in that it accounts for
both the non−spherical geometry and transport of angular momentum and
associated chemical mixing that are caused by rotation (see Endal & Sofia 1978
and Pinsonneault *et.al* 1989 for details.)

Running YREC with the rotational instabilities turned off displays the
effects that the non-spherical geometry has on the evolution of the model. As
can be seen in figure 1c, the effects on main sequence evolution are minimal
for a 15 M$_\odot$ model. The post−main sequence effects are more significant.

The lower luminosity of the blue supergiant branch and higher effective temperature of core He–ignition may help resolve the blue–to–red supergiant ratio discrepancies. Sizeable amounts of differential rotation may develop in the post–main sequence stages; this also points toward rotation producing significant changes in the later evolution of massive stars.

It is important to note, however, that much more work needs to be done on the rotating models. The rotating track shown here needs to be extended much further before any definite conclusions can be reached regarding the structural effects of rotation on this model. Also, the inclusion of rotational instabilities is very important, if the models are to be physically realistic. The results shown here are merely promising, rather than final.

PRELIMINARY CONCLUSIONS

The models to date suggest that even small amounts of convective overshoot may eliminate semiconvection. Convective overshoot seems necessary for explaining main sequence broadening; in the current runs, mass loss and rotation have little effect on main sequence evolution. The effects of mass loss and rotation are probably most important in the later stages of evolution, when the star has lost significant amounts of mass and developed sizeable amounts of differential rotation. Finally, the inclusion of rotation may help resolve the discrepancy between observed and predicted blue–to–red supergiant ratios

ACKNOWLEDGEMENTS

The author is deeply grateful to the following individuals: Pierre Demarque, for much advice and guidance, Mark Pinsonneault and David Guenther, for numerous modifications to YREC, and Bob Kurucz, for providing his most recent model atmospheres. This work has been supported in full by NASA grant NAGS–1486.

REFERENCES

Endal, A.S., Sofia, S. 1976 *Ap. J.* **210**:184
Endal, A.S., Sofia, S. 1978 *Ap. J.* **220**:279
Inglesias, C.A., Rogers, F.J. *Ap. J.* **37**:408
Kurucz, R.L. 1992 In *The Stellar Populations of Galaxies* ed. by B. Barbuy & A. Renzini, Kluwer, Dordrecht. In press.
Pinsonneault, M.H., Kawaler, S., Sofia, S., Demarque, P. 1989 *Ap. J.* **338**:424
Prather, M. 1976 Ph.D. Thesis, Yale University
Rogers, J.F., Inglesias, C.A. 1992 *Ap. J. Suppl.* **79**:507
Seidel, E., Demarque, P., Weinberg, D. 1987 *Ap. J. Suppl.* **63**:917
Waldron, W.L. 1984 In *The Origin of Non-Radiative Heating/Momentum in Hot Stars* ed. by A.B. Underhill, A.G. Michalitsianos, p. 2358 Washington:NASA

Massive Stars: Their Lives in the Interstellar Medium
ASP Conference Series, Vol. 35, 1993
Joseph P. Cassinelli and Edward B. Churchwell (eds.)

CNO ABUNDANCES IN THREE A-SUPERGIANTS

KIM VENN
U. Texas at Austin, Dept. of Astronomy, Austin, TX, 78712

ABSTRACT   Abundance results are presented for three intermediate mass ($\sim 10 M_\odot$), Pop.I A-type supergiants. We find these stars have roughly solar metal abundances (*i.e.*, Fe, Mg, Ti, Cr), with large enrichments of nitrogen (~ 0.8 dex). Slight carbon and possibly oxygen depletions are also present. We conclude the three A-supergiants presented here are most likely post-RGB stars that have experienced first dredge-up which has altered their surface CNO abundances. For the two A0Ib supergiants, the presence of *both* CN- and ON-cycled products is indicated.

INTRODUCTION

Stellar evolution calculations predict that the 8 to 12 $M_\odot$ mass, galactic A-supergiants are in a phase of helium-core burning, most likely following a red supergiant (RSG) phase (*c.f.*, Maeder and Meynet 1988). If the latter is true, then the presence of a deep surface convection zone that arises during the RSG phase will induce the first dredge-up of CNO-processed material, thus affecting the surface CNO abundances in a discernible way. The CNO abundances in normal, galactic A-supergiants have not previously been examined for their group characteristics. The three stars presented here are HD87737 ($= \eta$ Leo, A0Ib), HD46300 ($= 13$ Mon, A0Ib), HD13476 (A3Iab, an outlying member of Per OB1). These stars are MK standards and are not known binary systems.

ANALYSIS

Spectroscopic indicators have been used to determine the atmospheric parameters (T_{eff}, gravity), including the Hγ line profiles and ionization equilibrium using lines of MgI and MgII. Ionization equilibrium for FeI and FeII has also been calculated, but not used rigorously due to the uncertain, but estimated as large, non-LTE effects for the FeI atoms; only small non-LTE effects are predicted for MgI, MgII, and FeII. Chemical abundances and theoretical Hγ profiles are determined from model atmospheres analyses using Kurucz models generated from ATLAS6, and the WIDTH6 and BALMER codes. Thus, LTE is assumed throughout. Microturbulence (ξ) has been determined by forcing the FeII and TiII abundances to be independent of equivalent width.

The atmospheric parameters derived here are shown in Table I. The uncertainty in T_{eff} from ionization equilibrium of Mg varies from about ±150K (HD13476) to 250K (others), and the uncertainty in $\log g$ from an $H\gamma$ line profile fit at a given temperature is ±0.1 dex. Abundance results are also presented in Table I as unweighted averages per species ($\log\epsilon(x)$) and relative to solar ($[\epsilon] = \log\epsilon(x)_{star} - \log\epsilon(x)_\odot$), including the number of lines averaged. Solar abundances are from Anders and Grevesse (1989) and Grevesse et $al.$ (1990,1991). Weak lines only have been used to compute the averages (typically 10 mÅ$<$ $W_\lambda< 200$ mÅ). Estimated abundance uncertainties per species are ≤0.2 dex due to uncertainties in the atmospheric parameters (T_{eff}, $\log g$), the microturbulence, the measured W_λ's, and the line-to-line scatter. The final abundances for [Mg] are an unweighted average of MgI and MgII, yet the final abundances for [Fe] are considered as only the FeII abundances.

TABLE I   Elemental Abundances

elem	sun $\log\epsilon$	HD 13476 $\log\epsilon$	$[\,\epsilon\,]$	$(\#)$	HD 87737 $\log\epsilon$	$[\,\epsilon\,]$	$(\#)$	HD 46300 $\log\epsilon$	$[\,\epsilon\,]$	$(\#)$
C I	8.60	8.51	-0.09	(7)	8.33	-0.27	(5)	8.19	-0.41	(4)
N I	8.00	8.77	$+0.77$	(7)	8.95	$+0.95$	(2)	8.75	$+0.75$	(5)
O I	8.93	8.73	-0.20	(4)	8.95	$+0.02$	(4)	8.90	-0.03	(4)
Mg I	7.58	7.40	-0.18	(4)	7.50	-0.08	(3)	7.40	-0.18	(3)
Mg II	7.58	7.56	-0.02	(4)	7.56	-0.02	(1)	7.48	-0.10	(1)
Ti II	4.93	4.90	-0.03	(10)	4.72	-0.21	(12)	4.63	-0.30	(11)
Cr II	5.68	5.57	-0.11	(4)	5.62	-0.06	(1)	5.50	-0.18	(1)
Fe I	7.51	7.72	$+0.21$	(6)	7.59	$+0.08$	(2)	7.51	0.00	(2)
Fe II	7.51	7.51	0.00	(15)	7.64	$+0.13$	(12)	7.48	-0.03	(14)
[(C+N+O)/FeII]		$+0.03$			$+0.04$			$+0.08$		
T_{eff} , $\log g$, ξ		8600K,1.3,8km/s			9700K,2.0,4km/s			9900K,2.2,4km/s		

DISCUSSION

The first dredge-up calculations by Maeder and Meynet (1988) and Becker and Iben (1979) predict that significant CN-cycled material is mixed to the stellar surface. For the A0Ib stars the observed [N/C] and [N/O] ratios are larger than predicted, presumably due to the extreme overabundance of N. If only CN-cycling has occured, the N abundances are too large (converting all C to N from an initially solar composition results in a maximum $\log\epsilon(N) = 8.7$ dex). The sum of the nuclei (C+N+O)/Fe is roughly solar however, thus a significant amount of mixing of ON-cycled products is indicated. If CN- and ON-cycled material is present than N should scale with the sum C+O in these two stars, as seen in Fig. 1.

For the A3Iab star, it is odd that C is larger than O by 0.1 dex. The CN-cycle is the more efficient process reaching near equilibrium

before significant ON-cycling begins, thus C atoms should be more
depleted than O atoms in simple CNO-processed gas. The oddness of
the C and O abundances may be due to uncertainties in this analysis,
although the low O abundance is reminiscent of the low O abundances
in the F-K supergiants determined by Luck and Lambert (1985)
which was attributed to a presumably low primoridal O abundance.
Also, non-LTE effects cannot be ruled out as a source for non-solar
abundances in these stars.

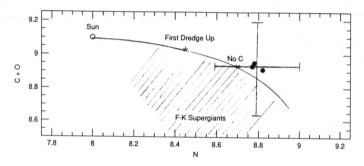

Fig. 1.   Plot of N vs (C+O) for the three A-supergiants
studied here (filled circles) with the estimated uncertainty
indicated. The solid line tracks abundances through CNO-
processing; along this line are marked the solar abundances,
C exhaustion, and first dredge-up predictions. The region of F-K
supergiant abundances (from Luck and Lambert 1985) is marked.

CONCLUSIONS

We conclude that these three, intermediate mass ($\sim$ 10 M$_\odot$) A-
supergiants are most likely post-RGB stars that have experienced first
dredge-up which has altered their surface abundances of CNO. These
stars appear to have significant CN- *and* ON-cycled products mixed in
their surface abundances which is unlike the predictions. The analysis
of a larger sample of stars is underway to confirm the results presented
here. Knowing the abundances for normal A-supergiants, then we may
more reliably study other populations of these stars such as the Pop.II,
the high galactic latitude, and the alleged Magellanic Clouds anomalous
A-supergiants.

REFERENCES

Anders E., Grevesse N. 1989, *Geochim. Cosmochim. Acta*, **53**, 197.
Becker S.A., Iben I. 1979, *Ap. J.*, **232** , 831 .
Grevesse N., *et al.* 1991, *Astr. Ap.* , **242** , 488 .
Grevesse N., *et al.* 1990, *Astr. Ap.* , **232** , 225 .
Luck R.E., Lambert D.L. 1985, *Ap. J.*, **298** , 782 .
Maeder A., Meynet G. 1988, *Astr. Ap. Suppl. Series*, **76** , 411 .

Massive Stars: Their Lives in the Interstellar Medium
ASP Conference Series, Vol. 35, 1993
Joseph P. Cassinelli and Edward B. Churchwell (eds.)

MASS LOSS FROM LATE TYPE SUPERGIANTS

MICHAEL WOODHAMS
Princeton University Observatory, Peyton Hall, Princeton NJ 08544-1001

ABSTRACT   Spectra are presented of four supergiants in the CO (J=3-2) line at 345 GHz. Mass loss rates are determined for the two that are members of OB associations, and comparison is made with mass loss rates determined by IR fluxes and OH maser shell sizes.

INTRODUCTION

A supergiant can lose over half its mass in stellar winds during its lifetime (Maeder & Meynet 1988.) The process of mass loss remains one of the greatest unknowns in the modeling of the evolution of massive stars. The lower initial mass limits for creating a supernova or a Wolf Rayet star and the return of matter processed by nuclear burning to the interstellar medium are among the more obvious matters relying on understanding mass loss.

I am conducting a CO $(J=3-2)$ (345 GHz) survey of sources from the IRAS point source catalog at the Caltech Submillimeter Observatory at Mauna Kea. Figure 1 shows spectra for four supergiants included in the sample.

From these spectra, the velocity of the stellar wind V_o and the radial velocity of the stars V_c, measured with respect to the local standard of rest, can easily be determined. The results are in Table I.

The spectrum of PZ Cas is particularly interesting. There are three lumps, with strength decreasing towards the red, plus a red wing. This clearly indicates non-spherical winds, and presumably asymmetric stellar envelope. This can have interesting implications when the star eventually goes supernova.

MASS LOSS RATES

PZ Cas is in Cas OB5 and S Per is in Per OB1, so accurate distances to these stars can be determined (Humphreys 1978, Garmany & Stencel 1992.) This information, along with an assumed CO mass fraction in the wind, is sufficient to make a radiative transfer model of the outflow. The mass loss rate is varied in the model until the calculated line profile matches that observed, giving a measure of the mass loss rate. This has been done for PZ Cas and S Per, and the resulting profile is shown by the dashed lines in Figure 1. Table II compares the mass loss rate derived by this method with that from IRAS fluxes and OH maser shell radii.

Due to the lack of a reliable distance, AFGL 2968 and T Sge are instead fitted with flattened parabolas. This is the theoretical result for unresolved spherical constant velocity outflow, with the degree of flattening depending on optical thickness. The best fit parabolas for PZ Cas and S Per are almost indistinguishable from the radiative transfer model profile at the scale of the figures.

For late type stars, the wind is probably driven by radiation pressure on dust. The dust in turn collides with the gas and drags it along. It is interesting

TABLE I   Supergiants detected in CO (J=3-2)

Star	Date observed	Integration Time (s)	T_{rms} (mK)	T_{max} (mK)	V_o km s^{-1}	V_c km s^{-1}
PZ Cas	Nov 91, June 92	2019	56	360	30	-38
S Per	Nov 91	777	55	190	20	-39
AFGL 2968	Nov 91	1553	45	240	29	-51
T Sge	June 92	466	132	370	18	26

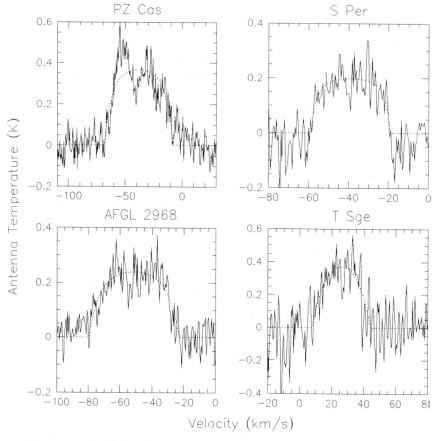

Fig. 1.   CO (J=3-2) line profiles of supergiants

TABLE II   Determinations of Mass Loss Rates

Star	OB Assn	Distance[a] (kpc)	Mass loss rate ($M_\odot\,\mathrm{yr}^{-1}$) This paper	OH maser shell[b]	IR excess[c]
PZ Cas	Cas OB5	2.0	4.7×10^{-5}	1.1×10^{-5}	6.4×10^{-6}
S Per	Per OB1	2.3	1.1×10^{-5}		7×10^{-6}

[a]Garmany & Stencel, 1992.
[b]Netzer & Knapp, 1987.
[c]Jura & Kleinmann, 1990, after correcting for different assumed distances.

to determine the amount of radiation absorbed by the dust by comparing the ratio of momentum in the wind with the momentum released in the form of stellar radiation, i.e. $\dot{M}v/(L/c)$. This ratio is 0.4 for PZ Cas and 0.1 for S Per using my mass loss rates, Humphreys (1978) for magnitudes, extinction and bolometric correction and the Garmany & Stencel (1992) distances.

Absorption of so much of PZ Cas's light should lead to half a magnitude of extinction, and indeed A_V for PZ Cas is 0.26 greater than the largest value for an early giant in the same association and 1 magnitude greater than the mean A_V for Cas OB5.

CONCLUSIONS

The evolution of massive stars cannot be understood without understanding the mass loss process. Submillimeter observations are a very valuable tool in coming to this understanding. Although current models assume spherically symmetric mass loss, there are examples where this is clearly not the case.

ACKNOWLEDGEMENTS

This work has been undertaken as part of a PhD thesis supervised by Jill Knapp. I thank Raoul Taco Machilvich for instruction on using the Caltech Submillimeter Observatory, and Antony Schinkel, Tom Philips, Larry Strom and the rest of the staff of the CSO for such an excellent telescope. Ken Young has given valuable advice, and also introduced me to the SIMBAD stellar database, which has been invaluable. This research was partly supported by NSF grant AST90-14689.

REFERENCES

Garmany, C. D. and Stencel, R. E. 1992, to appear in A&AS.
Humphreys, R. M. 1978, ApJS, 38, 309.
Jura, M. and Kleinmann, S. G. 1990, ApJS, 73, 769.
Maeder, A. and Meynet, G. 1988, A&AS, 76, 411.
Netzer, N. and Knapp, G. R. 1987, ApJ, 323, 734.

Massive Stars: Their Lives in the Interstellar Medium
ASP Conference Series, Vol. 35, 1993
Joseph P. Cassinelli and Edward B. Churchwell (eds.)

SPECTRAL INDICES OF WOLF-RAYET STARS AND THE INFRARED EXCESS

PATRICK W. MORRIS, KENNETH BROWNSBERGER,
PETER CONTI, and WILLIAM VACCA
Joint Institute for Laboratory Astrophysics, Campus Box 0440,
University of Colorado, Boulder, CO 80309-0440

PHIL MASSEY
KPNO, NOAO, P.O. Box 26732, Tucson, AZ 85726-6732

ABSTRACT   We find that the continuous energy distributions of Wolf-Rayet stars are well-represented by a power law $F_\lambda \sim \lambda^{-\alpha}$ over the wavelength region ~ 0.1 - 1.0 μm. The average spectral index value we obtain does not agree with a Rayleigh-Jeans representation of the optical-NIR continuum, negating the expectation for an inflection point near $1\,\mu$ due to free-free emission.

THE INFRARED EXCESS

The visible continuum of Wolf-Rayet (W-R) stars has often been approximated by a Rayleigh-Jeans (R-J) distrubution: $F_\lambda \sim \lambda^{-4}$. Early IR observations showed that fluxes at $\lambda \gtrsim 1$ μm were in excess of the R-J distribution, and could be attributed primarily to free-free encounters near He^+ and He^{++} ions in a circumstellar shell (Hackwell et al. 1974, Cohen et al. 1975, Hartmann & Cassinelli 1977). Postulating that such opacity dominates at longer wavelengths, Cassinelli & Hartmann (1977) qualitatively showed that an (approximately) inverse-square density distribution can produce the IR continuum shape observed in many W-R stars. This result is obtained from a power law flux distribution $F_\lambda \sim \lambda^{-\alpha}$, which rests on the validity of the approximation of the unerlying NIR continuum as an R-J distribution.

Using IUE satellite and ground-based optical and NIR spectra, we have examined the power law $F_\lambda \sim \lambda^{-\alpha}$ characteristics of 75 single W-R stars in the Galaxy and LMC over the wavelength range ~ 0.1 - 1.0 μm. This allows us to check the consistency of the shapes observed over this region with picture drawn in the IR excess models. The UV-NIR power law is justified from (1) an insensitivity of the representation to discrete portions of the total fitting range, and (2) predictions of power law behavior by the models of Schmutz et al. (1992). We also find that the color excesses obtained by nullifying the 2200-Å interstellar absorption "bump" are in good agreement with other published values.

THE SHAPE OF THE VISIBLE CONTINUUM AND BEYOND

The frequency distribution of slope values we obtain shows that the continua of these stars are much shallower than is demanded by the R-J function (Figure 1). A striking Gaussian-like distribution of values, centered around $\langle \alpha \rangle \simeq 2.85$, is understood chiefly in terms of a heterogeneity of mass-loss rates among these stars. There is some sensitivity of the actual values of $\dot{M}$ as a function of α to the form of the wind velocity law (cf. Hillier 1987, Hammann & Schmutz 1987). The WN and WC stars are matched to within 0.1 (or 1/2 the average uncertainty on α) of the total average.

To further examine how this behavior affects expectations for an inflection point near 1 μm (where free-free opacity is expected to be the dominant opacity source), we plot the extended distribution of HD 192163 (WR136, WN6) in Figure 2. Since this star is the central source of the bubble nebula NGC 6888, we do not plot IRAS 60 and 100 μm fluxes.

We see in Figure 2 that, apart from flux calibration uncertainties at the 10% level, no inflection points or departures from an $\alpha = 3.0$ power law are obvious. Similar behavior is observed for other W-R stars *without* detectable amounts of circumstellar dust.

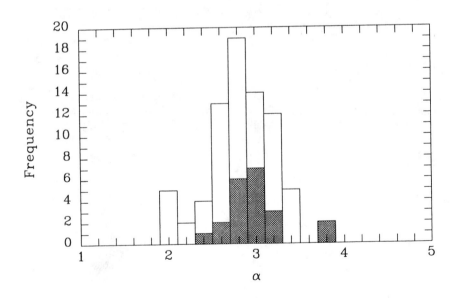

Fig. 1.   Frequency distribution of spectral index values for single Galactic and LMC WN and WC stars. Cross-hatching indicates WC stars. $\langle \alpha \rangle_{tot} = 2.85$ ($\sigma = 0.40$).

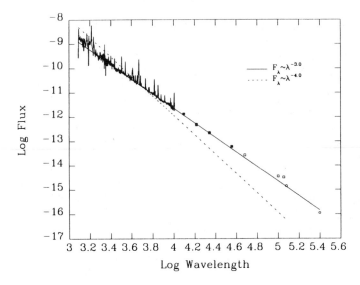

Fig. 2.   Extended distribution of HD 192163 (WR136, WN6). Open circles represent *JHKL* photometry of Mathis et al. (1992), filled circles are IRAS 12 and 25 μm fluxes. Squares are from IR photometry of Cohen, Barlow, & Kuhi (1975). All IR fluxes were dereddened using color excess determined from UV-NIR analysis, and normalized to $\alpha = 3.0$ power law at $\lambda 1.65$ μm.

ACKNOWLEDGEMENTS

We appreciate the comments of Drs. H. Lamers and P. Williams. We are also grateful to Dr. J.-M. Vreux for several NIR spectra, and (especially) to Dr. W. Schmutz for use of the model distributions and for helpful comments. Our work is being supported by NASA grant NAG5-1016 and NSF grant AST 9015240.

REFERENCES

Cassinelli, J.P., & Hartmann, L. 1977, Ap. J., 212, 488.
Cohen, M., Barlow, M.J., & Kuhi, L.V. 1975, Astr. Ap., 40, 291.
Hammann, W.-R., & Schmutz, W. 1987, Astr. Ap., 174, 173.
Hillier, J.D. 1987, Ap. J., 63, 947.
Hackwell, J.A., Gehrz, R.D., & Smith, J.R. 1974, Ap. J., 192, 383.
Hartmann, L., & Cassinelli, J.P. 1977, Ap. J., 215, 155.
Mathis, J.S., Cassinelli, J.P., van der Hucht, K.A., Prusti, T., Wesselius, P.R., & Williams, P.M. 1992, Ap. J., 384, 197.
Schmutz, W., Vogel, M., Hamann, W.-R., & Wessolowski, U. 1992, Astr. Ap., in preparation.

Massive Stars: Their Lives in the Interstellar Medium
ASP Conference Series, Vol. 35, 1993
Joseph P. Cassinelli and Edward B. Churchwell (eds.)

SPECTROPOLARIMETRY OF THE LUMINOUS BLUE VARIABLES AG CARINAE AND R127

REGINA E. SCHULTE-LADBECK
U. Pittsburgh, Department of Physics and Astronomy, 100 Allen Hall, Pittsburgh, PA 15260, USA

GEOFFREY C. CLAYTON
U. Colorado, CASA, Campus Box 389, Boulder, CO 80309, USA

MARILYN R. MEADE
U. Wisconsin, SAL, 1150 University Avenue, Madison, WI 53706, USA

ABSTRACT Many Luminous Blue Variables (LBV) are surrounded by extended nebulosity. It is thought that the nebulae originate from giant eruptions of the LBVs, caused by an unknown instability mechanism. Several of the resolved shells clearly display axisymmetric geometry. Is this geometry caused by the interaction of the ejecta with surrounding inhomogeneous interstellar medium, or is it a tracer of asymmetries in the LBV itself? We investigate this question with spectropolarimetry, a technique which is very sensitive in the detection of aspherical distributions. We review filter polarimetry available in the literature and discuss new spectropolarimetric data of the Galactic LBV AG Car and of R127 of the LMC. Both objects are found to be intrinsically polarized. In AG Car, the position angle of the intrinsic polarization is aligned with the major axis of the ring nebula and perpendicular to the "jet". In R127, the polarization is perpendicular to two bright features of the surrounding nebula. Our observations provide strong evidence that the asymmetries seen in the surroundings of these LBVs originate in the underlying stars.

INTRODUCTION

The radiation of LBVs is expected to be linearly polarized due to electron-scattering in their massive and highly ionized stellar winds. Electron scattering is most efficient within about $2R_*$ (Cassinelli, Nordsieck & Murison 1987, ApJ, 317, 290) while optical emission-line radiation originates from much larger volumes - hence the stellar continuum may be polarized more strongly than various lines. If a difference between the polarization of the continuum and a line is observed, we may infer the presence of intrinsic polarization and consequently of an asymmetry in the distribution of electrons very near the star.

237

From the vector difference between the polarization of the continuum and that in a line, the continuum-to-line vector, we can deduce the position angle (PA) of the electron distribution without having to know the interstellar foreground polarization. Spectropolarimetry can thus provide insight into the geometry of LBV winds at stellar distances which cannot be resolved by direct imaging or coronography.

ABOUT THE OBSERVATIONS

The linear polarization of AG Car and R127 was measured on November 27, 1991 with the half-wave, CCD polarimeter of the 3.9-m Anglo-Australian Telescope (AAT). A double-dekker with an aperature size of 2.7" was used to record simultaneously the star and sky spectra. Great care was taken in deriving a two-dimensional wavelength calibration. (This was of particular importance for the AG Car Hα-line results, in order to make sure that polarization changes across the Hα line were not introduced by slight misalignments of the line profiles in spectra taken at different wave-plate positions.) The observations cover a wavelength region from 4120Å to 6870Å; the sampling of the data was 2.7 Å/pixel. The images were corrected for bias and pixel-to-pixel variations. For each object, two sets of alternating star-sky spectra at 4 wave-plate positions were then combined to give the spectra of the three observed Stokes parameters. The "counts" spectra were not processed any further. The instrumental polarization was determined from an observation of the nearby bright star BS 1294 and the instrumental efficiency and zero-point of the position angle were measured by inserting an HN22 polaroid into the beam. The polarimetric calibration was checked against a measurement of the highly polarized standard star HD 298383, then it was applied to the program stars.

RESULTS

In both AG Car and R127, the polarization in Hα differs significantly from that of the continuum, thus giving evidence of intrinsic polarization. In Figure 1, we display the Hα continuum-to-line vectors of AG Car and R127. Intrinsic polarization was previously suspected to be present in AG Car because its UBV polarization varies as a function of time (Serkowski 1970, ApJ, 160, 1083). A change of polarization at Hα in R127 was noted by Drissen (1992, ASP Conf. Ser., 22, 3) relative to UBVRI–filter data.

The polarization across Hα of AG Car displays a variable PA with wavelength, indicating that the line itself is polarized. The Hα continuum-to-line vector points along a PA of ~145°. The continuum polarization is higher than when Serkowski observed it; the PA of this variation is also along ~145...150°. We interpret it as the constant intrinsic PA of the electron distribution; percentage polarization changes along this PA are then attributed to changes in the amount of free electrons in an axisymmetric stellar wind. The polarization PA is nearly co-aligned with the major axis of AG Car's elliptical ring nebula, ~135°, and perpendicular to the bright nebular regions at PAs ~35° and ~225°.

The polarization across Hα of R127 drops quickly to line center without PA variations across the profile. The continuum-to-line vector points along a PA of ~25°. This is almost perpendicular to two bright nebular patches whose axis has a PA of ~95...110° (cf., Schulte-Ladbeck et al. 1992, ApJ, subm.).

COMMENTS AND THOUGHTS

The intrinsic polarization in AG Car and R127 is caused by electron scattering. This requires a large density of free electrons in an aspherical distribution. Extended line wings of Hα previously attributed to electron scattering (Bernat & Lambert 1978, PASP, 90, 520) are found to be polarized in AG Car for the same reason, thus providing us with a new method of determining the electron temperature in a stellar wind (Schulte-Ladbeck et al. 1992, in prep.).

AG Car was near minimum, R127 was in maximum. A large electron-scattering optical depth must prevail in LBVs in either state.

The stellar winds are aspherical very near the stars, i.e., at distances of a few stellar radii, and any outflow mechanism must explain this property.

The nebulae surrounding AG Car and R127 both have two bright patches. In AG Car, these are located at the end of the bipolar AG Car "jet" (Paresce & Nota 1989, ApJL, 341, 83). The polarization is roughly perpendicular in both cases to an axis joining the bright nebular regions (polar or equatorial axis?).

The resolved nebulae are very old, ~10^4 yr. Yet the stars still have some "memory" of the nebular axes, evidenced by the polarization PA. A stable outflow mechanism is required. This points to either stellar rotation or binarity as the origin of the asymmetries seen in the ejecta of LBVs.

The Hα polarization of AG Car rotates across the profile. PA rotations in Be stars were shown by Poeckert & Marlborough (1978, ApJ, 220, 940) to originate from rotating, expanding, axisymmetric stellar winds. The PA rotation observed in AG Car may be interpreted with rotation.

The intrinsic polarization of R127 is in the same direction as the interstellar polarization in the LMC bar. Although LBVs are evolved, they are still relatively young objects. It could be possible that they have not lost all of their natal angular momentum and magnetic field during their short evolution.

This work was supported by NAS5-26777 and NAGW-2338. We thank T.L. Smith for her help in preparing the AG Car section of the poster.

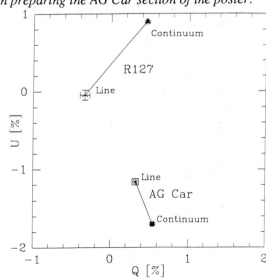

Fig. 1: QU diagram of the continuum-to-line Hα polarization vectors of AG Car and R127. Note: Angles in QU space are 2 x PA. The polarizations were measured in bandpasses centered on Hα. In R127, the filter was 8Å wide. For AG Car we used 20Å-wide band in order to obtain a better average over the PA rotation.

Massive Stars: Their Lives in the Interstellar Medium
ASP Conference Series, Vol. 35, 1993
Joseph P. Cassinelli and Edward B. Churchwell (eds.)

HE 3-519 : POST-LBV, PROTO-WR?

ARSEN HAJIAN
Cornell University, Department of Astronomy
Ithaca, NY 14853-6801

KRIS DAVIDSON AND ROBERTA HUMPHREYS
University of Minnesota, Department of Astronomy
116 Church St. SE, Minneapolis, MN 55455

He 3-519 is one of a handful of very luminous blue variable (LBV) and related Of/WN9 stars known in our galaxy. It is surrounded by a shell nebula similar to AG Car and is also spectroscopically very similar to AG Car, a known LBV,at minimum light. Although He 3-519 is not officially an LBV it has the Of/WN9 spectrum shown by several LBV's at minimum.

We have observed low-resolution IUE spectra of He 3-519 plus optical spectra and CCD photometry of it and its line of sight neighbors to better understand its physical nature and to make reliable estimates of its extinction, distance and luminosity.

The optical spectrum shows WN features and HeII emission with prominent P Cygni profiles in the hydrogen and He I lines. The most notable feature of the UV spectrum are two mysterious emission levels at 2753 and 3008Å. No similar features have been found in any other astronomical spectra although Sturve and Swings reported emission near 3010Å in P Cyg more than 50 years ago which they identified as FeIII. The feature may be FeIII but there are problems with this identification. We have no identification for the 2753Å feature.

The appearance of the spectrum with helium emission and WN features implies a temperature of about 30000K. Using a Zaustra-like method, we find that a 35000K blackbody continuum would provide just enough helium-ionizing photons to explain the HeI 4471 equivalent width. A 35000K Planck function fits the observed visual-UV continuum very well if E (B-V) = 1.4mag.

To estimate the distance to He 3-519 we have determined the relation between the extinction or reddening and distance for several stars in the direction of He 3-519 for which we have a spectral type and/or UBV photometry. This shows that He 3-519 with $A_V \equiv 4.5$mag must be at least as far away as AG Car at 6 kpc and is most likely farther away, perhaps at about 8 kpc in the Carina special feature. Its luminosity is then $M_{Bol} \equiv -11$ mag comparable to AG Car.

Massive Stars: Their Lives in the Interstellar Medium
ASP Conference Series, Vol. 35, 1993
Joseph P. Cassinelli and Edward B. Churchwell (eds.)

RELATIONSHIPS BETWEEN LUMINOUS BLUE VARIABLE,
OF, AND WOLF-RAYET STARS

ANNE B. UNDERHILL
Department of Geophysics and Astronomy, University of British
Columbia, Vancouver, B.C. V6T 1Z4, Canada

ABSTRACT   Key details of LBV, Of, and Wolf-Rayet spectra are
noted. It may be argued that the one kind of star does not evolve into one
of the others as a result of mass loss and normal nuclear reactions in the
core of the star. Rather in each case much of the emission-line spectrum
is generated in a disk. The observed light changes seem to be the result
of changes in the disk and of changes in the magnetically supported
filaments connecting the disk to the star.

INTRODUCTION

LBVs: A luminous blue variable is typically a relatively bright B0–B2
supergiant in the LMC or SMC. In the visible range, most show sharp emission
lines of Fe II, [Fe II], H, and sometimes He I on top of a continuous spectrum.
Normal B supergiant absorption lines are weak or absent. Slow changes of
apparent magnitude and colour may occur. The luminosity of a LBV remains
about constant and is typically in the range $M_{Bol} = -8$ to brighter than -10.
Those LBVs with $M_{Bol} \leq -10$ typically show no photospheric lines in the
normally observed optical region. Typical spectra are displayed in Stahl et al.
(1985) and in Zickgraf et al. (1986).

    Of stars: These are O stars with emission lines as well as the normal O-
type absorption lines in their spectra. Of stars show N III $\lambda\lambda4634,4640,4641$,
He II $\lambda4686$, and Hα strongly in emission; O(f) stars show the N III lines in
emission with a mixture of emission and absorption at He II λ 4686 and Hα
weakly in emission; O((f)) stars show the N III lines in emission but both the
He II $\lambda4686$ and Hα lines in absorption. The Ofpe/WN9 stars (Bohannan
and Walborn 1989) show sharp emission lines as in Of stars, on top of a blue
continuum; the H and He I lines are often in emission; the normal O-type
absorption lines may be weak.

    The normal O7 to O4 supergiants show an emission feature at Hα with
a central intensity between 1.20 and 1.35 as well as weak sharp emission lines
from N III, N IV, and Si IV. The O7((f)) to O4((f)) main-sequence stars show
Hα in absorption only. Weak sharp emission lines like those seen in supergiants
of the same spectral type are present. Spectra showing some of the weak sharp
emission lines in O-type spectra are displayed in Underhill and Gilroy (1990).

    WC stars: These show rather broad emission lines on a continuum; lines
of C III, C IV, O V, and O VI dominate; He I and He II are weakly in emission.
Most single WC stars show no absorption lines. A typical WC spectrum is
displayed in Underhill (1992).

WN stars: Most of these show rather broad emission lines on a continuum; lines of He II, N III, N IV, and N V dominate; He I and C IV emission lines are present. Some WN stars, particularly those of types WN7 and WN8 have sharp emission lines with the H Balmer series in emission. A few show intrinsic absorption lines typical of a star with T_{eff} near 30,000 K. Sections of the blue-violet spectrum of a WN5 star are shown in Underhill and Yang (1991).

LUMINOSITIES

O-type Iaf supergiants have $M_{Bol} \approx -9$; other Of stars have M_{Bol} in the range from ≈ -7 to -8; Ofpe/WN9 stars in the LMC have $M_{Bol} \approx -9$.

For most WC and WN stars $M_{Bol} \approx -6$ to -7, see, for instance, Underhill (1981, 1983). Most WN7 and WN8 stars have $M_{Bol} \approx -8$ to -9. This extra luminosity may be due to continuum radiation from an opaque disk in which the characteristic emission lines are generated. This source of continuum radiation does not appear to be strong for Wolf-Rayet stars of types other than WN7 and WN8.

DISCUSSION

Conti, Maeder, Humphreys, and Lamers, see, for instance, Lamers et al. (1991) and references therein as well as the papers by Langer and by Humphreys in *These Proceedings,* have suggested that the above types of star evolve with large rates of mass loss $(\approx 10^{-5}$ to $10^{-4} M_{\odot}$ yr$^{-1})$ from initial masses in the range 40–120 $M_{\odot}$ and that Wolf-Rayet stars show anomalous surface abundances of He, C, N, and O and are the final remnants of very massive stars.

These hypotheses are not true because (1) Wolf-Rayet stars have the properties of massive young stars just approaching the ZAMS and they have normal composition (Underhill 1991); (2) it may be questioned that the very large mass loss rates required to evolve massive model stars lying in the LBV part of the HR diagram rapidly actually occur; and (3) the excessive luminosity of some LBVs may be due to the presence of a disk which is optically thick and radiates a continuum plus edge-generated emission lines as is the case for cataclysmic variables.

REFERENCES

Bohannan, B. and Walborn, N. R. 1989, *P.A.S.P.*, **101**, 520.

Lamers, H. J. G. L. M., Maeder, A., Schmutz, W., and Cassinelli, J. P. 1991, *Ap. J.*, **368**, 538.

Stahl, O., Wolf, B., de Groot, M., and Leitherer, C. 1985, *Astr. Ap. Suppl.*, **61**, 237.

Underhill, A. B. 1981, *Ap. J.*, **244**, 963.

Underhill, A. B. 1983, *Ap. J.*, **266**, 718.

Underhill, A. B. 1991, *Ap. J.*, **383**, 729.

Underhill, A. B. 1992, *Ap. J.*, **397**, October 20 issue.

Underhill, A. B. and Gilroy, K. K. 1990, *Ap. J.*, **364**, 626.

Underhill, A. B. and Yang, S. 1991, *Ap. J.*, **368**, 588.

Zickgraf, F.-J., Wolf, B., Stahl, O., Leitherer, C., and Appenzeller, I. 1986, *Astr. Ap.*, **163**, 119.

Massive Stars: Their Lives in the Interstellar Medium
ASP Conference Series, Vol. 35, 1993
Joseph P. Cassinelli and Edward B. Churchwell (eds.)

OBN STARS AND BLUE SUPERGIANT SUPERNOVAE

NORBERT LANGER
Universitäts-Sternwarte Göttingen, Geismarlandstr. 11, W-3400 Göttingen, F.R.G.

It is known for a long time, that carbon and nitrogen absorption lines in the spectra of late O and early B stars show a large diversity. Stars with unusually strong carbon (OBC stars) or nitrogen (OBN stars) features are observed. Walborn (1976), who reviewed the OBC/OBN categories, suggested abundance differences as origin of the spectral anomalies in the sense that OBC stars may reflect the local "cosmic" CNO abundances, normal OB stars are slightly nitrogen enriched, and OBN stars should be extremely nitrogen rich and carbon depleted. The interpretation of OBN/OBC stars in terms of an abundance effect is confirmed by non-LTE abundance analyses of 4 OBN stars by Schönberner et al. (1988), who do not only find a very large nitrogen enrichment but additionally a considerable helium enrichment. Recently, Herrero et al. (1992) showed that even a large fraction of galactic O5-B0 stars appear to be helium enriched (which was occasionally found earlier; cf. Herrero et al., for references). Along with this, Herrero et al. find that spectroscopically determined masses of most OB stars are considerably smaller ($\sim 50\%$) compared to masses derived from evolutionary tracks. Note that these independent findings are quite consistent in the framework of stellar structure theory: an increased helium abundance decreases the mass to luminosity ratio of massive main sequence and supergiant stars (see below).

Further evidence in favour of CNO-cycled material at the surface of massive stars comes from the SN 1987A progenitor, which obviously was considerably nitrogen enriched (Fransson et al. 1989), while the progenitor star apparently did not show pronounced BN spectral characteristics (Walborn et al. 1989). Recently, also Gies and Lambert (1992) find evidence for CNO-products in early B-stars, while indications for N and He-enrichment in Luminous Blue Variables and Type II supernovae are reviewed by Walborn (1988).

It is very striking that the standard stellar evolution theory for massive stars does not predict any surface enrichments for early phases of luminous O and B stars with initial masses less that $\sim 60\,M_\odot$ (cf. Chiosi and Maeder 1986, Langer 1991, Schaller et al. 1992) except a slight CNO-signature for core-He burning blue supergiants in a post-red supergiant blue loop (see e.g. Langer 1991).

In this context, another fundamental discrepant observation concerning massive stars has to be mentioned: All stellar evolution calculations predict a gap between main sequence and post-main sequence stars (i.e. core hydrogen

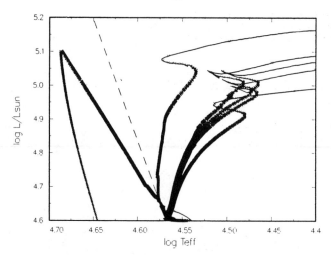

Fig. 1: *Evolutionary tracks of 20 $M_{\odot}$ sequences with $Z = Z_{\odot}/4$ in the HR diagram. From right to left, the following values for the rotational mixing parameter have been used: $\alpha = \infty$ (i.e. no rotational mixing), 0.09, 0.06, 0.03, 0.01, and 0.01. The left sequence has been computed with the Schwarzschild criterion for convection, while for the others the Ledoux criterion plus semiconvection has been invoked. A cross each 10^4 yr on the tracks indicates the speed of evolution. The theoretical zero age main sequence is drawn as a dashed line.*

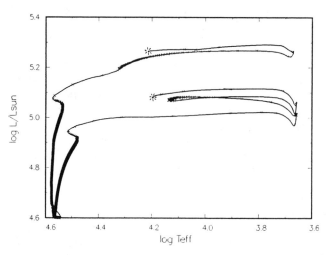

Fig. 2: *Tracks for the sequences computed with $\alpha = \infty$ and $\alpha = 0.01$ from the ZAMS to the pre-supernova stage in the HR diagram. Note the large differences in luminosity (and thus M/L) and core He-burning blue supergiant effective temperature. Both sequences explode as blue supergiants with $T_{eff} \simeq 16\,000\,K$, and leave the red supergiant branch $\sim 4\,10^4$ yr before they explode.*

and core helium burning) in the HR diagram (cf. Langer 1991), while the observed distribution appears to be rather continuous with no indication of a post-main sequence gap (cf. Blaha and Hymphreys 1989; Fitzpatrick and Garmany 1990; Garmany and Stencel 1992). This will be called the "TAMS problem", while the mass and helium discrepancy are named as mass and helium problem by Herrero et al. (1992).

In the following we want to present several theoretical stellar evolution sequences for an initial mass of $20 M_\odot$ and a metallicity of $Z_\odot/4$, where parametrized diffusive mixing in radiatively stable zones according to the turbulent diffusion model of Zahn (1983) for differentially rotating stars is invoked. We have incorporated this process in our code as described in earlier models of Maeder (1987), however, instead of the Schwarzschild criterion we have applied the Ledoux criterion for convection plus semiconvective mixing (as in Langer 1991a; cf. also Langer, this volume).

Due to this restriction of convection in the presence of molecular weight barrieres, we do not find — as Maeder (1987) — a bifurcation in stellar evolution for changing rotational mixing parameter α. Instead, the tracks of our sequences vary continuously with α (cf. Fig. 1), and moreover they always keep a large fraction of hydrogen in their envelopes even until core hydrogen exhaustion, and consequently avoid the branch of homogeneous evolution found by Maeder. The relevance of the Ledoux criterion for convection is demonstrated in Fig. 1, where we also included the evolutionary track of a sequence computed with the Schwarzschild criterion, which evolved at almost complete chemical homogeneity.

Fig. 2 compares the complete evolution of the "non-rotating" and "most rapidly rotating" sequences of Fig. 1 in the HR diagram. (Note that we did not include rotation in our calculations, but only parametrized rotationally induced mixing.) Most important in the present context are two features in Fig. 2: 1) The "rapidly rotating" sequence, which is greatly helium enriched in the whole envelope (the surface helium mass fraction rises from 0.25 to 0.35 during core hydrogen burning), shows a much hotter core helium burning blue supergiant stage as the "non-rotating" model, and 2) it explodes as a blue supergiant like the "non-rotating" model sequence (cf. Langer 1991a).

The folowing conclusions are drawn:
1. The mass-, helium-, and TAMS problem of luminous early type stars may be closely connected.
2. Helium enrichment in the whole stellar envelope may be the physical origin.
3. If rotationally induced mixing is responsible for the helium enrichment (other possibilities exist: e.g. mass transfer in a close binary system; cf. Ray and Rathnasree 1991), the choke-off of convection due to molecular weight gradients (semiconvection) is important.
4. Helium enrichment still allows for selfconsistent post-RSG blue supergiant supernova progenitor models in the semiconvection scenario.

ACKNOWLEDGEMENT The author is very grateful to the organizers of this conference, and especially to Prof. J. Cassinelli, for their generous support.

REFERENCES

Blaha C., Humphreys R.M., 1989, A.J. 89, 1598

Chiosi C., Maeder A., 1986, ARA&A 24, 329

Fitzpatrick E.L., Garmany C.D., 1990, ApJ , 363, 119

Fransson C., Cassatella A., Gilmozzi R., Panagia N., Wamsteker W., Kirshner R.P., Sonneborn G., 1989, ApJ 336, 429

Garmany C.D., Stencel R.E., 1992, A&A Suppl. 94, 211

Gies D.R., Lambert D.L., 1992, ApJ 387, 673

Herrero A., Kudritzki R.P., Vilches J.M., Kunze D., Butler K., Haser S., 1992, A&A 261, 209

Langer N., 1991, A&A , 252, 669

Langer N., 1991a, A&A 243, 155

Maeder A., 1987, A&A 178, 159

Ray A., Rathnasree N., 1991, M.N.R.A.S. 250, 453

Schaller G., Schaerer D., Meynet G., Maeder A., 1992, A &AS , in press

Schönberner D., Herrero A., Butler K., Becker S., Eber F., Kudritzki R.P., Simon K.P., 1988, A&A 197, 209

Walborn N.R., 1976, ApJ 205, 419

Walborn N.R., 1988, in *Atmospheric Diagnostics of Stellar Evolution*, IAU-Colloq. 108, K. Nomoto, ed., p. 70

Walborn N.R., Prevot M.L., Wamsteker W., Gonzales R., Gilmozzi R., Fitzpatrick E.L., 1989, A&A 219, 229

Zahn J.P., 1983, in *Astrophysical Processes in Upper Main Sequence Stars*, 13th Saas-Fee Course, B. Hauck, A. Maeder, eds.

Massive Stars: Their Lives in the Interstellar Medium
ASP Conference Series, Vol. 35, 1993
Joseph P. Cassinelli and Edward B. Churchwell (eds.)

IS THE W-R SYSTEM HD 5980 IN THE SMC UNDERGOING OUTBURST?

G. KOENIGSBERGER
Instituto de Astronomía, UNAM, México

L.H. AUER
Los Alamos National Laboratory, USA

O. CARDONA
INAOE, México

L. DRISSEN
Space Telescope Science Institute, USA

A.F.J. MOFFAT AND N. St. LOUIS
Dept. de Physique, U. Montreal, Canada

W. SEGGEWISS
Observatorium Hoher List, U Bonn, Germany

ABSTRACT   In this paper we report the spectacular changes which have occured in the SMC W-R system HD 5980=AB5 during the past decade. We present IUE data (1979-1991) which, along with complementary optical observations, show that the emission lines have enhanced greatly their equivalent widths as well as their line-to-continuum ratios over the past decade. The semiforbidden line of N IV at λ 1486 Å has shown the most spectacular increase, from being practically undetectable in 1981 to becoming one of the most intense lines in the 1989-1991 UV spectrum.

The recent optical spectrum present H and He emission lines, which now overwhelm the photospheric absorptions which were very evident in the 1975-1977 spectra. Furthermore, the relative strengths of N III λ 4640, N IV λ 4058 and N V λ 4603 and λ 4620, imply a classification of WN6, rather than the previous classification of WN4.5 (or WN3!) for the W-R component. This change from WN3 to WN6 is unprecedented.

The UV ($\lambda\lambda$ 1850-1860 A) continuum flux levels increased slightly (10%) between 1986 and 1991. However, the visual (FES) intensity of the system increased by 1.1 mag during this same period. This corresponds to an increase by a factor of almost three in flux, which is comparable with the increase by a factor of two of the equivalent width in the semiforbidden N IV λ 1486 Å line during this period.

Massive Stars: Their Lives in the Interstellar Medium
ASP Conference Series, Vol. 35, 1993
Joseph P. Cassinelli and Edward B. Churchwell (eds.)

A 37-DAY PERIOD IN THE PECULIAR WR STAR HD 191765?

A.F.J. MOFFAT
Département de physique, Université de Montréal, C.P. 6128, Succursale
A, Montréal, Québec, H3C 3J7 Canada, and Observatoire du mont
Mégantic

S.V. MARCHENKO
Instituto de Astronomía, Universidad Nacional Autónomia de México,
Apartado Postal 70-264, C.P. 04510, México, D.F., México and Main
Astronomical Observatory of the Ukrainian Academy of Sciences,
Goloseevo, 252127 Kiev, Ukraine

ABSTRACT   We report on two years of $\sim$ daily UBV monitoring of
the variable Wolf-Rayet star HD 191765 (WN6+?) at one of the 0.25m
automatic photometric telescopes on Mount Hopkins. Evidence is found
for a relatively long period (P = 37.2d) of the light curve in all three
filters, with amplitudes close to 0.03 mag. The 37-day period can be
interpreted as either: a) luminous blue variable - like microvariability, or
b) precession of the WR component in a close binary system of $P \sim 2d$.

INTRODUCTION

The strongly supersonic nature of WR star winds leads one to expect turbulent
outflow and hence time variability.   Yet the observed frequency of light
variability among WR stars is not significantly different from that among normal
stars (Moffat & Shara 1986). Clearly, the nature of the observed variability of
WR stars lacks a definitive explanation.
    In search of further clues to the origin of the variability, one important
aspect is periodicity. Its presence must be related to rotation,. pulsation or
binarity. Its lack would imply a stochastic nature intrinsic to the wind.
    The most frequently studied group of WR stars are the eight bright objects
in Cygnus (WR 133-140). Most variable among them is the well-known eclipsing
binary V444 Cygni (WR 139). Among the other seven, HD 191765 (WR 134,
WN6) stands out most for its variability. Previous attempts to find a periodicity
in this star on a short- and long-term basis have led to ambiguities. Of special
note is the recent intensive 10-night monitoring of its light on a 4-minute to day
timescale, by Antokhin *et al.* (1992). They found possible evidence for a period
close to 2 days, in addition to a long trend in the variation.

PHOTOMETRY

We report on two years of $\sim$ daily UBV monitoring of WR 134 at one of the 0.25 m automatic photometric telescopes on Mount Hopkins. This is the longest interval to date for monitoring of this star, coupled with good signal-to-noise (instrumental $\sigma(O - C) \simeq 0^m.006$ per data point in B and V). Fig.1 shows the overall light curve for 4 epochs of data in 1990 and 1991. The gaps are due to summer Monsoons and inaccessibility in winter.

A period search of these data was undertaken using standard sine-wave least squares fitting and phase dispersion minimization techniques. A significant period of 37.2 day was revealed with peak-to-valley amplitude $\sim 0.03 mag$ in U,B and V. The data are not suited to verifying the short 2-d period.

MECHANISMS FOR THE 37 D PERIOD

Traditionally, there are several mechanisms which one could propose for long-term (days-to-months), small-scale variability of massive, luminous hot stars:
1. Binarity: orbital revolution, eclipses, precession effects (Antokhin & Cherepashchuk 1989).
2. Rotation of a star with spotted surface (Harmanec 1987).
3. Intrinsic variability of the wind due to the stochastic nature of mass loss (Andriesse 1980) or line profile instability (Rybicki et al. 1990).
4. LBV-like microvariations (Sterken 1989; van Genderen 1989).

We can immediately reject 2. and 3. because of the long period (37 d). For example, a typical rotation velocity $V_e \sim 200 km \ s^{-1}$ for massive stars would lead to $P \sim 1 \ d$ for a star such as WR 134.

Thus, 1. and 4. appear to remain as possibilities. For example, we find a slope among all of our present data, of $\Delta(B-V)/\Delta V = -0.6$, compared to -0.25 for the LBV S Dor (Stahl et al., 1984). In this context, both values (hence also the physical process) can be considered similar. Nevertheless, such long-term variations in WR stars are, unlike LBV's, probably rare.

In case 1., we cannot be dealing with eclipses since the light amplitude implies a massive companion, which is not evident from the standpoint of radial velocity variations of WR 134. Thus, we must look at the case of a WR star with a low-mass (probably compact, e.g. neutron star, NS) companion. The 37 d period is then interpreted as the precession of a rapidly rotating WR star, whose extended equator is inclined to the orbital plane (cf. Antokhin & Cherepashchuk 1989). Following Brown & McLean (1977), Brown et al. (1978), Brown & Fox (1989) for electron scattering in an optically thin, axially symmetric, flattened envelope, we derive a model to reproduce the 37 d modulation of the light curve and the polarization (more details can be found in a longer paper: Moffat & Marchenko 1992). Fig.2 shows that a good fit is obtained for an inclination of the precession axis to the line of sight $i = 67.5°$ and precession cone angle $\Delta i = 12.5°$.

Is it feasible to have such a short precession period compared to a possible 2 d orbit? Following the pioneer work of Brouwer (1946 a,b), we calculate the flattening r_c/r_a for the WR star, for $M_{WR} = 15 \ M_{\odot}$, $R_{WR} = 7 R_{\odot}$, $P_{orbit} = 2 \ d$, $P_{precession} = 37.2 \ d$, and $M_{NS} = 1.4 \ M_{\odot}$. We obtain $r_c/r_a = 0.6 - 0.8$, depending on the density profile inside the WR star. For corotation, we found

$V_e = 200$ km s^{-1} for the WR star; hence it would be rotating at $\sim 30\%$ of break -up speed $(\sim 640\ km\ s^{-1})$. Furthermore, the orbital velocity of the WR star would be 38 $km\ s^{-1}$. After orbital projection, it is not surprising that such a small value has not been detected yet in such a broad-line, spectroscopically variable star.

CONCLUSIONS

Of the 4 conceivable mechanisms to explain the observed 37 day light variability in WR 134, only 2 of them survive:
- LBV-like microvariations, or
-precession of a flattened WR wind in a close binary system with a low-mass companion.
    The latter of these two is preferred. If such flattening actually exists, it could lead to equatorial wind enhancement in the WR star. This could be further amplified by magnetic acceleration (Poe $et\ al.$ 1989). Unfortunately, however, no observed rotational velocities are available for WR 134 (or any WR star) to check this.

REFERENCES

Andriesse,C.D.1980, Ap& SS,67,461.
Antokhin,I.I, and Cherepashchuk,A.M. 1989, Pis'ma AZh,15,701.
Antokhin,I.I., Irsmambetova,T.R., Moffat,A.F.J., Cherepashchuk,A.M., and Marchenko,S.V. 1992, ApJS, in press.
Brouwer,D. 1946, AJ, 51,223.
Brouwer,D. 1946, AJ, 52,57.
Brown,J.C., and McLean,I.S. 1977, A& A,57,141.
Brown,J.C., McLean,I.S., and Emslie,A.G. 1978, A& A, 68,415.
Brown,J.C., and Fox, G.K. 1989, ApJ, 347, 468.
Harmanec,P. 1987, BAC, 38,52.
Marchenko, S.V., Moffat,A.F.J., Annuk,K., Antokhin,I.I, Guralchuk,A.L., Irs-mambetova,T.R, Khalack,V.R., Morozhenko.A.V., and Vinogluadov,V.N. 1992, in preparation.
Moffat,A.F.J., and Shara,M.M. 1986, AJ, 92,952.
Moffat,A.F.J., and Marchenko, S.V. 1992, AJ, submitted.
Poe,C.H., Friend,D.B., and Cassinelli,J.P. 1989, ApJ, 337,888.
Rybicki,G.B., Owocki,S.P., and Castor,J.I. 1990, ApJ, 349,274.
Schulte-Ladbeck,R.E., Nordsieck,K.H., Taylor,M., Bjorkman,K.S., Magalha-es,A.M., and Wolff,M.J. 1992, ApJ, 387,347.
Stahl,O., Wolf,B., Leitherer,C., Zickgraft, F-J., Krautter, J., and de Groot,M. 1984,A& A, 140,459.
Sterken,C. 1989, in Physics of Luminous Blue Variables, Proc.   IAU Coll. N°. 113, eds.   K.Davidson, A.F.J.Moffat, H.G.J.L.M.Lamers (Dordrecht:Kluwer),p.59.
Van Genderen,A.M. 1989, A& A, 208,135.

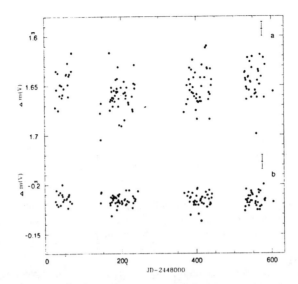

FIG.1 V-band light curves for (a) HD 191765 minus control star and (b) check minus control star. The 2σ instrumental error bars for individual data points are indicated. Note how , despite the horizontal compression, the 37 d period can be seen more clearly in the first, third and latter part of the fourth epochs.

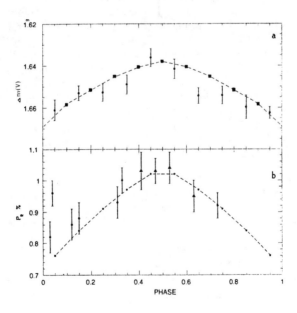

FIG.2 The model fit (squares) to the present mean V light curve and published polarization (Schulte-Ladbeck *et al.* 1992 : dots; Marchenko *et al.* 1992: triangles). Zero phase is arbitrarily chosen at $JD\ 2448000 + 37^d.2\ E$.

Massive Stars: Their Lives in the Interstellar Medium
ASP Conference Series, Vol. 35, 1993
Joseph P. Cassinelli and Edward B. Churchwell (eds.)

THE MOST EVOLVED WOLF-RAYET STARS

P.R.J. EENENS

Instituto de Astronomía, UNAM,
Apartado Postal 877, Ensenada, B.C., Mexico

ABSTRACT   Infrared spectra of two galactic WO stars were obtained with
CGS2 on UKIRT. Line identifications are given and new estimates of the He/C
ratio, of the stellar wind terminal velocity and of the mass-loss rate are derived.

Line identifications

Using CSG2 on UKIRT, we obtained infrared spectra of the two Oxygen-rich Wolf-Rayet
stars WR 142 = Sand 5 (WO2) and WR 102 = Sand 4 (WO1), covering wavelength
ranges of 1.0 – 3.3 μm and 1.0 – 1.3 μm respectively. The resolution was ~ 500 km/s.
The strongest lines in the infrared spectrum of WR 142 are due mainly to transitions
of C IV: 8–7 (1.191 μm), 11–9 (1.396), 9–8 (1.736), 10–9/13–11 (2.427), 14–12 (3.091),
11–10 (3.282). Other features are assigned to O V (9–8 near 1.111 and 14–11 near 1.153
μm) and O VI (10–9/13–11 near 1.079, 11–10 near 1.458 μm). Helium is clearly present
in the spectrum of that star, in particular He II 7–5 and 11–6 (near 1.163 μm), 10–6
(1.282), 9–6 (1.476) and 10–7 (2.189 μm). The presence of Helium in WR 102 is more
controversial. Our spectrum shows an emission feature near 1.16 μm (Fig. 1) coinciding
with He II 7–5 and 11–6, in apparent contradiction to the claim of Dopita *et al.* (1990)
that no Helium is present in this star. However the infrared spectrum published by
those authors is very noisy in this region, as well as near 1.282 μm, where He II 10–6
would fall. Those authors also searched for other He II lines in their infrared spectrum of
WR 102 (near 2.037, 2.164, 2.31-2.37 μm) but these are very faint in early WC spectra
(Eenens *et al.* 1991) and hence should be absent in WO spectra. Indeed, they are not
observed in our spectrum of WR 142.

The Carbon abundance

We derived relative abundances following the method described by Hummer *et al.* (1982).
We used recombination coefficients Q calculated for T_e = 30000 K and log N_e = 11,
for comparison with our analysis of WC types (Eenens & Williams 1992). Note that
an increase in temperature by a factor of two would not change our results beyond
the estimated uncertainties. Our analysis gives a ratio C IV/He II of **0.8 ± 0.2** (by
number) for WR 142. For WR 102 we derive a ratio of **2.0**, but the uncertainties are
larger, because only two lines were used in its estimate. From the absence of He I
and C III lines, we can assume that in the infrared line-emitting region Carbon and
Helium are mainly in the form of C^{++++} and He^{++}, and that the CIV/HeII ratio gives
a reasonable estimate for C/He. The lower values derived by Kingsburgh & Barlow
(1991) are probably affected by blending of the optical lines they used (see below). The
C/He ratio derived for the WO1–WO2 stars (0.8 – 2.0) is much higher than for WC
stars (0.07 – 0.41) (Eenens & Williams 1992), confirming that the WO stars are the
most evolved Wolf-Rayet stars.

The wind terminal velocity and mass-loss rate

The strong infrared lines of WR 102 have an asymmetrical profile (the flux in the blue wing is greater than in the red wing), similar to what we observed in the infrared spectrum of WR 142 (see Eenens & Williams 1991 and their Fig. 1). This indicates a heavier absorption of the continuous radiation from the wind beyond the star as it travels near the star. From half the full width at zero intensity (FWZI/2) of the infrared emission lines (and taking into account the instrumental profile), we obtain v_∞ $\sim$ 4500 $\pm$ 500 km/s for both stars. This is much higher than for WC or WN stars, but smaller than the values obtained from the FWZI of optical lines: 5500 km/s for WR 142 (Kingsburgh & Barlow 1991); and between 4600 (Kingsburgh & Barlow) and 5530 km/s (Dopita et $al.$ 1990) for WR 102. The discrepancies between those authors reflect the large dispersion of FWZI and of shape between optical lines (Kingsburgh, private communication), indicating possible problems with blending and continuum fitting in the optical region. The infrared spectra are less crowded and all the strong lines have similar FWHM, FWZI and shape, making them more suitable to derive v_∞. With our revised values of C/He and v_∞, a previous estimate for the mass-loss rate of WR 142 (Barlow 1991) must be revised upwards. Our new value of 5.5×10^{-5} $M_\odot$ yr^{-1} is close to the average for other Wolf-Rayet stars (6×10^{-5} $M_\odot$ yr^{-1}) and implies a strong metallic input into the stellar environment.

References

Barlow, M.J. 1991, in $Wolf\text{-}Rayet$ $Stars$ and $Interrelations$ $with$ $Other$ $Massive$ $Stars$ in $Galaxies$, IAU Symposium No 143, eds van de Hucht, K.A., & Hidayat, B., Reidel, Dordrecht, p. 281.

Dopita, M.A., Lozinskaya, T.A., McGregor, P.J. & Rawlings, S.J. 1990, $Astrophys.$ $J.$ 351, 563.

Eenens, P.R.J. & Williams, P.M. 1991, in The $infrared$ $Spectral$ $Region$ of $Stars$, Proc. Montpellier Conference, eds: C.Jaschek & Y.Andrillat, p. 158.

Eenens, P.R.J., Williams, P.M. & Wade, R. 1991, Mon. Not. R. astr. Soc. 252, 300.

Eenens, P.R.J. & Williams, P.M. 1992, Mon. Not. R. astr. Soc. 255, 227.

Hummer, D.G., Barlow, M.J. & Storey, P.J. 1982, in $Wolf\text{-}Rayet$ $Stars:$ $Observations,$ $Physics,$ $Evolution$, IAU Symposium No 99, eds de Loore C.W.H. & Willis A.J., Reidel, Dordrecht, p. 79.

Kingsburgh, R.L. & Barlow, M.J. 1991, in $Wolf\text{-}Rayet$ $Stars$ and $Interrelations$ $with$ $Other$ $Massive$ $Stars$ in $Galaxies$, IAU Symposium No 143, eds van de Hucht, K.A., & Hidayat, B., Reidel, Dordrecht, p. 101.

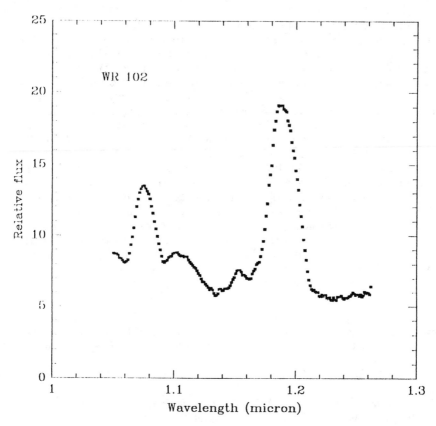

Fig. 1.   Near-infrared spectrum of WR 102. The main features are O VI near
1.079 μm, O V near 1.111 and 1.153, a He II+C IV blend near 1.163, and C IV
near 1.191 μm.

Massive Stars: Their Lives in the Interstellar Medium
ASP Conference Series, Vol. 35, 1993
Joseph P. Cassinelli and Edward B. Churchwell (eds.)

THE INFLUENCE OF CLOSE BINARY EVOLUTION ON THE THEORETICALLY PREDICTED NUMBER DISTRIBUTION OF WR STARS IN THE GALAXY AND IN THE MAGELLANIC CLOUDS.

D. VANBEVEREN
Dept. of Physics, Vrije Universiteit Brussel, 1050 Brussels, Belgium

C. DE LOORE
Astrophys. Inst., Vrije Universiteit Brussel, 1050 Brussels, Belgium.

ABSTRACT Binary computations are used in combination with single star evolutionary calculations in order to study the theoretically predicted number distribution of WR stars in the Galaxy and in the Magellanic Clouds.

INTRODUCTION

In order to explain the observed WR/O and WC/WN number ratio gradient as a function of metallicity Z, three suggestions appeared in litterature, i.e.

a. $\dot{M}$ during the WR phase is a function of Z (Vanbeveren and Conti, 1980)

b. Different IMF for different Z (Garmany et al., 1982)

c. $\dot{M}$ during the RSG phase is a function of Z (Maeder et al. 1981).

Using single star evolutionary computations only, quantitative results for the ratio's have been discussed by Maeder (1991) using suggestion C and it was concluded that they match reasonably well the observations. In the present paper we will combine binary and single star evolutionary computations and rediscuss the predicted ratios. We will demonstrate that it is very unlikely that suggestion C is the (only) reason for the observed gradient.
For single stars we will use the evolutionary computations of Maeder (1991). During the RSG phase, $\dot{M} \div Z^{0.5}$ whereas during the WR phase, $\dot{M}$ is independent from Z. The recent improvements of Schaller et al. (1992) lead to almost similar results. For massive close binaries we will use the Galactic results of Vanbeveren (1991). For the Magellanic Clouds we performed a complete set of binary computations which will be published in A.A. Supplements (de Loore and Vanbeveren, 1992). The binary computations use the same physical ingredients as the single star results, i.e. the same $\dot{M}$ during core hydrogen burning, the same convective core overshooting ($\alpha = 0.25$) and $\dot{M}$ during the WR phase is considered independent from Z. This set of single and binary computation will be called 'standard' evolution of massive stars.

THE RESULTS.

For the Galaxy and the MC's we adopt an IMF $\div M^{-\alpha}$ ($\alpha = 2.5 \pm 0.2$) and we assume that the IMF of primaries in massive close binaries is similar to that of single stars. Denoting by f the O + OB type binary frequency, it follows that the theoretically expected WR/O number ratio can be estimated as

$$\frac{WR}{O} = f\left(\frac{WR}{O}\right)_b + (1-f)\left(\frac{WR}{O}\right)_s$$

where the subscripts b and s stand for the predicted ratio for binaries, resp. single stars. It is obvious that the predicted ratio of WR single stars /WR binary components (with a normal companion) follows from

$$\frac{WR_s}{WR_b} = \frac{(1-f)\left(\dfrac{WR}{O}\right)_s}{f\left(\dfrac{WR}{O}\right)_b}$$

Using f = 0.35 (Garmany et al., 1980) for the Galaxy and for the MC's, the predicted ratios are given in table 1. The results are average values for $\alpha = 2.5$ and may differ by $\pm 10\%$ for α varying from 2.3 to 2.7.

TABLE 1 The theoretical WR distribution compared to the observed one; the first row (resp. the second row and the third row) holds for the Galaxy, resp. the LMC and the SMC.

WR/O				WR_s/WR_b	WC/WN	
single	binary	total	observed		theo	obs
0.072	0.12	0.09	0.116	1.12	1.3	
0.017	0.085	0.041	0.04	0.36	1.0	0.2
0.0027	0.045	0.018	0.015	0.11	1.3	0.06

The observed WR/O number ratio is given as well. We conclude

Standard evolution of single stars and close binaries predicts a WR/O number ratio for the Galaxy and for the MC's which corresponds to the observed values.

Standard evolution of single stars and close binaries predicts that there should be as many WR + OB close binaries as WR single stars in the Galaxy and that the number of WR + OB close binaries is at least 2.4 (resp. 8) times the number of WR single stars in the LMC (resp. the SMC), i.e.

Standard evolution of single stars and of close binaries predicts that the WR population in low metallicity environnements is largely dominated by close binaries.

In table 1 we also compare the theoretical WC/WN ratio predicted by standard evolution with the observed ratio. As can be noticed

Standard evolution of single stars and of close binaries predicts a WC/WN ratio which is far too large in the MC's compared to observations (i.e. a factor 5 in the LMC and a factor 20 in the SMC) and thus

Standard evolution of single stars and of close binaries does not reproduce the observed WR distribution in the Magellanic Clouds.

We have explored the model which has been proposed originally by Vanbeveren and Conti (1980) i.e. the $\dot{M}$ during the WR phase is a function of Z. The details of this model will be published in a forthcomming paper (Vanbeveren and de Loore, 1992). The quantitative results allow us to conclude that

only when the $\dot{M}$ during the WR phase is a function of Z the WC/WN number ratio predicted by stellar evolution corresponds to the observed value in the Galaxy and in the Magellanic Clouds.

REFERENCES.

Garmany, C.D., Conti, P.S., Massey, P.: 1980, Ap. J. **242**, 1063.
Garmany, C.D., Conti, P.S., Chiosi, C.: 1982, Ap. J. **263**, 777.
de Loore, C., Vanbeveren, D.: 1992, A.A. Suppl. (submitted).
Maeder, A.: 1991, A.A. **242**, 93.
Maeder, A., Lequeux, J., Azzopardi, M.: 1980, A.A. **90**, L17.
Schaller, G., Schaerer, D., Meynet, G., Maeder, A.: 1992, A.A. Suppl. (in press).
Vanbeveren, D.: 1991, A.A. **252**, 159.
Vanbeveren, D., Conti, P.S.: 1980, A.A. **88**, 230.
Vanbeveren, D., de Loore, C.: 1992, A.A. submitted.

Massive Stars: Their Lives in the Interstellar Medium
ASP Conference Series, Vol. 35, 1993
Joseph P. Cassinelli and Edward B. Churchwell (eds.)

X-RAY VARIABILITY IN V444 CYGNI - EVIDENCE FOR COLLIDING WINDS

M. F. CORCORAN, S. N. SHORE, J. H. SWANK, S. R. HEAP, G. L. RAWLEY, A. M. POLLOCK, I. STEVENS
Goddard Space Flight Center, Greenbelt MD 20771

ABSTRACT Phase-resolved ROSAT observations of the soft X-ray flux from V444 Cygni confirm the orbital dependence of the flux suggested by analysis of IPC observations. The X-ray behavior suggests that a region of X-ray emitting gas exists between the 2 stars, probably produced by a collision between the WR and O star winds.

INTRODUCTION

We used the ROSAT PSPC to examine the X-ray variability of the 4.2 day period WR+ O star binary system V444 Cygni. Our objectives were to:
1) compare our results with previous observations made with the IPC (Moffat et al. 1982)
2) explore the phase dependence of the X-ray variations, and
3) look for evidence of the generation of X-rays from colliding winds in the system.

The WR + O binary V444 Cygni is one of the best candidates systems in which to measure the importance of colliding winds on X-ray emission, since analysis of a long IPC observation by Moffat et al. (1982) suggested that the X-ray flux did indeed drop when the O star was eclipsed by the WR star. Unfortunately the IPC observation provided only 4 measures of the X-ray flux, only one of which was near O-star eclipse, so the significance of this variation could not be assessed.

OBSERVATIONS

ROSAT observed V444 Cygni with the PSPC 5 separate times, for a total integration time of about 20 ksec. Table 1 lists the journal of observations.

Table I Journal of Observations

Date	Phase	Exp. Time (sec)	Net Rate (cts/s)
6-May-91	0.71	1846	0.036±0.008
9-May-91	0.37	2073	0.010±0.010
9-May-91	0.45	1931	0.040±0.008
22-Oct-91	0.90	12584	0.015±0.007
3-Nov-91	0.70	1436	0.026±0.008

Figure 1 shows the lightcurve, in which the 22 Oct observation is broken up into 5 bins having lengths of 1500-2500 sec. The PSPC measured a significant decrease in counting rate near phase 0.45 (when the O star is moving in front of the WR star) and from phase 0.84– 0.96. Figure 2 shows a hardness ratio curve. The hardness ratio suggests a decrease at entrance to primary eclipse. We constructed an "outside eclipse" spectrum from the 6 May and 3 Nov observations along with the 9 May observation obtained at phase 0.37, and compared that to an "inside eclipse" spectrum constructed from the other observations. The "outside eclipse" spectrum was slightly harder (Fig. 3) with a nominally higher temperature (0.9 keV vs. 0.5 keV) but we could not adequately constrain the temperatures since each spectrum had fewer than 200 counts.

CONCLUSIONS

The observed change could be the result of a random (non-phase-locked) variation in the X-ray producing gas, but this is unlikely because of 1) the agreement between the 3 Nov and 6 May observations, both of which were obtained near phase 0.7, and 2) the IPC X-ray lightcurve. The fact that a decrease in counting rate is seen at both secondary and primary eclipse is strongly suggestive of generation of X-rays by colliding winds. Spectral modeling suggests a temperature near 1 keV, in accord with colliding wind models, and the derived luminosity, $L_x \approx 10^{31}$ ergs/s, is entirely consistent with the model predictions (Prilutskii and Usov 1978). Unfortunately our data does not allow us to resolve the dip in the X-ray emission, so we cannot reliably constrain the size of the X-ray emitting region between the stars. We note also that the flux near phase 0 shows at least 1 factor-of-two fluctuation, which may suggest a source of X-rays which is not fixed in the rotating frame. Further monitoring of the X-ray variability from V444 Cyg is needed.

REFERENCES

Moffat, A. F. J., et al., 1982, in *Wolf-Rayet Stars: Observations, Physics, Evolution*, (ed. C. de Loore and A. Willis), p. 577.

Prilutskii, O. F., and Usov, V. V., *Astron. Zh.* 53, 6.

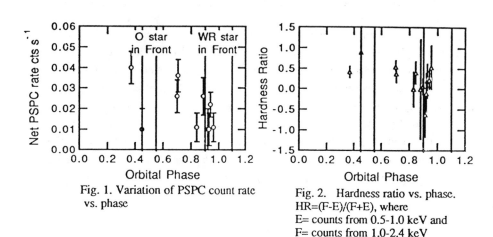

Fig. 1. Variation of PSPC count rate vs. phase

Fig. 2. Hardness ratio vs. phase. HR=(F-E)/(F+E), where E= counts from 0.5-1.0 keV and F= counts from 1.0-2.4 keV

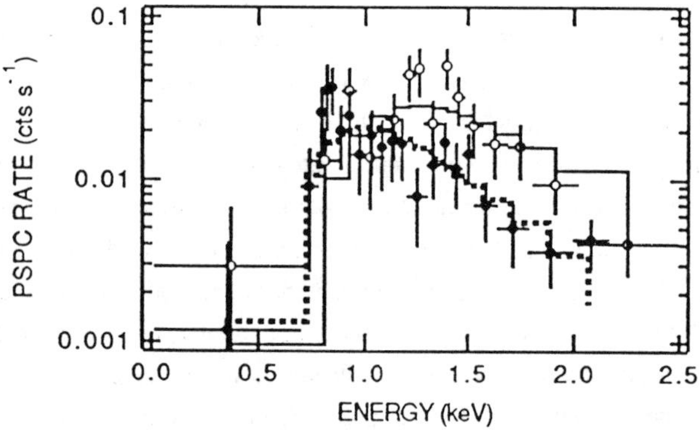

Fig. 3. Inside eclipse spectrum (filled circles) vs. outside eclipse spectrum (open circles). Best fit thermal models are shown (Solid line kT=0.9 keV, dashed line kT = 0.5 keV)

Massive Stars: Their Lives in the Interstellar Medium
ASP Conference Series, Vol. 35, 1993
Joseph P. Cassinelli and Edward B. Churchwell (eds.)

PROPER MOTIONS OF THE N CONDENSATIONS OF
ETA CARINAE

DENNIS EBBETS
Ball Aerospace, PO Box 1062, AR1, Boulder CO 80306

ELIOT MALUMUTH
CSC, NASA GSFC, Code 681, Greenbelt MD 20771

KRIS DAVIDSON
U. Minnesota Astronomy Dept, Minneapolis MN 55455

RICHARD WHITE, NOLAN WALBORN
STScI, 3700 San Martin Dr. Baltimore MD 21218

Eta Carinae is a luminous, massive star which has erupted with episodes of
greatly enhanced mass loss several times during the past few centuries. The
central object is now obscured by the debris from an outburst which occured
between 1830 - 1860. This irregular shell of material is known as the
Homunculus, which has been discussed by Hillier and Allen (1991),
Burgarella and Paresce (1991), and Hester et al. (1991.) Surrounding this is
an array of fainter condensations whose composition and kinematics strongly
suggest knots of ejecta. Some are contemporaries of the Homunculus while
others may have been expelled during earlier outbursts. Eta Carinae represents
a dramatic although possibly brief phase of the interaction between a massive
star and its environment.

Our study of the motions of the N Condensation was motivated by the work
of Walborn, Blanco and Thackeray (1978), and Walborn and Blanco (1988)
who showed that the two brightest knots, NN and NS have the fastest motions
of any of the outer condensations. We used their six reference stars to
determine the positions of twenty six additional stars on a CTIO CCD image
of the field. These were also visible on our Hubble Space Telescope WFC and
PC images, and served as our astrometric reference points. We measured the
positions of both knots of the N Condensation and the bright central star.
Images of the N Condensation give the visual impression of a jet like structure
directed radially outward. The measured offsets have clustered tightly around
a position angle of 27 degrees for over 40 years. Figure 1 shows our new data,

along with published positions from three earlier epochs. A linear least squares fit points back to within 0.14 arc seconds of the origin.

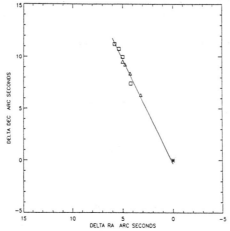

Fig. 1   Offsets of the N Condensations from the Central Object

An important question about the motions is whether the velocity has been constant with time, or whether the knots are being decelerated. We have investigated this using least squares fits to the time history of the radial positions. Figure 2 shows the offsets with linear and quadratic solutions. Both curves fit the data with maximum residuals of less than 0.1 arc seconds, and both are equally satisfactory in a chi-squared sense. The second order terms for both condensations are negative with nearly identical magnitudes. The possibility of decelerated motions, while not proven, is made plausible by the consistency of the two independently derived values.

Table I   Proper Motions and Physical Motions for d=2800pc

	NN Cond uniform	NN Cond decelerated	NS Cond uniform	NS Cond decelerated
T date AD	1862	1883	1870	1890
Mu (initial) " yr^{-1}	0.097	0.148	0.089	0.138
Mu (1991) " yr^{-1}	0.097	0.085	0.089	0.076
A " yr^{-2}	0	-5.8E-4	0	-6.2E-4
V initial km sec^{-1}	1288	1965	1181	1832
V (1991) km sec^{-1}	1288	1128	1181	1009
A km sec^{-1} yr^{-1}	0	-7.7	0	-8.2
V=0 date AD	never	2138	never	2113

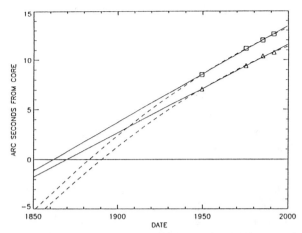

Fig. 2   Time History of Radial Positions With Linear and Quadratic Fits

Descriptions of the behavior of Eta Carinae date back to at least the mid-1830s (Walborn and Liller, 1977.) It was one of the brightest stars in the sky between 1835 and 1855. It declined in visual brightness until about 1888, whereupon a smaller brightening occurred. We would like to be able to associate the various knots of ejecta with the historical behavior, or at least to date the ejection events. Was all of the material now visible expelled during one outburst in the 19th century, or can we unambiguously identify debris from several such episodes? Our solutions for the motions of the NN and NS knots constrain their origin to the later half of the 19th century, as shown in Figure 2. The linear fits (uniform motion) extrapolate back to 1862 and 1870, which would associate their ejection with the decline from the great light maximum of the previous decades. On the other hand, the quadratic fits (constant deceleration) imply a slightly more recent origin, in 1883 and 1890. This could associate the N Condensation with the nova-like outburst of the early 1890s. It is difficult to visualize an interpretation of motions of the NN and NS knots that would place their origins before 1850 or after 1900 unless there had been an earlier epoch of acceleration of deceleration substantially different from what is observable in this century.

REFERENCES

Burgarella,D., Paresce,F. 1991, *A&A*, **241**, 595
Hester,J., et al. 1991, *A.J.*, **102**, 654
Hillier,D., Allen,D. 1992, *A&A* (in press)
Walborn,N., Liller,M. 1977, *Ap.J.*, **211**, 181
Walborn,N., Blanco,B. 1988, *PASP*, **100**, 797
Walborn,N., Blanco,B. Thackery,A. 1978, *Ap.J.*, **219**, 498

Massive Stars: Their Lives in the Interstellar Medium
ASP Conference Series, Vol. 35, 1993
Joseph P. Cassinelli and Edward B. Churchwell (eds.)

PROPER MOTIONS, MEMBERSHIP, AND PHOTOMETRY OF
CLUSTERS NEAR η CARINAE

KYLE CUDWORTH and STEVE MARTIN
Yerkes Observatory, The University of Chicago, P. O. Box 258,
Williams Bay, WI 53191

ABSTRACT Proper motions and photographic photometry
have been derived for nearly 600 stars with 7.5 < V < 15.5 in the
region of the very young open clusters Tr. 14, Tr. 16, and Cr.
232 based on 26 plates dating from 1893 to 1990. Cluster
membership probabilities have been derived from the proper
motions and color-magnitude diagrams of probable members
of each cluster are discussed. In contrast to some previous
studies we find all three clusters to lie at the same distance.

INTRODUCTION

The region in and around the Carina Nebula (NGC 3372) contains a
wealth of very young open clusters. Investigations in this region
of the sky have centered on the relation between the clusters, gas,
and dust (see, for example, Feinstein, Marraco, & Muzzio 1973;
Walborn 1973; Tapia *et al*. 1988) and the possibility of an
anomalous extinction law within the clusters (Smith 1987; Garcia
et al. 1988; see also Turner & Moffat 1980).
     Despite the large body of data from many observers (Feinstein
et al. 1973, Feinstein 1982; Turner & Moffat 1980; Forte & Orsatti
1981; Levato & Malaroda 1982; Walborn 1982 and references
therein; Smith 1987; Morrell, Garcia, & Levato 1988; Tapia *et al*.
1988), there appears to be some controversy over whether or not
Tr 14 and Tr 16 lie at the same distance. Walborn (1973) and
Morrell, *et al*. (1988) found that Tr 16 was located considerably
closer than Tr 14, with ~1 mag. difference in the CMD's. In
contrast, Turner & Moffat (1980) and Tapia *et al*. (1988) concluded
Tr 14 and Tr 16 are actually at the same distance. All of these
studies include stars which are intrinsic to the cluster as well as
some field stars along the same line of sight. To date, the probable
members of each cluster have been deduced based on an analysis
of color-color diagrams for the clusters.
     In order to better address questions concerning the
relationship among the clusters, it is necessary to systematically
determine the cluster membership of stars in these regions

independent of possible photometric biases. To this end, we obtained photographic B and V photometry plus proper motions and membership probabilities for nearly 600 stars in the regions of Tr 14, Tr 16, and Cr 232.

OBSERVATIONS, MEASUREMENTS, AND REDUCTIONS

In this study we have used 26 plates from 4 different telescopes spanning the epochs from 1893 to 1990. All plates were scanned on the PDS microdensitometer at MADRAF with image centroids and photometry derived using our standard software. Astrometric and photometric reductions and derivation of membership probabilities were then done at Yerkes using the software described by Cudworth (1985, 1986). We used only the new plates from Las Campanas telescopes for the photometry. The photometry was calibrated using the photoelectric observations by Feinstein *et al.* (1973) and Feinstein (1982), with stars noted by them as variable or peculiar excluded.

RESULTS

Color-magnitude diagrams of probable cluster members were constructed for the field as a whole and for the regions of each cluster. Although many field stars were eliminated via the proper motions, the CMD's strongly resemble those published by others.

Morrell, *et al.* (1988) argued that the HR diagram for Tr 14 lies approximately 1 mag. below that of Tr 16. We find that most of the stars observed by them which are in common with our study are indeed probable members, so we conclude that the possibility of contamination by foreground stars is probably not significant and the authors' conclusion of a shift in the diagrams is probably not due to field star contamination of the CMD's. However, their sample of stars was limited to the brightest members in each of the clusters ($V \lesssim 11$). Our survey contains considerably more stars and extends down to $V \sim 16$. If all the stars with high membership probabilities in our study are used to search for a shift in the CMD's of the clusters, no clear shift is evident (and certainly not at the level of about 1 mag.). We thus argue that the clusters are most probably at the same distance with reasonably similar reddening and extinction. The crucial difference between our study and Morrell, *et al.* is that our CMD's are not limited to the very bright stars where the main sequence is nearly vertical but rather reach faint enough that vertical shifts are much easier to investigate.

To investigate the kinematic association of these clusters we have derived the mean proper motions of the probable members ($P > 90\%$) in each of the three clusters. The means are not significantly different from zero, with errors of the mean ranging from about 3 to about 6 milliarcsec cent^{-1} (corresponding to velocities of 0.3 to 0.6 km/s at a distance ~2 kpc). This appears to be strong evidence for common kinematics (to a precision better than 1 km/s) among all 3 clusters, and hence for their association. We must caution, however, that the astrometric plate constant reductions could have removed real differences in proper motions.

ACKNOWLEDGMENTS

This project was begun at the instigation and with the assistance of Kathy Eastwood. Special thanks go to Nick Suntzeff and Steve Majewski for obtaining the Las Campanas plates used in this study and to Martha Hazen for the loan of the Harvard plates. As always, the hospitality of the University of Wisconsin during our use of the MADRAF facilities is much appreciated. This research has been partially supported by the NSF.

REFERENCES

Cudworth, K. M. 1985, AJ, 90, 65

Cudworth, K. M. 1986, AJ, 92, 348

Feinstein, A. 1982, AJ, 87, 1012

Feinstein, A., Marraco, H. G., & Muzzio, J. C. 1973, A&AS, 12, 331

Forte, J. C., & Orsatti, A. M. 1981, AJ, 86, 209

Garcia, B., Claria, J. J., & Levato, H. 1988, Ap & Sp Sci, 143, 317

Levato, H., & Malaroda, S. 1982, PASP, 94, 807

Morrell, N., Garcia, B., & Levato, H. 1988, PASP, 100, 1431

Smith, R. G. 1987, MNRAS, 227, 943

Tapia, M., Roth, M., Marraco, H., & Ruiz, M. T. 1988, MNRAS, 232, 661

Turner, D. G., & Moffat, A. F. J. 1980, MNRAS, 192, 283

Walborn, N. R. 1973, ApJ, 179, 517

Walborn, N. R. 1982, AJ, 87, 1300

Massive Stars: Their Lives in the Interstellar Medium
ASP Conference Series, Vol. 35, 1993
Joseph P. Cassinelli and Edward B. Churchwell (eds.)

PROPER MOTIONS OF HIGH-|b| OB STARS

K. M. CUDWORTH, J. BENENSOHN
Yerkes Observatory, The University of Chicago, P. O. Box 258,
Williams Bay, WI 53191

A. E. SCHWEITZER
Department of Astronomy, University of Wisconsin, 475 North
Charter St. Madison, WI 53706

ABSTRACT   A program is in progress at Yerkes Observatory to
measure proper motions for about 25 high-latitude stars with
O and early B spectral types. On classification spectra all
appear to be normal Pop. I stars located much further from
the Galactic plane than would be expected, often |z| ~1 kpc.
Preliminary proper motions and space velocities (with
distances based on assuming the stars are normal) have been
derived for the half of the program completed to date.

THE PROBLEM

The high-latitude OB stars have been recognized for some time as a
problem for our standard pictures of star formation and Galactic
structure. The difficulty is easily demonstrated by considering the
well-known specific case of HD 93521, an O9.5 V star with V = 7.04
that lies at b = +62°. If it is the normal Pop. I star implied by
Morgan's classification it lies about 2200 pc from the sun, and
nearly 2000 pc above the Galactic plane. Its radial velocity is -16
km/sec. For such a star one would normally infer a mass ~20 $M_\odot$
and a main sequence lifetime of ~3 x 10^6 years. To get where it is
now in this time would require a mean W velocity > 600 km/sec
over its lifetime. Since this is significantly greater than the local
Galactic escape velocity (~475 km/sec, Cudworth 1990), the star
cannot have simply reached its maximum z distance and now be
turning around. While some of the stars may well be runaways,
some like HD 93521 cannot be explained in this way. We are thus
left with the options that (1) occasional massive stars form well
out of the Galactic plane, *or* (2) these are blue stragglers with
greatly extended lifetimes, *or* (3) they are lower mass stars in a
late stage of stellar evolution in which their classification spectra
mimic normal massive Pop. I OB stars.

Since spectral classifications only tell us that these stars have the same T_e and log g as normal OB stars, one can easily concoct *ad hoc* lower mass stars with the proper T_e and log g that will have lower luminosities. These stars will then have both smaller distances (typically by a factor ~5) and much longer lifetimes, removing the problem. Ebbets & Savage (1982) indeed argued this for HD 93521, but a key detail of their chain of reasoning was refuted by Irvine (1989), who concluded this is a normal O9.5 V star.

Clearly these stars are worthy of attention in their own right, but it is their use in studying the "halo" ISM that drew our attention to them (Munch & Zirin 1961, Hobbs, et al. 1982). It should be pointed out, however, that *none* of these stars actually lies in the halo and that therefore the gas that can be studied in their foreground lies in the thick disk, not the halo.

PROPER MOTIONS

To improve the proper motion data currently available for these stars a program was begun with the Yerkes 40-in refractor in 1982 to obtain first epoch plates on about 25 of these stars. The goal was to investigate the kinematics of these stars using space velocities rather than only radial velocities, and to attempt a statistical parallax solution that should reveal whether these stars are significantly subluminous. The observations use standard techniques of photographic astrometry to reduce the likelihood of systematic errors due to colors and/or bright magnitudes. All measurements were done with the MADRAF PDS microdensitometer and reductions used our standard codes. Reductions from relative to absolute proper motion used a new technique developed by Cudworth & Hanson (1992).

RESULTS

It appears from the preliminary space velocities that we probably have a heterogeneous set of stars. Some have velocities ~50 to 100 km/sec characteristic of runaways. It may be significant that most have W velocity components directed away from the Galactic plane, and some appear to be large enough to have carried the stars to their present locations in acceptably short lifetimes.

HD 93521 has a very small velocity, comparable to low-|b| OB stars. If instead of being a massive star this is a star ~1 $M_\bullet$ its distance would be ~500 pc and its heliocentric tangential velocity only ~5 km/sec. In distinct contrast to the general pattern of the velocities, HD 137569 has a W component of 87 km/sec *toward* the Galactic plane.

HD 214080 has an enormous space velocity that is marginally
retrograde to Galactic rotation (in spite of 0 radial velocity!). It
may be difficult to reconcile such a velocity with a runaway
explanation. This star is classified as a supergiant, indicating that
its surface gravity is low compared to main sequence stars of
similar temperature. Perhaps this is one case where one should
seriously consider the possibility that the gravity is low because of
a low mass. If the assumed luminosity, and hence the distance,
were reduced to something appropriate for a 1 $M_\odot$ star the
velocity would be reasonable for a disk star. HD 215733 and 219188
also have space velocities $\geq$ 200 km/sec. They should perhaps be
considered with HD 214080 as candidates for stars of lower mass.

CONCLUDING REMARKS

These preliminary results indicate that the problem of the high-
|b| OB stars remains complicated. The final program results in a
few years should provide a good deal of useful additional basic
data, but probably not a definitive solution. If the population is
indeed heterogeneous, the planned statistical parallax solution
may be difficult to interpret.

ACKNOWLEDGMENTS

We thank Lew Hobbs and Elise Albert for drawing our
attention to these stars initially, and Lew Hobbs for continuing
useful discussions. This research has been partially supported by
the NSF.

REFERENCES

Cudworth, K. M. 1990, AJ, 99, 590

Cudworth, K. M. and Hanson, R. B. 1992, in preparation for AJ

Ebbets, D. C., & Savage, B. D. 1982, ApJ, 262, 234

Hobbs, L. M., Morgan, W. W., Albert, C. E., and Lockman, F. J. 1982,
    ApJ, 263, 690

Irvine, N. J. 1989, ApJ, 337, L33

Munch, G. and Zirin, H. 1961, ApJ, 133, 11

Massive Stars: Their Lives in the Interstellar Medium
ASP Conference Series, Vol. 35, 1993
Joseph P. Cassinelli and Edward B. Churchwell (eds.)

IRC+10420: A COOL HYPERGIANT NEAR THE TOP OF THE HR DIAGRAM

TERRY JAY JONES, ROBERTA M HUMPHREYS, ROBERT D GEHRZ, AND GEOFFREY LAWRENCE
University of Minnesota, Department of Astronomy, 116 Church St SE, Minneapolis MN, 55455

INTRODUCTION

IRC+10420 has the spectrum of a very luminous F supergiant (F8Ia+) and a very large infrared excess from circumstellar dust (Humphreys et al. 1973). A very high mass loss rate of $3 \times 10^{-4} M_\odot$/yr has been estimated from its CO emission with an assumed distance of 4kpc (Knapp and Morris 1985). It is also one of the warmest stellar OH masers known (Giguere et al. 1976). Recent observations show that the 1612 MHz feature is growing while the main line 1665 feature is weakening (Lewis et al. 1986). It has also brightened in the visual about one magnitude from 1930 to 1970 (Gottlieb and Liller 1978).

Although the underlying nature of IRC+10420 is far from clear, most authors tend to include the star in the loosely defined 'Proto-Planetary Nebulae" classification (e.g. Habing et al. 1989, Hrivnak, Kwok and Volk 1989). These are stars with $L \leq 10^4 L\odot$ that have recently left a high massloss phase on the upper Asymptotic Giant Branch, and are now evolving towards the planetary nebulae phase. The very low surface gravity in their tenuous photospheres produces a supergiant spectrum, even though these stars are not true supergiants with massive main sequence progenitors. Classic examples include 89 Her and the Egg Nebula.

We have obtained new observations of IRC +10420, including visual and IR spectroscopy, photometry, imaging and polarimetry that show that the star is a true supergiant near the upper limits of the HR diagram. Its evolutionary history may be analogous to intermediate mass red giants evolving across the HR diagram toward the planetary nebulae, but IRC+10420 must have had a massive ($M \geq 40 M\odot$) main sequence progenitor. In this short paper we present some of our new observations and discuss the distance and luminosity of

IRC+10420. The entire data set and a more complete discussion
is given by Jones et al. 1992.

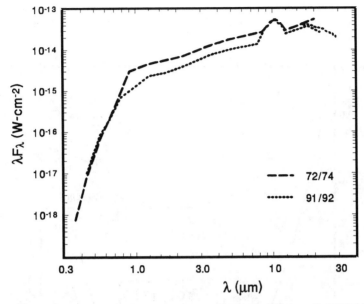

Figure 1. The spectral energy distribution of IRC+10420
at two different epochs.

DISCUSSION

With a B-V of ~2.7 mag, IRC+10420 is very highly reddened, but
what component is interstellar and circumstellar? Our visual
to infrared polarimetry and the polarimetry of Craine et al.
(1976) show that a major fraction of the polarization in
IRC+10420 is interstellar. The interstellar polarization
component in the K (2.2μm) band is about 1.5%, corresponding
to A_V~6 (Jones 1989). This is only slightly less than the
extinction one would compute assuming **all** of the E(B-V) color
excess is due to interstellar reddening.

The observed interstellar extinction of other stars in the
general direction of IRC+10420 (ℓ=47° b=-2.5°) reaches A_V = 4
mag at distances of 3 to 4 kpc. This places IRC+10420 ≥ 4kpc
from the Sun. IRC+10420 has a large positive radial velocity
of +75km-s^{-1} (LSR). The maximum observed HI velocity at this
longitude and latitude is also +75km-s^{-1} and IRC+10420 is
probably associated with this feature. Assuming that the
maximum radial velocity occurs at the tangent point, IRC+10420
must be at a distance of 5.8kpc (R_0 = 8.5kpc). Integrating the
observed energy distribution (see Figure 1) from 0.5 to 40μm
yields M_{bol} = -9.2 to -10.0 mag for a distance from 4 to 6kpc.

This places the star at or near the Humphreys-Davidson Limit at the top of the HR diagram (see Figure 2), well above the AGB limit and very much more luminous than typical proto-planetary nebulae. IRC+10420 could be evolving from a red supergiant back across the HR diagram, perhaps to become a Wolf-Rayet star.

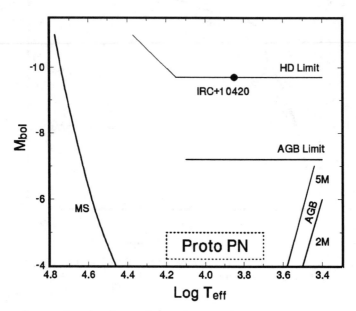

Figure 2. The location of IRC+10420 in the HR Diagram.

REFERENCES

Craine, E.R., Schuster, W.J., Tapia, S. and Vrba, F.J. 1976, Ap. J., 205, 802.

Giguere, P.T., Woolf, N.J., and Webber, J.C. 1976, Ap. J., 207, L195.

Gottleib, E.W. and Liller, W. 1978, Ap. J., 225, 488.

Habing, H.J., Hekkert, P. and Van der Veen, W.E.C.J. 1989, in Planetary Nebulae, IAU Symposium No. 131, Ed. S. Torres-Peimbert, p. 381.

Hrivnak, B.J, Kwok, S. and Volk, K.M. 1989, Ap. J., 346, 265.

Humphreys, R.M., Strecker, D.W., Murdock, T.L. and Low, F.J. 1973, Ap. J., 179, L49.

Jones, T.J., Humphreys, R.M., Gehrz, R.D., Lawrence, G., Zickgraf, F-J., Pina, R., Jones, B., Moseley, H., Casey, S., Glaccum, W.J., Staltzer, C. and Venn, K. in prep.

Jones, T.J. 1989, Ap. J., 346, 728.

Knapp, G.R. and Morris, M. 1985, Ap. J., 292, 640.

Lewis, B.M., Terzian, Y. and Eder, J. 1986, Ap. J., 302, L23.

Massive Stars: Their Lives in the Interstellar Medium
ASP Conference Series, Vol. 35, 1993
Joseph P. Cassinelli and Edward B. Churchwell (eds.)

DETERMINATION OF MAGNITUDE DIFFERENCES IN WC+ABS SYSTEMS

KENNETH R. BROWNSBERGER and PETER S. CONTI
Joint Institute for Laboratory Astrophysics, University of Colorado,
Campus Box 440, Boulder, CO 80309-0440

INTRODUCTION

Magnitude differences in binary or composite systems can be determined by comparing the equivalent width (EW) of nearly-constant lines with those of single WC stars. By assuming the flux of an emission line comes only from the WC star, a simple formula involving ratios of EW's leads directly to the magnitude difference between the WC star and it's companion.

EQUIVALENT WIDTH MEASUREMENTS

The EW measurements for this work include over 50 emission features stretching from 1550 - 10124 Å. We have used 45 single WCE stars from subtypes WC4 - WC7 to derive estimates of the mean line strengths for these emission features. Figure 1 shows a histogram of the mean EW of the 54 different emission features. Figure 2 shows a histogram of the STD of the mean EW. Note that 12 of the 54 lines show 1σ variation at or below the expected measurement uncertainty, which we estimate to be 30% given the difficulty of locating the continuum in WC stars.

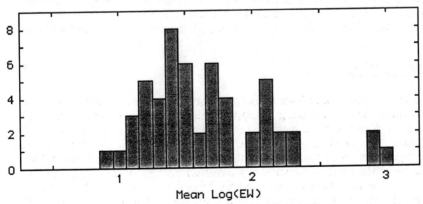

Fig. 1. Histogram of Mean values of Log(EW) for 54 different lines measured from 45 single WCE stars.

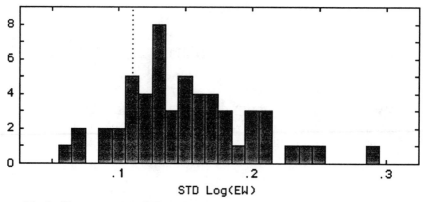

Fig. 2. Histogram of the STD of the Mean values of Log(EW) from figure 1. The dotted line marks the expected measurement uncertainty for individual EW measurements.

MAGNITUDE DIFFERENCE ESTIMATION

Assuming that all the flux within an emission feature comes from the WC star, and that the presence of a companion only increases the overall continuum level, it is easy to show that the magnitude difference, at wavelength λ, between the companion (OB) and the WC star can be written:

$$\Delta M = M_{OB} - M_{WC} = -2.5 \, [\, LOG(EW_{WC} - EW_{WC+OB}) - LOG(EW_{WC+OB}) \,]$$

where EW_{WC} is the mean EW at wavelength λ for all WC stars, and EW_{WC+OB} is the measured EW for the binary system.

Rather than limit ourselves only to those lines which show small variation, we have opted to compute the magnitude difference for ALL measurable lines in the WC+abs systems. Each estimate of ΔM is weighted by the measured STD of that particular line. The final magnitude difference for the system is computed by a weighted mean of all the estimated ΔM's. Magnitude Difference estimates for 24 WC+abs systems are listed in Table I. Note that with the exception of WR 50, the companions are always brighter than the WC star.

DISCUSSION

It is indeed overly simplistic to assume that the mean value of the EW for a particular line represents a good approximation of the line strength for all WC stars. However, for this method the assumption works fairly well for two reasons. First, several lines are fairly constant in strength across a wide range of subtypes. Second, even though many other lines change with subtype, the change is often small compared to the overall strength of the line.(Again, see Figures 1 and 2). Hence, the errors resulting from this assumption do not dominate the procedure and useful ΔM estimates can be obtained. A brief comparison with the literature (see Table II) shows that this method yields values of ΔM which agree reasonably well with published estimates.

TABLE I   Magnitude Difference Estimates

Star Name	Spectral Type	Mean	STD
Br 22	WC6+abs	-1.08	0.19
Br 28	WC4+abs	-1.43	0.44
Br 31	WC4+abs	-2.46	0.34
Br 32	WC5+abs	-1.89	0.36
Br 44	WC5+abs	-0.87	0.50
Br 67	WC4+abs	-2.15	0.37
Br 68	WC4+abs	-0.33	0.47
Br 70	WC5+abs	-0.73	0.56
Br 94	WC5+abs	-0.80	0.37
MG 01	WC4+abs	-0.82	0.46
MG 06	WC4+abs	-2.04	0.16
WR 009	WC5+abs	-0.13	0.50
WR 011	WC8+abs	-1.85	0.56
WR 030	WC6+abs	-0.56	0.45
WR 042	WC7+abs	-0.66	0.56
WR 048	WC6+abs	-2.39	0.38
WR 050	WC6+abs	0.28	1.07
WR 079	WC7+abs	-0.52	0.85
WR 086	WC7+abs	-0.72	0.79
WR 093	WC7+abs	-0.33	1.06
WR 113	WC8+abs	-1.34	0.80
WR 137	WC7+abs	-0.76	0.91
WR 140	WC7+abs	-0.92	0.34
WR 30a	WC4+abs	-1.88	1.09

TABLE II   Magnitude Difference Comparisons

Star Name	This Work	Lundström & Stenholm, 1984	Other
WR 011	-1.9	-1.3	-1.8[a], -1.4[b], -1.6[c]
WR 048	-2.4	-2.3	
WR 079	-0.5:	-1.0	
WR 093	-0.3:	-1.0:	
WR 113	-1.3:	0.0:	
WR 137	-0.8:	-1.0	

[a] Baschek & Scholz 1971, [b] Conti & Smith 1972, [c] Willis & Wilson 1976

REFERENCES

Baschek, B., Scholz, M. 1971, *A&A,* **11**, 83.
Conti, P.S., Smith, L.F. 1972, *ApJ,* **172**, 623.
Lundström, I., Stenholm, B. 1984, *A&AS,* **58**, 163.
Willis, A.J., Wilson, R. 1976, *A&A,* **47**, 429.

Massive Stars: Their Lives in the Interstellar Medium
ASP Conference Series, Vol. 35, 1993
Joseph P. Cassinelli and Edward B. Churchwell (eds.)

THE ISM STRUCTURES AND THE EARLY-TYPE STARS IN PUPPIS-VELA

M. SAHU AND A. BLAAUW
Kapteyn Laboratory, P.O.Box 800, 9700 AV Groningen, Netherlands

ABSTRACT   We identify three principal structures in the ISM in Puppis-Vela within ∼ 2 kpc: 1) the IRAS Vela Shell at ∼ 450 pc, 2) the Gum Nebula centred at a proposed distance of ∼ 800 pc and 3) the Vela Molecular Ridge (VMR) at ∼ 1 kpc. The early-type stars and associations most relevant for the interpretation of the ISM structures are identified and studied through a proper motion analysis. They include: 1) Vela OB2 at ∼ 450, 2) the young Vela R2 association at ∼ 800 pc and 3) the O-type supergiant, ζ Puppis. The main results are: 1) the IRAS Vela Shell is interpreted to be the remnant of the giant molecular cloud (GMC) around the Vela OB2 association and 2) the past projected path of ζ Puppis on the sky for various assumed distances suggests that the star originates in the VMR, close to the Vela R2 association. The Gum Nebula is interpreted to be a shell structure formed by the interaction of the Vela R2 stars with the surrounding ISM and is located at the near edge of the VMR.

1. INTRODUCTION

1.1 The ISM structures

Three principal structures can be identified in the ISM in the constellations of Puppis-Vela, extending from $l^{II} \sim 240°$ to 275° and $b^{II} \sim \pm 20°$, and within ∼ 2 kpc from the Sun:

1) The IRAS Vela Shell — the recently discovered ring-like complex of (which also includes several cometary globules (CGs) and dark clouds) with a radius of ∼ 7.°5, centred at about $(l^{II}, b^{II}) = (263°, -7°)$. This structure, which was discovered from the IRAS Sky Survey Atlas (ISSA, also referred to as the Super Sky Flux) maps, has been termed the IRAS Vela Shell (Sahu, 1992, Sahu and Blaauw, 1992). The stars of Vela OB2 lie within this shell. The mass and distance of this shell are estimated to be ∼ 10^6 M$_\odot$ and ∼ 450 pc respectively. The ionized gas in the region of the IRAS Vela Shell is seen as a bright emission patch which is superposed on the 36° diameter emission structure, the Gum Nebula (see below). A high resolution spectroscopic study (mainly in the [NII] λ 6584Å emission line) of the kinematics of the ionized gas in the Puppis-Vela region (Sahu, 1992, Sahu and Sahu, 1992) shows that the ionized gas in the region of the IRAS Vela Shell has an expansion of 10 ± 2 km s^{-1} with systemic velocities ranging from –10 to +12 km s^{-1} which are in fairly good agreement with molecular gas observations of the dark clouds and CGs.

2) The Gum Nebula — The appearance of the Gum Nebula, in H$_\alpha$ emission line surveys, is striking because of its large extent (diameter = 36°). Four principal attempts have been made to understand its nature and origin — 1) classical HII region model (Gum,

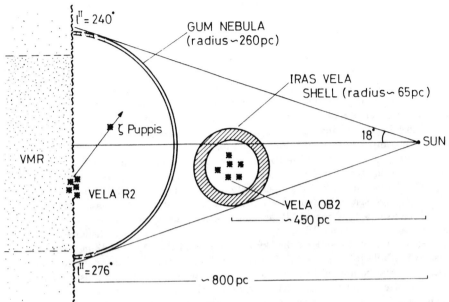

Fig. 1 The main conclusions drawn regarding the various ISM structures in Puppis-Vela – the Gum Nebula, the IRAS Vela Shell and the Vela Molecular Ridge – are schematically shown in this diagram as a projection onto the galactic plane (Note the diagram is not to scale). ζ Puppis and its past projected path, Vela OB2 stars and the Vela R2 stars are indicated in the figure together with the distances to the IRAS Vela Shell and the Gum Nebula. We have interpreted the Gum Nebula as a shell structure formed by the interaction of the Vela R2 stars with the surrounding ISM through their ionizing radiation, stellar winds and supernovae. The asymmetry ($\sim 8°$) between the H_α centre of the Gum Nebula and the Vela R2 stars maybe related to the displacement of ζ Puppis to lower and negative latitudes. The IRAS Vela Shell has been interpreted as the remnant of a GMC surrounding the relatively aged Vela OB2 association.

1956), 2) "Fossil" Stromgren sphere (FSS) model (Brandt et al. 1971), 3) old supernova remnant model (Reynolds, 1976) and 4) stellar wind model (Weaver et al., 1977).

In all these models the the intense emission region now associated with the IRAS Vela Shell was consodered to be part of the Gum Nebula since it appears that way on the H_α maps. However based on the recent ISSA maps and the study of the kinematics of the ionized gas in the region, we know that the IRAS Vela Shell and the Gum Nebula are two separate structures and our interpretation will be based on this.

3) The VMR — Recent wide-angle, extended CO emission observations of the Puppis-Vela region (Murphy and May, 1991), show that the most intense CO emission is confined to $l^{II} = 260°$ to 270°, $|b^{II}| < 2°$. This intense feature which can be most clearly seen on maps with velocities 5.9 km s^{-1} to 9.8 km s^{-1} was referred to by these authors as the Vela Molecular Ridge. Several population I objects lie within the CO contours defining the ridge. The young Vela R2 associations lie on a CO maximum in the ridge. The VMR

consists of four components whose radii range from 34 to 78pc and mass estimates are 10^5 to 10^6 M$_\odot$.

1.2 The early-type stars

We analysed newly determined proper motions of a sample of stars belonging to the Vela OB2 association and ζ Puppis (Sahu, 1992, Blaauw and Sahu, 1992) using modern meridian observations.

The proper motions of the Vela OB2 stars, after correcting for the solar reflex motion, suggest low internal velocity dispersion, indicating there is strong evidence for genetic coherence of the stars in this group. A comparison of the spectral-type luminosity diagram (using published Stromgren photometry – Brandt et al. 1971) with similar diagrams for other subgroup associations suggests that the Vela OB2 association is $\sim 2 \times 10^7$ years old. Therefore based on its age and low internal velocity dispersion, we conclude that Vela OB2 is an aged association on the verge of disintegration.

Combining the observed proper motion of ζ Puppis with the radial velocity, we worked out the past projected paths on the sky for the star for four assumed distances. For all the assumed distances, ζ Puppis has a space velocity over 70 km s^{-1} and belongs to the class of runaway OB stars. Independent of the assumed distance, the past projected paths of the star on the sky originate from the CO maximum in the VMR, close to the Vela R2 association. The time scales involved since the star's leaving this association was computed to be 1.5 $\times$ 10^6 years, which is entirely consistent with the kinematic ages found for massive runaway stars (Blaauw, 1992) and with the accepted age estimates for early-type supergiants (Maeder, 1987) and the estimated age of the Vela R2 (Herbst, 1975).

2. INTERPRETATION OF THE ISM STRUCTURES

The energy output of the Vela OB2 stars was estimated assuming that it is "standard" association subgroup. The observed kinetic energy of the IRAS Vela Shell can be explained by the expected energy output from the Vela OB2 stars through the combined effects of stellar winds and supernovae. In a similar approach, the observed kinetic energy of the Gum Nebula can be explained by the energy output from the Vela R2 association, whose members include ζ Puppis. The proposed relation between the ISM structures and the early-type stars in Puppis-Vela is shown in Figure 1.

REFERENCES

Brandt, J.C., Stetcher, T.P., Crawford, D.L., Maran, S.P.: 1971, Astrophys. J. **163**, L99

Blaauw, A.: 1992, in this volume

Blaauw, A. and Sahu, M.: 1992 (in preparation)

Gum, C.: 1956, Observatory **76**, 150

Herbst, W.: 1975, Astron. J. **80**, 503

Maeder, A.: 1987, Astron.Astroph. **178**, 159

Murphy, D.C., May, J.: 1991, Astron. Astrophys.**247**, 202

Sahu, M.: 1992, Ph.D. thesis, University of Groningen, The Netherlands

Sahu, M., Blaauw, A.: 1992 (in preparation)

Sahu, M., Sahu, K.C.: 1992 (in preparation)

Reynolds, R.J.: 1976, Astrophys. J. **206**, 679

Weaver, R., McCray, C.T., Castor, J., Shapiro, P., Moore, R.: 1977, Astrophys. J. **218**, 377

Massive Stars: Their Lives in the Interstellar Medium
ASP Conference Series, Vol. 35, 1993
Joseph P. Cassinelli and Edward B. Churchwell (eds.)

HIGHLY POLARIZED STARS IN CASSIOPEIA

DR. G. J. CORSO
Dep't. of Physics, De Paul University, Chicago, IL 60614

A. V. SHATZEL
Dep't of Math/CS, Barat College, Lake Forest, IL 60045

W. F. LANGE
Actinics, Inc., 1966 Raymond Dr., Northbrook, IL 60062

R. FOX
Dep't. of Physics, De Paul University, Chicago, IL 60614

ABSTRACT   A photographic survey of the x Cas region for highly polarized stars which is being conducted on the Yerkes Schmidt telescope is discussed. Preliminary results indicate that the stars in this area have an average polarization of about 3 to 5 percent but a few stars show polarization in excess of 6 percent.

A search for new highly polarized stars in the Cassiopeia region of the Milky Way is being conducted using the Schmidt telescope at Yerkes Observatory. The survey utilizes Kodak IIaO photographic plates and a Polaroid Corp. HN38 linear polarizer as well as a Sanyo CCD camera and an IBM compatible Hundai computer equipped with a Coreco Oculus 200 frame grabber and image analyzer.
    Observations by Hiltner (1956), Hall (1958), and Loden (1961) have established that the position angles of the polarization vector for stars in this region are clustered between 90 and 100 degrees and are parallel to the galactic plane. Our procedure consists in making two short (5 minute) closely spaced exposures on the same photographic plate with the Polaroid rotated between exposures so that the first exposure is made with the axis of the Polaroid parallel to the galactic equator and the second exposure made with the axis of the Polaroid at right angles to the galactic equator. All plates are developed in D19 and visually inspected using a 24 power microfiche reader for pairs of stars differing in brightness. These candidates for strong polarization are then scanned and digitized using the CCD camera and the Coreco frame grabber and

image analyzer. The percentage polarization is then obtained by numerical integration of the sky subtracted images. Because both images are recorded on the same plate and the exposure times are short, differences in the images due to processing or transparency and seeing changes are negligible.

A region covering approximately 24 square degrees which is located between galactic longitude 122 and 125 degrees and galactic latitude +4 and -4 degrees was surveyed for bright (apparent magnitude brighter than 11) highly polarized stars. This area includes Loden's x Cas region as well as several stars observed by Hiltner and Hall, so many calibration stars are available. In almost all cases where earlier measurements are available the difference between our results and previously published values was less than 1 1/2 %.

In general, the majority of the new highly polarized stars we have found show an average polarization of 3 to 5 percent with a few in excess of 6 percent. There is a tendency for the stars of highest polarization to occur south of the galactic equator in this region. Many, but not all, are early type stars. These results are consistent with Loden's suggestion that the degree of polarization in the region of SA8 is higher than that of the x Cas region and that in this direction stars within 100-150 pc of the sun show polarization.

REFERENCES

Hall, J. S. 1958, Publ. U.S. Nav. Obs., 17, VI
Hiltner, W. A. 1956, Vistas in Astronomy, 2, p. 1080
Loden, L. O. 1961, Stockhoms Observatorium Annaler, Band 21, 7

Massive Stars: Their Lives in the Interstellar Medium
ASP Conference Series, Vol. 35, 1993
Joseph P. Cassinelli and Edward B. Churchwell (eds.)

THE FALL AND RISE OF HR 8752'S [N II] EMISSION STRENGTH FROM 1961-1991

YARON SHEFFER
Department of Astronomy, The University of Texas at Austin, Austin, TX 78712

ABSTRACT McDonald spectra of HR 8752 from 1976-1991 reveal a monotonically increasing equivalent width of the [N II] emission at 6583 Å, thus confirming and expanding the trend first reported in Sheffer and Lambert (1987). By analyzing published B and V photometry we establish that the cause of [N II] equivalent width variability is the variable continuum of HR 8752. From the literature we uncover a remarkable minimum in [N II] strength around 1973 which possibly demonstrates a cyclical behavior of the continuum with a 30-year period, or longer.

INTRODUCTION

Very few yellow supergiants are known to exhibit the 6548 and 6583 ÅÅ [N II] lines superimposed on a photospheric absorption line spectrum. The enigmatic G0 Ia-0 star HR 8752 provided the first surprise when Sargent (1965) discovered the lines in its 1961 spectrum.

A B1 V companion is thought to ionize a local H II region which is believed to be the site for observed [N II] emission. It is clear that the angular extent of H II (and [N II]) is smaller than about 1", or roughly 3400 AU at the assumed distance of HR 8752 of 3.4 kpc (Humphreys, 1978). It is worthwhile, therefore, to check the behavior of this emission by monitoring its observables.

EQUIVALENT WIDTH OBSERVATIONS

Spectra at Hα that also show both [N II] lines were taken at coudé with the 2.7-m reflector of McDonald observatory. Since September 1976, the spectra have been recorded by Reticon detectors at frequencies of up to one observation a month. The slit width has been slightly oversampled by Reticons with different dispersions of 3 to 5 km s^{-1} diode^{-1}. Additional spectra were obtained at coudé with the 2.1 m Otto Struve reflector in 1986-7. Hereafter, we shall dwell only on the stronger [N II] line of HR 8752 at 6583Å.

We confirmed the 1977-86 monotonic increase in the 6583 Å equivalent width first reported in Paper II and discovered that the trend has persisted for the last 15 years, see Figure 1. The trend is obvious despite the scatter, which can be attributed to the following four causes: 1. Short-term continuum variations are caused by at least two radial pulsation modes on time scales of 300-400 days, see Paper II; 2. Continuum placement uncertainty of a few percent results in measurement noise; 3. Sporadic contamination of no more than 11 mÅ by the 6580.785Å telluric line of the blue wing of the profile; and 4. Possible variability of the underlying Si I photospheric line at 6583.707Å by less than 10 mÅ. Consequently, we find that the 6583 Å line has increased its equivalent width from 120 ±10 mÅ in 1976 to its present value of 230 ±10 mÅ.

R-CONTINUUM CALIBRATION

In order to investigate the continuum behavior, an analysis of R magnitudes is needed, since [N II] at 6583 Å lies very near the peak of the response curve for the R filter. But most photometrists report B and V magnitudes only. So we analysed V-R and B-V colors from the Arizona-Tonantzintla catalog, finding the following relation between the two unreddened indices:

$$(V\text{-}R)_0 = 0.751\ (B\text{-}V)_0^{0.732}. \tag{1}$$

We have adopted $E_{B\text{-}V} = 0.^m55$ and $E_{V\text{-}R} = 0.75 E_{B\text{-}V} = 0.^m41$. A comparison between 23 derived and observed R magnitudes yields an insignificant zero-point shift and a scatter of $\pm 0.^m026$.

Between 1976 and 1991 the R-continuum has faded by $0.^m62 \pm 0.^m11$. This fading can completely account for the increase of the [N II] equivalent width by $0.^m65 \pm 0.^m15$. From 9 yearly averages of overlapping extensive photometric and spectroscopic coverages we find the following relationship between the inferred R and the observed $W\lambda$:

$$W\lambda\ (6583\ \text{Å emission}) = 10^{0.4R+0.804} - 73, \tag{2}$$

where the largest residual from the fit is 12 mÅ. This convicingly demonstrates that the R-continuum induces the equivalent width changes of [N II], provided the intrinsic emission itself is of constant flux.

1961-1976 OBSERVABLES

We searched the literature for pre-1976 [N II] intensity values. In some cases the only way to extract the equivalent width was to measure it from published Figures. In this way we were able to find 6 more epochs from the 1960s and early 70s (see Sheffer and Lambert 1992 for details). As shown in Figure 1, the extended 1961-91 coverage establishes an equivalent width minimum around 1973 that was preceded by a decline.

We have also searched the literature for B, V, and R photometry for the 1961-1976 interval. Both emission equivalent width and inferred R brightness clearly show that the trend back to 1961 points to a rather prominent equivalent width, i.e. 200-odd mÅ, which certainly was a helpful factor in Sargent's (1965) discovery, who called the emission "moderately strong".

Figure 1

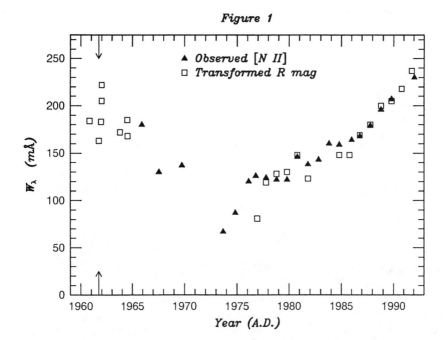

SUMMARY

The grand 1961-1991 compound plot of our [N II] equivalent widths and of inferred R magnitudes shows a unique history, see Figure 1. We see a 12-year long decline from 1961-72 leading to an equivalent width minimum of ca. 70 mÅ or less in 1972-3. We propose that it was caused by a maximum in R continuum brightness, in connection with the latest-ever spectral type (K, see Luck 1975) and associated temperature minimum that HR 8752 experienced ca. 1973. Thereafter, the [N II] emission has been increasing its equivalent width for the last 18 years, due to a rise in T_{eff} up to 6150 K (F8). The 1990-1 observables now have values similar to those they had around the 1961 discovery year of [N II] emission. The 30 years that have elapsed in the meantime constitute a discovery of the longest time scale of cyclic variations in HR 8752. It would be of great interest to stay tuned to 6583 Å in order to find out whether the T_{eff} variability is periodic over such long time scales.

REFERENCES

Humphreys, R. M. 1978, *ApJS*, **38**, 309
Luck, R. E. 1975, *ApJ*, **202**, 743
Sargent, W. L. W. 1965, *The Observatory*, **85**, 33
Sheffer, Y., and Lambert, D. L. 1987, *PASP*, **99**, 1277 (Paper II)
Sheffer, Y., and Lambert, D. L. 1992, *PASP*, in press.

Massive Stars: Their Lives in the Interstellar Medium
ASP Conference Series, Vol. 35, 1993
Joseph P. Cassinelli and Edward B. Churchwell (eds.)

HI BUBBLES AROUND O STARS

C.CAPPA DE NICOLAU [1a], V.NIEMELA [2b], P.BENAGLIA [1c]

[1] Instituto Argentino de Radioastronomia
C.C.5, 1894 Villa Elisa, Prov. Bs.As., Argentina

[2] Instituto de Astronomia y Fisica del Espacio
C.C. 67, Suc. 28, 1428 Buenos Aires, Argentina

INTRODUCTION

Recent HI 21-cm line observations of the interstellar medium surrounding WR stars have disclosed neutral gas bubbles around these stars (cf. Niemela and Cappa de Nicolau 1991, and references therein). Because of their low expansion velocities implying dynamical ages of a few million years, in all these bubbles the progenitor O stars of the WR stars have contributed substantially in the formation of the bubbles. Therefore, neutral gas bubbles should also appear around evolved O stars.

Here we report the discovery of three such bubbles around the O stars HD 112244, HD 135240 and HD 135591, HD 175754 and HD 175876.

HI DATA

In the analysis of the neutral gas distribution surrounding HD 112244, and HD 135240 and HD 135591, we have used the Strong et al.'s (1982) HI 21-cm line survey obtained with the Parkes radiotelescope. This survey has a regular grid of 0.5 in galactic longitude and 1^{O} in galactic latitude, with a

---

[a] Member of Carrera del Investigador,CONICET,Argentina

[b] Member of Carrera del Investigador, CIC, Prov.Bs.As. Argentina

[c] Fellow of CIC, Prov. de Buenos Aires, Argentina

half-power beam width of the antenna (HPBW) of $0\overset{\text{o}}{.}25$.
For the study of the neutral gas
distribution surrounding HD 175754 and HD 175876 we
obtained HI 21-cm observations with the 30-m
single-dish antenna at the Instituto Argentino de
Radioastronomia, in Villa Elisa, Argentina. This
antenna has a HPBW of $0\overset{\text{o}}{.}56$ at 21 cm.

RESULTS

The neutral gas distribution in the environs of the O
stars shows HI voids surrounded by thick envelopes
with stars located either near the center (HD 112244,
HD 175754) or near the higher density border (HD
135240 and HD 135591). The kinematical distances of
the voids agree with the spectrophotometric distances
of the associated O stars, whose parameters are
resumed in Table I.

TABLE I   Parameters of the O stars associated with the
           bubbles

Name HD	Spectral Type	Ref.	Spectr. parallax (kpc)	$-\log \dot{M}$ ($M_\odot$/y)	Ref.	Terminal veloc. (km/s) (3)
112244	O8.5Iab(f)	(2)	1.6	5.30	(5)	1575
135240	O7.5III((f))	(1)	1.1	6.93	(4)	2390
135591	O7.5III((f))	(1)	1.1	6.40	(4)	2180
175754	O8II((f))	(1)	2.3	5.90	(4)	2060
175876	O6.5III(n)(f)	(2)	2.3	5.22	(4)	2430

References: (1) Walborn 1972; (2) Walborn 1973; (3) Prinja
et al.1990; (4) Chlebowski & Garmany 1991; (5) Hutchings
1976.

The main observed and derived parameters of the
HI bubbles are listed in Table II. In each case, the
swept-up HI mass was obtained as a mean value between
the HI mass deficiency in the void and the HI mass in
the envelope. Both of them were derived from the HI
column density distribution. The ambient neutral gas
density was derived by distributing the swept-up mass
within the volume of the HI void. In all cases the

expansion velocities of the HI bubbles, taking into account the velocity interval where the structures are seen, are less than 10 km/s. The kinetic energy of the observed neutral gas bubbles is less than 10% of the kinetic energy of the stellar winds of the associated O stars. Therefore, the HI bubbles have most probably been blown by the strong stellar winds of the O stars.

TABLE II   Parameters of the HI bubbles

Related stars	HD 112244	HD 135240 HD 135591	HD 175754 HD 175876
Galactic coordinates (l,b) of the center	$303°.5,+3°.5$	$320°.3,-3°.5$	$16°.5,-9°.0$
Velocity interval v1,v2 (km/s)	$-17,-6$	$-33,-18$	$+5,+12$
Systemic velocity (km/s)	-15	-23	$+11$
Radius of the HI void (pc)	60	18	60
Radius of the HI envelope (pc)	75	27	85
Swept-up HI Mass ($M_\odot$)	11000	840	8300
Ambient gas density (cm^{-3})	0.5	0.1	0.4
Expansion velocity (km/s)	<10	<10	<10
Dynamical "age" (million years)	>4	>1	>4

REFERENCES

Chlebowski, T., and Garmany, C.D. 1991, Ap.J. 368, 241.

Hutchings, J.B. 1976, Ap.J. 203, 438.

Niemela, V.S., and Cappa de Nicolau, C.E. 1991, A.J. 101, 572.

Prinja, R.K., Barlow, M.J., and Howarth, I.D. 1990, Ap.J. 361, 607.

Strong, A.W., Riley, P.A., and Osborne, J.L. 1982, M.N.R.A.S. 201, 495.

Walborn, N. 1972, A.J. 77, 312.

Walborn, N. 1973, A.J. 78, 1067.

Massive Stars: Their Lives in the Interstellar Medium
ASP Conference Series, Vol. 35, 1993
Joseph P. Cassinelli and Edward B. Churchwell (eds.)

AN IUE SURVEY OF INTERSTELLAR H I Lyα ABSORPTION

ATHANASSIOS DIPLAS AND BLAIR D. SAVAGE
Washburn Observatory, University of Wisconsin-Madison,
475 North Charter Street., Madison, WI 53706

ABSTRACT   We study the Galactic interstellar neutral hydrogen distribution by analyzing existing interstellar absorption line data towards the complete sample of B2 and hotter stars observed at high resolution with the *IUE* satellite.   This study more than doubles the number of lines of sight previously observed. We have included the background correction algorithm of Bianchi and Bohlin in our data reduction and as a result, we achieve reduced errors in the derived HI column densities. We use the correlation between the [c1] index and the stellar Lyα absorption in order to correct the measurements for the effects of stellar Lyα blending. Approximately 30% of the stars are severely contaminated. We study the spatial distribution of the various phases of the ISM in the disk and in the galactic halo, and the relationship between interstellar gas and interstellar dust. By comparing the HI column densities with E(B-V) and E(Bump), several anomalous extinction regions are identified.

DATA ANALYSIS OVERVIEW

We analyze high dispersion (resolution $\Delta\lambda = 0.1$ Å), Short Wavelength Prime (SWP) camera images obtained with the *International Ultraviolet Explorer* (*IUE*) satellite, for **553** B2 and earlier stars. This number more than doubles the previously available sample (Shull & Van Steenberg 1985, Savage & Massa 1987, Bohlin, Savage & Drake 1978). The data were provided to us by the National Space Science Data Center (NSSDC) and span a 14 year period of observations with *IUE*. The stars cover all the sky, with 91 of the objects having |b| > 20°. The complete description of this work and the values of N(H I) are included in two upcoming papers (Diplas & Savage 1992, a, 1992, b).
The H I Lyα line lies in the short wavelength side of the SWP camera image where the Echelle order crowding is significant. As a result, there is a significant amount of scattered interorder background light present. We eliminate the effects of order overlap, by using the correction technique of Bianchi & Bohlin (1984). We also use the continuum reconstruction technique of Bohlin (1975), in order to derive the H I column density N(H I). Briefly, we multiply the spectrum by exp (τ_λ) = exp[$\sigma(\lambda)$ N(H I)], and the value of N(H I) is chosen so that the effects of the interstellar Lyα line are removed

from the reconstructed continuum. We assume that the absorption cross section $\sigma(\lambda)$ is Lorentzian.

Contamination of interstellar Lyα by stellar Lyα is often important for B stars. For stars cooler than B2.0 the contamination is so severe that we do not attempt to measure interstellar H I towards any significant number of such objects. Stellar Lyα absorption is mostly important for luminosity class IV and V stars, but it can occasionally be significant for more luminous stars as well. Shull & Van Steenberg (1985) did not account for the effects of stellar Lyα absorption in their analysis, so in many cases they overestimate the interstellar N(H I). The effect is especially important for nearby B2 V stars included in their sample. Savage & Panek (1974) established the following empirical method for estimating the stellar Lyα contamination. They note a positive correlation of the strength of the stellar Lyα line with the Stromgren narrow band photometry index $[c_1]$, for local stars with mainly stellar Lyα absorption. Furthermore, the observed values of N(H I) for the same group of stars provide a good fit to the NLTE calculations of the stellar Lyα line, by Mihalas (1972). We use these model calculations in order to estimate the value of stellar N(H I) for the stars in our sample. Whenever a $[c_1]$ measurement is available, we derive the interstellar value of N(H I) as the difference between the observed and the stellar N(H I). The stars with unreliable interstellar N(H I) represent 30% of the total sample and are rejected from any subsequent analysis. Our final "safe" sample includes 387 stars.

OBSERVATIONAL RESULTS

Distribution of the Gas away from the Galactic Plane
Figure 1 shows N(H I) sin |b| plotted versus |z| for the 387 stars. The solid line represents our fit to the data, with an exponential vertical gas distribution $n = n(0) \exp(-|z|/h)$, with mean density at the plane, n(0) = 0.42±0.08 cm^{-3} and scale height h = 156 ± 18 pc. A random scattering parameter σ (10^{σ} = 1.62) was added to the errors for each point, in order to force $\chi^2 = 1$. This indicates a very patchy distribution. We have identified a few stars with significant deviations from the assumed distribution. If we exclude the Sco - Oph and Orion stars that deviate significantly from the distribution, the derived parameters change to n(0)=0.38±0.06 cm^{-3}, h = 164 ± 15 pc, 10^{σ} = 1.41. The derived midplane density is smaller than the one similarly derived by previous studies. Shull & Van Steenberg (1985) report n(0)=0.55 cm^{-3} and h = 144 pc. The slightly higher value of n(0) their study can be interpreted as a result of stellar Lyα contamination. 21 cm observations indicate n(0) = 0.57 cm^{-3} (Dickey & Lockman 1990), for the sum of two gaussian distributions (n(0) = 0.395, 0.107 cm^{-3}, and FWHM = 212, 530 pc respectively) and of one exponential (n(0) = 0.064, h = 403 pc). The values of N(H I) sin |b| for the high |z| stars in our sample seem to keep rising with |z|, similarly indicating the presence of a more extended gas distribution than the one we are quoting here. A more detailed analysis can be found in Diplas & Savage (1992, b).

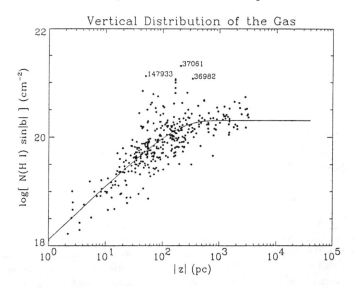

Fig. 1. N(H I) sin |b| plotted versus |z| for the 387 stars of our sample. The solid line indicates an exponential fit to the data. n(0) = 0.42 cm^{-3} and scale height h = 156 pc.

Gas versus Dust, and Anomalous sight lines

The derived mean ratio of gas to dust is <N(H I)/E(B-V)> = 4.91 10^{21} cm^{-2} mag and it remains fairly constant over a wide range of line of sight densities. For 200 of our stars, there are E(Bump) measurements of the 2175 Å extinction bump available from the ANS catalog of UV extinction measurements (Savage et al. 1985). The mean ratio is <N(H I)/E(Bump)> = 2.17 10^{21} cm^{-2} mag. We identify several sight lines with anomalous gas to dust ratio. The stars HD 15570, 24431, 24534, 36486, 37041, 37061, 55857, and 75222, have reliable N(H I) and E(B-V), but anomalous values of N(H I)/E(B-V).

REFERENCES

Bianchi, L., & Bohlin, R. C. 1984, A&A, 134, 31
Bohlin R. C. 1975, ApJ, 200,402
Bohlin R. C., Savage, B. D., & Drake, J. F. 1978, ApJ, 224, 132
Dickey, J. M., & Lockman, F. J. 1990, ARA&A, 28, 215
Diplas, A., & Savage, B. D. 1992, a, ApJ, submitted
Diplas, A., & Savage, B. D. 1992, b, in preparation
Mihalas, D. 1972a, NCAR Tech. Note NCAR - TN/STR-76 (Boulder: Nat. Center for Atmospheric Research)
Savage, B. D., & Massa, D. 1987, ApJ, 314, 380
Savage, B. D., Massa, D., Meade, M. & Wesselius, P. R. 1985, ApJS, 59, 397
Savage, B. D., & Panek, R. J. 1974, ApJ, 191, 659
Shull, J. M., & Van Steenberg, M. 1985, ApJ, 294, 599

Massive Stars: Their Lives in the Interstellar Medium
ASP Conference Series, Vol. 35, 1993
Joseph P. Cassinelli and Edward B. Churchwell (eds.)

SHOCK IONIZED GAS IN THE 30 DORADUS NEBULA

MICHAEL R. ROSA[1]
The Space Telescope-European Coordinating Facility, European Southern Observatory, Karl-Schwarzschild-Str. 2, D-8046 Garching, Federal Republic of Germany
[1] Affiliated to the Astrophysics Division, Space Science Department, European Space Agency

ABSTRACT   Considerable amounts of gas in the halo of the 30 Dor nebula seem to be ionized by shocks. Sources for the mechanical energy can readily be found in the hot interiors of large bubbles swept and maintained by fast stellar winds and SN explosions from the plentitude of massive stars in the 30 Dor starburst region. Including this additional interaction of massive stars with the ISM into photoionized model nebulae brings predictions of the ionization structure into agreement with observations.

INTRODUCTION

The interstellar medium in the prototype giant extragalactic H II region (GEHR) 30 Dor in the LMC is far from the quiescent case of a textbook Stroemgren sphere in isolation. Large ionized shells of expanding gas are revealed by deep imaging (eg. Meaburn 1984), and by emission line velocity mapping, as pioneered by Smith and Weedman (1971). More recently, X-ray images have demonstrated that hot gas at 5×10^6 K coexists with the H^+ gas at 10^4 K throughout the nebula. X-ray peaks are observed in rarefied bubbles presumably swept by stellar winds prior to SN explosions (Chu and MacLow 1990; Wang and Helfand 1991). Are the spectra of GEHRs contaminated by shocks as has been recently suggested by Peimbert, Sarmiento and Fierro (1991). If so, massive stars might keep these super-large volumes of gas ionized by all interactions with the ISM they are capable of during their entire life, namely: EUV photons, fast winds during Of and WR stages, and finally SN explosions.

30 DOR AND MODEL NEBULAE

Spectrophotometric data of the 30 Dor nebula have been analyzed for 70 positions covering a wide range of distances from the core region around R 136. In Figure 1 the ionic fractions of sulphur and oxygen are shown, overlayed with 5 model nebulae using Kurucz atmospheres and a blackbody of 100 kK as ionizing sources as described by Mathis and Rosa (1991). The bulk of the 30 Dor data is bracketed by model nebular tracks of 40 kK $\leq T_{eff} \leq$ 50 kK. However, about 20 % of the data seem to have too much S^+ at a given oxygen ionization, qualitatively similar to the situation in H II galaxies (Garnett

1989) where in some cases an ionizing source of infinite temperature would be implied (open triangles in Fig. 1).

THE SPREAD IN STELLAR TEMPERATURES

Stars of types O9 - O3 (33 kK $\leq T_{eff} \leq$ 55 kK) are definitively present in the 30 Dor cluster. The spread of data points perpendicular to model nebulae of differing ionizing T_{eff} at high ionization parameters, i.e. at low fractional O$^+$ and S$^+$ ionizations may therefore be due to different stars (open circles: lower T_{eff} than average, filled circles: higher T_{eff} than average) dominating the ionizing radiation at differing lines of sight. However, the pronounced deviation of data points from model tracks at low ionization parameters (star symbols), i.e. volumes where the ionization is dominated by the diffuse field, can not be explained that way.

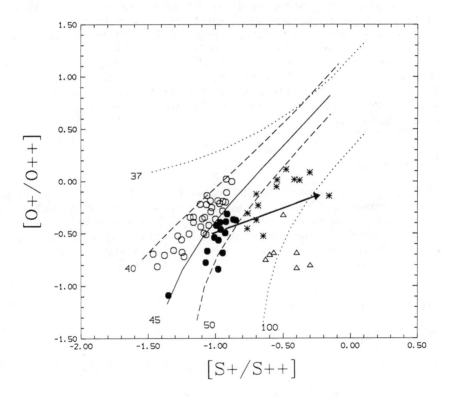

Fig. 1.   The sulphur – oxygen ionization of the 30 Dor nebula. Tracks of model nebulae are shown for differing ionizing T_{eff}. The data of 30 Dor are arbitrarily grouped to indicate low T_{eff} (open circles), high T_{eff} (filled circles), and positions not in accord with simple models (star symbols). Data for H II galaxies are shown by open triangles. The arrow indicates the shift of model tracks if shock ionized volumes are observed along the same line of sight.

ARE THE ATOMIC DATA APPROPRIATE ?

Garnett (1989) discussed the modification of S ionization cross–sections and dielectronic recombination rates as possible solutions for the H II galaxy case. It is clear that modification of a single atomic parameter for the sulphur ions in question will, to first order, simply result in a horizonrtal shift of the model tracks with respect to the observations. With such an approach data points representing integrated spectra of 30 Dor like complexes, i.e. the H II galaxies, might be brought into accord with model nebulae of reasonable ionizing sources. However, the good match of models with the bulk of the data for 30 Dor will be lost entirely.

SHOCK IONIZED VOLUMES

Adding the line emissivity of a shock with $v \leq 200$ km s^{-1} ionized volume to a photoionized model will reduce the observed degree of ionization much more in S than in O. Following Peimbert, Sarmiento and Fierro (1991) shock model emissivities have been added to emissivities from purely photoionized volumes in varying percentages. The arrow in Fig. 1 indicates how an average volume in 30 Dor would move in the diagram if a shock of 140 km s^{-1} contributed 50 % of the Balmer line emissivity. Note that most of the unusual data points in 30 Dor could actually be explained by shock contamination at levels between 5 and 20 % . Further support for this explanation comes from the fact that most of the positions probably contaminated by shocks are located in rarefied bubbles in the outer region of 30 Dor and/or are coincident with peaks in the X–ray images of the nebula.

IMPLICATIONS

Spatially integrated spectra of entire 30 Dor like nebulae (GEHRs) will be affected more by shock contamination than a pencil beam observation in a spatially resolved object, since the halo component usually contributes about 50 % of the total Balmer luminosity. Absolute chemical abundances might than be underestimated because of an overestimate of electron temperatures. Abundance ratios with respect to oxygen will however suffer very little as can be shown in the case of 30 Dor (Kinkel and Rosa, 1992).

REFERENCES

Chu, Y.–H., MacLow, M.–M. 1990, *Ap. J.*, **365**, 510.
Garnett, D.R. 1989, *Ap. J.*, **345**, 282.
Kinkel, U., Rosa, M.R. 1992, in *New Aspects of Magellanic Cloud Research*, ed. G. Klare, Springer, in press
Mathis, J.S., Rosa, M.R. 1991, *Astr. Ap.*, **245**, 625.
Meaburn, J. 1984, *M.N.R.A.S.*, **211**, 521.
Peimbert, M., Sarmiento, A., Fierro, J. 1991, *Pub. A.S.P.*, **103**, 815.
Smith, M.G., Weedman, D.W. 1971, *Ap. J.*, **169**, 271.
Wang, Q., Helfand, D.J. 1991, *Ap. J.*, **370**, 541.

SECTION 3

EFFECTS OF MASSIVE STAR WINDS
AND RADIATION ON THE ISM

Massive Stars: Their Lives in the Interstellar Medium
ASP Conference Series, Vol. 35, 1993
Joseph P. Cassinelli and Edward B. Churchwell (eds.)

WIND MASS AND ENERGY DEPOSITION

JOHN I. CASTOR
Lawrence Livermore National Laboratory, L-58, P. O. Box 808,
Livermore, CA 94550, U.S.A.

ABSTRACT   The subject of mass return to the interstellar medium by
the winds of massive stars is reviewed from the viewpoint of calculations
of stellar evolution with mass loss, combined with the birth rate function
for stars of different masses (IMF). The uncertainties due to the IMF and
in the assumed mass loss rates are discussed. The corresponding energy
deposition rates from stellar winds are also estimated.

I. INTRODUCTION

The cycling of material through massive stars—defined here as those stars that
start on the main sequence with about 9 or more solar masses, enough to form
a type II supernova at the end of their evolution—is a simple process. Almost
all the mass that first forms the star is returned to the interstellar medium in
one form or another by the end of the star's evolution; only perhaps one solar
mass of the original tens of solar masses is left as a compact remnant such as
a black hole or neutron star. Thus the gross rate of mass return is within a
few per cent of the total rate of formation of mass into stars. The *interesting*
question, of course, is how this mass return is distributed among the different
channels—main sequence mass loss in a stellar wind of O and B type stars,
mass loss in a red supergiant wind, very rapid mass loss during a transitory
phase as a luminous blue variable (LBV) similar to the prototypes P Cyg and
η Car, mass loss in an intense wind as a Wolf-Rayet (WR) star, and finally the
mass expulsion in a final supernova event.

The injection of energy into the interstellar medium through these mass
return processes is also of considerable interest. This is the energy that must
maintain the random velocities of the interstellar clouds—and therefore of
the stars that are formed from them—which would otherwise have damped
out early on in the history of the Galaxy from the dissipative effect of cloud
collisions. The energy deposition is easy to calculate together with the mass
deposition, since the *specific* energy (energy per unit mass) associated with
each channel of mass return is fairly well constrained observationally or
theoretically.

One way of calculating the mass injection to the ISM is to perform a
census of the stellar population within a fixed volume of space, and simply
sum up the injection rates that can be measured or estimated for each star.
This assumes that the stellar population is in a statistically steady state within

that volume, which requires that it be chosen large enough, and still additions must be made for phases of evolution so brief that the sample volume does not contain any representatives; *e.g.*, supernovae. This approach was taken by Abbott (1982b), who estimated for the first time the mass return rate through the winds of OB stars and WR stars, and found them to be sizeable, especially that for WR stars.

An alternative approach is to use the same stellar census information to derive the *initial mass function* (IMF) of the stars, that is, the birth rate function of stars as a distribution over initial stellar mass. This is found by combining star counts with theoretical calculations of the lifetimes of the stars in different phases of evolution, and their corresponding HR diagram locations. Assuming that the channels of mass return are properly represented in the evolutionary calculations, the rates of mass and energy return for the stellar ensemble are found by integrating over the IMF the mass and energy deposited in each channel for a given initial mass. The advantage of this indirect approach is that the channels that correspond to very rare or brief phases of evolution are included with much less statistical error than they would be by direct counting. The main disadvantage is the reliance on the theoretical evolutionary calculations, and the mass and energy yields that have been built into them. This kind of work has been done by Maeder and associates using their calculations of massive star evolution including mass loss (Maeder and Meynet 1987, 1988 and Maeder 1990). A thorough review of the mass and energy returns for synthesized stellar populations using this method has been done by Robert, Leitherer and Drissen (at this workshop and submitted to *Ap. J.*).

In §II the evolution calculations of Maeder (1990) will be used to illustrate how the mass yield through different mass return channels varies with the initial mass of the star. These yields will be combined with estimates of the initial mass function (IMF) to show the possible range of mass return rates. The evolution calculations have to be made with some assumption for the dependence of the rate of mass loss on the current structure of the star; this assumption leads directly to the mass yields for the various channels, and influences whether, for example, the star ever becomes a WR star, or a LBV or a red supergiant. Sometimes the mass-loss formula is based on theory, for example that of Castor, Abbott and Klein (1975), and sometimes an empirical relation such as those of Abbott (1982a) and de Jager, *et al.* (1988) is used. The range of predictions of some current mass loss formulae is considered in §IV. A summary of the results for the mass and energy injection rates, with warnings about the uncertainties, is given in §V.

II.   MASS BUDGET FOR MASSIVE STARS

The disposition of the material that forms into massive stars as the stars go through their evolution and shed bits from their surfaces can be illustrated using the evolutionary calculations of Maeder (1990). This work was a revision and extension of the calculations of Maeder and Meynet (1987, 1988), in which key changes were made to the assumed rates of mass loss, in addition to other changes. The mass loss rate for WR stars of type WNE (*i.e.*, WN4–WN6), WC and WO was assumed to vary in proportion to $\mathcal{M}^{2.5}$, following Langer (1989),

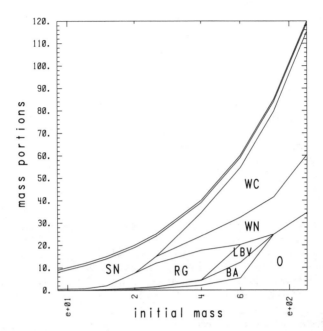

Fig. 1.   Mass lost by category *vs.* initial stellar mass. The distances
between successive curves upward are the mass lost to: O-star wind,
BA-supergiant wind, LBV mass loss, red supergiant wind, WN-type
Wolf-Rayet wind, WC-type Wolf-Rayet wind, and supernova ejecta. The
highest band represents the mass left in the compact stellar remnant,
assumed to be $1\,\mathcal{M}_\odot$.

rather than having the constant value $2.8 \times 10^{-5}\,\mathcal{M}_\odot\,\mathrm{y}^{-1}$ (Bieging, Abbott and
Churchwell 1982). The mass loss rate for most of the other evolutionary phases
was taken from the general formula of de Jager, Nieuwenhuijzen and van der
Hucht (1988), rather than the 1986 publication of the same authors.

These models are remarkable in that the Wolf-Rayet phase of evolution,
when He- or C-rich material is exposed on the surface, is quite extended in
time, and, as a result, large amounts of mass are lost in this period. This is
illustrated in Figure 1, which shows the amount of mass lost from the star in
each phase of evolution, as a function of the initial stellar mass. These amounts
are derived from the tables of total stellar mass at each evolutionary point in
Maeder (1990). I have assumed that each evolutionary sequence ends with a
supernova explosion, and for the figure I arbitrarily assumed that a compact
stellar remnant of $1\,\mathcal{M}_\odot$ was left; as can be seen in the figure, this amount of
mass is inconsequential.

The general trends of mass return with initial mass are easily seen
in Figure 1: At low initial mass all the mass returned is in the supernova
explosion, while at high mass the overwhelming mass return is in the hot-star
winds, especially those of the Wolf-Rayet phases. In the intermediate range,
$20\text{–}40\,\mathcal{M}_\odot$, red supergiant mass loss is quite important, and around $60\,\mathcal{M}_\odot$ the
loss as a LBV is significant. When we consider the energy deposition into the

ISM, the red supergiant and LBV phases almost drop out of consideration, since the velocities of these outflows (of order $10\,\mathrm{km\,s^{-1}}$ and $500\,\mathrm{km\,s^{-1}}$, respectively) are much less than for the hot star winds and the supernovae.

The mass yields in Figure 1 can be turned into mass return rates by folding in the IMF. The initial mass function $\xi(\log\mathcal{M})$ is defined (Scalo 1986) such that $\xi(\log\mathcal{M})d\log\mathcal{M}dV$ is the rate of star births in a spatial element dV and a range of logarithmic initial mass $d\log\mathcal{M}$. Let Q stand for one type of star, or for a star in a certain phase of evolution. Let $t_Q(\mathcal{M})$ be the lifetime of a star in the Q phase, given that its initial mass is $\mathcal{M}$. Likewise, let $\Delta\mathcal{M}_Q(\mathcal{M})$ be the mass returned in the Q phase for a star of initial mass $\mathcal{M}$. Then we have

$$\text{space density of } Q \text{ stars} = \int d\log\mathcal{M}\,\xi(\log\mathcal{M})\,t_Q(\mathcal{M}) \qquad (1)$$

$$\text{mass return rate of } Q \text{ stars} = \int d\log\mathcal{M}\,\xi(\log\mathcal{M})\,\Delta\mathcal{M}_Q(\mathcal{M}) \qquad (2)$$

Equation (1) is fitted to the observed counts of OB stars to derive the IMF for the massive stars. The masses of individual stars can be found by placing them on the evolutionary tracks, or the similarly-defined luminosity function can be found directly, then transformed statistically to the IMF; the former procedure should be more accurate.

The results of several determinations of the IMF for the massive stars are given by Scalo (1986, his Fig. 18). Particular note can be taken of the determinations by Garmany, Conti and Chiosi (1982) and by Humphreys and McElroy (1984). As summarized by Leitherer (1990), these two determinations are fitted by

$$\xi(\log\mathcal{M}) = 2.30 \times 10^{-3}\mathcal{M}^{-1.6}\,\mathrm{births\,y^{-1}\,kpc^{-2}\,dex^{-1}} \qquad \text{(GCC)}, \qquad (3)$$

and

$$\xi(\log\mathcal{M}) = 5.30 \times 10^{-2}\mathcal{M}^{-2.4}\,\mathrm{births\,y^{-1}\,kpc^{-2}\,dex^{-1}} \qquad \text{(HM)}. \qquad (4)$$

The determinations tend to agree best at the intermediate masses. At the low mass end, incompleteness of the volume-limited sample becomes a problem; at the high mass end the number of stars is very small, and the statistical error is large. Since an appreciable number of the most massive stars belong to giant OB associations, which are sparsely distributed, a sample volume $3\,\mathrm{kpc}$ in radius is none too large to represent them fairly.

A plot of the data in Figure 1 multiplied by $\xi(\log\mathcal{M})$ is shown for the GCC case in Figure 2. The axes are chosen so that an area on the figure is equal to the actual mass return rate. The dominance of supernovae, especially in the initial mass range below $25\,\mathcal{M}_\odot$, is apparent, as is the large contribution made by red supergiants. But just as remarkable, and not as expected, is the very substantial contribution of the Wolf-Rayet (WN and WC) stars. If the HM IMF is used in place of the GCC one, the results are as in Figure 3. Now the supernovae and the red supergiants are completely dominant, because the

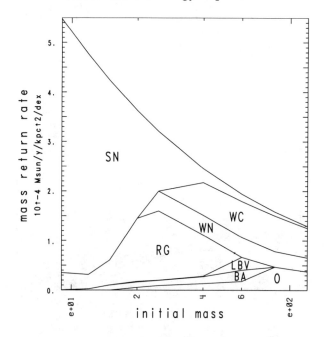

Fig. 2.   The data of Fig. 1 weighted by the GCC initial mass function.

steeper IMF in this case emphasizes the low masses over the high ones.

The space densities implied by the Maeder (1990) data, when combined with the GCC and HM initial mass functions, are shown in Table I. For comparison, the total rate of supernovae from the massive stars is also given. It is assumed that all stars with $\mathcal{M} \geq 9\,\mathcal{M}_\odot$ become supernovae, either type II or Ib (if the progenitor of a type Ib SN is a WR star). The O-star phase is assumed to end when $T_{\mathrm{eff}} < 33,000\,\mathrm{K}$. The BA phase ends and the LBV phase begins at core hydrogen exhaustion, and the LBV phase ends when $T_{\mathrm{eff}} < 5,000\,\mathrm{K}$ or the star becomes a WR star. All the integrations have a lower limit $9\,\mathcal{M}_\odot$, so only a minor fraction of all BA stars are included.

Table I   Star Densities *vs.* Evolutionary Phase for Two IMFs

IMF	O	BA	\(\text{star density kpc}^{-2}\)				SN rate \(\text{y}^{-1}\,\text{kpc}^{-2}\)
			LBV	RG	WN	WC	
GCC	27	259	0.62	44	0.65	0.99	1.86×10^{-5}
HM	47	879	2.03	146	0.81	0.98	5.10×10^{-5}

The corresponding results for the mass return rates, computed from equation (2), are given in Table II.

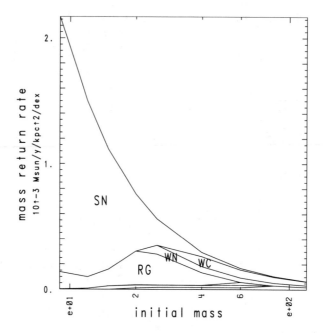

Fig. 3.   Like Fig. 2, but with the HM initial mass function.

Table II   Mass Return Rates for Two IMFs

		mass return rate $10^{-5}\,\mathcal{M}_\odot\,\mathrm{y}^{-1}\,\mathrm{kpc}^{-2}$					
IMF	O	BA	LBV	RG	WN	WC	SN
GCC	1.73	1.06	0.46	6.2	2.70	3.97	17.3
HM	1.54	1.56	0.43	11.9	2.96	3.47	48.0

The densities in Table I can be compared with those derived from star counts, as a check on the IMF and the evolutionary calculations. Abbott (1982b) gives the figures $20\,\mathrm{kpc}^{-2}$ and $1.7\,\mathrm{kpc}^{-2}$ for the densities of O and WR stars, respectively. These are in fair agreement with the GCC case in Table I. The often-used rate for type II SN is half of the total rate, which is one SN per galaxy per 30 years; this implies $2.4 \times 10^{-5}\,\mathrm{y}^{-1}\,\mathrm{kpc}^{-2}$ for SN II if the effective area of the galaxy is $\pi(15\,\mathrm{kpc})^2$. This also is in fair agreement with the GCC case. A thorough discussion of the comparative numbers of O and WR stars, of the WN/WC ratio, and of the populations of the WR subclasses has been given by Maeder (1991); a somewhat dissenting viewpoint of the same questions is offered by Vanbeveren (1990, also Vanbeveren and de Loore in this workshop). It remains unclear whether the O/WR ratio can be explained without assuming that a significant number of WR stars are made by mass exchange in binary systems.

The mass return from all types of hot star wind (O, BA, LBV and WR) is $9.9 \times 10^{-5}\,\mathcal{M}_\odot\,\mathrm{y}^{-1}\,\mathrm{kpc}^{-2}$ in the GCC case, and $1.00 \times 10^{-4}\,\mathcal{M}_\odot\,\mathrm{y}^{-1}\,\mathrm{kpc}^{-2}$ in the HM case, both in reasonable agreement with Abbott's figure $8.6 \times$

$10^{-5}\,\mathcal{M}_\odot\,y^{-1}\,kpc^{-2}$. The contribution of O stars is less and that of BA and LBV stars is greater, according to Table II, than Abbott found. The proportion of hot wind mass return to SN II+Ib mass return is 1:1.7 in the GCC case, and 1:4.8 in the HM case. The choice of IMF evidently makes a qualitative difference in the relative importance of hot wind mass input. This might be resolved by looking carefully at the O/B ratio, free of completeness problems.

III. SENSITIVITY TO MASS-LOSS RATES

The calculations of massive star evolution with mass loss by Maeder (1990) were made with the assumptions about wind mass loss rates that were mentioned earlier. The previous work by Maeder and Meynet (1987, 1988) used a uniform WR mass loss rate and an earlier version of the de Jager, et $al.$ formula for other stars. The results were qualitatively different: the mass lost in the WR phase was about $15\,\mathcal{M}_\odot$ for all stars that became WR stars, and these were the stars with $\mathcal{M} > 40\,\mathcal{M}_\odot$; this meant much less WR mass loss for the most massive stars. The typical WR lifetime was 5×10^5 y, while Maeder (1990) found WR lifetimes above 10^6 y for massive WR stars. The number ratio of O/WR stars could not be made to agree with star counts using any reasonable IMF if the Maeder and Meynet (1988) data were used, unless binaries played a major role, as Vanbeveren (1990) has suggested.

The great changes associated with a different set of mass loss rates underscores the importance of making an accurate determination of them. The means used to measure mass loss rates, and their accuracy, cannot be described here; for a review see Garmany (1988) and Conti (1988). It should be borne in mind, however, that for the large majority of OB stars the determination of $\dot{\mathcal{M}}$ is based on the strength of the P Cygni lines of CIV, NV and SiIV measured with the IUE observatory. These lines are not sufficient to fix $\dot{\mathcal{M}}$ without assuming an ionization model based on earlier Copernicus data, or making an empirical calibration to other $\dot{\mathcal{M}}$ measurements. Whichever of these routes is taken, the results cannot be considered very reliable.

The uncertainty of the predicted mass loss rates can be appreciated by examining Figure 4, in which mass loss rates on the zero-age main sequence, and at the end of the main sequence phase, are compared between the de Jager, et $al.$ (1988) empirical formula and the Kudritzki, et $al.$ (1989, KPPA) theoretical approximation. (The values of L, M and T_{eff} are taken from the tracks of Maeder [1990]. The KPPA formulae are evaluated using the force-law parameter set A of Kudritzki, et $al.$ 1992.) Also shown is the Howarth and Prinja (1989) empirical relation. What we see is a remarkably large spread in mass loss rate at $L = 10^6 L_\odot$, where the range between the de Jager ZAMS relation and the de Jager supergiant relation is a factor 27! These de Jager $\dot{\mathcal{M}}$ values are the same as those used and tabulated by Maeder (1990). The KPPA theoretical relations also predict a main sequence-supergiant $\dot{\mathcal{M}}$ difference in the same sense, but only a factor 3.4 at $L = 10^6 L_\odot$. Howarth and Prinja found that the relation illustrated in Figure 4 fit both main sequence and supergiant stars, and that there was no statistically significant dependence on luminosity class ($i.e.$, on T_{eff} for a given L). The contradiction with the de Jager, et $al.$,

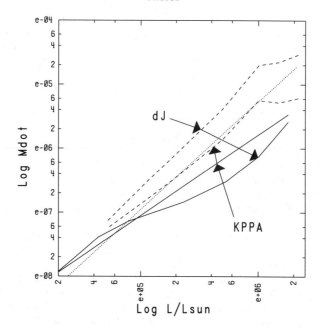

Fig. 4.   Mass loss rate *vs.* stellar luminosity. Solid lines: relation
along zero-age main sequence. Dashed lines: relation at core hydrogen
exhaustion. Curves labelled dJ use the de Jager, *et al.* (1988) formula;
those labelled KPPA use the Kudritzki, *et al.* (1989) formula. Dotted
curve: relation from Howarth and Prinja (1992).

relation is striking.

    Can the disparities shown in Figure 4 be understood? I think the answer
is, yes, in part. The strong T_{eff} dependence of $\mathcal{M}$ at large L according to
de Jager's formula is a consequence of fitting both the LBVs at low T_{eff} and
the O3 stars at high T_{eff}. The very high mass loss rate of the LBVs, and
their episodic behavior, is now thought to be related to a critical shift in the
ionization state of the wind (Pauldrach and Puls 1990, Lamers and Pauldrach
1991). The low $\mathcal{M}$ values ascribed to the O3 stars follow from their observed
weaker P Cygni absorptions in CIV, NV and SiIV; the true ionization state
in the O3 wind is unknown, and on the conservative assumption that it is like
that in the winds of cooler stars, a low $\mathcal{M}$ is indicated. The KPPA theoretical
mass loss rates, computed all with the same set of force parameters, have the
implicit assumption that the ionization state does not change very much across
the O star range. It is likely that if the true ionization state of the wind were
reflected in both the constants of the force law used to predict theoretical mass
loss rates, and in the ionization fractions used to infer mass loss rates from
observed line strengths, the discrepancies would be greatly reduced.

    The low ZAMS mass loss rates for the most massive stars, and the rapid
increase as T_{eff} declines, do influence the course of evolution of those stars.
This causes the O star mass loss to be smaller, and the WR mass loss, and the
duration of the WR phase, to be larger than otherwise. It is responsible for a

relatively large ratio of number of WR stars to number of O stars. This result is in agreement with observation, but until the mass loss rates receive further validation this might be regarded as fortuitous.

IV. SUMMARY OF MASS AND ENERGY INJECTION RATES

The estimated mass return rates to the ISM from massive stars are given in Table II. The part of this that is due to hot star winds (O+BA+LBV+WN+WC) should scale in proportion to the space density of WR and O stars, and this is given fairly well by the GCC IMF; thus the first line of Table II may be taken as a reasonable estimate of this part of the mass deposition. The SN part of the mass return varies more closely in proportion to the space density of those B stars with $\mathcal{M} \geq 9\mathcal{M}_\odot$. It is likely that the GCC IMF undercounts those stars, and the mass return rate from SN may lie between the GCC and HM values in Table II. The rates of energy return to the ISM by massive stars can be estimated from the mass return rates by multiplying them by $v_\infty^2/2$, where v_∞ is the appropriate average of the terminal speed of the wind, or, in the case of supernovae, the rms speed of the ejecta. This gives the return to the ISM in the form of kinetic energy. The overall conversion efficiencies of wind or SN energy into mechanical energy of the ISM will not be considered here.

In Abbott's (1982b) survey of mass and energy deposition he found average values of v_∞ for the O stars, WR stars and BA stars of $3000\,\mathrm{km\,s^{-1}}$, $2500\,\mathrm{km\,s^{-1}}$ and $1600\,\mathrm{km\,s^{-1}}$, respectively. Improved values of the energy deposition could in principle be found by fitting a KPPA wind model to each time point in Maeder's (1990) sequences. The KPPA theory is known to give a satisfactory fit to the v_∞ values of OB stars. For the WR stars, empirical values of v_∞ are still needed. Proceeding with Abbott's averages for all types, and also $v_\infty = 500$ and 10 for LBV and RG, and taking 10^{51} erg for the energy of all type II and Ib supernovae, leads to Table III.

Table III   Kinetic Energy Deposition

IMF	energy deposition $10^{38}\,\mathrm{erg\,s^{-1}\,kpc^{-2}}$						
	O	BA	LBV	RG	WN	WC	SN
GCC	0.49	0.085	0.004	2×10^{-5}	0.53	0.78	5.89
HM	0.44	0.13	0.003	4×10^{-5}	0.58	0.68	16.2

The total energy deposited by stellar winds is almost independent of the IMF: 1.90×10^{38} for the GCC IMF, and 1.83×10^{38} for the HM IMF. The energy deposited by SN II+Ib is more sensitive to the IMF. If SN I have an equal rate to SN II, and an energy of 5×10^{50} erg, then the proportion of wind energy to total SN energy is 1:4.65 for GCC and 1:13.3 for HM, with the correct figure likely to be between these limits. These results are essentially unchanged from those of Abbott (1982b).

Because a large fraction of massive stars are formed in OB associations, where the association members are nearly coeval, and because of the similar situation in a starburst, the history of the energy deposited by winds and supernovae becomes important, and the winds play a greater role than their total deposition would indicate. The reason is that the deposition of energy from the massive star winds begins promptly, and lasts through the 3–4 $\times 10^6$ y lifetime of these stars. The 9–20 $\mathcal{M}_\odot$ stars that dominate the SN energy deposition do not reach the SN stage until after 10^7 y. Thus the early evolution of the association or starburst is dominated by the winds, and the SN provide an "afterburner" effect later on. This is discussed in detail by Leitherer (1990) and Leitherer, et al. (1992).

I am grateful to N. Langer, H. Lamers, C. Garmany and C. Robert for helpful discussions during the workshop, and to R. Kudritzki for providing the subroutines to evaluate the KPPA formulae. This work was performed under the auspices of the U. S. Department of Energy by the Lawrence Livermore National Laboratory under Contract No. W-7405-ENG-48.

REFERENCES

Abbott, D. C. 1982a, *Ap. J.*, **259**, 282.
Abbott, D. C. 1982b, *Ap. J.*, **263**, 723.
Bieging, J. H., Abbott, D. C. and Churchwell, E. B. 1982, *Ap. J.*, **263**, 207.
Castor, J. I., Abbott, D. C. and Klein, R. I. 1975, *Ap. J.*, **195**, 157.
Conti, P. S. 1988, in *O Stars and Wolf-Rayet Stars*, eds. P. S. Conti and A. B. Underhill, NASA SP-497, 168.
de Jager, C., Nieuwenhuijzen, H. and van der Hucht, K. A. 1986, in *Luminous Stars and Associations in Galaxies*, IAU Symp. 116, eds. C. W. H. de Loore, A. J. Willis and P. Laskarides, 109.
de Jager, C., Nieuwenhuijzen, H. and van der Hucht, K. A. 1988, *Astr. Ap. Suppl.*, **72**, 259.
Garmany, C. D. 1988, in *O Stars and Wolf-Rayet Stars*, eds. P. S. Conti and A. B. Underhill, NASA SP-497, 160.
Garmany, C. D., Conti, P. S. and Chiosi, C. 1982, *Ap. J.*, **263**, 777.
Howarth, I. D. and Prinja, R. K. 1989, *Ap. J. Suppl.*, **69**, 527.
Humphreys, R. M. and McElroy, D. B. 1984, *Ap. J.*, **284**, 565.
Kudritzki, R.-P., Hummer, D. G., Pauldrach, A. W. A., Puls, J., Najarro, F. and Imhoff, J. 1992, *Astr. Ap.*, **257**, 655.
Kudritzki, R. P., Pauldrach, A., Puls, J. and Abbott, D. C. 1989, *Astr. Ap.*, **219**, 205 (KPPA).
Lamers, H. J. G. J. M. and Pauldrach, A. W. A. 1991, *Astr. Ap.*, **244**, L5.
Langer, N. 1989, *Astr. Ap.*, **220**, 135.
Leitherer, C. 1990, *Ap. J. Suppl.*, **73**, 1.
Leitherer, C., Robert, C. and Drissen, L. 1992, submitted to *Astrophysical Journal*.
Maeder, A. 1990, *Astr. Ap. Suppl.*, **84**, 139.
Maeder, A. 1991, *Astr. Ap.*, **242**, 93.
Maeder, A. and Meynet, G. 1987, *Astr. Ap.*, **182**, 243.
Maeder, A. and Meynet, G. 1988, *Astr. Ap. Suppl.*, **76**, 411.
Pauldrach, A. W. A. and Puls, J. 1990, *Astr. Ap.*, **237**, 409.
Scalo, J. 1986, *Fund. Cosmic Physics*, **VII**, 1.
Vanbeveren, D. 1990, *Astr. Ap.*, **234**, 243.

Massive Stars: Their Lives in the Interstellar Medium
ASP Conference Series, Vol. 35, 1993
Joseph P. Cassinelli and Edward B. Churchwell (eds.)

HIGH MASS LOSS RATE RED SUPERGIANTS: DO THEY EXHIBIT
NITROGEN-RICH CHEMISTRY?

M. JURA
Department of Astronomy, UCLA, Los Angeles CA 90024 USA

ABSTRACT   We consider the two red supergiants ($L > 10^5$ $L_\odot$) within
2.5 kpc of the Sun that are losing mass at an especially high rate (dM/dt
$\sim 10^{-4}$ $M_\odot$ yr^{-1}), NML Cyg and VY CMa. It is possible that the material
in the outflows from these stars is nitrogen-rich in the sense that [N]/[O]
> 1 and [N]/[C] > 1. In this case, we are witnessing the emergence onto
the surface of the star of CNO processing that occurred deep in the interior. If the material is nitrogen-rich, it may display hitherto-unrecognized
chemical properties; for example, NO may be detectable. Although these
stars may be significant sources of the Galactic nitrogen, they probably
contribute less nitrogen to the interstellar medium than do planetary nebulae. In an Appendix, we note that within the 2.5 kpc neighborhood of
the Sun, the most luminous red supergiants have $M_K = $ -11.8 mag. If we
adopt this as characteristic of red supergiants, this implies that a recent
determination of the Hubble constant should be 68 km s^{-1} Mpc^{-1} instead
of 86 km s^{-1} Mpc^{-1}.

INTRODUCTION

The material lost from massive ($M > 20$ $M_\odot$) stars is important for replenishing
the interstellar medium with processed material. These massive stars lose this
mass both as Wolf-Rayet stars and as red supergiants (Maeder & Meynet 1989).
Jura & Kleinmann (1990) have argued that in the hemisphere facing the Galactic
center there is much mass loss from M supergiants than from W-R stars, but in
the anticenter direction, the M supergiants return more mass than do the W-R
stars. Here, because of the restrictions on space and time, we concentrate on
the red supergiants.

One advantage in studying the red supergiants is that much of the outflowing material is molecular. Consequently, it is possible to study isotopes as
well as elemental abundances.

Jura & Kleinmann (1990) have catalogued all the luminous ($L > 10^5$ $L_\odot$)
red supergiants within 2.5 kpc of the Sun as part of their broader program to
characterize the local population of red giants. There are 21 supergiants stars

in their list. From the available data, they estimated the mass loss rates and found that the mass return from this set of stars is dominated by two objects: VY CMa and NML Cyg. Each of these stars appears to be losing $\sim 10^{-4}$ $M_\odot$ yr^{-1}. Here, we describe in more detail how studies of the circumstellar envelopes around these two stars can be used to improve our understanding of the return of processed material back into the interstellar medium.

Luminous red supergiants have been used as distance indicators to determine the Hubble constant (see, for example, Pierce, Ressler & Shure 1992). In an Appendix, we note that the absolute K-band magnitude of the most luminous red supergiants in the sample of Jura & Kleinmann (1990) is -11.8 mag. Therefore, following the analysis of Pierce et al. (1992), this suggests a value for the Hubble constant of 68 km s^{-1} Mpc^{-1} instead of 86 km s^{-1} Mpc^{-1}.

NML CYG AND VY CMA

The best estimate for the distance to NML Cyg is 2 kpc (Morris & Jura 1983), although a value as low as 0.5 kpc has also been suggested (Herman & Habing 1985). At near infrared wavelengths where the bulk of energy from the star is emitted, NML Cyg displays less than 0.2 mag of variation (Hyland et al. 1972). As far as can be determined, its luminosity is nearly constant, about 5×10^5 $L_\odot$. The mass of NML Cyg can be estimated from its location on the H-R diagram. From the tracks of Maeder & Meynet (1989), we estimate an initial main sequence mass of ~ 40 $M_\odot$.

Herbig (1970) estimates a distance of 1500 pc to VY CMa since it appears to be a member of the cluster NGC 2362 (see also Lada & Reid 1978). The derived luminosity is about 5×10^5 $L_\odot$.

Below we discuss the possibility that these two stars are nitrogen-rich. If the models for stellar evolution computed by Maeder (1987) pertain, then perhaps 1% of the mass of the atmosphere is nitrogen. This suggests that these stars within 2.5 kpc of the Sun are returning to the interstellar medium nitrogen at the rate of:

$$dM(N)/dt = 1.0 \times 10^{-7} M_\odot kpc^{-2} yr^{-1} \qquad (1)$$

In contrast, it seems that the total from the mass-losing AGB stars that are returning material to the interstellar medium is between 3 and 6×10^{-4} $M_\odot$ kpc^{-2} yr^{-1} (Jura & Kleinmann 1989). At least some of these mass-losing AGB red giants are nitrogen rich (see Jura 1991); it seems conservative to adopt that the typical nitrogen abundance by mass of these mass-losing red giants is 10^{-3} (Zuckerman & Aller 1986). Therefore, from mass-losing AGB red giants, it is likely that in the neighborhood of the Sun,

$$dM(N)/dt = 3 - 6 \times 10^{-7} M_\odot kpc^{-2} yr^{-1} \qquad (2)$$

Although there are significant uncertainties associated with these numbers, from the comparison of equations (1) and (2), it seems that the AGB stars are more important that the massive supergiants in returning nitrogen back to the interstellar medium.

CIRCUMSTELLAR ENVELOPES

Because they display powerful masers (Bowers, Johnston & Spencer 1983, Johnston, Spencer & Bowers 1985, Diamond, Norris & Booth 1984), the circumstellar
envelopes of NML Cyg and VY CMa have been extensively studied. They both
exhibit abundances that are usually described as being "oxygen-rich". However,
for both stars, the intensity of the $J = 1\text{-}0$ HCN line is comparable to the intensity of the $J = 1\text{-}0$ line of CO (Nercessian *et al.* 1989). This is very unusual for
oxygen-rich stars. More often, such stars display values of $I(HCN)/I(CO) \leq 0.2$
(Lindqvist *et al.* 1988, Omont *et al.* 1992). Therefore, it is possible that these
two stars may have unusually high abundances of nitrogen.

   Another reason to suspect that nitrogen has a high abundance in the atmospheres of these two stars is the relatively high abundance of $H^{13}CN$. Toward
NML Cyg, it is observed that $I(H^{13}CN)$ compared to $I(H^{12}CN)$ is 0.1 while the
same ratio toward VY CMa is about 0.03. It is realistic to imagine that the
emission in the HCN lines is optically thin and that the observed intensities of
these emission lines reflect $^{13}C/^{12}C$ (see, for example, Kahane *et al.* 1988, Sopka
et al. 1989) According to the detailed models of Maeder (1987) for a 25 $M_\odot$
star, the value of $^{13}C/^{12}C$ are substantially enhanced in zones where $[N] > [O]$.

   These calculations for the isotopic abundances are somewhat uncertain.
For example, Maeder (1987) also predicts that in the zones where $^{13}C/^{12}C$ is as
high as 0.25 then $^{17}O/^{16}O$ should also be nearly this high. However, there is no
known object where the value of $^{17}O/^{16}O$ is anywhere near 0.25. Landré (*et al.*
1990) have argued that the rates for the destruction of ^{17}O by nuclear processes
have been greatly underestimated and therefore calculations for the interiors of
stars may have overestimated by a large amount the synthesis of ^{17}O.

   Both stars exhibit strong infrared emission from circumstellar grains. VY
CMa also exhibits highly polarized optical reflection nebulosity (see Herbig 1972,
Jura 1975). Levan & Sloan (1989) have detected a broad weak absorption at
9.7 μm toward NML Cyg; there may be a narrower emission peak in the broad
absorption line profile of this star.

NITROGEN-RICH CHEMISTRY

A famous example of a nitrogen-rich massive star is η Car (Davidson, Walborn
& Gull 1982). There exist a number of other nitrogen-rich massive stars (see, for
example, Schonberner *et al.* 1988). While a few nitrogen-rich planetary nebula
are known (Nussbaumer *et al.* 1988, Zuckerman & Aller 1986), and while there
exists a class of apparently nitrogen-rich pre-planetaries which are also carbon-
rich (Jura 1991), perhaps the most prominent class of nitrogen-rich objects is
novae (Nussbaumer *et al.* 1988). Above, we have noted that VY CMa and NML
Cyg are nitrogen-rich. Here, we note possible observational consequences of this
hypothesis.

   The usual model is that the molecular composition of the gas-phase material that is ejected from a cool star "freezes-out" at some set of LTE molecular
abundances characteristic of some temperature near 1000 K (McCabe, Smith &

Clegg 1979; Lafont, Lucas & Omont 1982). In such a model, although there is some uncertainty, we expect that the bulk of the nitrogen is contained within N_2 (see, for example, Tsuji 1964). In the models presented by Nercessian *et al.* (1989) or Nejad & Millar (1988), the nitrogen in the outflow remains as N_2 until it is photo-dissociated by the ambient interstellar radiation field to produce atomic nitrogen. That is:

$$h\nu + N_2 \to 2N \tag{3}$$

The rate for reaction (3) is uncertain. Below, we use the "best estimate" by van Dishoeck (1987) of $2.3 \times 10^{-1-}$ s^{-1}, but the rate might be as low as 5×10^{-11}. The lower rate leads to a smaller total amount of NO since there is less N to react with OH in the zone where OH is largely concentrated as required in reaction (5).

A large amount of oxygen in the inner region of the circumstellar envelope is H_2O, and this molecule persists until it is photodissociated to form OH by the following:

$$h\nu + H_2O \to OH + H \tag{4}$$

In this outer region of the circumstellar envelope, it is thought that the OH and N can combine to form NO:

$$N + OH \to NO + H \tag{5}$$

Finally, the NO is removed photo-dissociation:

$$h\nu + NO \to N + O \tag{6}$$

This sequence of reactions may occur in the outflows around stars such as VY CMa and NML Cyg.

The NO molecule has been studied within the interstellar medium (Ziurys *et al.* 1991), but it has not yet been detected in the outflows from red giants (Olofsson 1992). However, this molecule has only been searched for in a few envelopes such as that of the mass losing star, TX Cam (Olofsson *et al.* 1991). It may be fruitful to search for NO around NML Cyg and VY CMa.

Nercessian *et al.* (1989) note that the rate for reaction (5) has only been measured near room temperature, and there may be a small activation energy barrier. However, at least in the case of VY CMa and NML Cyg, we expect that the gas temperature will be larger than 300 K in the region where the NO might be formed, $\sim 10^{17}$ cm from the star. When considering the balance of heating and cooling processes, Jura, Kahane & Omont (1988) have argued that in a circumstellar envelope, the gas temperature, T, asymptotically varies as:

$$T = r^{-1}[\sigma_{grain}n_{grain}/n](QL/c)^{3/2}(4\pi k)^{-1}(vdM/dt)^{-1/2} \tag{7}$$

where σ_{grain} is the mean cross section of the grains of space density n_{grain} compared to a hydrogen nuclei space density, n. The ratio of geometric to average optical cross section is Q, L is the luminosity of the star of mass loss rate dM/dt and outflow velocity v. As described by Jura *et al.* (1988), we can take $\sigma_{grain} n_{grain}/n = 10^{-21}$ cm^{-2}. With Q = 0.01, L = 5 10^5 L$_\odot$, v = 25 km s^{-1} and dM/dt = 10^{-4} M$_\odot$ yr^{-1}, then

$$T = 730 r_{17}^{-1} \qquad (8)$$

where r_{17} is the distance from the star in units of 10^{17} cm. Therefore, we anticipate that the gas temperature in the circumstellar envelopes around these stars can be quite substantial. It should be noted that this gas temperature is high compared to that in many other circumstellar envelopes because of the very high luminosities of these stars with the consequence that the grains stream through the gas very rapidly which results in a large amount of heating from gas-grain collisions.

To illustrate the chemistry in a nitrogen-rich environment, we display in Figure 1, a model calculation for the space density of circumstellar molecules around a mass-losing red supergiant. We employ the photo-rates used by Nercessian et al. 1989) which are similar to those in Nejad & Millar (1988), a mass loss rate of 10^{-4} M$_\odot$ yr^{-1}, an outflow velocity of 30 km s^{-1}, and an ultraviolet dust opacity of 1000 cm^2 gm^{-1} as is typical of the interstellar medium (see Spitzer 1978). Following Nercessian et al. (1989), we assume that reaction (5) has a rate of 7 × 10^{-11} cm^3 s^{-1}.

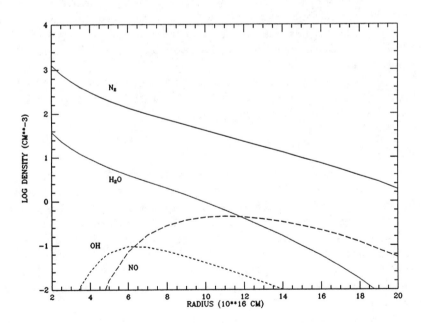

Figure 1. Plot of molecular densities for the model parameters described above.
We see that the density of N$_2$ and H$_2$O decrease substantially from the central star while we predict that there are shells of OH and NO around the

star. These shells of OH and NO are the result of the photo-dissociation of the parent molecules and the resulting chemical reaction described above. In the calculations displayed in Figure 1, the OH density peaks at about 6×10^{16} cm from the star. For VY CMa at a distance of 1500 pc, this distance corresponds to an angular separation of 2.7" in moderate agreement with the measured value of 2.1" (Bowers et al. 1983). For NML Cyg at a distance of 2000 pc, the predicted OH zone should lie at 2.0" from the star in agreement with the observed range of between 1.5" and 2.5" (Diamond et al. 1984). It should be noted that in these models, the rate of reaction of N with OH via reaction (5) is quite an important route for the destruction of OH. Therefore, the pure photo-chemical models for the formation and destruction of OH (Huggins & Glassgold 1982, Netzer & Knapp 1987) may be modified in especially nitrogen-rich environments.

Recently, SO has been discovered in the outflow from VY CMa (Sahai & Wannier 1992). Because this molecule may also be formed by reactions with OH (Nejad & Millar 1988), it is possible that NO also will be detected.

CONCLUSIONS

The mass returned to the interstellar medium within 2.5 kpc of the Sun by red supergiants is dominated by two stars, NML Cyg and VY CMa. On the basis of the high ratio of HCN to CO radio emission in their circumstellar envelopes and because it seems that $H^{13}CN$ is relatively strong compared to $H^{12}CN$, it is possible that these stars are nitrogen-rich in the sense that $[N] > [C]$ and $[N] > [O]$. The circumstellar chemistry in these possibly nitrogen-rich objects may display distinctive phenomena; for example radio emission from NO may be detectable.

This work has used the NASA/IPAC Extragalactic Database (NED) which is operated by the Jet Propulsion Laboratory, under contract with NASA. My research also has been supported by NASA.

APPENDIX

Jura & Kleinmann (1990) list 21 red supergiants within 2.5 kpc of the Sun. The values of M_K for these stars are derived from that paper. We have mostly used m_K from the *Two Micron Sky Survey* (Neugebauer & Leighton 1969), distances from cluster-membership or associations and most estimates of the extinction at K with the value of 0.15 mag/kpc. Under these assumptions, the four most luminous stars at K are BC Cyg, KY Cyg, μ Cep and VY CMa. They all have values near $M_K = -11.8$ mag. The exact value of M_K is uncertain because these stars are somewhat variable, and the extinction and distance are somewhat uncertain.

Pierce, Ressler & Shure (1992) use the brightest red supergiants to estimate the distance to the galaxy IC 4182 where supernova 1937C exploded. They then derive a value of the Hubble constant of 86 km s^{-1} Mpc^{-1}. However, their distance to IC 4182 may be too small. They assumed that the brightest red supergiants have M_K = -11.2 mag. Instead, as argued here, in the neighborhood of the Sun, the most luminous red supergiants are as bright as M_K = -11.8 mag. There also exist at least three red supergiants in the LMC, including HV 888, which also are this luminous (Elias *et al.* 1981, Wood, Bessell & Fox 1983). If we adopt M_K = -11.8 mag, the`arguments of Pierce, Ressler & Shure (1992) then imply a value for the Hubble constant of 68 km s^{-1} Mpc^{-1}.

Can we compare IC 4182 with the stars in the Solar neighborhood? One way to approach this subject is to argue that the solar neighborhood within 2.5 kpc of the Sun has a similar blue luminosity to that of the entire galaxy, IC 4182. That is, within the neighborhood of the Sun, the disk has a mass column density of 60 $M_\odot$ pc^{-2} (Bahcall 1984). With a mass to blue light ratio of 4.0 (extrapolating from the M/L_V ratio of 3.0 given in this same paper by Bahcall), this result predicts that for the galactic disk within 2.5 kpc of the Sun, L_B = 2.9 10^8 $L_{\odot,B}$ where $L_{\odot,B}$ is the luminosity of the Sun in the B band.

Measurements of the apparent magnitude of IC 4182 range from 11.4 mag (Tyson and Scalo 1988) to 12.9 mag (Corradi and Capaccioli 1990) with an intermediate value of 12.2 mag (extrapolating from the J mag given by van der Kruit 1987) and his equation (1).) Because he presents the most thorough discussion, we adopt van der Kruit's apparent magnitude of 12.2. Given the distance modulus of 27.0 mag to IC 4182 proposed by Pierce *et al.* (1992), this implies a value of M_B = -14.2 or L_B = 1.3 10^8 $L_{\odot,B}$. If the distance modulus is 27.6 mag, then L_B = 2.2 10^8 $L_{\odot,B}$. In either case, the blue luminosity of IC 4182 is comparable to the disk of the Milky Way within 2.5 kpc of the Sun, so it seems reasonable but not certain to adopt M_K = - 11.8 mag for the most luminous red supergiants within IC 4182.

REFERENCES

Bahcall, J. N. 1984, ApJ, 287, 926

Bowers, P. F., Johnston, K. J. & Spencer, J. H. 1983, ApJ, 274, 733

Corradi, R. L. M. & Capaccioli, M. 1990, A&A, 237, 36

Davidson, K., Walborn, N. R. & Gull, T. R. 1982, ApJ, 254, L47

Diamond, P. J., Norris, R. P. & Booth, R. S. 1984, MNRAS, 207, 611

Elias, J. H., Frogel, J. A., Humphreys, R. M. & Persson, S. E. 1981, ApJ, 249, L55

Herbig, G. H. 1970, ApJ, 162, 557

Herbig, G. H. 1972, ApJ, 172, 375

Herman, J. & Habing, H. J. 1985, A&AS, 59, 533

Huggins, P. J. & Glassgold, A. E. 1982, AJ, 87, 1828

Hyland, A. R., Becklin, E. E., Frogel, J. A. & Neugebauer, G. 1972, A&A, 16, 204

Johnston, K. J., Spencer, J. H. & Bowers, P. F. 1985, ApJ, 290, 660

Jura, M. 1975, AJ, 80, 227

Jura, M. 1991, ApJ, 372, 208

Jura, M., Kahane, C. & Omont, A. 1988, A&A, 201, 80

Jura, M. & Kleinmann, S. G. 1989, ApJ, 341, 359

Jura, M. & Kleinmann, S. G. 1990, ApJS, 73, 769

Kahane, C., Gomez-Gonzalez, J., Cernicharo, J. & Guélin, M. 1988, A&A, 190, 167

Lada, C. J. & Reid, M. J. 1978, ApJ, 219, 95

Lafont, S., Lucas, R., & Omont, A. 1982, A&A, 106, 201

Landré, V., Prantzos, N., Aguer, P., Bogaret, G., Lefebvre, A. & Thibaud, J. P. 1990, A&A, 240, 85

Levan, P. D. & Sloan, G. 1989, PASP, 101, 1140

Lindqvist, M. Nyman, L.-A., Olofsson, H. & Winnberg, A. 1988, A&A, 205, L15

Maeder, A. 1987, A&A, 173, 247

Maeder, A. & Meynet, G. 1989, A&A, 210, 155

McCabe, E. M., Smith, R. C. & Clegg, R. E. S. 1979, Nature, 281, 263

Morris, M., & Jura, M. 1983, ApJ, 267, 179

Nejad, L. A. M. & Millaer, T. J. 1988, MNRAS, 230, 79

Nercessian, E., Guilloteau, S., Omont, A. & Benayoun, J. J. 1989, A&A, 210, 225

Netzer, N. & Knapp, G. R. 1987, ApJ, 323, 734

Neugebauer, G. & Leighton, R. B. 1969, Two Micron Sky Survey (NASA SP-3047)

Nussbaumer, H., Schild, H., Schmid, H. M. & Vogel, M. 1988, A&A, 198, 179

Olofsson, H. 1992, in "Mass Loss on the AGB and Beyond", conference at La Serena, Chile, in press

Olofsson, H., Lindqvist, M., Nyman, L.-A., Winnberg, A., & Nguyen-Q-Rieu 1991, A&A, 245, 611

Omont, A., Loup, C., Forveille, T., te Lintel-Hekkert, P., Habing, H., & Sivagnanam, P. 1992, A&A, in press

Pierce, M. J., Ressler, M. E. & Shure, M. S. 1992, ApJ, 390, L45

Sahai, R. & Wannier, P. G. 1992, ApJ, 394, 320

Schonberner, D., Herrero, A., Becker, S., Eber, F., Butler, K., Kudritzki, R. P., & Simon, K. P. 1988, A&A, 197, 209

Sopka, R. J., Olofsson, H., Johansson, L. E. B., Nguyen-Q-Rieu & Zuckerman, B. 1989, A&A, 210, 78

Spitzer, L. 1978, Physical Processes in the Interstellar Medium, John Wiley: New York

Tsuji, T. 1964, Ann Tokyo Astr Obs 2nd Ser, 9, 1

Tyson, N. D. & Scalo, J. M. 1988, ApJ, 329, 618

van Dishoeck, E. F. 1987, in Astrochemistry, IAU Symp. No. 120, M. S. Vardya & S. P. Tarafdar eds., Reidel: Dordrecht

van der Kruit, P. C. 1987, A&A, 173, 59

Wood, P. R., Bessell, M. S. & Fox, M. W. 1983, ApJ, 272, 99

Ziurys, L. M., McGonagle, D., Minh, Y. & Irvine, W. M. 1991, ApJ, 373, 535

Zuckerman, B. & Aller, L. H. 1986, ApJ, 301, 772

Massive Stars: Their Lives in the Interstellar Medium
ASP Conference Series, Vol. 35, 1993
Joseph P. Cassinelli and Edward B. Churchwell (eds.)

STELLAR WIND BOW SHOCKS

Dave Van Buren, Theoretical Astrophysics, California Institute of
Technology, MS 130-33, Pasadena CA 91125

Abstract   Wind-blowing stars travelling through an ambient medium
produce bow shocks if their velocities are supersonic. The nearest
runaway O star, ζ Oph is prototypical. The bow shocks of runaway O
stars are *IRAS* objects due to reprocessing of stellar ultraviolet into the
infrared by swept-up dust. Wind bow shocks are manifested in several
other ways. In molecular clouds, newly formed O stars generate and
ionize bow shocks, giving rise to cometary ultracompact HII regions.
High velocity pulsars create bow shock nebulae when their relativistic
winds interact with the diffuse interstellar medium, examples being PSR
1957+20 and the pulsar in CTB 80. There is even a bow shock around
the high velocity cataclysmic variable 0623+72, presumably generated by
the x-ray induced wind of its accretion disk.

1. INTRODUCTION

Bow shocks are formed around supersonic objects as they push material out
of the way to make room for their passage. Natural examples of bow shocks
include meteors, comets and the heliopause at $r \approx 100$ A.U. Of special interest
in the context of this volume are the bow shocks formed by winds of massive
stars moving supersonically through the diffuse and molecular interstellar
medium.

Bow shocks in the interstellar medium are a recently recognized
phenomenon. One of the first bow shocks discovered was that around the
nearby runaway O star ζ Oph, found during a deep wide-field survey of line
emission by Gull and Sofia (1979). Since then a number of others have been
found using the IRAS satellite around other runaway O stars (Van Buren and
McCray, 1988). An early theoretical discussion of stellar wind bow shocks was
by Weaver et al. (1977), who considered "distorted" stellar wind bubbles.

Let a wind-blowing star be moving supersonically through the
interstellar medium. Because the stellar wind is divergent, its ram pressure
decreases with radius, while the ram pressure of the oncoming ISM (in the
frame of the star) is constant. Therefore there is a point at which they
balance, giving the stand-off distance of the bow shock,

$$\ell = \left(\frac{\dot{m}v_w}{4\pi\rho_0 v_\star^2} \right)^{1/2} \qquad (1)$$

315

written in terms of the star's mass-loss rate $\dot{m}$, wind terminal velocity v_w, the density of the ambient medium ρ_0, and the velocity at which the star is travelling through that medium v_*. Typical wind parameters are $v_w = 3000$ km s^{-1} and $\dot{m} = 10^{-6}$ M$_\odot$ yr^{-1}. For runaway O stars in the diffuse ISM $(n_H \approx 1$ cm$^{-3})$ the stand-off distance is of order 1 pc while for young O stars in a molecular cloud it is typically .01 pc.

Stellar wind bow shocks turn out to be quite common - just consider that the list of *IRAS* bow shocks contains mainly named stars. This common occurance follows from the shape of the velocity dispersion of early-type stars compared to the sound speed in the diffuse and molecular ISM. In the coronal medium only the few stars with $v_* \gtrsim 100$ km s^{-1} are capable of producing bow shocks. A simple estimate of the fraction of O stars with bow shocks starts with noting that the filling factor of the ISM with sound speed $\lesssim 10$ km s^{-1} is of order 20% while about 3/4 of the O stars have space velocities > 10 km s^{-1}, which is the maximum sound speed in the diffuse gas. The fraction of O stars, all of which have strong winds, that have bow shocks should then be about 15%.

In this review I have tried to focus on the rich diversity of stars which give rise to wind bow shocks rather than give a detailed account of the theory. My hope is that your interest in these objects will be awakened and that you will not only see them as interesting in their own right, but also view them as potentially valuable laboratories for studying a range of astrophysical processes.

IRAS BOW SHOCKS

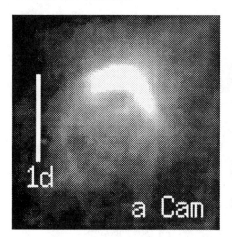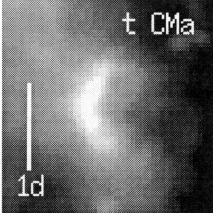

Figure 1. Thermal infrared emission at 60 μm from dust swept up in the α Cam and τ CMa bow shocks as observed by *IRAS*. The dust is heated by starlight and traces the shocked ISM. Positions of the stars are marked by dots. The bars are one degree long. Images are from the Infrared Sky Survey Atlas.

I term the parsec-sized bow shocks of runaway O stars in the diffuse ISM the *IRAS* bow shocks because they are most easily identified by the far infrared emission from dust in their shells. Figure 1 shows the 60 μm nebulae around α Cam and τ CMa from the *IRAS* Infrared Sky Survey Atlas. These are high contrast objects at 60 μm because the dust is several times hotter than average, resulting in a very large enhancement in emissivity even though the path lengths are small.

A cursory survey of the *IRAS* data for 60 μm bright arcuate nebulae was made by Van Buren and McCray (1988), and found a number associated with runaway O stars. We showed there that the bow shock model provides an adequate explanation for most of the data, including not only the observed stand-off distances ℓ, but also the far infrared luminosities and inferred dust temperatures. Several B stars could not plausibly be fit with the bow shock model owing to the weakness of their winds, although they have striking paraboloidal 60 μm nebulae. In these cases (for example δ Per) stellar radiation pressure acting on dust could provide the organizing force provided the stellar velocities are unusually small with respect to the gas.

Why *IRAS*?
On dimensional grounds we expect the swept-up column densities at the stagnation point of a bow shock to be of order $n_0\ell$. More detailed calculations bear this out (Mac Low et al., 1991). For $v_* \lesssim 125$ km s^{-1} and $\ell <$ the Strömgren radius, the bow shock itself is isothermal, so the gas is compressed by a factor $\mathcal{M}^2$, where $\mathcal{M}$ is the Mach number. The emission measure through the stagnation point is then

$$EM = \mathcal{M}^2\, n_0^2\, \ell, \tag{2}$$

which for typical values $v_* = 50$ km s^{-1}, $n_0 = 1$ cm^{-3} give $EM \approx 25$ cm^{-6} pc, small enough so that (even accounting for limb brightening) the Palomar Observatory Sky Survey provides inadequate survey material.

On the other hand, consider the dust mixed with the gas in the swept up shell. While the optical depth of the dust is small to the stellar continuum, the stellar luminosity is large, often $> 10^5$ L$_\odot$, resulting in a 60 μm surface brightness ≈ 10 MJy sr^{-1}, well above the *IRAS* detection limit. Since the stellar radiation field intensity declines as $1/r^2$ and the equilibrium dust temperature also declines with distance from the star, the shell stands out with some contrast against the confusing cirrus background, especially if it is seen edge on and limb brightened.

In retrospect this seems obvious, but the *IRAS* bow shocks were found serendipitously during a search for dust shells around OB associations. A hint that wind-blowing stars would be interesting in the *IRAS* data was provided a few years earlier by de Vries (1986), who interpreted the infrared nebula around α Cam as a stellar wind bubble. A directed survey is now underway using the Infrared Sky Survey Atlas (ISSA), targeting all known O stars from the list of Garmany, which contains ≈ 700 stars. If the estimate above is close to correct, then as many as 100 new bow shocks could be discovered. The ISSA images are sensitive enough to detect α Cam at a distance of 10 kpc but since most O stars, even runaways, are within a few 100 pc of the galactic plane, the survey is limited by source confusion rather than source strength.

Such a large survey is useful beyond the statistics it will provide since the large set of potential observables allows very detailed modelling of individual bow shocks.

Modelling *IRAS* Bow Shocks
The main ingredients of a model for the *IRAS* bow shocks are 2 dimensional hydrodynamics and radiative transfer. Mac Low and Van Buren (work in progress) are using a modified version of ZEUS 2D (Stone and Norman, 1992) to calculate the gas flow. Figure 2 shows some exploratory calculations with parameters chosen to be similar to the α Cam bow shock. The star moves to the right and emits a spherically symmetric wind. The wind cavity is a bell-shaped cavity with the star significantly offset upstream since the confining pressure is greatest in that direction. Further downstream the pressure drops via the Bernoulli effect because the flow around the cavity has accelerated. The end cap is spherical since the tail is full of isobaric shocked wind. Because of strong radiative cooling, the swept-up ISM forms a very thin shell only a few zones thick.

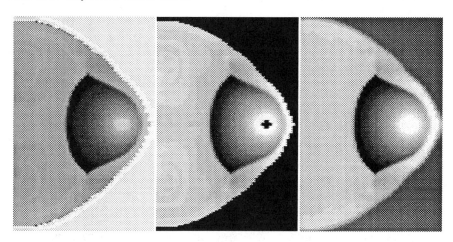

Figure 2. Left: density in a plane through the axis of symmetry of an exploratory ZEUS 2D model for the α Cam bow shock. Middle: temperature. Right: pressure. Courtesy of M.-M. Mac Low, University of Chicago.

To compare the hydrodynamical models with the *IRAS* data requires adding dust and making a radiative transfer calculation beginning with a model stellar atmosphere. As the stellar photospheric radiation propagates through the dust, it heats the dust, and is attenuated. The dust in turn radiates in the thermal infrared according to its optical properties, chosen to be those given by, for example, Draine and Lee (1985). The volume spectral emissivities calculated in this way for each grid zone then go into a visualization program that converts the 2 dimensional model into a 3 dimensional object and calculates how it would appear projected on the plane of the sky at an arbitrary angle.

Bow Shocks as Laboratories
Both the preshock and postshock material in a bow shock has significant
emissivity in a number of diagnostic lines and in the dust continuum, so
the physical effect of shocks on interstellar material can be assessed with a
minimum of assumptions in modelling. A rich laboratory for studying shock
physics is thus available and is certain to be exploited in much the same way
as supernova remnants have been.

One important question which is sure to be addressed is the shock
processing of interstellar dust. By combining *IRAS* or ISOPHOT broadband
photometry with KAO or *ISO* LWS spectroscopy, the line contributions to the
broadband fluxes can be removed from the pre and post shock dust emission,
allowing a direct measure of the change in the dust emissivity as the shock
is traversed. Modelling the infrared emission from various grain populations
and comparing with data of this sort could give the first before and after view
of grain destruction. Theoretical studies of shock processing using spherical
grains have yet to be empirically tested.

Hydrodynamically bow shocks are interesting too, since they break the
spherical symmetry of the best studied astrophysical flows, supernova

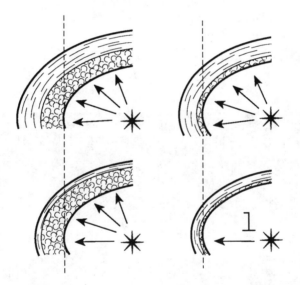

Figure 3. Schematic structure of a bow shock. Stellar wind undergoes
a terminal shock to form a hot shocked wind region (curls). As the
structure moves supersonically to the left, ambient ISM is accreted onto
a shell (layered), which flows around the shocked wind and is separated
from it by a contact discontinuity. Shown are four different bow shock
configurations, depending on effectiveness of post-shock cooling. From
top to bottom, left to right: both shocks non-radiative, outer non-
radiative and inner radiative, outer radiative and inner non-radiative
and both non-radiative.

remnants, but are still amenable to 2-d modelling. For example, consider
the case where magnetic fields are not important, but the effectiveness
of postshock cooling, which depends on a number of physical processes,
is varied. A bow shock is a two shock structure (like a colliding wind, or
a blast wave with reverse shock) giving four extreme possibilities: both
shocks can be radiative, both isothermal, or one of each (figure 3). Since
the structure is qualitatively different in each case, strong links between
models and observation are possible. As to the physics behind whether the
inner or outer shock is radiative or nonradiative, consider the effects of non-
equilibrium ionization in enhancing (or inhibiting) radiative cooling, thermal
conduction, and mixing via Kelvin-Helmholz "cat's eyes". Of course a full
description that can adequately treat the various instabilities which operate
requires 3-d hydro, or even MHD. The large set of observables guarantees
that the *IRAS* bow shocks will become an important test bed for hydro
codes.

ULTRACOMPACT HII REGIONS

The striking resemblance of ultracompact cometary HII regions (Reid and
Ho, 1985; Gaume and Mutel, 1987; Wood and Churchwell, 1989) to the *IRAS*
bow shocks leads directly to a bow shock model for their origin (Van Buren
et al., 1990; Mac Low *et al.*, 1991). The Wood and Churchwell survey and
it's continuation (in progress) provide a large number of contour maps of
ultracompact HII regions, a modest fraction of which have cometary

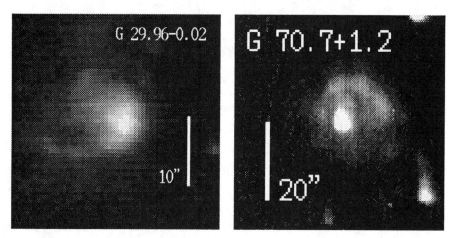

Figure 4. Right: A Br γ image of G29 taken with the Palomar 5-m
and Prime Focus Infrared Camera. The star is marked. In the nearby
continuum, the nebula is fainter than the star. Left: An [FeII] 1.64
μm image of G70 made with the same instrumentation. This line is
diagnostic of shocks energetic enough to destroy iron-bearing grains and
return the iron to the gas phase.

morphology. Van Buren et al. (1991) interpret G34.3+0.2 in terms of a bow shock model motivated by the cometary radio morphology. Wood and Churchwell (1991) present the case for G29.96-0.02 using recombination line maps. Figure 4 shows that the G29 bow shock is also visible in narrowband near IR line images where the extinction is low compared to the optical. Also shown is G70.7+1.2, a closely related but un-ionized compact cometary nebula.

The observed sizes, radio continuum fluxes, bolometric luminosities, radio recombination line kinematics and morphologies suggest that a typical ultracompact HII region is formed by an O star passing through a dense ($n_{H_2} \approx 10^5$ cm^{-3}) molecular cloud at a velocity of ≈ 10 km s^{-1}. Since the effective sound speed is $1 - 3$ km s^{-1}, bow shocks are expected to be common around wind-blowing stars in molecular clouds. As a class, the ultracompact cometary HII regions share these properties:

- paraboloidal radio continuum shape
- limb brightened, especially toward the coma
- very sharp upstream edge - often unresolved with the VLA
- kinematically broadened recombination lines, and
- inferred velocities of *approach* between radio recombination lines and molecular lines.

Recently Churchwell (1990) published a review on cometary ultracompact HII regions, focusing on their radio properties and the bow shock model, which accounts for many of their properties. A longstanding puzzle that the bow shock model solves is the lifetime problem of at least the cometary ultracompact HII regions: given their small sizes, their expansion timescales are very short, of order 10^3 years. But there are so many in the galaxy that it is implausible that they are formed at such a rapid rate. Consequently there

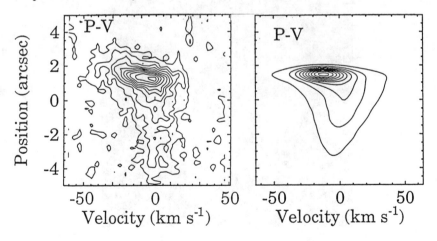

Figure 5. Comparison between velocity structure in G29 (left) and bow shock model for the same object (right). These are position-velocity diagrams corresponding to the slice of the three-dimensional data cube through the axis of symmetry. Model parameters are inclination 135°, $v_* = 20$ km s^{-1}, $n_H = 5.5 \times 10^5$ cm^{-3}, $v_w = 3370$ km s^{-1}, and $\dot{m} = 8.68 \times 10^{-6}$ M$_\odot$ yr^{-1}. G29 data courtesy D. Wood, NRAO.

must be a confining mechanism that increases the lifetimes by a large factor. The bow shock model solves this problem by providing strong ram pressure confinement, though at the cost of requiring rather large clumps of dense material in molecular clouds. Clumps of the appropriate size have been found by Churchwell, Walmsley and Wood (1992) towards several objects.

Application of the bow shock model to cometary ultracompact HII regions was made by Van Buren et al. 1990, Mac Low et al. 1991 and Van Buren and Mac Low (1992). The first of these papers gives an analytic theory for the region along the symmetry axis forward of the star while the following two construct a numerical model for interpreting VLA radio continuum and recombination line data. To show how well these models work, figure 5 shows a comparison between a position-velocity slice of the VLA observations of G29 (Wood and Churchwell, 1991) and the predictions of the bow shock model. Using wide field optical Fabry-Perot spectrometers, similar comparisons are possible for the *IRAS* bow shocks.

EXOTIC BOW SHOCKS

Cataclysmic Variables
There is a very beautiful bow shock around the cataclysmic variable CV0623+72 visible in optical emission lines. The accompanying figure (6) shows an [NII] image of this nebula, obtained by Hollis et al. (1992) at Kitt Peak. Other lines were also observed. Particularly noteworthy is that the

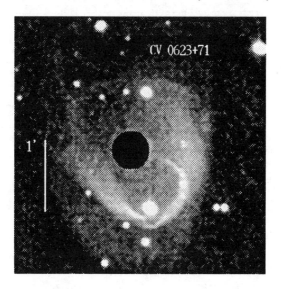

Figure 6. [NII] image of the cataclysmic variable CV0623+71, courtesy of J. M. Hollis. The dark circle near the center of the field of view is an occulting disk in the optical path.

[OIII] emission, characteristic of higher excitation and hence stronger shocks, is localized to only the coma of the bow shock, just as expected if the shock is oblique elsewhere. *IUE* low resolution spectroscopy of the star shows strong P Cygni lines, presumably arising in a wind from the accretion disk around the degenerate star. The mass loss rate inferred from the ultraviolet, $\approx 10^{-11} M_\odot$ yr^{-1}, is consistent with the observed stand-off distance for an ISM density of $\gtrsim 1$ cm^{-3}. By observing the consequences of the disk wind at a distance, as it interacts with the ISM, the accretion disk wind model can now be subjected to new tests.

Pulsars
Pulsars, in light of the opportunities for asymmetry in the explosions that create them, are intrinsically high space velocity objects. They also have strong relativistic winds. The interaction between pulsar winds and an external medium, aside from the larger scale supernova remnant - ISM interaction, has been seen in a number of pulsars (Fesen, Shull and Saken, 1988; Kulkarni and Hester, 1988; Frail and Kulkarni, 1991). Figure 7 shows the Hα bow shock nebulae of PSR 1957+20 and PSR 1951+32 in CTB 80. The kinematics of CTB 80 show the signature of a bow shock viewed head on (J. Hester, private communication).

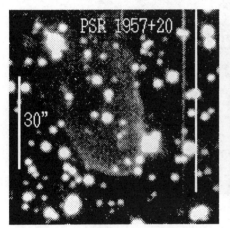 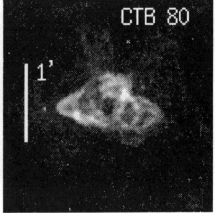

Figure 7. Left: PSR 1957+20, the eclipsing binary millisecond pulsar, has an arcminute scale bow shock visible in Hα. Right: The small nebula around the PSR 1951+32 in CTB 80 is almost certainly a bow shock seen nearly straight on. To understand the morphology visualize a badminton birdie with a slightly squeezed tail. The [OIII] image is confined to the central 30", showing that the shock is strongest there, where it is less oblique. Both images are courtesy J. Hester, Arizona State University.

CTB 80 is thought to be interacting with its own much larger supernova remnant, as is PSR1757-24. As such they provide interesting tools for probing the density stratification of the shell. PSR 1957+20, a recycled pulsar, is likely interacting with the diffuse ISM many kpc and millions of years from it's remnant. One interesting aspect of the pulsar bow shocks is that the winds that blow them are powered by the rotational energy of the neutron star, a quantity that is related to the star's moment of inertia, structure and the equation of state at nuclear densities.

Planetary Nebulae

The planetary nebula Abell 35 (figure 8) shows a striking bow shock morphology when viewed in [OIII] 5007 Å(Jacoby 1981). The Hα emission is on a larger scale and has very low surface brightness except for several wisps reminiscent of cirrus. The space motion of the central star, a binary, is along the axis of symmetry with $v_* = 150$ km s^{-1}. Inferred parameters are $n_0 = 11$ cm^{-3}, $v_w = 185$ km s^{-1} and $\dot{m} = 3 \times 10^{-9}$ M$_\odot$ yr^{-1}. Borkowski, Sarazin and Soker (1990) list 6-7 large proper motion planetaries showing evidence for ISM-PN interaction, including Abell 35 and the Helix nebula.

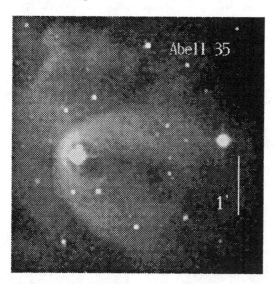

Figure 8. [OIII] image of Abell 35, courtesy of G. Van de Steen, Kapteyn Institute, taken with the KPNO 0.9m. Note the cavity around the central star which extends tailward.

The interaction of the ISM with high velocity planetary nebulae was studied theoretically by Smith (1976), Isaacman (1979), and more recently by Borkowski, Sarazin, and Soker (1990), and Soker, Borkowski, and Sarazin (1991). These models do not yet include the potentially strong winds from the central stars which would stabilize the calculated structures against the Rayleigh Taylor instability that is found in the calculations.

ACKNOWLEDGEMENTS

I'd like to thank the taxpayers of the US, UK and Netherlands for making the *IRAS* mission possible. I would also like to thank the Observatoire de Grenoble for their hospitality and use of facilities during part of the time this review was written. Much of my work on this topic has been supported by a series of grants from the NASA ADP program, NSERC (Canada) and the Davis Foundation. Of particular value to me were many discussions about bow shocks with Sue Terebey, Shri Kulkarni, Jeff Hester, Sterl Phinney, Mike Hollis, Howard Bond, Ed Churchwell, Doug Wood, Dick McCray and Mordecai Mac Low. The near infrared images were obtained with the capable assistance of Juan Carrasco of Palomar Observatory.

REFERENCES

Borkowski, K. J., Sarazin, C. L., and Soker, N. 1990, ApJ, 360, 173
Churchwell, E., 1990, AARev, 2, 79
Churchwell, E., Walmsley, C. M., and Wood, D. O. S. 1992, AA, 253, 541
de Vries, C. P. 1985. AA, 150, L15
Draine, B. T., and Lee, H. M. 1985, ApJ, 285, 89
Fesen, R. A., Shull, J. M., and Saken, J. M. 1988, Nature, 334, 229
Frail, D. A., and Kulkarni, S. R. 1991, Nature 352, 785
Gaume, R. A.,. and Mutel, R. L. 1987, ApJ Suppl., 65, 193
Gull, T. R., and Sofia, S. 1979, ApJ, 230, 782
Hollis, J. M., Oliversen, R. J., Wagner, R. M., and Feibelman, W. A. 1992,
     ApJ, 393, 217
Isaacman, R., 1979, AA, 77, 327
Jacoby, G. H. 1981, ApJ, 244, 903
Kulkarni, S. R., and Hester, J. J. 1988, Nature, 335, 801
Mac Low, M.-M., Van Buren, D., Wood, D. O. S., and Churchwell, E. 1991,
     ApJ, 369, 395
Reid, M. J., and Ho, P.T. P. 1985, ApJ Letters, 288, L17
Soker, N., Borkowski, K. J., and Sarazin, C. L. 1991, AJ, 102, 1381
Smith, H., 1976, MNRAS, 175, 419
Van Buren, D., Mac Low, M.-M., Wood, D. O. S., and Churchwell, E. 1990,
     ApJ, 353, 570
Van Buren, D., and Mac Low, M.-M. 1992, ApJ, 394, 534
Van Buren, D., and McCray, R. 1988, ApJ, 329, L93
Weaver, R., McCray, R., Castor, J., Shapiro, P, and Moore, R. 1977, ApJ 218,
     377
Wood, D. O. S., and Churchwell, E. 1989, ApJ Suppl., 69, 831
Wood, D. O. S., and Churchwell, E. 1991, ApJ, 372, 199

Massive Stars: Their Lives in the Interstellar Medium
ASP Conference Series, Vol. 35, 1993
Joseph P. Cassinelli and Edward B. Churchwell (eds.)

HOT PHASES OF THE INTERSTELLAR MEDIUM: STELLAR WINDS, SUPERNOVAE, AND TURBULENT MIXING LAYERS

J. MICHAEL SHULL
Department of Astrophysical, Planetary, and Atmospheric Sciences,
University of Colorado, Campus Box 391, Boulder CO 80309; and
Joint Institute for Laboratory Astrophysics, University of Colorado and
National Institute of Standards and Technology

ABSTRACT   I discuss the hot "third phase" of the interstellar medium together with mechanisms for depositing energy in it through supernovae, stellar wind bubbles, and superbubbles. A young (< 1 Myr) superbubble with an active OB association is illustrated by the $2° \times 5°$ infrared (IRAS) supershell around Cyg OB1. The thermal energy deposited by SNe and winds can be transferred to cold or warm interstellar gas through "turbulent mixing layers", which radiate characteristic diffuse UV, EUV, and optical emission lines at $\bar{T} \approx 10^{5.3}$ K (peak of cooling curve).

I. INTRODUCTION

In this review, I will discuss three topics related to the production and detection of the "hot phase" of the interstellar medium (ISM). First, after a brief discussion of "thermal phases", I describe the standard theory of stellar wind bubbles, including the effects of magnetic fields on thermal evaporation and the interior temperature. Second, I discuss the extension of this bubble theory to "superbubbles" driven by winds and supernovae from associations of OB stars. I conclude by covering the new subject of "turbulent mixing layers", which may provide a new, simplified model for understanding the transfer of energy from hot phases to cold, warm, and intermediate-temperature gas. These mixed regions of the ISM appear ubiquitously during interactions between supernova remnants (SNRs) and interstellar clouds and between superbubbles and the gas layer at the disk – halo interface. More generally, turbulent mixing layers may provide a paradigm for modeling the "interstellar froth" and churned gaseous structures seen in the Milky Way and selected external spiral and irregular galaxies.

II. THERMAL PHASES OF THE INTERSTELLAR MEDIUM

Interstellar gas may exist in a variety of forms depending on its local heating, ionization, and past history. Parcels of gas at a stable specified temperature and density are termed "phases", although the term is often used somewhat loosely. The two-phase interstellar medium was discussed by Field, Goldsmith, & Habing (1969) and was updated by Shull and Woods (1985) to include metallicity changes, grain photoelectric heating, and X-rays. The so-called "three-phase model" of the ISM was introduced by Cox & Smith (1974) and McKee & Ostriker (1977) to account for new observations of hot gas at temperatures between 10^5 K and 10^6 K suggested by UV absorption lines (O VI, N V, C IV, Si IV) and soft X-ray emission. Although gas at such temperatures is probably thermally unstable (see review by Shull 1987) hot, low-density interstellar gas has a long cooling time and is referred to as the "hot third phase" of the ISM. In fact, the gas seen in UV absorption lines is probably in a transient state of ionization and temperature, produced by thermal conduction, blast-wave heating, a galactic fountain, or turbulent mixing layers. I will return to this subject in § V.

III. THEORY OF STELLAR WIND BUBBLES

The standard theory of a wind-driven bubble developed by Castor, McCray, & Weaver (1975) and Weaver et al. (1977) is based on a structure with an inner free-wind jacketed by a bubble of hot shocked gas. During a brief early phase, the fast stellar wind from a hot star impacts directly on the surrounding medium and the bubble is momentum-driven. Soon, a radiative outer shell forms and the fast wind shocks at an interior radius, leaving a large bubble of hot gas separating the inner and outer shocks. The shell is then driven by the bubble's thermal pressure, P_b.

   The equations of momentum and energy conservation for this bubble structure can be used to relate the shell's radius, R_s, and velocity, $V_s = \dot{R}$, to the (time-dependent) rate of input of wind energy, $L_w(t)$,

$$\frac{d}{dt}(M_s \dot{R}_s) = 4\pi R_s^2 P_b \tag{1}$$

$$\frac{d(E_b)}{dt} = L_w(t) - 4\pi R_s^2 P_b \dot{R} - (\text{Radiative Cooling}) . \tag{2}$$

It is generally assumed that once a radiative shell forms, all the swept-up interstellar mass remains in the shell, $M_s = (4\pi R_s^3/3)\rho_o$, where $\rho_o = (1.4 m_H n_o)$ is the mass density of the ambient medium. The bubble's thermal pressure and internal energy E_b are connected by the thermodynamic relation $P_b = (E_b/2\pi R_s^3)$, and $L_w(t) = \dot{M}_w V_w^2/2$ is the mechanical luminosity of the stellar wind, which for a typical O-star with mass loss rate $\dot{M}_w \approx 10^{-6}\ M_\odot$ yr^{-1} and wind velocity $V_w \approx 2000$ km s^{-1} may be approximated,

$$L_w = (1.3 \times 10^{36}\ \text{ergs s}^{-1})(\dot{M}_w/10^{-6})(V_w/2000)^2 . \tag{3}$$

If we neglect radiative cooling of the hot bubble interior (a valid assumption for the low densities characteristic of the diffuse ISM and during all but the late stages of evolution of the bubble) we may combine equations (1) and (2) into a single equation for shell radius,

$$\frac{d}{dt}\left[R_s^3 \frac{d^2}{dt^2}(R_s^4)\right] = \frac{6 R_s^2}{\pi \rho_o} L_w(t) . \tag{4}$$

For a constant wind luminosity, $L_w(t) = L_o$, a self-similar solution exists,

$$R_s = \left[\frac{125 L_o t^3}{154 \pi \rho_o}\right]^{1/5} = (26.2 \text{ pc}) L_{36}^{1/5} n_o^{-1/5} t_6^{3/5} \tag{5}$$

$$V_s = \left(\frac{3 R_s}{5t}\right) = (15.4 \text{ km s}^{-1}) L_{36}^{1/5} n_o^{-1/5} t_6^{-2/5}, \tag{6}$$

where $L_{36} = (L_o/10^{36} \text{ ergs s}^{-1})$ and $t_6 = (t/10^6 \text{ yr})$. In this solution, 5/11 of the wind's luminosity goes to increasing the bubble's internal energy and 6/11 goes to $P\,dV$ work on the shell. Thus, $E_b(t) = 5L_o t/11$ and the bubble pressure is

$$\frac{P_b}{k} = \left(\frac{E_b}{2\pi R_s^3 k}\right) = (3.12 \times 10^4 \text{ cm}^3 \text{ K}) L_{36}^{2/5} n_o^{3/5} t_6^{-4/5}, \tag{7}$$

which is around 10 times the mean interstellar thermal pressure.

The characteristic density n_b and temperature T_b of the bubble interior are determined by the "mass loading" from evaporated gas off the shell and other sources (entrainment, clouds that penetrate the shell, etc.). If one considers solely conductive evaporation from the shell, the mass loss rate is

$$\dot{M}_b = \frac{16 \pi \mu R_s (6 \times 10^{-7} T_b^{5/2}) \kappa_o}{25k}, \tag{8}$$

where $\mu = 0.609 m_H$ is the mean particle mass for fully-ionized gas with He/H = 0.1 by number. We have assumed the classical conductivity (Spitzer 1962) in cgs units, multiplied by a dimensionless scaling factor $\kappa_o \leq 1$ to account for possible magnetic suppression. For interior temperature and density profiles appropriate for conductive flow, $T(r) = T_b[1 - (r/R_s)]^{2/5}$ and $n(r) = n_b[1 - (r/R_s)]^{-2/5}$, one finds $\dot{M}_b = (125\pi/39)(1.4 m_H n_b R_s^3)$ and $T_b = (P_b R_s^3/\dot{M}_b)(\mu/k)(125\pi/39)$. By integrating $\dot{M}_b$ together with the previous expressions for $R_s(t)$, $P_b(t)$, and $T_b(t)$ we arrive at the expressions,

$$M_b(t) = (46 \text{ } M_\odot) L_{36}^{27/35} n_o^{-2/35} t_6^{41/35} \kappa_o^{2/7} \tag{9}$$

$$T_b(t) = (1.85 \times 10^6 \text{ K}) L_{36}^{8/35} n_o^{2/35} t_6^{-6/35} \kappa_o^{-2/7} \tag{10}$$

$$n_b(t) = (7.3 \times 10^{-3} \text{ cm}^{-3}) L_{36}^{6/35} n_o^{19/35} t_6^{-22/35} \kappa_o^{2/7}, \tag{11}$$

where we have used $n_b = (P_b/2.3kT_b)$ for the density of hydrogen nuclei. Notice that the effects of magnetic suppression of conductivity would need to be severe ($\kappa_o \ll 1$) to produce a large increase in the interior temperature, since $T_b \propto \kappa_o^{-2/7}$. More likely, the interior density and temperature are regulated by the entrainment of clouds and the penetration of ambient interstellar clouds that survive photoevaporation by the O star (McKee, Van Buren, & Lazareff 1984). Clouds will penetrate the radiative shell if their column densities exceed that of the swept-up gas in the shell, $N_{sh} = n_o R_s/3 \approx (3 \times 10^{19} \text{ cm}^{-2}) L_{36}^{1/5} n_o^{4/5} t_6^{3/5}$. Clouds have been found interior to the Cygnus Loop (Fesen, Kwitter, & Downes 1992). The detection of clouds inside wind-driven bubbles would be an equally important discovery, owing to their potentially strong radiative effects on the surrounding hot plasma.

IV. SUPERBUBBLES AROUND OB ASSOCIATIONS

An obvious modification of the McKee-Ostriker 3-phase theory occurred
during the 1980's when modelers took into account the spatial and temporal
correlation of OB stars in associations. These associations typically contain 10
to 100 upper-main-sequence stars with powerful winds and the potential for
ending their lives as Type II or Type Ib supernovae (McCray & Kafatos 1987;
Heiles 1987, 1990). The dynamics of the "superbubble" can be approximated
by scaling up the wind-bubble theory for a continuous energy input. If the
association's stars are formed roughly coevally, then the rate of energy input
can be computed by coupling the initial mass function (IMF) with evolutionary
tracks for the massive stars ($m \geq 25\ M_\odot$).

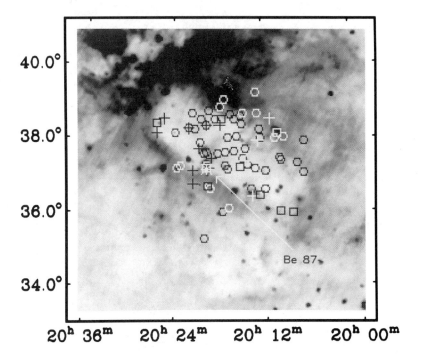

Fig. 1.   The IRAS supershell (Saken et al. 1992) surrounding the Cyg
OB1 association at (1.5 ± 0.3) kpc overlaid with 77 member OB stars
(Garmany & Stencel 1992) and 9 Wolf-Rayet stars (Conti & Vacca 1990).
Post-main-sequence stars are plotted as crosses and Wolf-Rayet stars as
squares. The open cluster Be 87 at 950 pc containing the star WR 142 is
shown as an asterisk near the southern cavity boundary.

Superbubbles and supershells have been identified in 21-cm emission
maps (Colomb, Poppel, & Heiles 1980) and in X-ray surveys (Cash et al.
1980). Models of supershell dynamics and "blowout" of the disk gas layer
have been made by many authors (e.g., Tomisaka & Ikeuchi 1986; McCray &
Kafatos 1987; Tenorio-Tagle & Bodenheimer 1988; Mac Low et al. 1989). The

Colorado group has recently identified the "smoking gun" of superbubbles – a clear example of a young superbubble and its parent OB association in the "Cygnus X" region (Saken et al. 1992). Figure 1 shows the hot stars in the Cyg OB1 association centered on a $2° \times 5°$ cavity in the IRAS $60/100$ μm ratio image. At the association distance of (1.5 ± 0.3) kpc, this cavity has dimensions 50×150 pc and agrees well with an Hα filamentary boundary (Y.-H. Chu, private communication). The shell's elongated morphology is consistent with OB-star subclustering over the 10^6-yr age of the bubble. With the parent OB association still visible (10 O stars between 25 and 45 $M_\odot$, 3 Wolf-Rayet stars inside the cavity, and possibly ~ 3 past supernovae from stars more massive than 45 $M_\odot$) this object is ideal for analyzing the formation, energy input, and evolution of a supershell.

The first 4×10^6 years of (coeval) superbubble evolution are dominated by stellar winds from O stars, which increase in strength as the stars evolve off the main sequence and eventually become Wolf-Rayet stars. During later stages, when stellar winds from the most massive stars have subsided, Type II supernovae dominate the energy input down to the point when stars with $m \approx 8$ $M_\odot$ end their lives. An approximate formula for the mechanical luminosity in the SN phase may be written,

$$L_{\text{SN}}(t) = \left(\frac{dN(> m)}{dm} \right) \left(\frac{dm}{dt} \right) E_{\text{SN}} \, , \tag{12}$$

which scales as $L \propto t^{[(\alpha/\beta)-1]}$, where the IMF is taken to be $N(> m) \propto m^{-\beta}$ and the mass-age relation for upper-main-sequence stars is parameterized as $t(m) \propto m^{-\alpha}$. For stars in the mass range $8 - 30$ $M_\odot$, the mass-age relation may be fitted with $\alpha \approx 1.6$ (Maeder & Meynet 1989) and measurements of IMF slopes for upper-main-sequence stars typically yield $\beta = 1.6 \pm 0.3$ (Garmany, Conti, & Chiosi 1982; Massey et al. 1989). Thus, since $\alpha \approx \beta$, the energy input, $L_{\text{SN}}(t)$, tends to be approximately constant with time during the SN phase. Slight changes in the cluster IMF slope and the mass-age relation cause $L_w(t)$ to have a more complex behavior in more sophisticated cluster models.

The actual energy input from an OB association varies with time due to the switchover from O-star winds to Wolf-Rayet star winds and SNe. If we scale up the stellar wind energy input by a factor of 100, to $(10^{38}$ ergs s$^{-1})L_{38}$, we may write the shell radius, velocity, and bubble pressure,

$$R_s = (65.9 \text{ pc})L_{38}^{1/5} n_o^{-1/5} t_6^{3/5} \tag{14}$$

$$V_s = (38.6 \text{ km s}^{-1})L_{38}^{1/5} n_o^{-1/5} t_6^{-2/5} \tag{15}$$

$$\frac{P_b}{k} = (1.24 \times 10^6 \text{ cm}^3 \ K)L_{38}^{2/5} n_o^{3/5} t_6^{-4/5} \, , \tag{16}$$

after $t = (10^6 \text{ yr})t_6$. If the interior mass is set by thermal conduction, as in the isolated bubble, the bubble interior temperature and density are given by,

$$T_b = (5.3 \times 10^6 \ K)L_{38}^{8/35} n_o^{2/35} t_6^{-6/35} \kappa_o^{-2/7} \tag{17}$$

$$n_b = (1.6 \times 10^{-2} \text{ cm}^{-3})L_{38}^{6/35} n_o^{19/35} t_6^{-22/35} \kappa_o^{2/7} \, . \tag{18}$$

An interesting test of this theory would be to measure the X-ray temperature of the hot interiors. The superbubble's observability is determined by the X-ray surface brightness, proportional to $n_b^2 R_b$, and by the temperature T_b, which is governed by the mass loading of the interior. Magnetic fields can suppress the rate of conductive mass loss into the interior by a factor $\kappa_o \leq 1$, thereby raising T_b and reducing n_b. Counteracting this effect is the increase of interior mass through entrainment of gas off the "chimney" walls or cloud penetration of the shell, which will increase the cloud surface area subject to evaporation. Penetrating clouds or turbulent mixing layers allow an efficient transfer of energy to gas at temperatures $T \approx 10^5$ K, which can radiate efficiently in the UV and EUV.

The Hertzsprung-Russell diagram for Cyg OB1 is shown in Figure 2. A discrepancy between the superbubble dynamical age $\leq$ 1 Myr and the main-sequence turnoff age (5 Myr) probably requires non-coeval star formation.

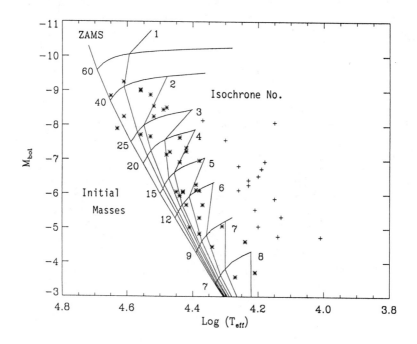

Fig. 2.   H-R diagram of 53 stars in Cyg OB1; post-main-sequence stars are plotted as crosses. Evolutionary tracks for solar metallicity are from Maeder & Meynet (1988) and Maeder (1990), labeled to left of ZAMS with initial mass. Isochrones (# 1-8) are labeled to right of stars corresponding to ages of 2.5, 4.0, 6.5, 8.25, 11.8, 16.7, 26, and 45 Myr.

This interpretation is consistent with the sub-clustering hypothesis, with the number of post-main-sequence stars of initial mass lower than the main-sequence turnoff, and with the likelihood that few SNe can have occurred inside the relatively small cavity. Thus, if the 3 Wolf-Rayet stars came from stars with initial masses between 45 and 50 $M_\odot$, extremely massive stars may not have formed in this association, and the superbubble may have been driven by stellar winds alone. To test this hypothesis, we show in Figure 3 the energy input from winds and SNe from a model Cyg OB1 association. The wind input rises quickly during the first 4 Myr, peaks near 10^{39} ergs s^{-1}, and then slowly declines as the most massive stars complete their Wolf-Rayet phases and produce SNe. This energy input into the relatively small IR cavity requires that the superbubble be less than 1 Myr in age.

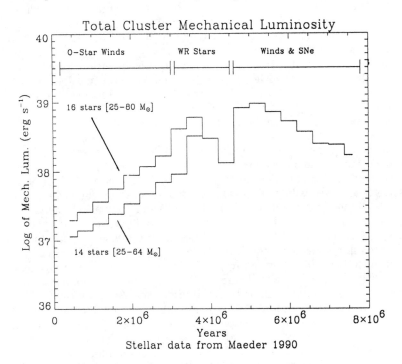

Fig. 3.   Energy input to cavity from stellar winds and SNe in two model Cyg OB1 associations (Saken & Shull 1992) with upper mass cutoffs of 64 and 80 $M_\odot$. Stars form coevally with IMF $N(> m) \propto m^{-1.6}$ and evolve with mass-loss prescription and tracks of Maeder (1990).

V. TURBULENT MIXING LAYERS IN THE INTERSTELLAR MEDIUM

As noted in § II, the ISM of galaxies is a complex system containing several phases, ranging from hot (10^6 K) to cold (10^2 K) or warm (10^4 K) gas. A major task of any theory of the ISM is to explain how mass and energy are transferred from stars to the gas and between gaseous phases. The relative proportions in each phase are significant, since the radiative cooling rate is low for $T \leq 10^4$ K and $T \geq 10^6$ K, and peaks at $\sim 10^5$ K. It is widely believed that the hot phase is created and maintained by SNe and stellar winds, while the cold, cloudy phase may arise from radiative cooling aided by the compaction of diffuse gas by shocks. The primary means proposed to convert hot gas to cold/warm gas has been through radiative cooling, while cold/warm gas is converted to hot gas by thermal conduction.

It is likely, however, that a dynamically active ISM can find more effective means to transfer energy from the point of deposition (SNe and winds) to intermediate-temperature gas. We have proposed (Slavin, Shull, & Begelman 1993) that "turbulent mixing" of hot and cold gas provides such a mechanism for understanding the energetics and line ratios in a frothy medium. On a broader theoretical level, turbulent layers initiated by shear flows at the boundaries of hot and cold gas produce intermediate-temperature gas at $\bar{T} \approx 10^{5.0-5.5}$ K that radiates strongly in the optical, ultraviolet, and extreme ultraviolet (at the peak of the cooling curve). These layers convert thermal energy to ionizing radiation. Expanding on the idea of Begelman & Fabian (1990), we have modeled these layers under the assumptions of rapid mixing and mean steady flow. By including the effects of non-equilibrium ionization and self-photoionization of the gas as it cools after mixing, we predict the intensities of numerous optical, infrared, and ultraviolet emission lines, as well as absorption column densities of C IV, N V, Si IV, and O VI. Spectroscopically, mixing layers include aspects of both hot, collisionally ionized gas and photoionized gas.

We believe that turbulent mixing layers are common in the ISM of our Galaxy and selected external galaxies. Prime locations for turbulent mixing layers include: (1) Rayleigh-Taylor (RT) and Kelvin-Helmholtz (KH) instabilities generated by shock waves overtaking interstellar clouds (Klein, McKee, & Colella 1990); (2) RT instabilities produced when superbubbles break through the gas layer above the Galactic disk (Tenorio-Tagle & Bodenheimer 1988; Mac Low & McCray 1990); and (3) instabilities generated by shear flows along the walls of chimneys; or (4) supernova remnants or stellar wind shocks interacting with density inhomogeneities in the ISM. The equilibrium state of the ISM is strongly influenced by dynamical events, leading to a frothy or churned morphology. The diffuse infrared cirrus seen in IRAS images has a ragged appearance that suggests sheets and fragments of clouds torn apart by shock waves (Shull 1987). Several authors (Kennicutt & Hodge 1986; Hunter & Gallagher 1990; Rand, Kulkarni, & Hester 1990; Dettmar 1990) have pointed to the filamentary appearance of the diffuse ionized gas in external spiral and irregular galaxies. These filaments exhibit large velocity widths and an ionization level lower than that in classical giant H II regions, as evidenced by strong [S II], [N II], and [O II] relative to Hα, but weak [O III]/Hα.

Mixing layers can produce the line ratios [S II]/H$\alpha \approx 0.5$, [N II]/H$\alpha \approx 0.3 - 0.7$, and [O II]/[O III] $\approx 2 - 3$, in good agreement with observations of diffuse ionized gas and "interstellar froth" in Magellanic irregulars and some edge-on spirals. Selected line-ratio diagrams (Figure 4) illustrate the characteristics of the optical spectra produced by mixing layers. Because the cooling occurs in many EUV emission lines, mixing layers develop features of both radiative shocks and photoionized nebulae, as the ionizing radiation is absorbed by adjoining neutral gas. However, the [O I] and [S II] ratios are separated from shocks in these diagrams, and can be used to analyze the collective spectra of large emission-line regions in external galaxies.

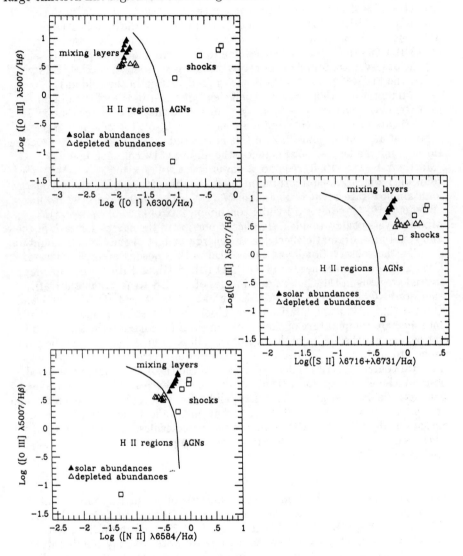

Fig. 4.   Optical line-ratio diagrams for [S II], [N II], and [O I] versus [O III] (each normalized to Hα or Hβ). Models are for mixing layers (Slavin et al. 1993) and radiative shocks (Shull & McKee 1979). The dividing line between AGNs and H II regions (Veilleux & Osterbrock 1987) also is shown.

The absolute line intensities from mixing layers scale in proportion to gas pressure. In the Galaxy, at a pressure $p/k_B \approx 3000$ cm^{-3} K, mixing layers can account for: (1) between 2% and 10% of the diffuse Hα in the Galactic disk; (2) between 30% and 100% of the the diffuse Hα at high latitude, with correct line ratios of [S II], [N II], [O II], and [O III]; (3) all of the high-latitude C IV $\lambda1550$ emission observed by Martin & Bowyer (1990), with the observed ratios of O III] $\lambda1663$ to C IV $\lambda1549$ emission and C IV emission to absorption; and (4) the observed column density ratios of the highly ionized species, C IV, N V, O VI, and Si IV if one allows for some grain depletion in the cold gas.

Mixing layers thus provide an efficient amplifier of the effects of star formation. On average, $(1-3)$ $M_\odot$ yr^{-1} of star formation over the Galaxy leads to a SN rate of ~ 0.025 yr^{-1} and a SN energy input of $(1-2) \times 10^{42}$ ergs s^{-1} (Slavin et al. 1993). Some 20% of this power and 75 $M_\odot$ yr^{-1} may be processed through mixing layers. For reference, the diffuse C IV emission-line flux (Martin & Bowyer 1990) requires a power source of $(4 \times 10^{40}$ ergs s$^{-1})R_{12}^2$ for collisional ionization equilibrium in a disk of radius (12 kpc)R_{12} and a Galactic fountain flow rate between 10 and 20 $M_\odot$ yr^{-1}. The Galactic diffuse Hα flux (Reynolds 1984) requires a Lyman continuum flux of $(4 \times 10^{41}$ ergs s$^{-1})R_{12}^2$. Mixing layers could provide a significant portion of the energy for each of these, and could have dramatic effects on the energy budget of the Galactic fountain.

Further observational and theoretical work is needed on mixing layers. Optical line ratios are needed of [O I], [O III], [S II], and Hα in the same high-latitude regions in the Galaxy, and images of the Hα structure might clarify the source. Observations of the emission lines O III] $\lambda1663$, [O III] $\lambda5007$ and O VI $\lambda1032, 1038$ at high latitude are important to test our models for the intermediate temperature of the layers. In external galaxies, the studies of line ratios, line widths, and morphology of the diffuse ionized gas are just beginning to be provided; more detailed spatial information is needed to separate out hot-star sources from turbulent sources. On the theoretical side, unresolved issues include pinning down the entrainment efficiencies, the potential role of magnetic fields, dust grain processing, and "slow mixing layers" into molecular gas (mixing rates less than 20 km s^{-1}) that might affect the chemistry of such species as H_2, CH^+, C I/CO and more exotic molecules. An overall goal of these studies should be to understand the nature and life history of "frothy" diffuse ionized gas in the ISM.

I am grateful to the ASP conference organizers for the invitation to speak on the subject of winds and SNe. This work was supported, in part, by the NASA Astrophysical Theory Program (NAGW-766) and by various NASA data analysis grants, which have allowed me to work with my collaborators Rob Fesen, Jon Saken, Jon Slavin, Dick McCray, and Mitch Begelman. Many thanks to their advice and help over the past several years.

REFERENCES

Begelman, M. C., & Fabian, A. C. 1990, MNRAS, 244, 26P
Cash, W. C., et al. 1980, ApJL, 238, L71
Castor, J., McCray, R., & Weaver, R. 1975, ApJL, 200, L107

Colomb, F. R., Poppel, W. G. L., & Heiles, C. 1980, A&AS, 40, 47

Conti, P. S., & Vacca, W. D. 1990, AJ, 100, 431

Cox, D. P., & Smith, B. W. 1974, ApJL, 89, L105

Dettmar, R.-J. 1990, A&A, 232, L15

Fesen, R. A., Kwitter, K. B., & Downes, R. A. 1992, AJ, in press

Field, G., Goldsmith, D., & Habing, H. 1969, ApJL, 281, L25

Garmany, C. D., & Stencel, R. 1992, A&AS, in press

Garmany, C. D., Conti, P. S., & Chiosi, C. 1982, ApJ, 263, 777

Heiles, C. 1987, ApJ, 315, 555

Heiles, C. 1990, ApJ, 354, 483

Hunter, D., & Gallagher, J. 1990, ApJ, 362, 480

Kennicutt, R., & Hodge, P. 1986, ApJ, 306, 130

Klein, R. I., McKee, C. F., & Colella, P. 1990, in The Evolution of the
      Interstellar Medium, ed. L. Blitz, (San Francisco, PASP), 117

Mac Low, M.-M., McCray, R., & Norman, M. L. 1990, ApJ, 337, 141

Maeder, A. 1990, A&AS, 84, 139

Maeder, A., & Meynet, G. 1988, A&AS, 76, 411

Maeder, A., & Meynet, G. 1989, A&A, 210, 155

Martin, C., & Bowyer, S. 1990, ApJ, 350, 242

Massey, P., Garmany, C., Silkey, M., & DeGioia-Eastwood, K. 1989, AJ, 97,
      107

McCray, R., & Kafatos, M. 1987, ApJ, 317, 190

McKee, C. F., Van Buren, D., & Lazareff, B. 1984, ApJL, 278, L115

McKee, C. F., & Ostriker, J. P. 1977, ApJ, 218, 148

Rand, R., Kulkarni, S., & Hester, J. 1990, ApJL, 352, L1

Reynolds, R. J. 1984, ApJ, 282, 191

Saken, J. M., & Shull, J. M. 1992, ApJ, in preparation

Saken, J. M., Shull, J. M., Garmany, C. D., Nichols-Bohlin, J., & Fesen, R.
      1992, ApJ, 397, in press (Oct. 1)

Shull, J. M. 1987, in Interstellar Processes, ed. D. J. Hollenbach & H. A.
      Thronson, (Dordrecht: Kluwer), 225

Shull, J. M., & McKee, C. F. 1979, ApJ, 227, 131

Shull, J. M., & Woods, D. T. 1985, ApJ, 288, 50

Slavin, J., Shull, J. M., & Begelman, M. C. 1993, ApJ, in press

Spitzer, L. 1962, Physics of Fully Ionized Gases, (New York: Wiley-
      Interscience)

Tenorio-Tagle, G., & Bodenheimer, P. 1988, ARAA, 26, 145

Tomisaka, K., & Ikeuchi, S. 1986, PASJ, 38, 697

Veilleux, S., & Osterbrock, D. E. 1987, ApJS, 63, 295

Weaver, R., McCray, R., Castor, J., Shapiro, P., & Moore, R. 1977, ApJ, 218,
      377 (errata 220, 742)

Massive Stars: Their Lives in the Interstellar Medium
ASP Conference Series, Vol. 35, 1993
Joseph P. Cassinelli and Edward B. Churchwell (eds.)

THE RADIATIVE IONIZATION OF THE INTERSTELLAR MEDIUM--
THE WARM IONIZED GAS

RONALD J. REYNOLDS
Department of Physics, University of Wisconsin,
1150 University Avenue, Madison, WI 53706

ABSTRACT   Faint   optical   emission   lines   and   pulsar
dispersion measures reveal warm (10^4 K) ionized gas to be
a major component of the interstellar medium of the Milky
Way and other galaxies.   The origin of this ionization
has   not   been   determined,   but   the   temperature   and
ionization state of the gas appear similar to that of a
low excitation HII region.   Of the known sources of
ionization in our Galaxy only the ionizing flux from O
stars and   the kinetic   energy of supernovae meet or
surpass the power requirements of this global ionization.
However, if the source is O stars, then ionizing photons
must be able to travel hundreds of parsecs through the
Galaxy's HI layer, and if it is supernovae, the
ionization mechanism must be very energy efficient. From
the   available   data   it   is   clear   that   an   accurate
understanding of interstellar matter and processes will
require a thorough investigation of this warm ionized
medium.

INTRODUCTION

The mechanisms by which massive stars influence the global
properties of the interstellar medium are the subject of
debate and controversy.   It has been proposed, for example,
that the kinetic energy deposited in the interstellar medium
by supernovae and stellar winds produces a pervasive hot (10^6
K) gas that determines the large scale structure and dynamics
of the medium (Cox and Smith 1974; McKee and Ostriker 1977;
Shapiro and Field 1976).   But these global effects of the hot
gas have not yet been confirmed by observations, and a recent
rethinking of some of the original assumptions has put the
theoretical basis for such models in doubt (Cox 1989; Slavin
and Cox 1992).   Another unresolved question is the extent to
which massive stars are responsible for the large scale
ionization of the interstellar hydrogen; and, if they are,
what the ionization mechanism is.   Within the last few years,
warm ($\sim 10^4$ K) ionized hydrogen has been recognized as one of

the interstellar medium's major components (e.g., Kulkarni and Heiles 1987, Cox 1989, McKee 1990). Its surprisingly large mass, high extent above the midplane, and enormous power requirements have modified our understanding of the composition and structure of the interstellar medium and the distribution of ionizing radiation within the Galactic disk and halo. Proposed sources of this ionization have ranged from O stars (e.g., Mathis 1986) to dark matter (Sciama 1990).

The following is a brief review of the observed properties of this ionized component and the constraints on the source of its ionization.

WARM IONIZED HYDROGEN: TRADITIONAL PICTURE VS ACTUAL SITUATION

In the traditional picture, regions of warm, fully ionized hydrogen are confined to "Stromgren spheres" of radius R_s around O stars. Within each of these "HII regions" the Lyman continuum luminosity L_* of the star balances the hydrogen recombination rate, so that $L_* = \alpha\, n_e^2\, R_s^3\, 4\pi/3$, where α is the hydrogen recombination rate coefficient and n_e is the gas density. For an average interstellar gas density $n_e \approx 1\ \mathrm{cm}^{-3}$, and a typical O star ionizing photon luminosity of $10^{49}\ \mathrm{s}^{-1}$, $R_s \approx 60$ pc. Since O stars tend to be located near regions of higher than average density, the expected value of R_s is even smaller. The transition at the edge of the Stromgren sphere between the fully ionized hydrogen and the neutral hydrogen is very sharp ($\lesssim 1$ pc), a few mean free path lengths of an ionizing photon in HI. Therefore, the warm ionized hydrogen in the interstellar medium should be confined to isolated regions near the Galactic midplane and within 60 pc of an O star. This is the picture most of us still hold to some extent because the interstellar medium actually appears this way on Hα filter surveys (e.g., Palomar Sky Survey red prints).

The actual distribution of ionized gas in the interstellar medium is very different from the traditional picture, however. From measurements of faint optical interstellar emission lines and pulsar dispersion measures we now know that:

- most (90%) of the HII is located in warm ($\sim 10^4$ K), low density ($\sim 0.1\ \mathrm{cm}^{-3}$), fully ionized regions located far from O stars;

- half of the HII is more than 600 pc from the Galactic midplane;

- the HII mass is approximately 1/3 the HI mass;

- these HII regions fill more than 20% of the volume

within a 2 kpc thick layer about the midplane.

Therefore, although most of the hydrogen recombinations (i.e., Hα luminosity) in the Galaxy occur within traditional O star HII regions, nearly all of the HII mass is associated with a much more extended, low density component that is not detected on broad band, limited sensitivity Hα photographs.

THE DATA

Many diverse observations of the interstellar medium have provided evidence for substantial ionization of the interstellar hydrogen outside the bright, traditional HII regions. These include free-free absorption at very low radio frequencies (Hoyle and Ellis 1963), ultraviolet interstellar absorption lines (Gry, York, and Vidal-Madjar 1975), Galactic rotation measures, pulsar dispersion measures, and faint optical interstellar line emission. The last two provide most of the information about the nature of diffuse ionized gas.

Dispersion Measures.
Pulsar dispersion measures have revealed that the ionized gas is widespread and massive. Distances have been determined for approximately 10% of the known pulsars through 21 cm absorption measurements (e.g., Weisberg, Rankin, and Boriakoff 1980), VLBI parallax (Gwinn et al. 1986), and the fact that some pulsars reside in globular clusters and supernova remnants of known distance. Distances range from approximately 130 pc to 16 kpc and sample a large portion of the Galactic disk. Because the dispersion measure is equal to the column density of electrons (and HII) along the line segment to the pulsar, these data provide a direct measurement of the space averaged electron density near the midplane ($\langle n_e \rangle \approx 0.025$ cm^{-3}; Weisberg, Rankin, and Boriakoff 1980). In addition, the pulsars in globular clusters (e.g., Wolszczan et al. 1989) located far ($|z| > 3$ kpc) from the midplane reveal the total vertical extent ($H_z \approx 900$ pc) and column density ($N_{HII} \approx 7 \times 10^{19}$ cosec $|b|$ cm^{-2}) of the HII above the midplane (Reynolds 1991a). Multiple pulsars within some of the clusters (e.g., Manchester et al. 1991) also indicate that any dispersion measures intrinsic to the pulsars and the clusters are small ($\lesssim 10^{18}$ cm^{-2}).
The mass surface density of this diffuse HII is about one third that of the HI, outweighing the traditional HII regions by about 20 to 1 (Reynolds 1991a). A comparison of dispersion measures with 21 cm HI column densities toward the five high $|z|$ globular clusters reveals N_{HII}/N_{HI} column density ratios that range from 0.26 to 0.63 (Reynolds 1991a; Lockman, Langston, and Reynolds 1992). The pulsar data also indicate that the HII is distributed much more smoothly within the

interstellar medium than the HI (compare Fig. 1 in Reynolds 1991a with Fig. 1a in Edgar and Savage 1989, for example).

There is, of course, some contribution to the dispersion measures by the hot (10^5–10^6 K) component of the interstellar medium. However, this contribution is relatively small. For example, the Local Bubble responsible for much of the soft x-ray background (Sanders et al. 1977) has a density near 5×10^{-3} cm^{-3}, extends ≈ 100 pc from the Sun, and thus contributes about 1.5×10^{18} cm^{-2} ($\lesssim 2\%$) to the HII column densities at high latitudes. The amount of hot gas located beyond the Local Bubble is a subject of debate; however, the distribution of dispersion measures as a function of distance from the Galactic midplane shows no evidence for a significant contribution to HII column densities from any component with a scale height much larger than 1 kpc (Reynolds 1991a). Therefore, even if the hot gas occupied <u>all</u> of the volume up to 1 kpc, it would only contribute 20% to the observed HII column densities. The substantial HII column densities revealed by the globular cluster pulsars must therefore be associated with a higher density and cooler component.

Pulsar dispersion measures can provide little additional information about the ionization state, density, filling fraction, kinematics, and temperature of the gas. That information is obtained through the detection and study of faint optical emission lines.

Optical Emission Lines
Faint emission lines from the diffuse ionized gas have been explored in our Galaxy with a large throughput, high spectral resolution Fabry-Perot spectrometer developed at the University of Wisconsin (Reynolds 1990a; Roesler et al. 1978). This instrument is capable of detecting and measuring the radial velocities and line widths of HII emission regions having an Hα surface brightness of 5×10^{-8} ergs cm^{-2} s^{-1} sr^{-1}, which is 70 to 120 times fainter than the faintest regions detectable on Sivan's (1974) Hα survey and the Palomar Sky Survey red prints, respectively, and 1000 times more sensitive than radio continuum and recombination line techniques (e.g., Lockman 1976). The Hα observations with this instrument have revealed a complex emission pattern that covers the sky with intensties of $(1-4) \times 10^{-7}$ cosec $|b|$ ergs cm^{-2} s^{-1} sr^{-1} at $|b| \gtrsim 15°$ (e.g., Reynolds 1992). These H-recombination line intensities imply an average hydrogen recombination (and thus ionization) rate of $r_G = 4.4 \times 10^6$ s^{-1} per cm^2 of Galactic disk (Reynolds 1984).

The presence of collisionally excited forbidden lines of [SII] and [NII] provides a lower limit on the electron temperature $T_e > 5500$ K (Reynolds 1989), while the widths of the Hα and [SII] lines give an upper limit of $T_H < 20,000$ K with a best fit value of 8000 K (Reynolds 1985a). The line width analysis also indicates that non-thermal motions within the gas have speeds typically between 10 km s^{-1} and 30 km s^{-1}.

The absence of detectable [OI] λ6300 and [NI] λ5201 from the warm ionized gas implies a hydrogen ionization ratio $n(H^+)/n(H^\circ) > 2$ (e.g., Reynolds 1989). Thus the ionization is associated with "fully" ionized rather than mostly neutral gas. This rules out penetrating radiation such as x-rays or cosmic rays as the primary ionization mechanism. However, the ionization/excitation conditions differ significantly from conditions in the traditional photoionized HII regions. The very low intensity of [OIII] λ5007 (Reynolds 1985b) and the anomalously strong [SII] λ6716 (Reynolds 1985a) relative to Hα suggest a low state of excitation with few ions present that have ionization energies greater than 23 eV.

A comparison of emission measures ($\int n_e^2 \, ds$) derived from the Hα intensity with dispersion measures ($\int n_e \, ds$) reveals that the HII is clumped into regions that have an average density $n_e \simeq 0.08$ cm^{-3} and occupy at least 20% of the volume within a 2 kpc layer about the Galactic midplane (Reynolds 1991b). If the density decreases exponentially with increasing $|z|$, the average density within an ionized region near the midplane is 0.16 cm^{-3}. The volume filling fraction of the gas may increase with $|z|$ (Kulkarni and Heiles 1987; Reynolds 1991b).

In contrast to the close-up, from-the-inside view of our Galaxy provided by the Wisconsin, large beam Fabry-Perot, relatively broad band imaging and echelle spectroscopy of other galaxies have begun to provide a more global perspective of diffuse ionized gas (e.g., Rand, Kulkarni, and Hester 1990; Keppel et al. 1991; Hunter and Gallagher 1990; Walterbos and Braun 1992; Bland-Hawthorn, Sokolowski, and Cecil 1991; Dettmar 1992). Although the number of galaxies that have been investigated so far is small and the sensitivity to Hα is not at the level attained for our Galaxy, these observations have confirmed the presence of warm ionized gas in other galaxies and provided important, new information.

In the edge-on spiral NGC 891, for example, the most thoroughly studied galaxy, the ionized gas is found to have a scale height and [SII]/Hα line intensity ratio similar to that in our Galaxy, an HII mass surface density twice that of our Galaxy (nearly equal to that of the HI), and a radial extent of more than 9 kpc from the galactic center (Rand, Kulkarni, and Hester 1990). The [SII]/Hα and [NII]/Hα line intensity ratios appear to increase with distance from the galactic midplane (Dettmar and Schulz 1992). Excitation conditions inferred from the forbidden line emission appear to vary from galaxy to galaxy. Regions of diffuse emission in some irregular galaxies, for example, have [OIII]/Hα intensity ratios (Hunter 1984) that are much higher than in NGC 891 and our Galaxy. The presence of detectable, high $|z|$, "extra planar" ionized gas appears to be associated with galaxies that have thick disks of radio continuum emission. The extended radio and Hα emission also seem to correlate with star formation activity in the underlying disk (see review by Dettmar 1992). Of course such correlations do not necessarily imply a cause and effect relationship.

SOURCE OF THE IONIZATION

0 Stars

In our Galaxy the hydrogen ionization rate of 4.4×10^6 s^{-1} per cm^2 of disk implied by the high latitude Hα intensity places a tight constraint on the possible sources of the ionization. B stars, planetary nebula nuclei, hot white dwarfs, cosmic rays, the x-ray background, and the extragalactic radiation field are not capable of accounting for this rate, individually or collectively (Reynolds 1984; Kutyrev and Reynolds 1989). Of the known sources only 0 stars, which emit 3×10^7 Lyman continuum photons s^{-1} per cm^2, produce a sufficient amount of ionizing radiation. However, getting this radiation to the diffuse gas is problematical (see below). At 13.6 eV per hydrogen ionization the minimum power required is 1×10^{-4} ergs s^{-1} cm^{-2} or 10^{51} ergs per 70 yrs for the entire Galaxy. Therefore, supernovae also are a potential source, but only if the kinetic energy they inject into the interstellar medium is somehow converted with high efficiency into the ionization of hydrogen.

The crucial question raised by the hypothesis of 0 star ionization is how the ionizing photons travel from the stars into the extended HII layer. With an exponential scale height of 900 pc one half of the ionized hydrogen is at $|z| > 600$ pc and half of the ionizing photons ($\approx 7\%$ of the total 0 star Lyman continuum photons) are absorbed at $|z| > 300$ pc. These photons must leak out of the traditional HI cloud layer as well as penetrate the "Lockman layer" of HI, which extends to $|z|$ of 500 pc or more (Lockman, Hobbs, and Shull 1986).

Moreover, 0 star photons must travel not only large distances above the plane, but also within the HI cloud layer. This was shown by an examination of the known sources of ionization around the line segment to the pulsar PSR 0823 + 26 (Reynolds 1990b). The distance to this pulsar, determined by VLBI parallax (Gwinn et al. 1986), is 360 ± 80 pc. At $l = 197°$, $b = +32°$ this line segment extends from the midplane (Sun) to $z = + 190$ pc (pulsar) and intersects no traditional HII regions. The pulsar's dispersion measure implies an HII column density of 6.0×10^{19} cm^{-2}. None of the B stars (fully sampled at these distances) or any of the known hot white dwarf stars, individually or collectively, comes even within a factor of ten of being able to produce this HII column density. Only three 0 stars in the solar neighborhood were found to be capable of producing the ionization, the closest

(ξ Per, 07.5 III) at 270 pc from the line segment on the opposite side of the Galactic midplane.

O star ionization thus requires at least a 300 pc radius HII region with a mean density n $\leq$ 0.1 cm^{-3} between ξ Per and the line segment, which subtends an angle of 67° as seen from the star (the dispersion measure implies n $\simeq$ 0.054 cm^{-3} along the segment). An investigation of the Hα background in a large region of the sky near ξ Per (Reynolds 1988) showed no evidence for an HII region extending more than about 90 pc from the star. In fact, the available data suggest that most of the star's ionizing flux is absorbed within a radius of 90 pc. If the ionization of the PSR 0823+26 line segment were assigned to another O star, an even larger (radius $\gtrsim$ 400 pc) HII region would be required (Reynolds 1990b).

The large radii that are required seem to be inconsistent with the generally depicted distribution of HI clouds. For example, a single cold HI cloud (10 cm^{-3}) located 30 pc or more from ξ Per would totally absorb the incident ionizing radiation if its column density were 3 × 10^{19} cm^{-2} or larger. T - τ statistics of clouds (Kulkarni and Heiles 1987; Payne, Salpeter, and Terzian 1983) suggest that there should be approximately 3-4 such clouds along a 300 pc line of sight near the midplane. Moreover, the number of clouds is projected to rise very rapidly with decreasing column density (Dickey and Garwood 1989). At a distance of 100 pc from the star, a cloud (10 cm^{-3}) with a column density of only 3 × 10^{18} cm^{-2} will absorb all of the incident ionizing photons. In the McKee and Ostriker (1976) model the mean free path is 88 pc between cloud cores and much less between the warm neutral (0.3 cm^{-3}) cloud envelopes.

There is evidence that ionizing radiation can penetrate 100 pc or more from some O associations (Leisawitz and Hauser 1988; Reynolds and Ogden 1979). Norman (1991) has suggested that superbubbles or "chimneys" around O associations could let ionizing photons penetrate the HI layer, and Miller and Cox (1992) propose that O star radiation can ionize HI free channels between the clouds and through a low density intercloud HI layer. That these processes actually produce the diffuse ionization, such as that toward PSR 0823+26, has not been demonstrated, however. In any case, if O stars are the source, then the distribution of the HI must differ from the random cloud models commonly envisioned when interpreting 21 cm data.

Alternative Sources

If O star radiation is not the source, then there must be another, more widely distributed source of ionization within the Galactic disk and halo. For example, Slavin, Shull, and Begelman (1992) and Benjamin and Shapiro (1992) have investigated the cooling of hot 10^6 K gas through turbulent mixing with cooler gas and Galactic fountain flows, respectively, and concluded that these processes could account for much of the observed diffuse optical line emission. The ultimate source of power in these models is the kinetic energy of supernovae. Raymond (1992) has proposed that Galactic magnetic field reconnection (microflares) could provide significant heating and ionization in the halo. Raymond points out that the stirring of the magnetic field by supernovae could power the hot, x-ray and UV emitting component, but that, if this mechanism also powered the cooler, optically emitting component, efficiency considerations would require Galactic rotation to be the ultimate power source.

The most exotic source proposed is the photodecay of neutrinos (Sciama 1990). Under this hypothesis the dark matter in galaxies is composed of massive (28 eV) neutrinos, which decay into a 14 eV photon and a lower mass neutrino. Sciama found consistency between the properties of the neutrino required to account for the diffuse galactic ionization and a variety of other galactic and extragalactic observations, including the electron density of the diffuse gas, the intergalactic radiation field, and galactic rotation curves. Attempts to detect this 14 eV radiation have not been successful, however (Davidsen et al. 1991).

CONCLUSIONS

Warm ionized gas is an important constituent of the interstellar medium of our Galaxy and others. Its large surface density, scale height, and power requirements challenge traditional views of the composition and structure of the interstellar medium as well as the distribution and source of ionizing radiation within galactic disks and halos. This component also may exert significant global influences on the galaxy through its contribution to the pressure of the interstellar medium (Cox 1989) and its effect on the dynamics of hot gas above the galactic midplane (Heiles 1990).

The source of the ionization has not yet been established. Neither is its relationship to the other components of the medium known. Does this gas constitute the outermost layers of HI clouds (McKee and Ostriker 1977), or the ionized portion of a pervasive HI intercloud medium (Cox 1989), or the inside walls of superbubbles and "worms" (Heiles 1992; Norman 1991)? Is the warm ionized gas the principal constituent of the lower halo (Reynolds 1991a)? Answers to these questions will require more detailed observations both of our Galaxy and others, where different conditions can provide valuable new perspectives on the nature of this gas. Because warm ionized gas can be explored primarily through its optical line emission, a more accurate understanding of interstellar matter and processes is going to require a

continuing ground based observational effort, including the
development of higher sensitivity spectroscopic
instrumentation.

This work has been supported by the National Science
Foundation, currently through AST 91-15703.

REFERENCES

Benjamin, R. A. and Shapiro, P. R. 1992, McDonald Observatory
    preprint No. 178
Bland-Hawthorne, J., Sokolowski, J., and Cecil, G. 1991, ApJ,
    375, 78
Cox, D. P. 1989 in IAU Colloquium No. 120, Structure and
    Dynamics of the Interstellar Medium, ed. G.
    Tenorio-Tagle, M. Moles, and J. Melnick (New York:
    Springer-Verlag), 500
Cox, D. P. and Smith, B. W. 1974, ApJ, 189, L105
Davidsen, A. F., et al. 1991, Nature, 351, 128
Dettmar, R.-J. 1992, preprint, Radioastronomisches Institut,
    Bonn
Dettmar, R.-J. and Schulz, H. 1992, A&A, 254, L25
Dickey, J. M. and Garwood, R. W. 1989, in IAU Colloquium No.
    120, Structure and Dynamics of the Interstellar Medium,
    ed. G. Tenorio-Tagle, M. Moles, and J. Melnick (New
    York: Springer-Verlag), 511
Edgar, R. J. and Savage, B. D. 1989, ApJ, 340, 762
Gry, C., York, D. G., and Vidal-Majar, A. 1985, ApJ, 296, 593
Gwinn, C. R., Taylor, J. H., Weisberg, J. M., and Rawley, L.
    A. 1986, AJ, 91, 338
Heiles, C. 1990, ApJ, 354, 483
Heiles, C. 1992, preprint
Hoyle, F. and Ellis, G.R.A. 1963, Australian J. Phys., 16,1
Hunter, D. A. 1984, ApJ, 276, L35
Hunter, D. A. and Gallagher, J. S. 1990, ApJ, 362, 480
Keppel, J. W., Dettmar, R.-J., Gallagher, J. S., and Roberts,
    M. S. 1991, ApJ, 374, 507
Kulkarni, S. R. and Heiles, C. 1987, in Interstellar
    Processes, ed. D. J. Hollenbach and H. A. Thronson,
    Jr. (Dordrecht:Reidel), 87
Kutyrev, A. S. and Reynolds, R. J. 1989, ApJ, 334, L9
Leisawitz, D. and Hauser, M. G. 1988, ApJ, 332, 954
Lockman, F. J. 1976, ApJ, 209, 429
Lockman, F. J., Hobbs, L. M., and Shull, J. M. 1986, ApJ, 301,
    380
Lockman, F. J., Langston, G., and Reynolds, R. J. 1992, in
    preparation

Manchester, R. N., Lyne, A. G., Robinson, C., D'Amico, N.,
    Bailes, M., and Lim, J. 1991, Nature, 352, 219
Mathis, J. S. 1986, ApJ, 301, 423
McKee, C. F. 1990 in The Evolution of the Interstellar Medium,
    ed. L. Blitz (ASP Conf. Series Vol. 12), 3
McKee, C. F. and Ostriker, J. P. 1977, ApJ, 218, 148
Miller, W. and Cox, D. P. 1992, in preparation
Norman, C. A. 1991, in IAU Symposium No. 144, The Interstellar
    Disk-Halo Connection in Galaxies, ed. H. Bloemen
    (Dordrecht:Kluwer), 337
Payne, H. E., Salpeter, E. E., and Terzian, Y. 1983, ApJ, 272,
    540
Rand, R. J., Kulkarni, S. R., Hester, J. J. 1990, ApJ, 352,
    L1 (Erratum 362, L35)
Raymond, J. C. 1992, ApJ, 384, 502
Reynolds, R. J. 1984, ApJ, 282, 191
Reynolds, R. J. 1985a, ApJ, 294, 256
Reynolds, R. J. 1985b, ApJ, 298, L27
Reynolds, R. J. 1988, AJ, 96, 670
Reynolds, R. J. 1989, ApJ, 345, 811
Reynolds, R. J. 1990a, in IAU Symposium No. 139, Galactic and
    Extragalactic Background Radiation, ed. S. Bowyer, and
    C. Leinert (Dordrecht:Kluwer), 157
Reynolds, R. J. 1990b, ApJ, 348, 153
Reynolds, R. J. 1991a, in IAU Symposium No. 144, The
    Interstellar Disk-Halo Connection in Galaxies, ed. H.
    Bloemen (Dordrecht:Kluwer) 67
Reynolds, R. J. 1991b, ApJ, 372, L17
Reynolds, R. J. 1992, ApJ, 392, L35
Reynolds, R. J. and Ogden, P. M. 1979, ApJ, 229, 942
Roesler, F. L., Reynolds, R. J., Scherb, F., and Ogden, P.
    M. 1978, in High Resolution Spectroscopy, ed. M. Hack
    (Trieste:Observatorio Astronomico), 600
Sanders, W. T., Kraushaar, W. L., Nousek, J. A., and Fried, P.
    M. 1977, ApJ, 217, L87
Sciama, D. W. 1990, ApJ, 364, 549
Shapiro, P. R. and Field, G. B. 1976, ApJ, 205, 762
Sivan, J.-P. 1974, A & A Suppl., 16, 163
Slavin, J. D. and Cox, D. P. 1992, ApJ, 392, 131
Slavin, J. D., Shull, J. M., and Begelman, M. C. 1992,
    preprint
Walterbos, R. A. M. and Braun, R. 1992, A&A Suppl., 92, 625
Weisberg, J. M., Rankin, J. M. and Boriakoff, V. 1980, A&A,
    88, 84
Wolszczan, A., Kulkarni, S. R., Middleditch, J., Backers, D.
    C., Fruchter, A. S., and Dewey, R. J. 1989, Nature, 337,
    531

Massive Stars: Their Lives in the Interstellar Medium
ASP Conference Series, Vol. 35, 1993
Joseph P. Cassinelli and Edward B. Churchwell (eds.)

O STAR GIANT BUBBLES IN M33

M.S. OEY
Steward Observatory, University of Arizona, Tucson, AZ 85721

P. MASSEY
Kitt Peak National Observatory, National Optical Astronomy
Observatories, P.O. Box 26732, Tucson, AZ 85726

ABSTRACT   We have identified two giant (110 pc), Hα bubble nebulae
in M33, and have spectroscopically identified a single, dominant O9–B0
star centrally located in each. Despite their large values, the nebular sizes
of our objects are consistent with the terminal radius of the model for
wind-blown bubbles created by individual massive stars. We suggest that
this is indeed the case.

Giant Hα shells and bubbles in the ISM are usually related to OB associations
which are thought to give rise to the bubble structure through the action of
stellar winds and/or supernovae (*e.g.* Weaver *et al.* 1977, McKee *et al.* 1984,
MacLow & McCray 1988). Investigations of such objects (*e.g.* Meaburn 1980,
Georgelin *et al.* 1983, Rosado 1986) have been carried out mostly on nearby
extragalactic examples since our views of Galactic shells tend to be highly
obscured. However, understanding of their physical nature remains sketchy
because evaluation of their stellar contents is difficult, and the effects of stellar
multiplicity complicate interpretation of the gas dynamics and morphology.

With the exception of the Milky Way (*e.g.* Saken *et al.* 1992), all studies of
such bubbles until now have examined objects in dwarf irregular galaxies
with turbulent and complex ISM. On the other hand, the nearby Local Group
Sc galaxy, M33, offers the advantage of a more uniform, better-understood
environment in which to study these objects. The presence of Hα bubble-
like structures has already been noted in M33 by Courtès *et al.* (1987). From
large-scale Hα maps of this galaxy taken with the NOAO 0.6-m Schmidt
telescope, we have identified a number of such giant ($\sim$50–200 pc) loops. We
then identified the hottest stars contained in these bubbles from U images
kindly provided by W. Freedman, and unpublished photometry obtained by
PM on the KPNO 2.1-m telescope. The former were obtained at the prime
focus of the CFHT. On the nights of 6–7 October 1991 at the Multiple-Mirror
Telescope, we obtained spectra satisfactory for rough classification of the stars
in three of the bubbles, which are identified in Table I.

TABLE I   Bubble Positions

Bubble	α (1950)	δ (1950)
#1	01 32 13.9	30 21 50
#2	01 32 11.8	30 22 58
#3	01 30 09.2	30 20 27

Bubbles #1 and #2 are of particular interest because we found only a single dominant O9–B0 star centrally located in each bubble. We classify the central star of Bubble #1 to be O9–B0, with an additional, fainter B star of mid to late spectral type enclosed. In Bubble #2, we classify the central star to be O9. The two nebulae are nearly identical, virtually spherical, and are situated near each other at large galactocentric radii (6.5 and 6.2 kpc). They each measure ~110 pc in diameter, which is larger than typically observed around individual stars by roughly a factor of 4 (Chu 1991, Lozinskaya 1982)! The diameters of Wolf-Rayet ring nebulae in M33 identified by Drissen *et al.* (1991) range from 10–40 pc in size. Nevertheless, we believe that the stellar winds of the observed stars are the most likely cause of these enormous bubble structures. Using the analytical model for wind-driven bubbles given by Weaver *et al.* (1977) and McCray (1983), we find that the observed size agrees well with the idealized terminal radius of the shell expected from an O9 V star (see Table II). Furthermore, the size of our objects agrees well with the radii of HI shells around individual Of stars observed by Cappa de Nicolau, Niemela, and Benaglia (1992, this volume). Due to their uniform morphology and agreement with the stellar wind model, these nebulae are likely to be the largest examples known of wind-blown bubbles due to individual, unevolved stars.

Table II presents theoretical parameters for bubbles blown by stars of similar spectral type to those found in Bubbles #1 and #2, following the Weaver *et al.* (1977) and McCray (1983) model for wind-driven bubbles. The first two columns show the spectral type and Strömgren radius of the assumed stars, while the third and fourth columns give the radius and age of the shells at the stall time. The fifth column gives the bubble age for the observed radius of 57 pc, while the sixth column gives the expected main-sequence lifetime of the given star from Maeder & Meynet (1988). All radii are given in pc, and all times are in Myr.

We also studied the stellar contents of a third nebula of the same size, Bubble #3, which is a typical complex object linked to the association OB21a cataloged by Humphreys and Sandage (1980). We classify the stellar spectral types to be O6–7, B2 I, and the WC star MC6, identified by Conti and Massey (1983). Two other massive stars of spectral type O8.5 and B1 I are located nearby but on the exterior of the shell.

TABLE II   Theoretical Bubble Parameters

Type	R_s	R_{stall}	t_{stall}	t_{57}	t_{ms}
O8 V	117	71	4.2	2.9	6.0
O9 V	77	51	3.0	–	8.0
B0 V	57	38	2.2	–	10.5
O8 I	189	121	3.7	1.1	3.7
O9 I	161	120	4.4	1.3	4.4
B0 I	139	121	5.0	1.4	5.0

ACKNOWLEDGEMENTS

We acknowledge T. Armandroff and C. Neese for their roles in obtaining the Schmidt Hα frames.

REFERENCES

Cappa de Nicolau, C., Niemela, V.S., Benaglia, P. 1992, this volume

Chu, Y.-H. 1991, in *Wolf-Rayet Stars and Interrelations with Other Massive Stars in Galaxies*, IAU Symp. 143, K.A. van der Hucht and B. Hidayat, eds., p.349.

Courtès, G., Petit, H., Sivan, J.-P., Dodonov, S., Petit, M. 1987, A&A 174, 28

Drissen, L., Shara, M., Moffat, A. 1991, AJ, 101, 1659

Georgelin, Y.M., Georgelin, Y.P., Laval, A., Monnet, G., Rosado, M. 1983, A&AS, 54, 459

Humphreys, R. & Sandage, A. 1980, ApJS, 44, 319

Lozinskaya, T.A. 1982, Ap&SS, 87, 313

MacLow, M-M. & McCray, R. 1988, ApJ, 324, 776

Maeder, A. & Meynet, G. 1988, A&AS, 76, 411

Massey, P. & Conti, P.S. 1983, ApJ, 273, 576

McCray, R. 1983, in *Highlights of Astron.* 6, R.M. West, ed., p.565

McKee, C., Van Buren, D., Lazareff, B. 1984, ApJL, 278, L115

Meaburn, J. 1980, MNRAS, 192, 365

Rosado, M. 1986, A&A 160, 211

Saken, J.M., Shull, J.M., Garmany, C.D., Nichols-Bohlin, J., Fesen, R.A. 1992, ApJ, in press

Weaver, R., McCray, R., Castor, J., Shapiro, P., Moore, R. 1977, ApJ, 218, 377

Massive Stars: Their Lives in the Interstellar Medium
ASP Conference Series, Vol. 35, 1993
Joseph P. Cassinelli and Edward B. Churchwell (eds.)

THE ERIDANUS SOFT X-RAY ENHANCEMENT: A NEARBY HOT BUBBLE

DAVID N. BURROWS, JOHN A. NOUSEK, GORDON P. GARMIRE
Department of Astronomy and Astrophysics, Penn State University
525 Davey Lab, University Park, PA 16802

K. P. SINGH
X-ray Astronomy Group, Tata Institute of Fundamental Research
Homi Bhabha Road, Bombay 400 005, India

JOHN GOOD
Infrared Processing and Analysis Center, California Institute of Technology, M/S 100-22, Pasadena, CA 91125

ABSTRACT   We present soft X-ray maps of the Eridanus soft X-ray enhancement. Comparison with 100μm and N_H maps shows that the X-ray enhancement consists of two distinct components: a hook-shaped component (EXE1) and a circular component (EXE2) at different temperatures. EXE1 appears to be a nearby stellar wind bubble, possibly reheated by supernovae blast waves.

X-RAY MAPS

The Eridanus enhancement $((\ell, b) \approx (200°, -42°))$ is a well known feature of the soft X-ray background with diameter $\sim 20°$ in the 1/4 keV band. Naranan *et al.* (1976) interpreted it as the hot interior of an old supernova remnant. It is surrounded by a partially ionized shell (Heiles 1976, Reynolds & Ogden 1979).

We present X-ray maps (from the HEAO-1 A2 experiment) of the Eridanus enhancement with the highest spatial resolution and sensitivity to date (Figure 1). EXE1 is the large hook-shaped feature in both the 1/4 keV data

(L1 band) and the 0.6 keV data (M1 band) that dominates these maps. The smaller enhancement (5° diameter) at $(\ell, b) = (210°, -43°)$ is seen only in the 1/4 keV data; we refer to it as EXE2. Comparison of the X-ray maps with

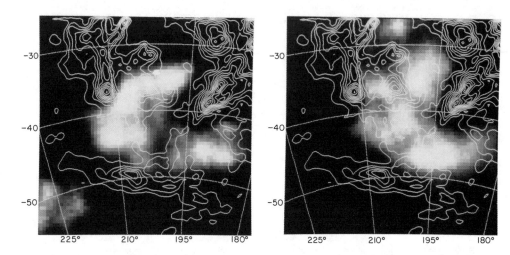

Figure 1: MEM deconvolution of the raw X-ray maps with IRAS 100μm contours overlaid. (a) The L1 map (15 – 30 cps). (b) The M1 map (3.3 – 6.7 cps).

N_H maps suggests that the two X-ray features are associated with different velocity components of the neutral gas. The smaller object, EXE2, is outlined by receding neutral gas spanning ~ 40 km/s in velocity and has an X-ray temperature of $\sim 1.5 \times 10^6$ K. Its distance is not known. The larger object, EXE1, is associated with approaching neutral gas spanning up to ~ 55 km/s in velocity, and is anticorrelated in a fairly detailed way with a convoluted cavity apparent in 60μm and 100μm IR maps. It has an X-ray temperature of $\sim 2.2 \times 10^6$ K. Absorption by L1569 $(185°, -35°)$ places this cavity beyond 130 pc. The far edge is presumed to be located near the Ori OB1 association at ~ 450 pc.

THE STELLAR WIND BUBBLE INTERPRETATION

From the X-ray observations, we can derive the model independent parameters of these enhancements given in Table 1. If the X-ray enhancements are interpreted as stellar wind bubbles, we obtain the results given in Table 2. The parameters

Table 1: Model independent parameters of X-ray enhancements

	EXE1		EXE2	
$\mathcal{F}_x$ (ergs s^{-1} cm^{-2})	2.9×10^{-8}		1.4×10^{-9}	
T_x (10^6 K)	2.13		1.57	
d (pc)	130	400	130	300
$\mathcal{L}_x$ (ergs s^{-1})	6.7×10^{34}	6.3×10^{35}	2.8×10^{33}	1.5×10^{34}
R_s (pc)	22.6	69.5	5.7	13.1
n_x (cm^{-3})	0.011	0.0064	0.016	0.010
P_{th} (cm^{-3} K)	23400	13600	24800	16300

derived for EXE2 are within the range expected for a typical stellar wind bubble. However, since no star with sufficient wind luminosity is located within this object, its nature remains undetermined.

EXE1 requires a somewhat more energetic wind, which is comparable to the total stellar wind luminosity of the Ori OB1 association. This is consistent with the interpretation that the X-ray bubble fills a cavity created by this association (Reynolds & Ogden 1979).

Table 2: Stellar wind bubble parameters

	EXE1		EXE2	
d (pc)	130	400	130	300
n_0 (cm^{-3})	18.9	33.2	20.2	30.7
L_w (10^{36} ergs/s)	3.1	9.6	0.28	0.65
t_6 (10^6 yrs)	2.3	12.5	0.56	1.98

ACKNOWLEDGEMENTS

This work was supported by NASA grants NAG5-941 and NAG8-478 and by JPL contract 657268.

REFERENCES

Heiles, C. 1976, ApJ, 208, L137

Naranan, S., Shulman, S., Friedman, H., & Fritz, G. 1976, ApJ, 208, 718

Reynolds, R. J., & Ogden, P. M. 1979, ApJ, 229, 942

Massive Stars: Their Lives in the Interstellar Medium
ASP Conference Series, Vol. 35, 1993
Joseph P. Cassinelli and Edward B. Churchwell (eds.)

WIND-BLOWN BUBBLES IN EJECTA MEDIUM

GUILLERMO GARCIA-SEGURA
University of Illinois at Urbana-Champaign, U.S.A.
Instituto de Astrofisica de Canarias,Tenerife,Spain

MORDECAI-MARK MAC LOW
NASA Ames Research Center, Moffett Field, U.S.A.
University of Illinois at Urbana-Champaign, U.S.A.
University of California at Berkeley, U.S.A.

ABSTRACT   We analytically predict the observed X-ray emission from
Wolf-Rayet bubbles using a two-wind model. In this model, the fast wind
from the Wolf-Rayet star expands into a previously ejected, slow, red
supergiant wind. An ellipsoidal, self-similar solution gives the shape of
the bubble. An energy conserving bubble with isotropic internal pressure
expands in each direction at a constant velocity in an inverse-square
density distribution. Our approximation of a self-similar shape for the
shell is justified because the red wind has such an inverse-square density
distribution in each direction. This solution accounts for departures
from spherical symmetry caused by a red supergiant wind with a higher
mass loss rate at the equator than at the pole. Our solution also includes
conductive evaporation of mass into the hot interior from the radiatively
cooled shell of swept-up red wind. We compare our model to Einstein
observations of the Wolf-Rayet bubble NGC 6888. Bubble models with a
homogeneous ambient medium cannot explain the observed flux, but our
two-wind model yields excellent agreement with the observed flux.

CALCULATIONS

We take the shape of the swept up shell of slow wind to be a prolate spheroid
with the minor axis in the plane of the stellar equator. The prolate spheroid is
formed by the rotation of an ellipse about its major axis. The major axis has a
length $2a$, and the minor axis has a length $2b$. The eccentricity of the bubble is
then

$$\varepsilon = \sqrt{1 - \frac{b^2}{a^2}} \ . \tag{1}$$

The red wind has an angular variation that we model with the density function

$$\varrho(b) = [(\varrho_{min} - \varrho_{max})\sin\phi + \varrho_{max}]\left(\frac{b}{r_i}\right)^{-2} \ , \tag{2}$$

where ϱ_{min} and ϱ_{max} are the polar and equatorial densities at the base of the slow wind, at a radius of r_i. (We take $r_i = 2$ A.U.) Because the wind has a inverse-square power law dependency in each direction, the shell expands at constant velocity in each direction, and so retains its shape. We calculate the dynamics of a constant eccentricity bubble in this density distribution analytically.

In a future paper we will describe this calculation in detail. Here we present some results of the model. The age of the bubble is given by

$$t = (9.8 \times 10^5 \text{yr}) \ \frac{b_{pc}}{v_5} , \tag{3}$$

where b_{pc} is the semi-minor axis in parsecs, and v_5 is the expansion velocity along that axis in km s^{-1}. Note that inclination effects must be included in deriving this velocity from the observed expansion velocity. Inverting the dynamical equations of our model, we find the mechanical luminosity of the wind in terms of observable quantities

$$L_w = (2.5 \times 10^{35} \text{ergs}^{-1}) \ f(\varepsilon)^{-1} \ g(\varepsilon)^{-2} \ F(\varrho)_{-13} \ b_{pc}^3 \ t_4^{-3} , \tag{4}$$

where $t_4 = t/10^4$ yr, the function $g(\varepsilon) = b/a$ and

$$f(\varepsilon) = 1 + \frac{\sin^{-1} \varepsilon}{\varepsilon \ g(\varepsilon)} . \tag{5}$$

The function

$$F(\varrho) = (4\pi - \pi^2) \varrho_{max} + \pi^2 \varrho_{min} . \tag{6}$$

We define $F(\varrho)_{-13} = F(\varrho)/10^{-13}$; it can be expressed in terms of the shell mass M_s by

$$F(\varrho)_{-13} = (7.196) \ g(\varepsilon) \ b_{pc}^{-1} \ M_s . \tag{7}$$

In the Einstein and ROSAT bands, the X-ray emissivity of hot gas remains nearly independent of temperature down to 5×10^5 K where there is a sharp cutoff (Chu and Mac Low 1990). For the emissivity of the plasma (Raymond, Cox, & Smith 1976, Chu and Mac Low 1990) we have used a constant value $\Lambda_x = 9.0 \times 10^{-24}$ ergs cm^3 s^{-1}. With the above asumptions, the X-ray luminosity of the hot gas is

$$L_x = (1.437 \times 10^{33} \text{ergs}^{-1}) \ g(\varepsilon)^{-25/21} \ f(\varepsilon)^{-17/21} \ F(\varrho)_{-13}^{17/21} \ L_{36}^{13/21} \ t_4^{-3/7} \ I(\tau) , \tag{8}$$

where the dimensionless integral is

$$I(\tau) = \frac{125}{33} - \frac{5}{11} \tau^{11/2} + \frac{5}{3} \tau^3 - 5 \tau^{1/2} \tag{9}$$

and the dimensionless temperature

$$\tau = (0.16) \ g(\varepsilon)^{-2/21} \ f(\varepsilon)^{2/21} \ F(\varrho)_{-13}^{-2/21} \ L_{36}^{-4/21} \ t_4^{2/7} . \tag{10}$$

The central temperature of the hot bubble is $T_c = (5 \times 10^5 \text{K})/\tau$.

We can recover the densities at the base of the slow wind using (6) and $\varrho_{min} = g(\varepsilon)\varrho_{max}$. The mass loss rate of the red supergiant phase can be calculated by

$$\dot{M} = (1.43 \times 10^{-7} M_\odot \text{yr}^{-1}) \ v_{\text{red}} \ F(\varrho)_{-13} \ , \tag{11}$$

v_{red} is the velocity of the red supergiant wind; typically $v_{red} = 10\text{--}15 \ \text{km s}^{-1}$.

RESULTS AND CONCLUSIONS

We predict the soft X-ray emission from the wind-blown bubble NGC 6888 in our two-wind model. This nebula has a size of 12×18 arc min, and a distance of 1.3 Kpc. This mean that we have for the semi-major and semi-minor axes of the ellipsoid $a = 3$ pc and $b = 2$ pc, and for the eccentricity $\varepsilon = 0.74$. The H_α expansion velocity observed is 80 km s^{-1} . High resolution radio observations give a shell mass of 5.3 $M_\odot$. With the above data we have a luminosity in the Einstein band $L_x = 1.22 \times 10^{34}$ erg s^{-1}. To compare with the observed flux of the Einstein satellite, we have calculated the flux using the tables of Bochkarev(1985) for the X-ray absorption, knowing that the extinction $A_v = 2.01$. The total calculated flux in the Einstein band is finally $F_x = 7.34 \times 10^{-13}$ erg s^{-1} cm^{-2}. This result is in good agreement with the observed flux (Bochkarev 1988) $F_x(0.2 - 3.0 \text{ KeV}) = 10^{-12}$ erg s^{-1} cm^{-2}. The calculated mass loss rate for the red supergiant wind from the central star (HD 129163) is $\dot{M} = 2.72 \times 10^{-5}$ $M_\odot$ yr^{-1}, for a red wind velocity of $v_{red} = 15$ km s^{-1} Within the observed range for red supergiant stars (Humphreys 1990).

By taking into account the density distribution of a red supergiant wind, we can explain the shape and X-ray emission of a Wolf-Rayet bubble. The two-wind model supports a natural evolutionary track for Wolf-Rayet stars coming after a red supergiant phase. We note that departures from spherical symmetry in the shape of the bubble cannot be caused by a variation from pole to equator of the strength of the fast wind, because an energy conserving bubble is driven by the isotropic internal pressure. Asymmetries can only exist if a density gradient is present in the external medium. When the angular gradient in the red supergiant wind is not present, the spherical solution is recovered, independently of any gradient in the fast wind. Bubble models using a homogeneus ambient medium have difficulty explaining Wolf-Rayet bubbles containing stellar ejecta. They disagree with observations of X-rays and the shape of the bubbles.

It is a pleasure to thank You-Hua Chu for her helpful discussions, encouragement, and support during this work. Support for this work came from the NASA Astrophysical Data Program.

REFERENCES

Bochkarev, N.G. 1988, *Nature*, 332, 518
Bochkarev, N.G. 1985, *Soviet Astr.*, 29, 509-515
Chu, Y.-H., and Mac Low, M.-M. 1990, *Ap. J.*, , 365, 510
Humphreys, R. M. 1990, IAU symp. 143, Wolf-Rayet Stars, 485
Raymond, J. C., Cox, D. P., and Smith, B. W. 1976, *Ap. J.*, , 204, 209
Weaver, R., McCray, R., Castor, J., Shapiro, P. and Moore, R. 1977, *Ap. J.*, ,
218, 377

Massive Stars: Their Lives in the Interstellar Medium
ASP Conference Series, Vol. 35, 1993
Joseph P. Cassinelli and Edward B. Churchwell (eds.)

SI IV AND C IV ABSORPTION TOWARD ζ OPHIUCHI AT HIGH RESOLUTION:
EVIDENCE FOR PHOTOIONIZED AND COLLISIONALLY IONIZED GAS

Kenneth R. Sembach and Blair D. Savage
Washburn Observatory, University of Wisconsin-Madison,
475 North Charter Street., Madison, WI 53706

Edward B. Jenkins
Princeton University Observatory, Princeton, NJ 08544

OBSERVATIONS

We obtained high-resolution GHRS echelle mode observations of the Si IV, C IV, and N V lines in the spectrum of ζ Ophiuchi as a part of the HST science verification program during May 1991. The observations were obtained with the light of ζ Oph positioned in the small (0.25"x0.25") entrance aperture. Comb-addition and FP-split procedures were used to reduce fixed pattern noise introduced by the digicon detectors. The combined data have S/N ≈ 30 to 60, a resolution (FWHM) of ~3.5 km s^{-1}, and an uncertainty of ~3.5 km s^{-1} (1σ) in heliocentric velocity.

OBSERVATIONAL RESULTS

We detect Si IV λλ1393, 1402 and C IV λλ1548, 1550 absorption toward ζ Oph. The Si IV profiles contain a strong narrow component (b = $[2kT/m]^{1/2}$ = 5.3 km s^{-1}, log N = 12.74±0.02) centered at -15 km s^{-1} and a broad shallow component (b ~ 16 km s^{-1}, log N = 11.82±0.16) centered at ~ -26 km s^{-1}. In contrast, the C IV profiles reveal only a broad shallow component (b ~ 19 km s^{-1}, log N = 12.73±0.14) centered at ~ -26 km s^{-1}. Using the N V λ1238 line, we set an upper limit of log N(N V) < 12.15 (Wλ(1238) < 3 mÅ). Previous Copernicus and IUE observations of these lines were not of high enough quality to detect the broad weak absorption.

H II REGION GAS

Measurements of Hα and [N II] arising in the H II region surrounding ζ Oph imply n$_e$ ≈ 4 atoms cm^{-3} and T ≈ 7000 K (Reynolds & Ogden: 1982, AJ, 87, 306). The single component emission lines are centered on a velocity of -13±1 km s^{-1}.

The narrow Si IV absorption is well-described by a Doppler parameter corresponding to a thermal width at T ~ 45,000 K in the absence of turbulence. If one assumes $T = 7000$ K for the H II region, then $b_{turb} \approx 4.8$ km s^{-1}, smaller than the value of $b_{turb} \approx 6.6$ km s^{-1} found by Reynolds & Ogden upon examination of the Hα and [N II] emission lines. The emission line values of b_{turb} may be larger due to the finite beam size of the instrument used to obtain them.

Si IV peaks in abundance at $T = 8 \times 10^4$ K in a plasma in collisional ionization equilibrium (Shapiro & Moore: 1976, ApJ, 207, 406). The $b = 5.3$ km s^{-1} observed for the narrow Si IV component is smaller than the 6.9 km s^{-1} width predicted for gas at $T = 8 \times 10^4$ K. At $T = 8 \times 10^4$ K, N(C IV)/N(Si IV) ~ 6 in collisionally ionized gas with solar abundances. No noticeable enhancement of C IV absorption is present at -15 km s^{-1}. Combined with the small width of the Si IV component, the possibility that the narrow Si IV component is being produced in gas at $T \geq 8 \times 10^4$ K is unlikely. The width and central velocity of this component indicate that it is produced by photoionization within the warm gas of the ζ Oph H II region. We estimate an upper column density limit of 1.2×10^{12} cm^{-2} for C IV in the H II region, assuming the C IV has a profile described by $b = 5.3$ km s^{-1} centered on -15 km s^{-1}. The observed Si IV abundance is a factor of ~30 lower than predicted using the photoionization models of Cowie, Taylor, & York (1982, ApJ, 248, 548), while the C IV upper limit is a factor of ~10 lower than predicted.

CORONAL GAS

The large width and smoothness of the broad Si IV and C IV component indicate that it is probably being formed in gas at high temperature ($T \sim 10^5$ K). At temperatures where these ions peak in abundance in collisional equilibrium, Si IV and C IV have thermal Doppler spread parameters $b = 6.9$ and 11.8 km s^{-1}, ~2x smaller than observed. This leads us to believe: 1) The temperature of the gas in the single component is a factor of 4 larger than the peak abundance temperatures; or 2) Multiple components contribute to the absorption; or 3) Large scale gas motions dominate the line widths.

Using the non-equilibrium cooling gas results of Shapiro & Moore, we expect N(C IV)/N(N V) = 0.3 and N(C IV)/N(O VI) = 0.08 for collisionally ionized gas of solar abundance with $T = 2 \times 10^5$ K. The observed C IV abundance would imply $\log N(N\ V) = 13.26$ and $\log N(O\ VI) = 13.83$; these columns yield $W_\lambda(N\ V\ \lambda 1238.821) = 38$ mÅ and $W_\lambda(O\ VI\ \lambda 1037) = 43$ mÅ, in disagreement with the observed limits of 3 mÅ and 12 mÅ (Table I; Morton: 1975, ApJ, 197, 85). Thus, we conclude $T < 2 \times 10^5$ K.

Weaver et al. (1977, ApJ, 218, 377) predict $\log N(Si\ IV) = 11.18$ and $\log N(C\ IV) = 12.38$ for a sight line through a standard interstellar bubble. The observed C IV column density for the broad component requires that ~3 of these bubbles be encountered along the ζ Oph sight line. The inability of the interface models to reproduce the observed value of N(C IV)/N(Si IV) may be due to the neglect of ionizing photons originating in the cooling gas itself. If the absorption originating in the interfaces is overlapping, the factor of 2 difference in the observed and predicted values of b can easily be reconciled.

Perhaps the individual components occur at the boundary of the nearby diffuse cloud seen in absorption near -27 km s^{-1}.

Nonthermal motions such as flows at a cloud-intercloud boundary may govern the shape of the broad absorption Shells of expanding gas surrounding OB associations have expansion velocities of 100 km s^{-1} in some cases (Cowie et al.: 1981, ApJ, 250, L25) and could account for the large observed width of the broad feature. The [O III] emission bubble seen by Gull & Sofia: (1979, ApJ, 230, 782; see also Van Buren, this volume) could prove to be a site for the absorbing gas.

In conclusion, we detect two distinctly different types of gas containing highly ionized carbon and silicon toward ζ Oph: a photoionized warm (T ~ 10^4 K) phase associated with the H II region near the star and a collisionally ionized hot (T ~ 10^5 K) phase that may be related to an interface near ζ Oph or the diffuse cloud complex seen in the low ionization lines at -27 km s^{-1}.

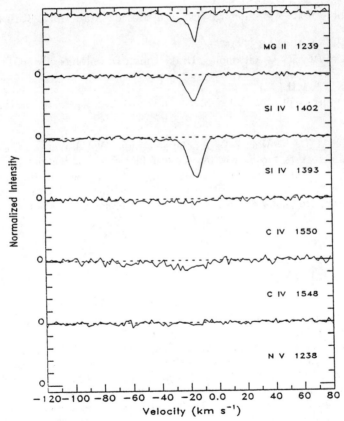

Normalized intensity versus heliocentric velocity for selected lines toward ζ Oph. The Mg II line traces the diffuse cloud complexes at v ~ -27 km s^{-1} and v ~ -15 km s^{-1}. The Si IV component at v = -15 km s^{-1} arises within the ζ Oph H II region. Both Si IV and C IV exhibit broad shallow absorption that likely arises in T ~ 10^5 K collisionally ionized gas along the sight line.

Massive Stars: Their Lives in the Interstellar Medium
ASP Conference Series, Vol. 35, 1993
Joseph P. Cassinelli and Edward B. Churchwell (eds.)

NEW SURVEYS OF NEBULAE AROUND WOLF-RAYET STARS

YOU-HUA CHU, GUILLERMO GARCIA-SEGURA
Astronomy Dept, University of Illinois, Urbana, IL 61801

MICHAEL A. DOPITA, J. F. BELL
MSSSO, The Australian National University, Private Bag, Weston Creek
P. O., ACT 2611, Australia

TATIANA A. LOZINSKAYA
Sternberg State Astronomical Institute, Moscow State Univ., Russia

ANTHONY P. MARSTON
Dept. of Physics & Astronomy, Drake Univ., Des Moines, IA 50311

GRANT J. MILLER
Astronomy Dept., San Diego State University, San Diego, CA 92182

ABSTRACT   New surveys of nebulae around Wolf-Rayet stars, using
digital detectors and narrow interference filters, are carried out for the
Galaxy and the Magellanic Clouds. Many new ring nebulae are found
in these surveys. Most importantly, small WR ring nebulae (a few pc in
diameter) are finally discovered in the Large Magellanic Cloud.

1. INTRODUCTION

Wolf-Rayet (WR) stars are often surrounded by ring-shaped nebulae, which
suggest dynamical interactions between the stars and the ambient medium.
Previous systematic surveys for nebulae around WR stars were made more
than 10 years ago, which all used existing photographic material that had low
sensitivity and/or low spatial resolution (the Galaxy: Chu 1981, Heckathorn
et al. 1982; the LMC: Chu & Lasker 1980). The apparent lack of small WR
ring nebulae in the LMC clearly indicated the incompleteness in the LMC
survey (Chu & Lasker 1980; Chu 1991); the galactic survey might also be quite
incomplete.
    During the IAU Symposium of "WR Stars and Interrelations with Other
Massive Stars in Galaxies" held in Bali, 1990, M. A. Dopita suggested to
initiate a new nebular survey for the WR stars in the Magellanic Cloud. In
the 1990-1991 season, all WR stars in the SMC and 50% of the WR stars
in the LMC were surveyed; many new small ring nebulae were discovered.
Encouraged by these results, we surveyed the galactic WR stars in the northern

sky in June 1991, and extended the galactic survey to the southern sky in 1992. Here we report the progress and preliminary results from these surveys.

2. WOLF-RAYET STARS IN THE GALAXY: I. NORTHERN SKY

The 1-m telescope at Mount Laguna Observatory was used for this survey, with a focal reducer and a TI 800x800 CCD. The field of view was about 13', and the 15 μm pixel size corresponded to about 1''. The FWHM of the seeing was typically $\leq 2''$ during the observing run in June 1991. 62 galactic WR stars were observed, representing the complete list of van der Hucht et al. (1981) down to a declination of -26°. All 62 stars were observed at least once in the Hα+[N II] filter (FWHM = 62 Å); 28 were also observed in the [O III] filter (FWHM = 49 Å).

We found probable new ring nebulae around WR stars number 113, 116, and 132, and possible new rings around WR stars number 133 and 153. Mosaic images of NGC 6888 (associated with WR136) and Anon (MR100 = WR134) were made in both Hα+[N II] and [O III]. The shell of NGC 6888 appears complete in the [O III] line, contrary to the "leaky bubble" hypothesis suggested to explain its low X-ray luminosity (Bochkarev 1988).

This survey forms part of the master's thesis of G. J. Miller, and the results have been reported in a paper submitted to Ap. J. Suppl. (Miller & Chu 1992).

3. WOLF-RAYET STARS IN THE GALAXY: II. SOUTHERN SKY

The Curtis Schmidt Telescope at Cerro Tololo Inter-American Observatory was used for this survey. A Thomson 1024x1024 CCD was used. The field of view was about 31', and the 19μm pixel size corresponded to about 1''.8. During the service observing run in April 1992, 37 stars were imaged with an Hα filter (FWHM = 17 Å), and 32 of them were also imaged with an [O III] filter (FWHM = 22 Å). We intend to continue this survey for the remaining 60 stars in 1993.

We found possible new ring nebulae around WR stars number 65 and 68. The improved spatial resolution and sensitivity helped to clarify the physical structure of RCW104 (around WR75) and the morphology of RCW78 (around WR55) and RCW118 (around WR85). RCW104 consists of a sharp filamentary region in the west, and an amorphous region in the east; the relation between the amorphous region and the WR star is not clear (Chu 1982). Our new [O III] image shows clearly a small ring nebula centered on WR75 and the sharp filaments form the western rim of the ring; the amorphous region is not detectable in the [O III] image. The Hα image of RCW78 shows a filamentary shell structure, as opposed to the "amorphous HII region" described previously (Chu & Treffers 1982). RCW118, on the other hand, appears quite amorphous in the new Hα image; no shell structure is obvious, although it was described as a shell HII region (Chu, Treffers, & Kwitter 1983).

4. WOLF-RAYET STARS IN THE MAGELLANIC CLOUDS

The 2.3 m Advanced Technology Telescope of Australian National University
was used in the 1990-1991 and 1991-1992 seasons to image the SMC WR stars
(Azzopardi & Breysacher 1979) and the LMC WR stars (Breysacher 1981). All
stars were imaged in both Hα and [O III] filters (FWHM = 16 Å). In the 1990-
1991 season, the Double-Beam Spectrograph was used in the imaging mode
with its Photon Counting Array, which gave a field of $5'.1$ at a scale of $0''.65$
pixel^{-1}. For the 1991-1992 season, the new focal reducing imager was used
with a Tektronix 1024x1024 CCD, which gave a field of $6'.7$ at a scale of $0'.6$
pixel^{-1}. Since the S/N ratio of the images taken by the CCD imager was much
better, all stars with associated nebulosity observed in the 1990-1991 season
were re-observed in the 1991-1992 season.

The most remarkable result of the LMC survey is that many small WR
ring nebulae, a few pc across, are discovered for the first time. The high spatial
resolution of this survey helps to reveal the true ring nebulae around Br2
and Br12, which are embedded in the previsouly identified large rings of N79
and DEM45, respectively (Rosado 1986; Chu & Lasker 1980). Several small
WR rings may be good candidates for ejecta-type nebulae; two of them are
morphologically similar to NGC 6888. About 16% of the LMC WR stars
are found in ring nebulae. It may be statistically significant that 18% of the
WN3,4 stars are in ring nebulae while only 10% of the later type WN stars are
in ring nebulae. Preliminary results of this survey have been presented in the
European Workshop "New Aspects of Magellanic Cloud Research" (Dopita *et
al.* 1992).

5. REFERENCES

Azzopardi, M. & Breysacher, J. 1979, *Astr. Ap.*, **75**, 120.
Bochkarev, N. G. 1988, *Nature*, **332**, 518.
Breysacher, J. 1981, *Astr. Ap. Suppl.*, **43**, 203.
Chu, Y.-H. 1981, *Ap. J.*, **249**, 195.
Chu, Y.-H. 1982, *Ap. J.*, **254**, 578.
Chu, Y.-H. 1991, in IAU Symposium No. 143, *Wolf-Rayet Stars and Their
    Interrelations with Other Massive Stars in Galaxies*, ed. K. A. van
    der Hucht and B. Hidayat (Dordrecht: Kluwar), p.349.
Chu, Y.-H. & Lasker, B. M. 1980, *Pub. A.S.P.*, **92**, 730.
Chu, Y.-H. & Treffers, R. R. 1982, *Ap. J.*, **249**, 586.
Chu, Y.-H., Treffers, R. R., & Kwitter, K. B. 1983, *Ap. J. Suppl.*, **53**, 937.
Dopita, M. A., Bell, J. F., Lozinskaya, T. A., & Chu, Y.-H. 1992, in
    proceedings of "*New Aspects of Magellanic Cloud Research*",
    held in Heidelberg, Germany, June 15-17, 1992.
Heckathorn, J. N., Bruhweiler, F. C., & Gull, T. R. 1982, *Ap. J.*, **252**, 230.
Miller, G. J. & Chu, Y.-H. 1992, submitted to *Ap. J. Suppl.*
Rosado, M. 1986, *Astr. Ap.*, **160**, 211.
van der Hucht, K. A., Conti, P. S., Lundstrom, I., & Stenholm, B. 1981,
    Space Sci. Rev., **28**, 227.

Massive Stars: Their Lives in the Interstellar Medium
ASP Conference Series, Vol. 35, 1993
Joseph P. Cassinelli and Edward B. Churchwell (eds.)

HIGHLY IONIZED GAS IN THE LARGE MAGELLANIC CLOUD

YOU-HUA CHU
BART WAKKER
GUILLERMO GARCIA-SEGURA
Astronomy Dept, University of Illinois, Urbana, IL 61801

ABSTRACT   We intend to use interstellar absorption by highly ionized gas as a tool to search for signatures of SNR shocks in superbubbles. The high dispersion, short wavelength spectra of LMC objects in the IUE archives are examined. It is found that most of the detected highly ionized gas appears associated with hot early-type stars. Large velocity shifts between high ionization species and low ionization species are seen in several superbubbles. The origin of this velocity offset is not known, but could be related to SNR shocks. We also find some evidence for hot halo gas around the LMC.

1. INTRODUCTION

X-ray emission has been detected in several superbubbles around OB associations in the Large Magellanic Cloud (LMC). Pressure-driven superbubble models predict X-ray luminosities that are much lower than observed. Therefore, it has been suggested that off-center supernova remnants (SNRs) hitting the superbubble shells produce the excess X-ray emission (Chu & Mac Low 1990; Wang & Helfand (1991). Hα emission-line observations of the X-ray bright superbubbles usually are not sensitive enough to detect the high-velocity, shocked material (Chu & Kennicutt 1993); therefore, we resort to the interstellar absorption lines of highly ionized species, such as Si IV and C IV, to search for evidence of SNR shocks.

Unfortunately, photoionization by hot stars, stellar winds inside superbubbles, and the possible existence of a hot gaseous halo around the LMC could all influence the interstellar Si IV and C IV lines and confuse the shock signatures we are searching for. We therefore include sightlines through different interstellar environments in order to empirically sort out the confusion. This poster reports the preliminary results from our examination of available IUE archival data of LMC stars in a variety of environments.

2. DATABASE AND MEASUREMENTS

In the IUE archives, high-dispersion spectra are available for about 50 objects in the LMC. However, data of only 28 of them have a sufficiently high S/N ratio for further analysis. These objects are found in the following environments:

12 – in superbubbles (5 are in X-ray bright superbubbles)
8 – in classical HII regions
5 – in diffuse nebulosity
2 – without any nebulosity
1 – other (questionable association with HII regions)

We expect shocked material to exhibit higher velocity and ionization than its ambient medium, hence we chose as our primary lines C IV $\lambda\lambda$1548, 1550 and Si IV $\lambda\lambda$1393, 1402. Our reference lines are the least confused and least saturated low-ionization lines of S II $\lambda\lambda$1250, 1253, Si II $\lambda\lambda$1526, 1808, and C II* $\lambda\lambda$1335. We measured the centroid velocities of all these interstellar lines. To confirm that features were interstellar, we compared each spectrum to the stellar spectrum of a galactic standard star with similar spectral type; these reference spectra were taken from Walborn $et\,al.$ (1985) and Walborn & Nichols-Bohlin (1987). For the C IV and Si IV lines we also measured their equivalent widths to derive column densities.

3. HIGHLY IONIZED GAS

The amount of information we can derive from the observations of highly ionized gas in the LMC is limited by the small number of objects available. Nevertheless the first results look encouraging and interesting.

Interstellar C IV and Si IV absorption was detected in all stars inside superbubbles, except Sk−69 209a (in 30 Dor C) for which only a very noisy spectrum was available. For stars in classical H II regions (without superbubble structure) interstellar C IV and Si IV absorption is sometimes detected. For example, the O3 III star Sk−67 211 in N 59A (DEM 241) shows clear interstellar C IV and Si IV, while the O9.7 Ib star Sk−67 05 in N 3 (DEM 7) shows hardly any trace of interstellar C IV or Si IV absorption despite the good S/N ratio of its spectrum. These results suggest that the highly ionized absorbing gas might be strongly concentrated near the hot, early-type stars, as opposed to being distributed uniformly over the entire LMC. The column density ratio of C IV to Si IV ranges from 1 to 3, indicating that both photoionization and collisional ionization are needed (Savage 1984).

Four of the objects in diffuse nebulosity are within the boundary of the supergiant shell LMC 3 (Meaburn 1980). All of them show interstellar absorption at $V_\odot \sim 230 \, \mathrm{km \, s^{-1}}$ in the low ionization lines; one of them (NGC 1984) also shows clear C IV absorption and another (NGC 1994, or Sk−69 147a) shows Si IV and possibly C IV absorption, all at similar radial velocities. It is tempting to conclude that the supergiant shell LMC 3 contains hot, highly ionized gas; future work is needed to confirm this.

The two stars without any nearby nebulosity are good candidate probes to study the hot gaseous halo of the LMC. Sk−69 221 (B 2Ia) shows Si IV absorption at $\sim 230 \, \mathrm{km \, s^{-1}}$. However, this may not be due to a gaseous halo, since this star is close to the 30 Dor complex and the Si IV velocity is very similar to those detected in the supergiant shell LMC 3. The best evidence for highly ionized gas in a LMC halo comes from SL 360, which is a cluster superposed on the H II region N 199. A physical association between SL 360 and the H II region is doubtful because of the absence of interstellar S II absorption. An interstellar C IV component at $175\,\mathrm{km\,s^{-1}}$ is clearly detected in the spectrum of SL 360, however.

4. RADIAL VELOCITIES OF THE INTERSTELLAR LINES

To interpret the radial velocities of the low ionization lines, we correlated them with the velocities of the H I and Hα emission lines. The H I velocities are taken from Rohlfs *et al.* (1984); Hα velocities are taken from Chu & Kennicutt's (1993) survey of 30 H II regions in the LMC. For expanding superbubbles, the velocity of the approaching side is used in this correlation. We find that the velocities of the low ionization lines correlate very well with the Hα velocities. The H I velocities generally are red-shifted from the Hα and low ionization line velocities by $10\text{--}20\,\mathrm{km\,s^{-1}}$. This is probably due to the fact that H I samples the full depth of the LMC along the line of sight, while the interstellar absorption lines sample only the front part of the LMC.

While in general the average radial velocity of C IV and Si IV agrees with that of the low ionization lines to within $\pm10\,\mathrm{km\,s^{-1}}$, for five of the stars in superbubbles we find high ionization lines blue shifted from the low ionization lines by $25\text{-}45\,\mathrm{km\,s^{-1}}$. Three of these are in X-ray bright regions, *i.e.*, 30 Dor and N 51D. Similar velocity offsets have been seen in the I Per OB association in the Galaxy (Phillips & Gondhalekar 1981). Is the velocity offset between the high ionization species and low ionization species an indication of (SNR) shocks? To answer this question we are currently investigating the X-ray properties of the superbubbles that show large velocity offsets, and we plan to study interstellar UV lines for more stars in X-ray bright superbubbles.

This research is supported by NASA grant NAG5-1755. We thank Dr. N. Walborn for providing information about spectral classification of the stars studied. We also thank Drs. B. Savage, K. de Boer, and D. Bomans for enlightening discussions.

5. REFERENCES

Chu, Y.-H. & Kennicutt, R. C. Jr., 1993, in preparation.
Chu, Y.-H. & Mac Low, M.-M. 1990, *Ap. J.*, **365**, 510.
Meaburn, J. 1980, *M.N.R.A.S.*, **192**, 365.
Phillips, A. P. & Gondhalekar, P. M. 1981, *M.N.R.A.S.*, **196**, 533.
Rohlfs, K., Kreitschmann, J., Siegman, B. C., & Feitzinger, J. V. 1984,
    Astr. Ap., **137**, 343.
Savage, B. D. 1984, in *Future of UV Astronomy Based on Six Years of
    IUE Research*, NASA Conference Publication 2349, p. 3.
Walborn, N. R. & Nichols-Bohlin, J. 1987, *Pub. A.S.P.*, **99**, 40.
Walborn, N. R., Nichols-Bohlin, J., & Panek, R. J. 1985, *IUE Atlas of O-Type
    Spectra from 1200 to 1900 Å*, NASA Reference Publication 1155.
Wang, Q. & Helfand, D. J. 1991, *Ap. J.*, **373**, 497.

Massive Stars: Their Lives in the Interstellar Medium
ASP Conference Series, Vol. 35, 1993
Joseph P. Cassinelli and Edward B. Churchwell (eds.)

COLD IONIZED GAS AROUND HOT HII REGIONS

J.S. ONELLO
Dept. of Physics, State University of New York, Cortland, NY, 13045

J.A. PHILLIPS
Owens Valley Radio Observatory, Caltech 105-24, Pasadena, CA 91106

ABSTRACT   We report the initial results of a survey for H168α recombination lines emanating from cold gas in the vicinity of three HII regions: NRAO 584, W48, and G70+1.2. Each of the nebulae exhibited narrow lines from cold hydrogen in photodissociation regions outside their Strömgren spheres. The coldest gas was found in G70+1.2. The width of the hydrogen line from that source indicated a kinetic temperature less than 280 K and a heating time of only $\sim$ 100 yr.

INTRODUCTION

Radio recombination lines are powerful diagnostic probes of cold ionized gas in photodissociation regions around hot young stars. Broad hydrogen lines arise in hot fully ionized HII regions while narrow lines of hydrogen and heavier elements originate in adjacent, partially ionized gas with considerably lower temperatures (see Onello, Phillips & Terzian 1991 and references therein). The source of ionization in the cold gas is controversial. Krügel & Tenorio-Tagle (1978) proposed a model in which stellar winds from early-type stars produce a soft X-ray flux that ionizes cold gas outside the Strömgren sphere. Another possibility was given by Hill (1977) who modeled weak D-type ionization fronts and showed that hydrogen emission features can form in the outermost regions of the front where the temperature is $\sim$ 100 K.

Although the narrow hydrogen lines are relatively strong they have proven difficult to detect because they are always superposed with the broad and much stronger hydrogen line from the nearby HII region. To understand better the partially ionized medium we have begun a survey for narrow recombination lines from a number of HII regions. In the course of this work we have detected an unusual radio recombination line from the molecular globule G70+1.2 that we believe harbors a very young HII region around an early type star.

OBSERVATIONS

In this paper we report H168α recombination line observations of 3 HII regions: W48, NRAO 584, and G70+1.2. The W48 observations were made in 1990 October at the Arecibo Observatory and originally reported by Onello et al. (1991). Observations of the other two sources were conducted in March and April 1992

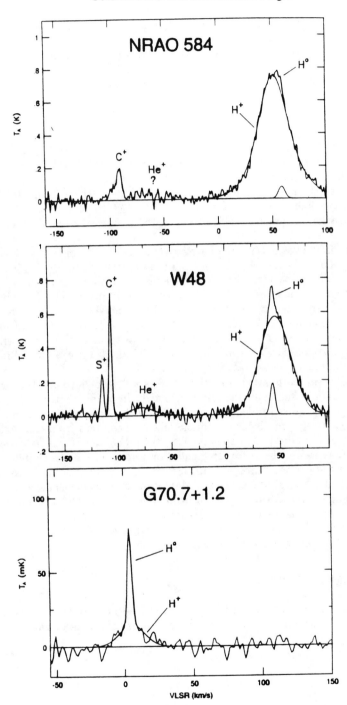

FIGURE I   H168α recombination line spectra of three HII regions.

using the same observing procedures and data reduction techniques. Two 1024-channel autocorrelation spectrometers were operated with a total bandwidth of 2.5 MHz for NRAO 584 and W48, while a bandwidth of 5.0 MHz was used for G70, providing a velocity resolution of 0.53 and 1.06 km/s, respectively. The hanning-smoothed profiles are shown with their gaussian fits in Figure 1.

DISCUSSION

W48 and NRAO 584 exhibit broad lines from hot ionized hydrogen (H^+) and helium (He^+), as well as narrow lines from cold carbon (C^+), sulfur (S^+), and hydrogen (H^o). For both sources the H^o line is weaker than that of H^+ *and* that of carbon. If we assume that the excitation conditions of the C^+ and H^o lines are the same, the line temperature ratios $T_l(H^o)/T_l(C^+) < 1$ imply that the fractional ionization in the cold gas is less than the cosmic abundance of carbon $x_C \simeq 4 \times 10^{-4}$.

Unlike the first two spectra, the narrow hydrogen line from G70+1.2 is relatively strong. For NRAO 584 and W48 the temperature ratio $T_l(H^o)/T_l(H^+)$ is 0.1 and 0.3. The same ratio for G70+1.2 equals 4.6, more than an order of magnitude larger. From the size of the HII region (Phillips, Onello & Kulkarni 1992) and the electron density inferred from its low-frequency spectrum (Kulkarni et al 1992), we conclude that the fractional ionization in the cold gas is large and probably exceeds $\sim 90\%$. The 3.6 km/s line width of G70 is considerably narrower than the values of 7.0 km/s for NRAO 584 and 5.1 km/s for W48, and corresponds to a kinetic temperature of only 280 K.

The equilibrium temperature of HII regions is well-known to be of order 10^4 K. The energy gain in such nebulae arises primarily from the photoionization of hydrogen. Figure 2 shows the heating function $\Gamma(T_e)$ corresponding to the Lyα flux from a B0 star. Energy is lost to the nebula via radiation from collisionally excited states of abundant ions such as C^+, Si^+, Fe^+ and O^+ (Dalgarno & McCray 1972), as well as from free-free emission and electronic recombination. Figure 2 shows the cooling function, $\Lambda(T_e)$, for a fully ionized nebula assuming standard values for the abundances and depletion of heavy elements (Spitzer 1978). For cold gas such as that observed from G70, the cooling is dominated by C^+ 158μ line emission and electronic recombination in hydrogen atoms. $\Gamma \gg \Lambda$ in the cold gas and the time required to reach thermal equilibrium is only ~ 100 years. The cold material is therefore very young.

ACKNOWLEDGMENTS

We wish to thank Shri Kulkarni for bringing G70+1.2 to our attention, and John Carlstrom, Shri Kulkarni, and Dave van Buren for enlightening discussions.

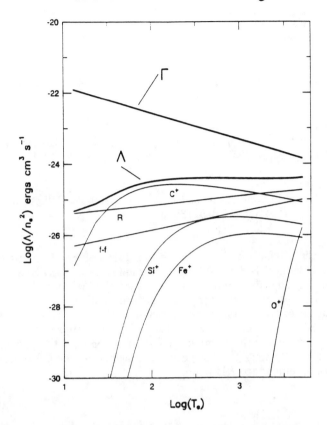

FIGURE II   Heating and cooling functions for a fully-ionized HII region. The curves labeled "f-f" and "R" represent energy losses due to free-free emission and electronic recombination, respectively. Curves labeled "X⁺" denote cooling from the collisional excitation of ionic species X.

REFERENCES

Dalgarno, A. & McCray, R.A. 1972, Ann. Rev. Astr. Ap., 10, 375.

Hill, J.K 1977, ApJ, 212, 692.

Krügel, E. & Tenorio-Tagle, G. 1978, A&A, 70, 51.

Kulkarni, S.R., Vogel, S., Wang, Zh. & Wood, D. 1992, Nature, submitted.

Onello, J.S., Phillips, J.A. & Terzian, Y. 1991, ApJ, 383, 693.

Phillips, J.A., Onello, J.S. & Kulkarni, S.R. 1992, in preparation.

Spitzer, L. 1978, Physical Processes in the Interstellar Medium (New York: Wiley), 4.

Massive Stars: Their Lives in the Interstellar Medium
ASP Conference Series, Vol. 35, 1993
Joseph P. Cassinelli and Edward B. Churchwell (eds.)

THE INITIAL MASS FUNCTION IN THE OUTER GALAXY

Michel Fich
Harvard-Smithsonian Center for Astrophysics and Physics Department,
University of Waterloo, Waterloo, Ontario, Canada N2L 3G1

Susan Terebey
Infrared Processing and Analysis Center, M/S 100-22, California Institute
of Technology, Pasadena, CA 91125

ABSTRACT   The Initial Mass Function (IMF) of massive stars located
in the outer Galaxy is derived from published infrared (IRAS) and CO
observations. The (preliminary) result is that the IMF has a steep slope,
$\Gamma = -2.1 \pm 0.3$. However this result is strongly dependent on two
assumptions: (1) the infrared luminosity in each object is dominated by
the luminosity of the most massive star and (2) the appropriate timescale
to use in normalizing the Luminosity Function counts is the sum of the
Kelvin-Helmholtz time plus 10^5 years, assumed to be the infrared emitting
lifetime of an embedded HII region.

INTRODUCTION

The results of most IMF studies is that the function is a power law for the
massive stars. Numerous studies of stars in the Solar Neighborhood have found
slopes between $\Gamma = -1.2$ and -1.8 (see Miller and Scalo (1979) and Scalo
(1986) for reviews of this subject). Garmany, Conti, and Chiosi (1982) found
a slope of -1.3 in the inner Galaxy and -2.1 in the outer Galaxy. In this paper
we discuss preliminary results on measuring the IMF of massive stars in the
outer Galaxy by counting pre-main sequence massive stars that are still deeply
embedded in the molecular cloud cores where they form and are therefore only
seen in the infrared. Most IMF studies are based on optically selected samples
where most of the stars are main or post-main sequence stars.

THE LUMINOSITY FUNCTION AND THE INITIAL MASS FUNCTION

The data set used in this study is the IRAS selected (from the Point Source
Catalog) bright infrared sources observed by Wouterlout and Brand (1989)
in CO (J=1 $\rightarrow$ 0). These objects are bright at 60 and 100 microns and span
15 to 20 degrees in galactic latitude in the longitude range $85° < l < 280°$.
The Wouterlout and Brand (1989) catalog includes estimates of the distance

and infrared luminosities of these objects and Figure 1 shows these quantities plotted for the entire data set.

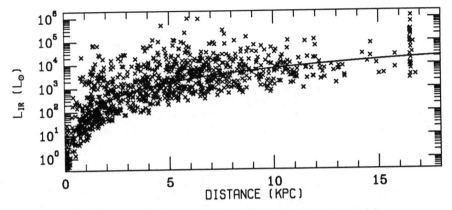

Fig. 1.   The infrared luminosities and distances estimated by Wouterlout and Brand (1989) to the objects in the data set. Our estimate of the completeness limit of the set is shown with the solid line.

The line through Figure 1 is our estimate of the completeness limit of the data set. We estimate that it is complete to a luminosity of $10^3 L_\odot$ only to ≈ 3.5 kpc and to $10^4 L_\odot$ at a distance of ≈ 15 kpc. We have counted objects to produce Luminosity Functions for the infrared sources for these two distance limits and these are shown in Figure 2. The data points for the near distance limit (3.5 kpc) are shown with crosses in the middle of the error bars. Note the turnover in the counts of the 3.5 kpc set at $\approx 10^3 L_\odot$ and at $\approx 10^4 L_\odot$ for the 15 kpc set, as expected from the completeness limits estimated above.

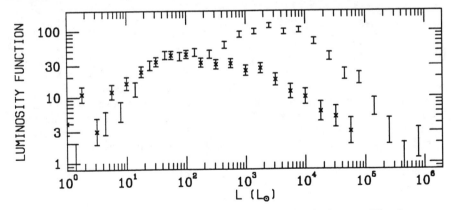

Fig. 2.   The Luminosity Function of the infrared objects. The data points shown with crosses are for objects at distances of 0.1 to 3.5 kpc. The other data points are for those objects between 0.5 and 15 kpc.

The IMF is obtained from the Luminosity Function by converting luminosity to mass and normalizing the counts by the appropriate timescale. Unlike intermediate to low mass stars, the pre-main sequence evolution of massive stars is entirely along radiative tracks where there is little change in the luminosity. Thus we have chosen to use the main-sequence mass-luminosity relation. The timescale for massive stars to evolve along radiative tracks is the Kelvin-Helmholtz time. However massive stars remain embedded for a time even after they reach the main-sequence. Ultimately they blow away the surrounding molecular material through the action of their winds and through the formation of an HII region, and then they cease to emit all of their luminosity in the infrared. We have used 10^5 years as this embedding timescale. The IMFs derived for our two distance limits are shown in Figure 3.

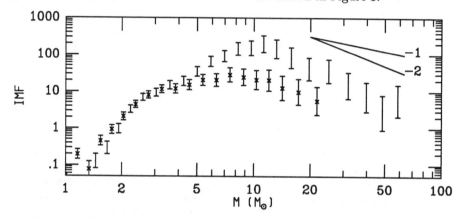

Fig. 3.   The IMF for the 3.5 kpc data set (shown with crosses) and for the 15 kpc data set. Power law slopes of -1 and -2 are shown in the top right hand corner.

CONCLUSIONS

We have binned the data in various ways and looked at a variety of distance limited sub-samples. The power-law slopes of the IMF derived from these has ranged from $\Gamma = -1.8$ to -2.8 and we estimate that on average the slope is -2.1 ± 0.3. We have begun looking at radio continuum surveys of the outer Galaxy to use them to examine HII regions in each infrared source. This will aid us in constraining our assumptions about the relationship between the massive star content and the infrared luminosity of each object.

REFERENCES

Garmany, C. D., Conti, P. S., and Chiosi, C. 1982, *Ap. J.*, **263**, 777.
Miller, G. E. and Scalo, J. M. 1979, *Ap. J. Suppl.*, **41**, 513.
Scalo, J. M. 1986, in *Fund. Cosmic Phys.*, **11**, 1.
Wouterlout, J. G. A. and Brand, J. 1989, *Astr. Ap. Suppl.*, **80**, 149.

Massive Stars: Their Lives in the Interstellar Medium
ASP Conference Series, Vol. 35, 1993
Joseph P. Cassinelli and Edward B. Churchwell (eds.)

MASS, MOMENTUM, AND ENERGY FROM MASSIVE STARS

CARMELLE ROBERT, CLAUS LEITHERER[1], AND
LAURENT DRISSEN
STScI, 3700 San Martin Dr., Baltimore, MD 21218
[1] Affiliated with the Astrophysics Division of ESA

ABSTRACT   We present theoretical models for the integrated wind properties of a population of massive stars formed in a typical starburst galaxy. The release of mass, momentum, and energy from the integrated winds is compared to corresponding quantities from supernovae. The influence of the metallicity and the slope and upper cut-off mass of the IMF are also discussed. In general, winds are more important than supernovae at higher metallicity and during earlier phases of the burst. More results related to this work can be found in the paper by Leitherer *et al.* 1992 (*Ap. J.*, Dec.).

INTRODUCTION

Massive stars, OB and Wolf-Rayet (W-R) stars (descendents of the most massive O stars), display strong stellar winds which are believed to be principally due to radiation pressure (Cassinelli & Lamers 1987, *Exploration of the Univers with IUE Satellite*, 139). The mass-loss rates ($\dot{M}$) observed for these objects exceed 10^{-6} $M_\odot$ yr^{-1} with terminal velocity: $v_\infty \geq 10^3$ km s^{-1}. An important contribution to the evolution of the interstellar medium (ISM) is therefore attributed to winds from massive stars. Supernova (SN) explosions occuring at the latest evolutionary phase of massive stars will also create an important interaction with the ISM. Many observational manifestations of the interaction of massive stars with the ISM can be found in the literature, *e.g.* rings around O and W-R stars (Chu 1991, *IAU Sym. 143*, 349).

POPULATION SYNTHESIS MODELS

We use Maeder's (1990 *A&AS*, 84, 139) evolutionary tracks given for various metallicities (Z). An interpolation is done for a mass interval of 1 $M_\odot$ and a time resolution of 10^4 yr. The mass-loss rate and the wind velocity are known at each evolutionary step according to observational databases and wind models which include the Z-dependence. We adopt a zero age main sequence mass for the SN progenitor of 8 $M_\odot$ (Maeder 1990), an energy of 10^{51} erg for each SN explosion (McKee 1990, *The Evolution of the ISM*, 3) and a remnant mass of 1.4 $M_\odot$. The integrated wind parameters for a stellar population are calculated given a certain slope (α) and cut-off masses (M_l and M_u) for the initial mass function (IMF).

RESULTS

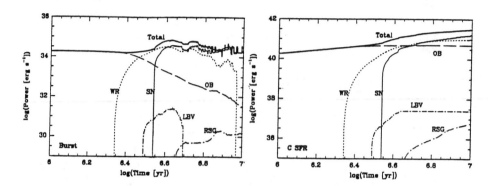

FIGURE 1. Stellar wind power of individual stellar phases and SN'e power during a burst of 1 $M_\odot$ and a constant SFR of 1 $M_\odot$ yr^{-1}. The burst duration is 0.5 Myr. $Z = Z_\odot$, $\alpha = 2.35$, $M_l = 1\ M_\odot$ and $M_u = 120\ M_\odot$.

Figure 1 illustrates the relative importance of individual stellar phases for the wind power in the case of a burst and a constant SFR. In both cases, the contribution from the luminous blue variables (LBV) and the red supergiants (RSG) are negligible as a consequence of their short life-time and low v_∞. For a constant SFR, OB and W-R stars are of almost equal importance after equilibrium between stellar birth and death. For a burst, at $t > 10^{6.5}$ yr, W-R stars will be the dominant contributors over OB stars. At solar metallicity ($Z_\odot$) and $t > 10^{6.7}$ yr, SN'e become the largest individual source of power.

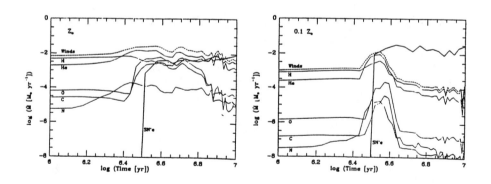

FIGURE 2. Mass-loss rate from an instantaneous burst of total mass 10^6 $M_\odot$. $\alpha = 2.35$, $M_l = 1\ M_\odot$ and $M_u = 120\ M_\odot$.

The influence of Z on the output mass deposition in a burst is presented in figure 2. At $Z \geq Z_\odot$, the stellar winds are the most important contributors to $\dot M$. SN'e give the largest $\dot M$ for $Z < Z_\odot$. At low metallicity fewer W-R stars are formed (these are responsible for the narrow bump seen in the winds curve at low Z). Similar correlations are found for the wind power and Z (not demonstrated here) as expected since $\dot{M}$ and v_∞ are correlated with Z.

The relative contribution to the mass deposition from the individual elements H, He, C, N, and O is also shown in figure 2. At $t = 10^6$ yr, we see the initial surface abundances of O stars. With time, there is an enrichment in processed material due to mixing and removal of H-rich layers by winds; N increases at the expense of C and O. At $t \geq 10^{6.5}$ yr, C and O abundances are the largest due to He burning in W-R stars.

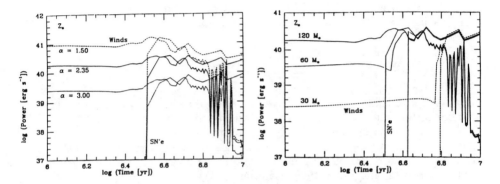

FIGURE 3. Power from an instantaneous burst of total mass 10^6 $M_\odot$. $Z = Z_\odot$, $\alpha = 2.35$ in (b), $M_l = 1$ $M_\odot$ and $M_u = 120$ $M_\odot$ in (a).

Figure 3 presents the effect of the IMF parameters on the wind power in a burst. For a flat IMF slope (e.g. $\alpha = 1.5$), the power increases due to the presence of more massive stars. Stellar winds are favored over SN'e for a longer time. In the case of a steep IMF (e.g. $\alpha = 3.0$), SN'e will dominate later since the typical progenitors are less massive. The relative contributions from stellar winds and SN'e as a function of Z are unchanged for $1.5 < \alpha < 3$.

The upper cut-off mass has a tremendous effect on the output power. If M_u is of the order of 60 $M_\odot$, the O star contribution is reduced. The W-R stars input will also be affected if $Z < Z_\odot$. The SN'e appear later since the lifetime is longer for lower mass stars.

CONCLUSIONS

There are a few points which remain to be verified: the theoretical prediction of the stellar mass-loss rate in different chemical environments, the effect of mass loss on stellar evolution, the increasing number of W-R stars due to inclusion of binaries, the large number of SN'e from W-R stars... Despite these uncertainties we feel that stellar winds can be an important contributor to the deposition of energy into the ISM.

L. D. and C. R. gratefully acknowledge CRSNG (Canada) for a Post-Doctoral Fellowship and STScI's Director's Research Fund for travel support.

Massive Stars: Their Lives in the Interstellar Medium
ASP Conference Series, Vol. 35, 1993
Joseph P. Cassinelli and Edward B. Churchwell (eds.)

EFFECTS OF SHOCK-PRODUCED X-RAYS ON THE IONIZATION
BALANCE IN THE WIND OF ε Ori

J.J. MacFARLANE,[1,2] M. WOLFF,[1] and P. WANG,[2]
Department of Astronomy,[1] Fusion Technology Institute,[2]
University of Wisconsin, Madison, WI 53706

ABSTRACT   We report on our initial series of calculations to study the
effects of shock-produced X-rays on the ionization state of hot star winds.
For the wind of ε Ori (B0 Ia), we investigate the dependence of the
ionization distribution on X-ray source characteristics.

INTRODUCTION

Hot stars have high mass loss rates and are known to be X-ray emitters. The
X-ray emitting plasmas typically have characteristic temperatures of $10^6 - 10^7$
K and luminosities of $L_x \sim 10^{-7} L_{bol}$. The source of the X-rays is not currently
known. The most likely candidates are strong shocks propagating through
their winds (Lucy 1982, Owocki et al. 1988) or coronae at the base of their
winds (Cassinelli and Olson 1979, Waldron 1984).

Surprisingly little is known about the physical characteristics of hot star
winds, such as the temperature structure and ionization distribution. Previous
notable studies of wind ionization include Klein and Castor (1981), Pauldrach
(1987), and Drew (1989). Drew pointed out the importance of heavy element
line cooling in influencing the temperature structure of hot star winds. This
effect leads to significantly lower temperatures compared to those found in
previous studies. However, comparison of calculated ionization distributions
and those deduced from observations of UV P-Cygni profiles (Groenewegen and
Lamers 1991) finds significant discrepancies between theory and observation.

In this paper, we describe our initial series of calculations aimed at
achieving a better understanding of the physical state of hot star winds. Our
calculations are in many ways similar to those reported by Drew (1989), but
with one important exception: we include the effects of Auger ionization
resulting from the X-rays. The location of the X-ray source in our models is
variable; it can be positioned at the base of the wind to simulate a corona, or
can be embedded in the cooler wind to simulate a shock. Below we describe
results for the particular case of ε Ori (B0 Ia). We shall focus in particular on
the sensitivity of the wind ionization distribution to the X-ray source location.

MODELS

Multilevel statistical equilibrium calculations were performed to predict the
distribution of atomic level populations throughout the wind. The following
effects are included in our model: photoionization from photospheric radiation
($h\nu < 54$ eV), diffuse radiation from the He II Lyman continuum (54 eV
$< h\nu \lesssim 100$ eV), and X-ray radiation ($h\nu \gtrsim 100$ eV); photoexcitation and

spontaneous decay; radiative and dielectronic recombination; and collisional excitation, deexcitation, ionization, and recombination. Calculations were performed for multicomponent plasmas (H, He, C, N, O, and Si) consisting of a total of 25 ions and 87 atomic levels. The levels and ions were selected so that a direct comparison with the results of Drew could be made.

The radiation field is modelled as follows. Below the He II photoionization edge (54 eV), the photospheric radiation flux is taken from non-LTE model atmosphere tables of Mihalas (1972), and corrected by the geometric dilution factor. Between the He II Lyman edge and about 100 eV, it is assumed the wind is optically thick. At these frequencies we assume the radiation field is given by the local continuum source function. Above 100 eV, the radiation field is constrained from X-ray satellite observations (see, e.g., Cassinelli et al. 1981). The frequency-dependent X-ray flux is computed using the XSPCT code (Raymond and Smith 1977). In the calculations described below, we assume an X-ray temperature of $T_x = 10^{6.5}$ K and emission measure of 10^{55} cm^{-3}. The attenuation of X-rays by inner-shell absorption is included in our calculations.

Photoexcitation rates are computed using the Sobolev approximation (see Castor 1970). Photoionization out of excited states is included. Bound-free cross sections for all levels and subshells were obtained from Hartree-Fock calculations (Wang 1991). Low-temperature dielectronic rate coefficients are based on the work of Nussbaumer and Storey (1983).

RESULTS
Results for calculations for the wind of ε Ori are shown in Fig. 1. Relevant parameters used in the calculations are shown in Table I. Results for ionization distributions for N and O are shown for 3 cases: (i) no X-ray source, (ii) an X-ray source at the base of the wind (coronal model), and (iii) an X-ray source at $r = 3R_*$ (shock model). In each case we use a wind temperature distribution based on the results of Drew.

Figure 1 shows the ionization fractions of N III, N IV, N V, O III, O IV, and O VI as a function of radius for each of the three calculations. In each case N IV and O III are the dominant stage of ionization. It is also seen that the N III, N IV, and O III distributions are almost identical in each case; i.e., their abundance is unaffected by the X-rays. Also note that the N V and O VI fractions in the absence of an X-ray field are predicted to be less than 10^{-6}. For the coronal X-ray case the N V and O VI fractions decrease with radius — gradually in the case of N V and more dramatically for O VI. For the shock X-ray case the N V and O VI fractions peak at the shock location ($r = 3R_*$).

TABLE I ε Ori Parameters

$R_x = 34$ R$_\odot$	$v_\infty = 2,010$ km/s
$T_{eff} = 27,500$ K	$T_x = 10^{6.5}$ K
$g = 1,000$ cm^2/s	$EM_x = 1 \times 10^{55}$ cm^{-3}
$\dot{M} = 1 \times 10^{-6}$ M$_\odot$/yr	

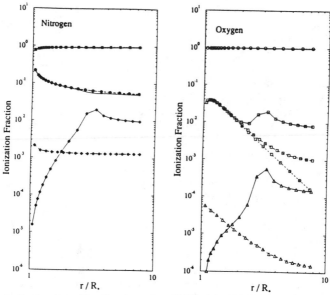

Fig. 1. Ionization distributions for N III (●), N IV (■), N V (◆), O III (○), O IV (□), and O VI (△). Results are given for 3 cases: shock X-ray model (solid curve), coronal X-ray model (dashed curve), and no X-rays (dotted curve). Note the radically different dependence of the N V and O VI fractions on the X-ray source location.

The rapid decrease in abundance in the direction of the star results from inner-shell attenuation by the intervening wind.

A detailed analysis of the influence of X-rays on the ionization state of hot stars winds will be presented in a forthcoming paper.

ACKNOWLEDGEMENTS
The authors thank J. Cassinelli for helpful discussions. This work was supported in part by NASA Grant NAGW-2210.

REFERENCES
Cassinelli, J.P., and Olson, G.L. 1979, *Ap. J.*, **229**, 304.
Cassinelli, J.P., Waldraon, W.L., Sanders, W.T., Hernden, F.R., Rosner, R., and Vaiana, C.S. 1981, *Ap. J.*, **250**, 677.
Castor, J.I. 1970, *M.N.R.A.S.*, **149**, 111.
Drew, J.E. 1989, *Ap. J. Suppl.*, **71**, 267.
Groenewegen, M.A.T., and Lamers, H.J.G.L.M. 1991, *Astr. Ap.*, **243**, 429.
Klein, R.I., and Castor, J.I. 1981, *Ap. J.*, **220**, 902.
Lucy, L.B. 1982, *Ap. J.*, **255**, 286.
Mihalas, D. 1972, NCAR Tech. Note, STR-76.
Nussbaumer, H., and Storey, P.J. 1983, *Astr. Ap.*, **126**, 75.
Owocki, S.P., Castor, J.I., and Rybicki, G.B. 1988, *Ap. J.*, **335**, 914.
Pauldrach, A. 1987, *Astr. Ap.*, **183**, 295.
Raymond, J.C., and Smith, B.W. 1977, *Ap. J. Suppl.*, **35**, 419.
Waldron, W.L. 1984, *Ap. J.*, **282**, 256.
Wang, P. 1991, Ph.D. Dissertation, Univ. of Wisconsin-Madison.

Massive Stars: Their Lives in the Interstellar Medium
ASP Conference Series, Vol. 35, 1993
Joseph P. Cassinelli and Edward B. Churchwell (eds.)

THE RING AROUND SN1987A AND ITS CONSTRAINTS ON THE POST-MAIN-SEQUENCE EVOLUTION OF THE SUPERNOVA PROGENITOR

CRYSTAL L. MARTIN
Steward Observatory, University of Arizona, Tucson, AZ 85721

DAVID ARNETT
Departments of Physics and Astronomy, University of Arizona, Tucson, AZ 85721

ABSTRACT   A two dimensional hydrodynamical calculation of colliding stellar winds is presented. We discuss present work using SN1987A's ring to constrain the time scales of the progenitor's post-main-sequence evolution.

INTRODUCTION

Stars in the 9 - 40 $M_\odot$ range play a prominent role in the hydrodynamical and chemical evolution of galaxies. Their stellar winds and supernova explosions are believed to create the hot component of the interstellar medium (ISM). In some galactic disks, the kiloparsec sized superbubbles formed around clusters of massive stars may blow out of the disk plane and release hot, metal enriched gas into the galaxy's halo. Additionally, the expanding shock front of a superbubble in the disk may trigger additional star formation. Furthermore, similar processes probably drive the galactic winds associated with starburst nuclei that enrich the intracluster and intergalactic mediums.

Nonetheless, the explosion of a blue supergiant in the Large Magellanic Cloud (LMC), SN1987A, illuminated the incompleteness of our understanding of massive stars. Evolutionary models of massive stars do not synthesize the observed supergiant populations in either the Milky Way or LMC. Our modeling of the formation of SN1987A's ring will improve our knowledge of both the post-main-sequence evolution of massive stars and their coupling to the ISM in galaxies.

SN1987A'S RING

It is now generally agreed that SN1987A's progenitor was a blue supergiant (BSG) (Arnett *et al.* 1989) of spectral type B3 Ia, and there is considerable evidence that this star passed through a red supergiant (RSG) stage earlier in its life. Theoretical tracks for low metallicity $20M_\odot$ stars (e.g. Arnett 1991) suggest the progenitor passed through two RSG stages and one BSG stage prior to the the final BSG phase. Observationally, the RSG phase is inferred from the narrow UV emission lines seen in spectra of SN1987A taken after May 24, 1987 (Fransson 1989). These line widths and strengths were shown to be consistent with an origin in a low density, (1-4) x 10^4 cm^{-3}, photoionized gas having CNO abundance ratios suggestive of a nucleosynthetic origin via the CNO cycle (Lundqvist & Fransson 1991). Hence, the emerging picture suggests that ashes from the nuclear burning in the interior were mixed into the progenitor's outer layers during a RSG phase and expelled in a slow stellar wind. During the subsequent BSG stage, a faster stellar wind from the hotter, smaller star collided with this circumstellar material. The resulting strong shock front swept up a dense shell (relative to the ISM) of ambient RSG wind which was subsequently photoionized by the UV/soft x-ray burst of the supernova explosion. A shell expansion rate of 15 km/s was derived by Lundqvist and Fransson (1991) from the UV emission lines.

High resolution imaging of the supernova remnant (SNR) has further constrained the morphology of this circumstellar structure. The image taken with ESA's FOC on HST through a narrow [OIII] λ 5007 filter (Panagia et al. 1991) reveals a clumpy ellipse of emission with a major axis a bit more than one light year across surrounding the SNR. The radial velocity gradient along the minor axis of the emission strongly suggests that the ellipse on the plane of the sky is actually a ring tilted relative to our line of sight rather than a limb brightened shell (Crotts 1991; Meikle et al. 1991). Crotts and Heathcote (1991) derived a ring expansion speed of 10.3 km/s from high-resolution spectra of the nebulosity. Luo and McCray (1991) have shown that the interacting winds scenario can produce an hourglass shaped circumstellar shell with a ring at the waist if the RSG wind has an equator to pole density contrast. Although the cause of such a wind asymmetry is unclear, a low mass secondary in a common envelope phase with the RSG can cause a highly asymmetrical mass-loss profile (Soker, 1992). A binary companion to the progenitor is also one possible cause for the observed asymmetric envelope expansion (Chevalier & Soker 1989).

CALCULATION

We use the PROMETHEUS code (Fryxell et al. 1989) to calculate the hydrodynamical interaction of a BSG wind overtaking an asymmetric RSG wind. We calculate one quadrant of a 50 x 100 grid oriented perpendicular to the plane of the ring with an origin at the progenitor star. This calculation is a step toward a 3D calculation which will investigate the stability of the ring. Only the last BSG and RSG phases of the progenitor's wind are modeled since the termination shocks of the winds blown during any previous blue loops and on the main sequence should be at much larger distances than the observed ring. The RSG wind asymmetry is described by a latitudinal density contrast in the RSG wind. We change the inner boundary condition from a RSG wind to a BSG wind abruptly since the acceleration mechanisms for the RSG and BSG winds are believed to be quite different and the stellar models indicate the progenitor evolves from the red to the blue on a time scale that is short in comparison to the time spent at either envelope solution. By choosing a RSG wind density toward the upper end of the estimated range and a BSG wind velocity toward the lower end of the observed range, we can place an upper limit on the duration of the final BSG stage.

RESULTS

The density contours and velocity vectors in Figure 1 illustrate the ring and constricted bubble resulting from an adiabatic calculation of a weak BSG wind crashing into an asymmetric, dense RSG wind with a 10:1 equator to pole density contrast. The ring has reached a radius of 1.85 lt. yr. when the progenitor has been in the final BSG stage for $\tau_{BSG} = 20,000$ yr. This distance is a factor of three larger than the observed position of the ring, and τ_{BSG} is only 80% of the value predicted by the stellar model of Arnett (1991). Hence we have a discrepancy between the interacting winds model for the ring and the stellar model. We are presently repeating the calculation with the inclusion of additional microphysics. The addition of radiative cooling and ionization losses will work to slow the ring down and increase its density. The results of this calculation will reveal whether or not there really is any discrepancy between the ring formation model and the stellar evolutionary models.

REFERENCES

Arnett, D. 1991, Ap.J. **383**, 295.

Arnett, W. D. *et al.* 1989, Annu. Rev. Astron. Astrophys. **27**, 629.

Chevalier, R. A. and Soker, N. 1989, Ap.J. **341**, 867.

Crotts, A. P. S. and Heathcote, S. R. 1991, Nature **350**, 685.

Fransson, C. *et al.* 1989, Ap.J. **336**, 429.

Fryxell, B. A. *et al.* 1989, *in Numerical Methods in Astrophysics*, P. R. Woodward, ed., (Academic Press: New York).

Lundqvist, P. and Fransson, C. 1991, Ap.J. **380**, 575.

Luo, D. and McCray, R. 1991, Ap.J. **379**, 659.

Meikle, W. P. S. *et al.* 1991, in *in Proc. ESO/EIPC Workshop on Supernova 1987A and Other Supernovae* (Marciana Marina, Elba, Italy, 1990 September 17 - 22), ed. I. J. Danziger & K. Kjär (Garching: ESO), 595.

Panagia, N. *et al.* 1991, Ap.J. **380**, L23.

Soker, N. 1992, Ap.J. **386**, 190.

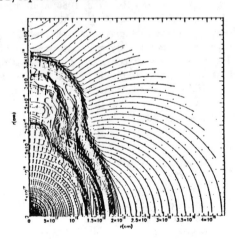

Figure 1. Density contours and velocity vectors of a spherically symmetric, fast wind running into an a slow wind with a density enhancement in the equatorial plane.

Massive Stars: Their Lives in the Interstellar Medium
ASP Conference Series, Vol. 35, 1993
Joseph P. Cassinelli and Edward B. Churchwell (eds.)

RADIATION DRIVEN STELLAR WINDS WITH ALFVÉN WAVES IN HOT STARS

L. C. dos Santos, V. Jatenco-Pereira, and R. Opher
Universidade de São Paulo
Instituto Astronômico e Geofísico
Caixa Postal 9638, 01065 São Paulo, SP, Brazil

ABSTRACT We present a model to explain the mass loss in Wolf-Rayet stars. The possibility that matter can be accelerated in these stars by Alfvén waves, acting in conjunction with radiation pressure, is considered. It is shown that this joint mechanism can explain the observed flow velocities and mass loss rates. The correction factor, which allows for the finite size of the stellar disk, is also considered.

INTRODUCTION

Observations of Wolf-Rayet stars in the ultraviolet first showed that these stars loose a considerable part of their mass in the form of a stellar wind (Morton, 1967, 1976; Snow and Morton, 1976). The wind expansion velocities reach 1000 − 3000 km/sec. The mass loss rate determined in the ultraviolet, infrared, as well as radio, indicates $\dot{M} \sim 10^{-5} M_\odot/yr$ (Prinja and Howarth, 1986; Walborn and Nichols-Bohlin, 1987; Howarth and Prinja, 1989; de Jager et al., 1988).

Many investigations have been made to explain the stellar winds from these stars. Until now, the models generally assume only radiation pressure to accelerate the stellar wind. The mass loss rate, in models considering only radiation pressure, is a problem. It is worth recalling that current empirical mass loss rates of massive stars is much greater than the theoretically predicted mass-loss rates. For these stars we have a very large wind momentum flux $\dot{M}u_\infty$. This gives rise to the "wind momentum problem" (Barlow et.al., 1981; Cassinelli, 1982).

ALFVÉN WAVE RADIATION DRIVEN WINDS

The radiative driven wind theory does not correctly predict the observed values of the terminal velocity and the mass-loss rate of early-type stars. Our model is a fusion of the Alfvén wave wind model of Jatenco-Pereira and Opher (1989a, 1989b, 1989c) and the radiation pressure model of Castor et al. (1975)(CAK) to provide a possible solution to the momentum problem. We assume that a magnetic field exists in Wolf-Rayet stars, and that the geometry of this field is similar to that of a solar coronal hole.

The observed mass-loss rate determines the initial velocity by the relation, $u_0 = \dot{M}/4\pi R_*^2 \rho_0 f_{max}$, where R_* is the stellar radius, and ρ_0 is the initial density (at the stellar surface), f_{max} is the opening solid angle for the coronal hole. The equation of motion (i.e., Eq. (52) of Holzer et al., 1983) is:

$$\frac{1}{u}\frac{du}{dr}\left[u^2 - v_{th}^2 - \frac{1}{4}\left(\frac{1+3M_A}{1+M_A}\right)<\delta\,v^2>\right] = \frac{S}{r}\left[v_{th}^2 + \right.$$

$$\left. +\frac{1}{4}\left(\frac{1+3M_A}{1+M_A}\right)<\delta\,v^2> + \frac{1}{4}\frac{r}{L}<\delta v^2> - \frac{1}{2S}v_e^2\right] \tag{1}$$

where u is the velocity of the wind at the radius r, $V_A = B/(4\pi\rho)^{1/2}$, the Alfvén velocity, $M_A = u/V_A$, the Alfvén Mach number, L the damping length, S a parameter describing the divergence of the geometry, v_{th} the ion thermal velocity and v_e the escape velocity.

We use an effective escape velocity

$$-\frac{1}{2}v_e^2 = -\frac{G(1-\Gamma_{Edd})M_*}{r} + \frac{K\,L_*\,F_{CA}CF}{(1-\alpha)r}\left(\frac{1}{\rho\,c}\frac{du}{dr}\right)^\alpha , \tag{2}$$

where L_* is the stellar luminosity, α the CAK power index expressing the effect of all lines, M_* the mass of the star, c the speed of light, Γ_{Edd} the usual Eddington factor, and ρ the density. K depends on the normalization of the line number distribution (Owocki et al., 1988). The factor $F_{CA} = (X/10^{-3})^{1-\alpha}$ describes chemical abundance, where X is the fraction of all atoms in the gas which are able to absorb effectively in the spectral region where most of the flux is emerging (CAK). For normal abundance, we assume that the CNO number abundance is $X = 10^{-3}$ (e.g. CAK) and for solar chemical composition without hydrogen we use $X = 10^{-2}$, and the factor CF takes into account the finite size of the disk of the star (see Kudritzki et al. (1990)).

RESULTS AND DISCUSSION

We solved the wind equation (1) for the Wolf-Rayet star with $M_* = 16M_\odot$, $R_* = 12R_\odot$, $T_* = 3.0 \times 10^4 K$, $\dot{M} = 10^{-4.7}M_\odot/yr$, and $L_* = 10^{5.1}L_\odot$. Our results for solar chemical composition, and non solar chemical composition are presented in Table 1.

In our model, the area of the flux tube increases by a factor $f_{max}(R_*/r)^2$ at large radii. If a fraction of the surface occupied by the flux tubes is $F_A \equiv (1/f_{max})$, then the initial velocity is given by u_0 below, and $F_A f_{max} = 1$. The average magnetic field is $\bar{B} = B_0 F_A$. We also investigate the case where $F_A f_{max} > 1$ and u_0 is supersonic. The value of $\bar{B}$ and $F_A f_{max}$ are also shown in Table 1.

We note that $\bar{B}_0$ in Table 1 is $\bar{B}_0 < 1000\ G$, appreciably less than $\bar{B}_0 \geq 10000\ G$ found necessary in the models of Hartmann and Cassinelli (1981) for a pure Alfvén wave driven Wolf-Rayet wind and that of Poe et al. (1989) for the magnetic rotator model. We also note in Table 1 the importance of the Alfvén waves when the Alfvén wave flux is reduced, for example, for $\alpha = 0.5$ and $f_{max} = 5$ we have $u_\infty = 2016\ km/sec$ for $\phi_{M0} = 6.7 \times 10^{12}\ erg\ cm^{-2}s^{-1}$ but only $u_\infty = 1366\ km/sec$ when $\phi_{M0} = 1.6 \times 10^{12}\ erg\ cm^{-2}s^{-1}$. For this example the relation, $u_\infty = 3v_{e0}$, indicated by observations (Barlow, 1982) is obtained.

Table 1. Results for solar chemical abundance (upper part) and non-solar chemical abundance (bottom part)

f_{max}	α	ϕ_{M0} erg $cm^{-2}s^{-1}$	$\bar{B}_0$ G	$F_A f_{max}$	u_0 km/sec	u_∞ km/sec	η
2	0.50	1.3×10^{12}	500	6.5	177	1781	14
2	0.50	2.4×10^{12}	950	1.0	27	1800	14
2	0.50	2.4×10^{12}	950	8.2	223	2010	14
5	0.50	1.6×10^{12}	142	1.0	67	1366	10
5	0.50	6.7×10^{12}	680	1.0	67	2016	15
5	0.60	5.2×10^{12}	491	1.0	67	1400	10
2	0.55	4.9×10^{12}	1245	1.0	27	2180	16
5	0.50	1.6×10^{12}	142	4.2	279	3866	29
2	0.55	6.5×10^{11}	221	7.0	189	2262	17

Acknowlegments

The authors would like to thank the brazilian agencies FAPESP and CNPq for support.

REFERENCES

Barlow, M. J. 1982, in "Wolf-Rayet Stars: Observations, Physics, Evolution, eds. C. W. H. de Loore and A. J. Willis (Dordrecht: Reidel)", p. 185.
Barlow, M. J., Smith, L. J., and Willis, A. J. 1981, M.N.R.A.S., 196, 101.
Cassinelli, J.P. 1982 in "Wolf-Rayet Stars: Observation, Physics, Evolution, eds. C. W. H. de Loore and A. J. Willis (Dordrecht: Reidel)" p. 173
Castor, J.I., Abbott, D.C. and Klein, R.I. 1975, Ap. J. 195, 157 (CAK).
de Jarger, C. Nieuwenhuijzen, H., and Van der Huch, K.A. 1988, AA 72, 259
Hartmann, L.W., and Cassinelli, J.P. 1981 Bull. A.A.S. 13, 785
Holzer, T.E., Fla, T., and Leer, E. 1983, Ap. J. 275, 808
Howarth, I.D. and Prinja, R.K. 1989, Ap. J. Suppl. Ser. 69, 527
Jatenco-Pereira, V. and Opher, R., 1989a, Ap. J. 344, 513
Jatenco-Pereira, V. and Opher, R., 1989b, Astr. Ap. 209, 327
Jatenco-Pereira, V. and Opher, R., 1989c, M.N.R.A.S., 236, 1
Kudritzki, R.P., and Hummer, D.G. 1990, Ann. Rev. Astron. Astrophys. 28, 303
Morton, D.C. 1976, Ap.J. 203, 386
Morton, D.C. 1967, Ap. J. 150, 535
Owocki, S. P., Castor, J. I., and Rybicki, G. B. 1988, Ap. J. 335, 914
Poe, C.H.,Friend, B., Cassinelli, J.P. 1989, Ap. J. 337, 888
Prinja, R.K. and Howarth, I.D. 1986, Ap. J. Suppl. Ser. 61, 357
Snow, T.P., and Morton, D.C. 1975 AP. J. Suppl. Ser. 32, 357
Walborn, N.R., and Nichols-Bohlin, J. 1987, P.A.S.P. 99, 40

Massive Stars: Their Lives in the Interstellar Medium
ASP Conference Series, Vol. 35, 1993
Joseph P. Cassinelli and Edward B. Churchwell (eds.)

Be STAR ENVELOPES DRIVEN BY OPTICALLY THIN LINES

Haiqi Chen and J. M. Marlborough
Department of Astronomy, The University of Western Ontario, London,
Ontario, N6A 3K7, Canada

ABSTRACT   In this paper we investigate the driving of the envelopes of Be
stars by a radiation force arising from optically thin lines. We introduce two
parameters, ϵ and η, to describe the force and calculate the radial velocity
distribution for various values of ϵ and η. The terminal velocity lies between
50 and 330 km s^{-1} for the values of ϵ and η of interest.

INTRODUCTION

In a previous study (Chen, Marlborough and Waters 1992), we excluded radiation
pressure due to optically thick lines as an important driving force for Be star
envelopes. We then discussed several other mechanisms, one of which is the
radiation force due to the optically thin lines. In this investigation, we consider
the driving of these envelopes by this radiation force.

The importance of the weak line force in driving stellar winds has been
discussed by Lamers (1986). After investigating the dynamics of the wind of
P Cygni, he concluded that the wind was best explained by the radiation force
arising from a large number of optically thin lines. Although no detailed
calculation has ever been done to show how large the weak line force could be,
the presence of a very large number of optically thin lines in IUE spectra for Be
stars suggests that this force might play a important role in driving the envelopes
of Be stars.

THE FORCE DUE TO OPTICALLY THIN LINES

If the total number of optically thin lines is N, the radiation force due to all weak
lines will be

$$F_w = \Sigma \, \kappa_l L_l \, / \, 4\pi c r^2 \,, \tag{1}$$

where c is the speed of light, L_l is the luminosity at the wavelength of the weak
line, κ_l is line opacity integrated over the line, and Σ sums over all N weak lines.

If we express the line opacity κ_l in terms of the electron scattering coefficient, σ, i.e., $\kappa_l = \delta_l \sigma$, and introduce an average

$$<\delta>=1/N\Sigma\delta_l(L_l/L_*), \qquad (2)$$

where L_* is the total luminosity of the star, equation (1) becomes

$$F_w=W(r)GM(1-\Gamma)/r^2, \qquad (3)$$

where $W(r)=N(r)<\delta>\Gamma/(1-\Gamma)$. From equation (3), $W(r)$ is the ratio of the force due to optically thin lines divided by the gravitational force.

The factor $W(r)$ may have a very complicated dependence on r. At present, we are not able to evaluate it from first principles. Thus we assume $W(r)$ to have a simple dependence on r and adjust the free parameters so that a reasonable radial velocity distribution will be produced. We therefore assume

$$W(r)=\eta(r/R)^\epsilon, \qquad (4)$$

where η and ε are constants; the weak line force will have a dependence upon r similar to that derived earlier (Chen, Marlborough and Waters 1992). The parameter η measures the strength of the force due to the optically thin lines at the surface of the star, and ε measures the dependence of the force on r.

VELOCITY DISTRIBUTION CALCULATIONS

In our calculation, the circumstellar envelope is assumed to be symmetric about the rotation axis ($\theta=0,\pi$) and the equatorial plane. We restrict our attention to the dynamics in the equatorial plane only.

The calculations are for a star like γ Cas: $M = 16\ M_\odot$, $R = 10\ R_\odot$, $T = 25,000$ K, $L = 5.4\times10^4\ L_\odot$. Since both η and ϵ are free parameters, we have calculated the wind model for different values of ϵ and η. Here we only show models with $\epsilon=0.01$ and different values of η (Table 1 and Fig. 1). In Table 1, columns 1 and 2 give ϵ and η values, respectively; column 3 gives sonic points; column 4 gives the initial velocities in km/s; and Column 5 gives V_{100}, the velocity at 100R. Although V_{100} is not the terminal velocity, V_∞, it is a good indication of V_∞. Column 6 gives R_h, the radius at which $V_{100}/2$ is reached; R_h is an indication of how fast the radial velocity increases.

In Fig. 2, we give the velocity distribution produced by a strong line radiation driven model with rotational velocity of $0.7V_{break-up}$ (indicated by "CAK" above the curve) and the velocity distribution produced by a weak line model with $\epsilon=0.01$ and $\eta=0.495$. The velocity distribution for a strong line radiation driven model reaches half of its terminal velocity at only 1.5 stellar radii compared with 4.4 for the weak line model.

TABLE I   Model Parameters for $\epsilon=0.01$

ε	η	$x_c(R)$	$V_0(km/s)$	$V_{100}(km/s)$	$R_h(R)$
0.01	0.495	1.71	2.6	69.7	4.3
0.01	0.490	3.16	0.04	59.6	7.2
0.01	0.480	7.46	$<10^{-6}$	47.9	13.5

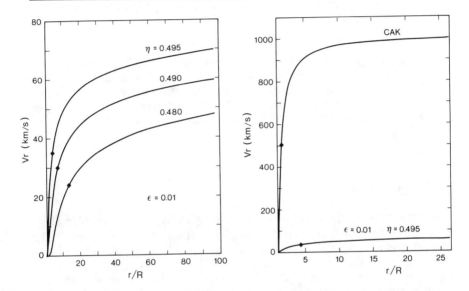

Figure 1.   The radial velocity distribution for $\epsilon=0.01$ and $\eta=0.495$, 0.490 and 0.480. The diamonds indicate the location of R_h, at which the velocity reaches half of V_{100}.

Figure 2.   The comparison of the radial velocity distribution for a strong line driven wind model and for our model with $\epsilon=0.01$ and $\eta=0.495$. The diamonds indicate the location of R_h, at which the velocity reaches V_{100}.

ACKNOWLEDGEMENT

This research was supported by The Natural Sciences and Engineering Research Council of Canada.

REFERENCES

Chen, H., Marlborough, J. M., & Waters L. B. F. M. 1992, ApJ, 384, 605
Lamers, H. J. G. L. M. 1986, A&A, 159, 90

Massive Stars: Their Lives in the Interstellar Medium
ASP Conference Series, Vol. 35, 1993
Joseph P. Cassinelli and Edward B. Churchwell (eds.)

HE I LINES FROM THE WINDS OF LUMINOUS YOUNG STELLAR OBJECTS

MELVIN G. HOARE
Department of Physics, Astrophysics, University of Oxford, Keble Road, Oxford, OX1 3RH

ABSTRACT   We present some initial results from NLTE modelling of the H I and He I IR emission lines from the winds of young massive stars. These models can be used to constrain the effective temperatures of these stars, and hence whether they are on the main sequence or not, as well as testing the viability of a spherically symmetric wind.

INTRODUCTION

The early stages in the life of a massive star, when it is still embedded in its natal molecular cloud, are characterized by a strong ($\dot{M} \approx 10^{-6}$ $M_\odot yr^{-1}$), slow ($v_\infty \approx 300$ km s^{-1}) ionized stellar wind. This is made manifest by broad near-IR H I emission lines (e.g. Persson *et al.* 1984) and radio continuum spectra like $S_\nu \propto \nu^{0.6}$ (e.g. Simon *et al.* 1983). The driving mechanism and geometry of these winds and their relation to the larger scale bipolar molecular outflows (Lada 1985) is unknown.

These winds are different to the faster, radiatively driven, winds well known in MS OB stars, which may mean that active accretion via a disk is still going on, and/or, that the star has not reached the MS. In order to study the evolutionary state of these objects we need to place them on the H-R diagram. The luminosities are reasonably well known from far-IR fluxes, but the heavy obscuration also precludes direct observation of the star to determine its effective temperature. If the ionized wind originates from the star and not from an accretion disk we can use the excitation state of the wind to infer T_{eff}. This requires detailed NLTE modelling of which the most comprehensive attempt to date was by Höflich & Wehrse (1987) who only considered H I lines. Here we include He I lines motivated by the observations of such lines in S106 IRS3 by Garden & Geballe (1986) and the expectation that they will provide further constraints on T_{eff}.

Many of the He I lines appear blended with the H I lines giving rise to asymmetric line profiles which have been interpreted as evidence for bipolarity. Detailed modelling will tell us if a spherically symmetric model can explain all the observational data. If it can then is radiative driving a plausible explanation? If not then the wind must originate from the disk or some other source.

MODELLING

Our modelling is performed using the NLTE radiative transfer code developed by Dr Peter Höflich (MPI) and described by Höflich & Wehrse (1987). This code calculates the level populations and temperature structure for both the expanding envelope and underlying photosphere. The initial model we present here treats 14 H I, 8 He I singlet and 7 He I triplet levels in NLTE with 134 NLTE transitions. In addition a further 230 higher level He I transitions are included for which the upper level is assumed to be in LTE and similarly the lower level unless it has $n \leq 5$. We present an illustrative model which loosely matches the luminosity and radio flux of S106 IRS3. Model parameters are:

Effective temperature	28 000 K
Luminosity	8.0×10^4 L$_\odot$
$\log g$	3.5
Photospheric radius	12 R$_\odot$
Envelope radius	300 R$_*$
Mass loss rate	1.4×10^{-6} M$_\odot$yr^{-1}
Velocity law	Castor & Lamers $\beta = 1.0$, $v_\infty = 200$ km s^{-1}

RESULTS

Figures 1 and 2 show model spectra around Brα and Pfγ. The He I $5g^{1,3}$G-$4f^{1,3}$F 4.049μm line was observed in S106 IRS3 by Garden & Geballe and the $5f^{1,3}$F-$4d^{1,3}$D blend has recently been detected in the same source by Bunn & Drew (priv comm). Note the blending on the blue wing of Pfγ as discussed by Garden & Geballe although, the pure H profile also appears asymmetric due to blue-shifted absorption.

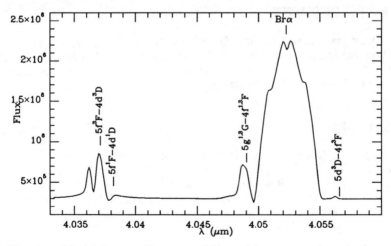

Fig. 1.   Model line profiles around Brα with rest wavelengths (vacuum)

marked. Some structure in the line profiles is due to inadequate gridding.

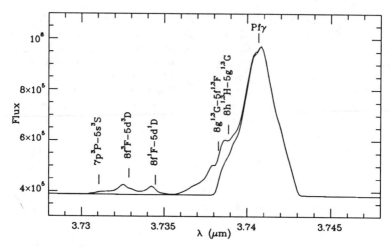

Fig. 2.   As Fig. 1 for Pfγ with a pure H model also shown.

FUTURE WORK

We will use the NLTE models to attempt a match to all the observed line profiles for S106 IRS3 in order to constrain the T_{eff} and $\log g$ of the star and the velocity law in the wind. This should enable us to determine if a radiatively driven spherically symmetric wind is a viable explanation for the phenomena we see.

ACKNOWLEDGEMENTS

We are very grateful to Dr. Peter Höflich (MPI) for his generous help in providing and implementing the NLTE code and to Dr. Pete Storey (UCL) for providing the He I atomic data. Discussions with Dr. Janet Drew and Ms. Jenny Bunn have contributed greatly.

REFERENCES

Garden R. P. & Geballe, T. R. 1986, *M.N.R.A.S.*, **220**, 611.
Höflich, P & Wehrse, R 1987, *Astr. Ap.*, **185**, 107.
Lada, C. J. 1985, *Ann. Rev. Astr. Ap.*, **23**, 267.
Persson, S E, Geballe, T R, McGregor, P J, Edwards, S & Lonsdale, C J 1984, *Ap. J.*, **286**, 289.
Simon, M., Felli, M., Cassar, L., Fischer, J. & Massi, M. 1983, *Ap. J.*, **266**, 623.

SECTION 4

SUPERNOVAE FROM MASSIVE STARS

Massive Stars: Their Lives in the Interstellar Medium
ASP Conference Series, Vol. 35, 1993
Joseph P. Cassinelli and Edward B. Churchwell (eds.)

SUPERNOVA EXPANSION, SHOCKS, AND REMNANTS

ROGER A. CHEVALIER
University of Virginia, Department of Astronomy, P.O. Box 3818,
Charlottesville, VA 22903

ABSTRACT   Most massive stars are believed to end their lives in
supernova explosions. The properties of the supernova and its interaction
with the surrounding medium give information on the progenitor star and
its previous evolution. Supernova light curves give a measure of the radius
of the progenitor star. Interaction with a stellar wind from the progenitor
star has been observed and suggests mass loss rates from red supergiant
progenitors up to $2 \times 10^{-4} M_\odot \mathrm{yr}^{-1}$ for a wind velocity of 10 km s^{-1} .
Supernova remnants show evidence for interaction with circumstellar
shells or rings on a variety of size scales.

INTRODUCTION

From their positions in galaxies, their light curves, their spectra, and their
rates, SN II (Type II supernovae) are inferred to be the explosions of most
massive stars that have retained their hydrogen envelopes (Woosley and
Weaver 1986). The positions of SN Ib (Type Ib supernovae) in galaxies also
suggest a massive star origin, but hydrogen is absent in their spectra. Wolf-
Rayet stars or massive stars in binary systems have been suggested as possible
progenitors for SN Ib. Even if they do not end their lives as SN Ib, Wolf-Rayet
stars may explode and produce remnants that can be observed in our Galaxy
and nearby galaxies.

The massive star progenitors of SN II and SN Ib are known to lose mass
during their evolution. For O-type main sequence stars, strong winds are
clearly present while for the B stars, effects of a wind are usually unobservable
(Abbott 1982). An exception is the Be stars, which are observed to have
winds with velocities 600 to 1100 km s^{-1} and mass loss rates of 10^{-11} to 3 $\times$
$10^{-9} M_\odot \mathrm{yr}^{-1}$ (Snow 1981). The wind luminosity is much lower than that of
the O-type stars. The radii of the bubbles created by the winds in a uniform
medium can be many 10's of pc (Weaver *et al.* 1977).

The mass range for SN II ($\geq 8\ M_\odot$) covers both O4–B0 stars with large
bubbles and the early B stars which are likely to have smaller regions affected
by winds. The nature of the interstellar medium is likely to be important for
the surroundings of the early B stars. If SN Ib have very massive progenitors
that evolve to a Wolf-Rayet phase, they are expected to be in large bubbles
created by the main sequence wind and photoionizing radiation. If they result

from the binary evolution of lower mass stars, an extended cavity may not be present.

During the next evolutionary phase for a massive star, the red giant or supergiant phase, the star has a slow wind with velocity 5–50 km s^{-1} and a mass loss rate of $10^{-7} - 10^{-4} M_{\odot} \mathrm{yr}^{-1}$ (Zuckerman 1980). The total duration of the red giant phase is about 10% of the main sequence lifetime. The rate of mass loss may evolve during this phase; the highest rates of mass loss have been observed in OH/IR stars at the tip of the red giant branch with absolute magnitudes M < −6. Most Type II supernovae are expected to explode at this point, but extreme mass loss can complicate the evolution. Loops can occur in the HR diagram (Chiosi and Maeder 1986); while the star is relatively blue, a faster, lower density wind is expected. Loss of the hydrogen envelope leads to a blue Wolf-Rayet star; these stars have typical mass loss rates of $10^{-5} - 10^{-4} M_{\odot} \mathrm{yr}^{-1}$ and wind velocities of 1000–2000 km s^{-1} (Chiosi and Maeder 1986). Massive stars with their hydrogen envelopes but low metallicity may also explode as blue stars (*e.g.* SN 1987A).

The interaction of the fast wind from the blue star with the red supergiant wind creates a shocked, cool shell of the dense wind and a hot shell of shocked fast wind. Ring nebulae have been observed around Wolf-Rayet stars. These typically have radii of 3–10 pc, velocities of 30–100 km s^{-1}, and densities of 300–1000 cm^{-3} (Chu et al. 1983). Some of the nebulae show evidence for abundance enhancements of N and He (Kwitter 1984). Many of the wind blown nebulae are asymmetrically distributed about the Wolf-Rayet star (Chu *et al.* 1983). Interaction with the interstellar medium is needed if the asymmetry is due to stellar motion. Asymmetries in the winds may also be a factor.

Here, I discuss three areas where properties of supernovae and their remnants give information on the progenitor stars and their previous evolution. The areas are supernova light curves, interaction with the progenitor wind, and interaction with a circumstellar shell. The emphasis is on cases where detailed information is available and there are significant constraints on the nature of the progenitor star.

SUPERNOVAE

The explosion of a supernova heats the stellar layers that can then give rise to photospheric emission from the expanding gas. The internal, radiative energy drops because of adiabatic expansion (approximately as 1/radius), so that a star with a small initial radius ends up being a weaker luminosity source than one with a large initial radius. For parameters appropriate to SN II, the luminosity at optical maximum is roughly proportional to initial radius (Arnett 1980). The observed luminosities of most SN II require an initial radius $\geq 3 \times 10^{13}$ cm, i.e. a red supergiant star (Chevalier 1976a; Falk and Arnett 1977). Explosions of red supergiants are consistent with expectations from stellar evolution theory.

The peak luminosity of a supernova actually occurs when the shock front breaks out of the stellar surface and a hard ultraviolet/soft X-ray burst is emitted (Klein and Chevalier 1978; Falk 1978). The peak luminosity for a SN II is expected to be $\sim 10^{45}$ ergs s^{-1} and to last ~ 30 minutes. Such bursts have not yet been directly observed and would require a wide field satellite detector to have a good chance for discovery. If such bursts could be detected, they would give useful information on the initial stellar radius because the star would not yet have expanded. Although SN 1987A was more compact than a typical SN II, it did have a strong ultraviolet burst, as inferred by the photoionization of the surrounding gas (Fransson and Lundqvist 1989).

SN 1987A has especially complete observations and the light is believed to have been dominated by shock heated gas for the first few weeks (Arnett *et al.* 1989). The initial light from the supernova was unusually faint and the progenitor star was deduced to have a radius $\sim 3 \times 10^{12}$ cm, consistent with the explosion of Sk-69 202, which was at the position of the supernova. Further confirmation of the progenitor radius came from the lag time between the neutrino burst and the first light observed from the supernova; the time delay of $2-3$ hours implies a compact star. The time for the supernova shock front to traverse a red supergiant star is close to 1 day.

Because of the small initial radius of SN 1987A, after day 40 some other source of power was needed for the supernova light and ^{56}Co decays were found to be the source. Although early models for the light curve were able to reproduce some of the basic observed properties, more detailed fits required the inclusion of mixing of the supernova gas (Woosley 1988; Shigeyama and Nomoto 1990). The models have assumed smooth, spherically symmetric mixing and have not yet taken account of the complex multi-dimensional mixing that probably actually occurs (Fryxell, Müller, and Arnett 1991; Herant and Benz 1992). The detailed light curve of a supernova might be expected to give detailed information on the structure of the progenitor star. However, mixing processes that occur during the supernova explosion may limit at present the information that can be deduced from a light curve.

It has long been recognized that the explosions of massive stars that lost their hydrogen envelopes would be relatively faint because of their small initial radii (Chevalier 1976b). SN 1985F was initially identified as such an object because it was faint, appeared to be associated with massive stars, and appeared to be composed of the heavy elements that are believed to be synthesized in massive stars (Filippenko and Sargent 1985). It was later recognized that this was the late stage of a SN Ib, which are powered by ^{56}Co decays near maximum light. The low initial ultraviolet luminosity of the Type Ib SN 1983N suggests that its initial radius was $< 10^{12}$ cm (Panagia 1985). The nature of SN Ib progenitors is still unclear. Explosions of Wolf-Rayet stars tend to give too broad a light curve near maximum (Ensman and Woosley 1988), but there are uncertainties in the opacity of the gas. Lower mass stars which lose their hydrogen envelopes by binary evolution are another possibility, and the nature of the light curve may be affected by whether hydrodynamic instabilities occur during the explosion (Shigeyama *et al.* 1990). The mixing out of ^{56}Co may be important for the production of He lines that are observed in SN Ib (Lucy 1992). While SN Ib appear to be powered by radioactivity, other small radius explosions could take place with less production of ^{56}Ni, giving rise to especially faint supernovae.

Another area where supernova studies are relevant to massive stars is the composition of the expanding gas. Near maximum light, the problem is a complicated one in radiative transfer, especially because non-LTE effects can be important (*e.g.* Eastman and Kirshner 1989). At late times, the radiative transfer is simplified and the problem is one of solving for the emission from a gas heated by γ-rays (Fransson and Chevalier 1989; Kozma and Fransson 1992). The late times are of special interest to nucleosynthesis because it is possible to see in to the central regions of the star. Unfortunately, these studies are also affected by uncertain mixing and clumping of the gas and definitive results on abundances have not yet been possible, even for SN 1987A.

Extreme mass loss from the progenitor star can affect the the supernova light curve near optical maximum (Falk and Arnett 1977; Grasberg and Nadëzhin 1987). SN 1988Z may well be a case where circumstellar interaction powered the light curve soon after maximum light, if not at maximum itself (Stathakis and Sadler 1991). For most massive star explosions, circumstellar interaction has shown up in non-optical wavelength bands, or at very late times, as discussed in the next section.

STELLAR WIND INTERACTIONS

Most SN II are expected to initially interact with the slow, dense wind from the red supergiant phase and there is excellent evidence for interaction with a dense circumstellar wind for the Type II supernovae SN 1979C and SN 1980K (see reviews by Fransson 1986 and Chevalier 1990). The evidence includes radio emission from the interaction region for both supernovae, infrared dust echoes for both supernovae, thermal X-ray emission from the interaction region in SN 1980K, and the ultraviolet line emission from highly ionized atoms in SN 1979C. The radio emission is a particularly good diagnostic because the early absorption of the radio emission can be interpreted as free-free absorption by the preshock gas and an estimate of the circumstellar density is obtained. Lundqvist and Fransson (1988) have made a detailed study of the temperature and ionization of the circumstellar gas in the radiation field of the supernova and have been able to reproduce detailed features of the radio light curves. They derive mass loss rates of $12 \times 10^{-5} M_\odot \text{yr}^{-1}$ and $3 \times 10^{-5} M_\odot \text{yr}^{-1}$ for a wind velocity $v_w = 10 \text{ kms}^{-1}$ for SN 1979C and SN 1980K respectively. The optical emission from SN 1980K appears to be dominated by circumstellar interaction at an age of 8 years (Fesen and Becker 1990).

TABLE I   Radio Supernovae

SN	Type	Ratio to Cas A at 6 cm	t_{20} (days)	$\dot{M}/v_w$ $(M_\odot/\text{yr})/(\text{km/s})$
1979C	II	~ 250	950	1×10^{-5}
1980K	II	~ 15	190	3×10^{-6}
1983N	Ib	~ 125	30	5×10^{-7}
1986J	II	~ 700	1600	2×10^{-5}
1987A	II	~ 0.02	2	1×10^{-8}

There are presently 5 radio supernovae with fairly extensive data, including the rising part of the radio light curve. They are SN 1979C, SN 1980K (Weiler *et al.* 1986), SN 1983N (Sramek, Panagia, and Weiler 1984), SN 1986J (Rupen *et al.* 1987; Weiler, Panagia, and Sramek 1990), and SN 1987A (Turtle *et al.* 1987). Table I lists the supernova type, the radio luminosity ratio to Cas A, the time of optical depth unity at 20 cm, t_{20}, and the circumstellar density given in terms of the presupernova mass loss rate divided by the wind velocity. SN1986J was probably not observed near maximum light and the Type II designation given here is based solely on the presence of hydrogen line emission. Because of their low density circumstellar environment, SN 1983N and SN 1987A did not show clear evidence for circumstellar interaction outside of radio wavelengths.

The results show that the winds around SN 1979C, 1980K, and 1986J are consistent with the dense slow winds expected around red supergiant stars. Confirmation of this picture for SN 1986J has come from the discovery of X-ray emission from the source at a level of a few 10^{40}ergs s^{-1} (Bregman and Pildis 1992), close to the prediction of the circumstellar interaction model. The density around the Type Ib event SN 1983N is considerably lower than that inferred around the above three Type II events, but is higher than the value expected for a typical Wolf-Rayet star. A wind velocity of 1000 km s^{-1} and $\dot{M} = 10^{-4}M_\odot$yr^{-1} leads to a value of $\dot{M}/v_w$ that is a factor of 5 below the estimated value. SN 1987A had an even earlier turn-on and was a faint radio supernova, but the estimated value of $\dot{M}/v_w$ is roughly consistent with the density expected around a B3 I star like the Sk-69 202 progenitor star (Chevalier and Fransson 1987). The observational estimate is a factor of a few larger than the expected value. If there is clumping in the circumstellar wind, the observational estimates for the mass loss rates are reduced.

In the circumstellar interaction model for the radio emission, the radio luminosity at a given age should be correlated with the density of circumstellar material. This is observed. It appears that the circumstellar interaction does give information on the properties of the supernova progenitor.

SN1979C has now been observed at radio wavelengths for more than a decade since the explosion (Weiler *et al.* 1991), implying more than $1M_\odot$ of mass loss. The interaction with the very dense wind is expected to last no longer than tens of years. During the next phase of evolution the supernova may approach free expansion in a lower density wind. When SN 1979C enters this phase, an accelerated rate of decline of the radio emission is expected. The supernova SN 1957D has been detected as both a radio (Cowan and Branch 1985) and optical (Long, Blair, and Krzeminski 1989) source. The broad optical lines are probably due to heating of the supernova gas by energetic photons from the circumstellar interaction region. The optical flux has recently been found to decline by a factor of 5 over a few years (Long, Winkler, and Blair 1992); this is plausibly an object where the shock front has run through the dense wind.

CIRCUMSTELLAR SHELL INTERACTIONS

After the supernova shock front has traversed the stellar wind from the immediate progenitor star, the supernova remnant can be expected to

brighten as it interacts with circumstellar shells, formed as discussed in the
Introduction. Table II lists the properties of some supernova remnants that
appear to be interacting with circumstellar shells.

TABLE II   Supernova Remnants with Shell Interaction

Remnant	Appearance	Shell radius (pc)	Nitrogen overabundant
Cas A	shell	1.7	yes
N132D	ring	3	?
Kepler	bow shock	3	yes
Cygnus Loop	shell	20	no
SN 1987A	ring	0.25	yes

Of the young supernova remnants, Cassiopeia A is the most likely to be
related to Wolf-Rayet stars and possibly to SN Ib. It has fast moving oxygen-
rich gas and is interacting with dense nitrogen-rich circumstellar gas; Fesen,
Becker, and Blair (1987) have suggested that the progenitor was a Wolf-Rayet
WN star. The problems with the Type Ib identification are that Cas A has
recently been found to have fast-moving hydrogen-rich gas (Fesen *et al.* 1987,
Fesen and Becker 1991) and the supernova was probably too faint to be a
typical Type Ib event, even if it was observed by Flamsteed in 1680 (Ashworth
1980). Although the presence of hydrogen would appear to rule out a Type
I supernova, it is perhaps possible that the hydrogen would not have been
detectable spectroscopically near maximum light. If the progenitor was a Wolf-
Rayet star, it is likely that the slowly moving, nitrogen-rich gas is in a shell
that was initially surrounding the supernova.

   Another remnant with oxygen-rich ejecta which is likely to be interacting
with circumstellar gas is N132D in the Large Magellanic Cloud. Optical studies
of this remnant show a faint outer shell of radius 40 pc, an inner disk of radius
16 pc, and a ring of expanding oxygen-rich ejecta at a radius of 3 pc (Lasker
1978, 1980). X-ray emission from the remnant has approximately a 16 pc
radius (Mathewson et al. 1983). The inner oxygen knots give an age of only
1300 years, which implies that the X-ray emitting gas has been expanding in a
low density medium if it is a normal supernova (Hughes 1987). Hughes (1987)
suggests that the cavity was created by an HII region. A fast stellar wind may
also play a role. The present X-ray remnant is then the result of interaction
with the swept up red supergiant wind. The faint outer shell may be a remnant
of the stellar wind and photoionizing radiation while the progenitor star was on
the main sequence.

   An interesting recent suggestion is that Kepler's supernova remnant is
the result of a Type Ib supernova (Bandiera 1987). An analysis of the X-ray
emission indicates that the initial stellar mass was $> 7M_\odot$ (Hughes and Helfand
1985), although this is rather uncertain. Bandiera (1987) has argued that the
progenitor star was a runaway Wolf-Rayet star from the galactic plane. Proper
motion studies of the dense optical knots do imply a high space velocity (van

den Bergh and Kamper 1977) and the asymmetry of the supernova remnant may be due to the interaction of presupernova mass loss with an ambient medium (Bandiera 1987).

There is evidence that some larger supernova remnants are interacting with circumstellar gas. The Cygnus Loop, with a radius of 20 pc, is interacting with gas density ≥ 5 cm^{-3} over much of its surface area, yet the appearance of the remnant shows circular symmetry. The X-ray properties imply that the remnant has been expanding in a low density medium until recently (Charles, Kahn, and McKee 1985). The implication is that the remnant is interacting with a spherical shell of radius 20 pc. Since this radius is smaller than the bubbles expected around O stars, Charles, Kahn, and McKee (1985) suggest that the progenitor star was an early B star.

Although circumstellar interaction around SN1987A did not give large effects during its initial evolution, there is evidence for dense gas close to the supernova (Fransson et al. 1989). Chevalier and Liang (1989) and Luo and McCray (1991) estimated that shock interaction with the dense gas should occur at an age of 15–20 years and will lead to a brightening of the supernova at radio through X-ray wavelengths. Broad optical and ultraviolet emission lines are expected. A rising radio flux has been observed beginning in July 1990 (Staveley-Smith et al. 1992); this could be due to supernova shock interaction with the termination shock of the blue supergiant wind (Chevalier 1992). Imaging observations of the dense line emitting gas with the ESO New Technology Telescope (Wampler et al. 1990) and with the Hubble Space Telescope (Jakobsen et al. 1991) show that the gas is in a ring around the supernova. The interaction of the supernova with the ring may give rise to the appearance of ejecta in a ring, as is observed in N132D.

The young supernova remnants contain many clues regarding the nature of the massive star that exploded, including the stellar mass, composition, and properties of the surroundings. However, the complexity of the interaction makes it difficult to obtain definitive results. First, a detailed hydrodynamic model is needed for the remnant in question (e.g. Borkowski, Blondin, and Sarazin 1992 for Kepler's remnant). Then the emission properties of the model must be compared with the observations. Line emission at optical through X-ray wavelength provides especially powerful diagnostics; non-equilibrium effects may well be important. Our ability to model young remnants may again be limited by incertain factors such as mixing and clumping. SN 1987A will provide a particularly good test of supernova remnant theory, because we have detailed information on both the supernova and its surroundings before the interaction takes place.

ACKNOWLEDGEMENTS

This research was supported in part by NASA grant NAGW-2376.

REFERENCES

Abbott, D. C. 1982, Ap. J., 263, 723.
Arnett, W. D. 1980, Ap. J., 237, 541.

Arnett, W. D., Bahcall, J. N., Kirshner, R. P., and Woosley, S. E. 1989, Ann. Rev. Astr. Ap., 27, 629.

Ashworth, W. B. 1980, J. Hist. Astr., 11, 1.

Bandiera, R. 1987, Ap. J., 319, 885.

Borkowski, K. J., Blondin, J. M., and Sarazin, C. L. 1992, Ap. J., in press.

Bregman, J. N. and Pildis, R. A. 1992, Ap. J. (Letters), submitted.

Charles, P. A., Kahn, S. M., and McKee, C. F. 1985, Ap. J., 295, 456.

Chevalier, R. A. 1976a, Ap. J., 207, 872.

Chevalier, R. A. 1976b, Ap. J., 208, 826.

Chevalier, R. A. 1990, in Supernovae, ed. A.G. Petschek (New York: Springer), p. 91.

Chevalier, R. A. 1992, Nature, 355, 617.

Chevalier, R. A. and Fransson, C. 1987, Nature, 328, 44.

Chevalier, R. A. and Liang, E. P. 1989, Ap. J., 344, 332.

Chiosi, C. and Maeder, A. 1986, Ann. Revs. Astr. Ap., 24, 329.

Chu, Y.-H., Treffers, R. R., and Kwitter, K. B. 1983, Ap. J. Suppl., 53, 937.

Cowan, J. J. and Branch, D. 1985, Ap. J., 293, 400.

Eastman, R. G. and Kirshner, R. P. 1989, Ap. J., 347, 771.

Ensman, L. M. and Woosley, S. E. 1988, Ap. J., 333, 754.

Falk, S. W. 1978, Ap. J. (Letters), 226, L133.

Falk, S. W. and Arnett, W. D. 1977, Ap. J. Suppl., 33, 515.

Fesen, R. A. and Becker, R. H. 1990, Ap. J., 351, 437.

Fesen, R. A. and Becker, R. H. 1991, Ap. J., 371, 621.

Fesen, R. A., Becker, R. H., and Blair, W. P. 1987, Ap. J., 313, 378.

Filippenko, A. V. and Sargent, W. L. W. 1985, Nature, 316, 407.

Fransson, C. 1986, in Radiation Hydrodynamics in Stars and Compact Objects, eds. D. Mihalas and K. H. A. Winkler (Berlin: Springer), p. 141.

Fransson, C. and Chevalier, R. A. 1989, Ap. J., 343, 323.

Fransson, C. and Lundqvist, P. 1989, Ap. J. (Letters), 341, L59.

Fransson, C., Cassatella, A., Gilmozzi, R., Kirshner, R. P., Panagia, N., Sonneborn, G., and Wamsteker, W. 1989, Ap. J., 336, 429.

Fryxell, B. A., Müller, E., and Arnett, W. D. 1991, Ap. J., 367, 619.

Grasberg, E. K. and Nadëzhin, D. K. 1987, Sov. Astr. AJ, 31, 629.

Herant, M. and Benz, W. 1992, Ap. J., 387, 294.

Hughes, J. P. 1987, Ap. J., 314, 103.

Hughes, J. P. and Helfand, D. J. 1985, Ap. J., 219, 544.

Jakobsen, P. et al. 1991, Ap. J. (Letters), 369, L63.

Klein, R. I. and Chevalier, R. A. 1978, Ap. J. (Letters), 223, L109.

Kozma, C. and Fransson, C. 1992, Ap. J., 390, 602.

Kwitter, K. B. 1984, Ap. J., 287, 840.

Lasker, B. M. 1978, Ap. J., 223, 109.

Lasker, B. M. 1980, Ap. J., 237, 765.

Long, K. S., Blair, W. P., and Krzeminski, W. 1989, Ap. J. (Letters), 340, L25.

Long, K. S., Winkler, P. F., and Blair, W. P. 1992, Ap. J., in press.

Lucy, L. B. 1991, Ap. J., 383, 308.

Lundqvist, P. and Fransson, C. 1988, Astr. Ap., 192, 221.

Luo, D. and McCray, R. 1991, Ap. J., 379, 659.

Mathewson, D. S., Ford, V. L., Dopita, M. A., Tuohy, I. R., Long, K. S., and Helfand, D. J. 1983, Ap. J. Suppl., 51, 345.

Panagia, N. 1985, in <u>Supernovae as Distance Indicators</u>, ed. N. Bartel (Berlin: Springer), p. 14.

Rupen, M. P., van Gorkom, J. H., Knapp, G. R., Gunn, J. E., and Schneider, D. P. 1987, A. J., <u>94</u>, 61.

Shigeyama, T. and Nomoto, K. 1990, Ap. J., <u>360</u>, 242.

Shigeyama, T., Nomoto, K., Tsujimoto, T., and Hashimoto, M. 1990, Ap. J. (Letters), <u>361</u>, L23.

Snow, T. P. 1981, Ap. J., <u>251</u>, 139.

Sramek, R. A., Panagia, N., and Weiler, K. W. 1984, Ap. J. (Letters), <u>285</u>, L59.

Stathakis, R. A. and Sadler, E. M. 1991, M.N.R.A.S., <u>250</u>, 786.

Staveley-Smith, L. *et al.* 1992, Nature, <u>355</u>, 147.

Turtle, A. J. *et al.* 1987, Nature, <u>327</u>, 38.

van den Bergh, S. and Kamper, K. W. 1977, Ap. J., <u>218</u>, 617.

Wampler, E. J., Wang, L., Baade, D., Banse, K., D'Odorico, S., Gouiffes, C., and Tarenghi, M. 1990, Ap. J. (Letters), <u>362</u>, L13.

Weaver, R., McCray, R., Castor, J., Shapiro, P. R. and Moore, R. T. 1977, Ap. J., <u>218</u>, 377.

Weiler, K. W., Panagia, N., and Sramek, R. A. 1990, Ap. J., <u>364</u>, 611.

Weiler, K. W., Sramek, R. A., Panagia, N., van der Hulst, J. M., and Salvati, M. 1986, Ap. J., <u>301</u>, 790.

Weiler, K. W., Van Dyke, S. D., Panagia, N., Sramek, R. A., and Discenna, J. L. 1991, Ap. J., <u>380</u>, 161.

Woosley, S. E. 1988, Ap. J., <u>330</u>, 218.

Woosley, S. E. and Weaver, T. A. 1986, Ann. Rev. Astr. Ap., <u>24</u>, 205.

Zuckerman, B. 1980, Ann. Rev. Astr. Ap., <u>18</u>, 263.

Massive Stars: Their Lives in the Interstellar Medium
ASP Conference Series, Vol. 35, 1993
Joseph P. Cassinelli and Edward B. Churchwell (eds.)

LATE STAGES IN SUPERNOVA REMNANT EVOLUTION

DONALD P. COX
Department of Physics, University of Wisconsin-Madison, Madison,
Wisconsin, U.S.A.

ABSTRACT   In the late stages of SNR evolution ($t \geq 10^6$ years),
the bubble–and–shell configuration is transformed by dispersal of the
shell (which reexpands to ambient density). The relic bubble lives on,
slowly cooling and shrinking in rough pressure equilibrium with the
surroundings. The SNR population in the solar neighborhood creates
only a modest porosity ($q \sim 0.1$) and need not disrupt a warm intercloud
component ($n \sim 0.2$ cm^{-3}) or generate a pervasive hot interstellar phase.
A large fraction of the high stage ion content in the Galactic plane
probably resides in the collection of cooling SNR bubbles, with a similar
fraction possible in superbubbles.

EARLY COMPLICATIONS

A typical discussion of the zeroth order model of supernova remnant (SNR)
evolution begins with an explosion in a uniform medium. Thereafter,
several phases of the evolution are identified: free expansion of the ejecta,
thermalization of the energy via the outward and reverse shocks, approach
to the Sedov structure for adiabatic evolution of a point explosion, further
nearly adiabatic evolution, sudden onset of cooling at the outer boundary, and
development of a bubble–and–shell configuration. In the latter, a slowly cooling
bubble of hot gas is surrounded by a dense cool shell and, at the outer edge, a
radiative shock. Subsequent evolution has been described as a pressure driven
snowplow in which the hot bubble pushing on the interior of the thin shell
gradually raises the radial momentum of the latter. Of course, reality has not
been kind enough to make this simple picture of SNR evolution very useful.

Massive stars do a great deal of prestructuring of their circumstellar
environment prior to explosion. Isolated stars have a sequence of outflow epochs,
winds varying between fast low density and slow high density, winds whose
characteristics vary rapidly with initial stellar mass. These winds, together
with the pressure of the H II regions create a complex density structure
consisting of relic wind zones, interaction shells, and wind blown cavities in
the interstellar medium. In addition, such stars find themselves in a variety of
ambient environments, within dense molecular clouds, in the diffuse intercloud
environment, and within the collective bubbles generated by a cluster of such
stars acting in concert. (See papers by John Castor, Roger Chevalier, and
Nino Panagia, among others, in this volume for further detail.) One supposes,
however, that most supernovae deriving from upper main sequence stars begin
their lives as relatively docile B stars in which these complications are not too
extreme.

Even supernovae arising from low mass configurations (Type Ia) are not without complications. They may once have produced a planetary nebula shell, and may many times have produced nova outbursts.

Although this paper is not about early complications in the life of an SNR, we have to confront the possibility that the later evolution depends somewhat on what happened early. More precisely, I shall be assuming that for isolated remnants in diffuse interstellar environments these complications no longer leave an appreciable signature by the time a remnant reaches the bubble and shell configuration. I do not know this to be true, but it is the current strategy in exploring late time evolution.

QUESTIONS AND FURTHER COMPLICATIONS

The major questions one hopes to answer about the late evolution of a supernova remnant appear to be:

(1) How sensitive is the bubble and shell configuration of the post cooling epoch to the early time complications discussed above, and to inhomogeneities in the ejecta or interstellar medium?

(2) How big does a remnant get?

(3) How long does it last?

(4) What makes it go away?

(5) What does it look like at late times?

(6) How does the population of such remnants affect the structure and state of the diffuse interstellar medium (DISM)?

There are, however, even more complications which must be addressed before reliable answers to these questions are available. One of those complications is that until the answer to question (6) above is available, we cannot know for certain how to approach studying the evolution of an individual remnant. We flat out do not know the most probable density or temperature of the DISM, to within an order of magnitude. A second complication is that we do not have a very complete understanding of hot gas dissipation in diffuse environments. Does it lose energy mainly by radiative processes, by collisional losses to dust, by thermal conduction, by interphase mixing and the formation of turbulent boundary layers, or by processes not yet named? We play with 1D models, using a hydrodynamic approximation when particle mean free paths are larger than the system, invoking poorly understood magnetic effects to enforce locality, and then argue about whether those effects will efficiently suppress thermal conduction. A third complication is our poor understanding of the interaction of old remnants with the nonthermal components of interstellar pressure. Are they largely transparent to cosmic rays or are localization and/or acceleration important? How does a remnant interact with the real interstellar magnetic field? Finally, one needs to know more about the three dimensional structure of the interstellar cloud configuration to begin to understand how these gross density perturbations might affect remnant evolution. Is it safe to assume that remnants evolve around and between clouds, as suggested by McKee and Ostriker, 1977? Is it more likely that remnants will evaporate or ablate the clouds, incorporating at least some of their mass into the hot interior, also as suggested by McKee and Ostriker, 1977? Or are clouds large scale walls of material which constrain remnants from evolving in some directions?

These and other complications may prevent the formulation of reliable answers to the questions posed, but they have not altogether stopped theoretical progress on some elementary sample evolutions. And recently, such investigations have brought some surprises.

COMPLETING THE EVOLUTION OF THE THEORISTS' SNR

Parameter Choices
For part of his Ph. D. thesis, Jon Slavin (1990) constructed a 1D hydrocode to explore the late time evolution of an SNR in a homogeneous medium. The ionic structure and gas cooling coefficient were followed using the Raymond, Cox and Smith (1976) and Edgar (see Gaetz, Edgar and Chevalier, 1988) rate codes.

The principal goal of the project was to complete the evolution outlined in the opening paragraphs of McKee and Ostriker (1977), to see whether those authors were correct in their assertion that supernovae in the DISM would certainly disrupt it into a hot frothy state with all cool material in the cloud configuration. As a result, an ambient density of 0.2 cm^{-3} was chosen, with a temperature of 10^4 K. These parameters derive from assuming that the warm neutral and warm ionized interstellar gas have equal densities and are uniformly distributed (apart from some being ionized) in intercloud space.

The only significant difference in assumption from McKee and Ostriker, was that we felt that the nonthermal pressures could not safely be ignored. The model neglected cosmic rays and magnetic tension, but included a magnetic pressure term proportional to density squared. The nominal value was 5 μG at $n = 0.2$ cm^{-3}. Many workers in ISM theory seem to feel that this value of B is too large, but there is by now a significant body of observation to the contrary. (See, for example, Boulares and Cox, 1990, for a discussion.) In any case, the qualitative features discussed below do not depend on the magnitude of B.

Shell Dissipation
One of Slavin's most interesting results (Slavin and Cox, 1992) is that the bubble/shell configuration is not a good description of a remnant at late times. Just after the cooling epoch, there is a dense shell, with a pressure roughly 100 times ambient. But even then, the magnetic pressure limits the compression factor to less than 10 in the radiative shock. Later, as the ram pressure decreases, the shell reexpands, returning the stored magnetic energy to the flow. At very late times when the remnant is approaching pressure equilibrium with its surroundings, the "shell" is extremely thick (the outer edge is a weak magnetosonic wave, the inner is the boundary of the uncooled bubble), and is essentially indistinguishable from the ambient medium. Our inference is that the shell of such a remnant would dissipate and disappear, leaving an essentially naked bubble.

Naked Bubble Evolution
The hot bubble of an SNR forms as a specific entity when cooling of the remnant edge differentiates it from the shell. When first formed the bubble contains roughly 200 solar masses. It is much reduced in density from the ambient medium, and that contrast grows as the bubble expands further.

During the expansion (and later contraction), the bubbles's mass steadily drops as the denser outer parts cool and accrete on to the outer wall, or shell. Expansion continues until the interior pressure is roughly 1/3 of ambient (pressure undershoot) after which contraction begins. The bubble returns to the ambient pressure. Thereafter it contracts only because it radiates, remaining at nearly constant pressure. When it finishes radiating, its volume has reduced to zero and the remnant consists of nothing but a few ripples in a largely ambient-like medium.

The quantitative aspects of this evolution are less certain than the qualitative ones given above. The details depend on how rapidly the bubble is able to cool, which in turn depends on the density and temperature structure within the inner 100 $M_\odot$ of the remnant. If that material had been shocked during a Sedov-like phase, and avoided thermal conduction and all other entropy-mixing processes, then the inner parts would be extremely hot and rarefied and take nearly forever to cool (e.g. see Cui and Cox, 1992). If, on the other hand, thermal conduction proceeded at an unquenched (but possibly saturated) rate, there would be a density and temperature plateau in the remnant center which would end the bubble life with finality as that isentropic core cooled. In what follows, I shall assume the latter is more appropriate.

ANSWERS WITHOUT COMPLICATIONS

Slavin's standard case with $n_o = 0.2$ cm^{-3}, $E_o = 5 \times 10^{50}$ ergs, $B_o = 5$ μG generated a bubble which reached a maximum radius of 59 pc at an age of 1.2×10^6 years. It finished cooling and disappeared at an age of 5.8×10^6 years and had a four-volume integral $\int V_B(t)dt = 2.2 \times 10^{12}$ pc^3 yr. With a supernova rate per unit volume of $S = 0.4 \times 10^{-13}$ pc^{-3} yr^{-1} outside of associations (e.g. Heiles, 1987), the implied porosity factor for the solar neighborhood would be
$$q = S \int V_B dt = 0.09$$
suggesting that remnant bubbles might well occupy only about 10% of interstellar space. This is in strong contrast to the result of McKee and Ostriker, 1977, who felt certain that q exceeded 3.

One cannot be certain from these results that supernovae would not disrupt a warm intercloud medium into a hot froth; but one can be certain that it is not certain that they would. The many caveats of this investigation and details of the standard case are discussed in Slavin and Cox, 1992. The dependencies of these results on n_o, E_o, and B_o are discussed in a second paper (Slavin and Cox, 1993).

A second major result of Slavin's modeling is that the bubbles might well make a major contribution to the high stage ion content of the ISM. As an example, the O VI dosage of the standard case is $D(O\ VI) = \int N(OVI)\,dt = \int \int n(OVI)\,dV\,dt = 1.3 \times 10^{61}$ ion yr, implying that remnants should contribute $\bar{n}(O\ VI) = S\ D(O\ VI) = 1.8 \times 10^{-8}$ ion cm^{-3} to the mean O VI density within the SNR-active interstellar disk in the solar neighborhood. Jenkins (1978) inferred the observational average OVI density to be 2.8×10^{-8} cm^{-3}, while a reanalysis taking into account significant contributions from the Local Bubble finds 1.4 to 2×10^{-8} cm^{-3} (Shelton and Cox, 1992). The details depend on details, but it is certainly true that the predicted contribution of the SNR bubbles is essentially identical to that actually found. The same is true of N V and C IV.

There is one major difference between Jenkins' O VI analysis results and the SNR model, the characteristic column density per bubble (and therefore the mfp for encounters). Jenkins found there to be many small bubbles with typical column density 1×10^{13} cm^{-2} and mfp 160 pc. Slavin's standard case predicts a characteristic column density of 4×10^{13} cm^{-2} with a mean free path of 750 pc. This was in fact foreshadowed by Jenkins' further result that there was an uncommon population of larger column density features that did not fit into his analysis. In the reanalysis of Shelton and Cox, the low O VI column densities are absorbed into the Local Bubble contribution, after which the larger features fit nicely into the analysis and provide results consistent with Slavin's.

EVEN MORE COMPLICATIONS

Within our group we have a working hypothesis that hot gas in the ISM derives entirely from regions affected by discrete events. We have shown that the high stage ion content in the midplane at the solar meighborhood could derive entirely from SNR bubbles. In the process, however, we have attributed an O VI column density of roughly 2×10^{13} cm^{-2} to the Local Bubble, an object which does not correspond to one of our single SNR bubbles. There are also indications of specific contributions from the Orion–Eridanus superbubble and the Gum Nebula. One needs accurate modeling of large or collective OB association bubbles, and a census to learn what overall fraction of the high ions derives from those.

In addition, the other high stage ions are measured to extend far out of the plane of the Galaxy (e.g. Savage and Massa, 1987) and it is not clear whether SNR bubbles will be sufficiently common at high z to account for that.

A theoretical complication, soon to be explored, is 2D MHD modeling of SNR bubbles so that magnetic effects will be better described. Such models may or may not improve our understanding, however, because they may be too smooth, ignoring shock induced turbulence and amplification of the interstellar wave field. They will certainly suggest that SNR bubbles collapse preferentially across the field into prolate regions, due to the anisotropy of the magnetic force. What needs to be clearly understood is that, although the magnetic field is certainly too important to ignore, it is also probably too irregular to treat naively.

A SURVEY OF TENTATIVE TRUTHS

There may not be a hot phase in the interstellar medium. The soft x–ray background may derive almost entirely from the single bubble of hot gas surrounding the Sun, the Local Bubble, (e.g. Cox and Reynolds, 1987) possibly last reheated by the explosion of Geminga's precursor (Bertsch, et al. 1992, Halpern and Holt, 1992). The high ion content of the Galaxy may derive almost exclusively from the bounded bubbles of OB associations and SN remnants.

The ISM is a very much thicker (4 to 6 kpc) and higher pressure ($p_o/k \sim$ 25,000 cm^{-3} K) environment than usually imagined (Boulares and Cox, 1990), and greatly interferes with the expansion of SNR and OB association bubbles. As a consequence, concepts such as breakout, blowout, and fountains may have no usefulness in the solar neighborhood. The response of this medium to spiral density waves may be sufficient to provide the abundance of high velocity material far off the galactic plane.

It will probably be a long time before we know for sure. On the other hand, EUVE is up and working and may see the EUV emission predicted in Slavin and Cox, 1992 for one or more bubbles. A spectrum of the soft x-ray background would greatly help in constraining models (is it even really thermal emission), particularly clarifying the dust content of the hot gas. One should be available soon from DXS and or the Wisconsin calorimeter rocket payload. More careful analysis of the high ion data looking for a radial gradient in the Galaxy will help, as will more complete Hα mappings of that diffuse background and a more complete comparison with H I data. The development of a Spatial Heterodyne Interferometer for operation in the UV also promises the possibilty of spectral imaging of bubble boundaries in the foreseeable future.

ACKNOWLEDGEMENTS

I would like to thank Jon Slavin for not listening to me when I told him not to get involved in the morass of hydrocode building, and Robin Shelton for a helpful reading of the manuscript. I have been supported by the National Aeronautics and Space Administration under grants NAGW–2532 and NAG 5-629 during the preparation of this manuscript.

REFERENCES

Bertsch, D. L. et al. 1992, Nature, 357, 306.
Boulares, A., and Cox, D. P. 1990, ApJ, 365, 544.
Cox, D. P., and Reynolds, R. J. 1987, ARAA, 25, 303.
Cui, W., and Cox, D. P. 1992, ApJ, 401, (Dec. 10), in press.
Gaetz, T. J., Edgar, R. J., and Chevalier, R. A. 1988, ApJ, 329, 927.
Halpern, J. P. and Holt, S. S. 1992, Nature, 357, 222.
Heiles, C. 1987, ApJ, 315, 555.
Jenkins, E. B. 1978, ApJ, 220, 107.
McKee, C. F., and Ostriker, J. P. 1977, ApJ, 218, 148.
Raymond, J., Cox, D. P., and Smith, B. W. 1976, ApJ, 224, 94.
Savage, B. D, and Massa, D. 1987, ApJ, 314, 380.
Shelton, R. L., and Cox, D. P. 1992 in proceedings of EIPC Conference on Star
    Forming Galaxies and their Interstellar Media, ed. J. Franco, in press.
Slavin, J. D. 1990, PhD thesis, University of Wisconsin–Madison.
Slavin, J. D., and Cox, D. P. 1992, ApJ, 392, 131.
Slavin, J. D., and Cox, D. P. 1993, ApJ, (submitted).

Massive Stars: Their Lives in the Interstellar Medium
ASP Conference Series, Vol. 35, 1993
Joseph P. Cassinelli and Edward B. Churchwell (eds.)

RECENT DEVELOPMENTS IN SN1987A

J. H. ELIAS, M. M. PHILLIPS, N. B. SUNTZEFF, A. R. WALKER, and
B. GREGORY
Cerro Tololo Inter-American Observatory, National Optical Astronomy
Observatories, Casilla 603, La Serena, Chile

D. L. DEPOY
Department of Astronomy, Ohio State University, 174 W. 18th St.,
Columbus OH 43210

ABSTRACT   Recent observations of SN1987A are reviewed.   The
principal results presented here are: (1) Continuing observations of
the bolometric light curve, although increasingly difficult, continue to
suggest the presence of significantly higher-than-solar abundances of ^{57}Co
relative to ^{56}Co in the expanding remnant. (2) Visible and near-infrared
observations of the two companion stars show that Star 2 has spectral type
B2III and Star 3 is probably a Be star and is variable. (3) Imaging of
the nebulosity surrounding the supernova, in the HeI 1.083 μm line shows
evolving structure, which can be interpreted as emission in response to
the initial uv light peak of the supernova.

INTRODUCTION

Supernova 1987A was the brightest and closest supernova to have gone off in
the last three centuries. It has been possible to observe its development for far
longer than for other recent supernovae, and to do so with much higher spatial
resolution. More than 5 years after it exploded, SN 1987A continues to show
interesting developments. We discuss here three main areas: the evolution of
the bolometric luminosity, the properties of the companion stars, and additional
"light echo" phenomena still being observed. The latter area in particular is
also covered by other papers presented at this Symposium (Panagia 1992, Wang
1992, Cummings 1992).

BOLOMETRIC LUMINOSITY

Measurements
Measurements of the bolometric luminosity of the supernova are essential to
an understanding of the initial explosion and its products. However, these
measurements are becoming increasingly difficult as the supernova continues to
fade, for a variety of reasons.

At shorter wavelengths, it is necessary to separate the emission from the supernova itself from that from the surrounding nebulosity, as well as nearby companion stars, two of which are already considerably brighter than the supernova itself in the visible and near-infrared. Imaging with reasonably good spatial resolution is sufficient to do these observations. In fact, sensitivity, both in the near-infrared and the visible, is such that continued measurements are likely to be possible for several years in the future. Some of the light curves obtained to date are shown in Figures 1 and 2.

Most of the observed luminosity from the supernova is not emitted at these wavelengths, but rather at wavelengths beyond 5 μm (see for example Suntzeff and Bouchet 1990). This leads to several problems. First, the longest wavelength at which measurements are currently being made is 20 μm; which corresponds roughly to the peak in the enrgy distribution. The luminosity beyond this wavelength must be estimated, usually by fitting a blackbody to the 10 and 20 μm data (see Suntzeff and Bouchet 1990, Suntzeff et al 1991,1992). Second, measurements at 10 and 20 μm are by now extremely difficult, and in fact the most recent CTIO observations are no more than upper limits. Finally, there also appears to be a faint 10 μm source (but not detected at 20 μm) near the supernova, which appears to have contaminated the reference positions used for the CTIO observations. A correction for this source has been applied where it affects the data (those taken after day 1000). These results are shown in Figure 2. One should note that the CTIO data are now in better agreement with the ESO data taken after day 1000 because of this correction (see Suntzeff et al 1991).

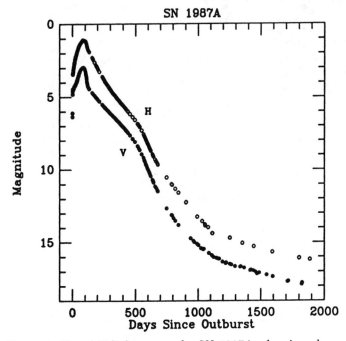

Figure 1: V and H light curves for SN 1987A, showing observations up to roughly 1850 days after the initial explosion.

Despite the problems with the long wavelength data, the observations continue to provide useful constraints on the bolometric luminosity evolution. This is because the upper limits set by the long wavelength data provide an upper limit to the total luminosity, while the observations at shorter wavelengths determine a lower limit to the luminosity.

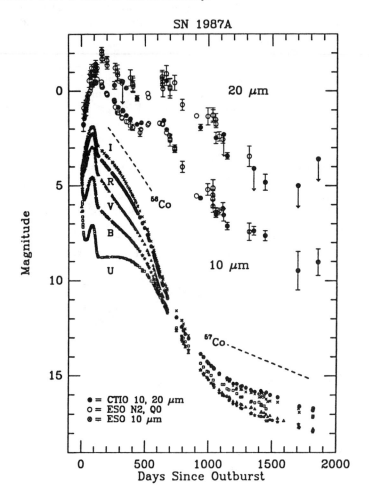

Figure 2: UBVRI and 10 and 20 μm light curves for SN 1987A, for the same period as Figure 2. For reference, dashed lines corresponding to the decay rates of ^{56}Co and ^{57}Co are also shown.

Model Comparisons

A comparison with model predictions is shown in Figure 3 (see Suntzeff et al 1991, 1992 for more details). The comparison shows a good fit to a model where is the ^{57}Co abundance, relative to ^{56}Co, is five time solar, as was found by Suntzeff et al. As discussed therein, this abundance is higher than that inferred from other observations. [One suggestion made at this conference (R.

Chevalier) is that the slow decay may in fact reflect longer recombination times
in the expanding remnant more than [57]Co decay.]

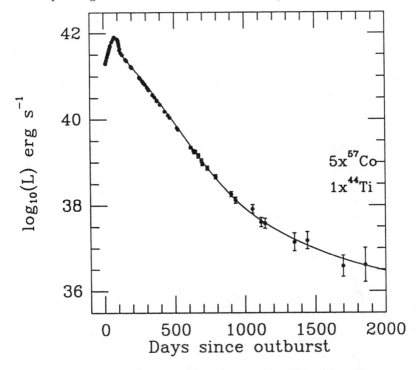

Figure 3: Model fit to observations. Details given in Suntzeff et al (1992).

COMPANION STARS

There are several stars in the vicinity of SN 1987A. These are of interest because
they may be physically associated, in which case information on their properties
may aid in the understanding of the supernova progenitor. Even if there is not
a physical association, they are still of interest because they provide probes of
lines of sight close to the supernova, and will thus provide information on the
expanding remnant, when it eventually crosses these lines of sight, and, until
then, on whatever material is present from earlier phases in the stars' evolution.

The two brightest companions are Star 2, to the NW, and Star 3, to the
SE. Another much fainter star is present to the SW, and other still fainter
objects may be present as well. Of these, only Stars 2 and 3 can presently be
usefully observed. Spectra of Stars 2 and 3 are presented in Figure 4, together
with several spectra of standard stars. The nebular spectrum was removed
from those of Stars 2 and 3 by adjusting the subtraction until the forbidden
lines disappeared; while this should have largely removed the Balmer lines as
well, it is not possible to be sure that the removal was exact, particularly for
Star 3.

Nevertheless, examination of the helium lines in the spectra clearly shows that both stars have a temperature class of B2. The signal to noise is not sufficient to determine a luminosity class directly, but the spectra in combination with the known distance and brightness are consistent with Star 2 being B2III.

Although Star 3 shows the Balmer lines in emission, these could be an artifact of the nebular subtraction. Observations at better spatial and spectral resolution by Wang et al (1992) show that Star 3 has a broad, non-nebular component of Hβ emission, suggesting that it is a Be star.

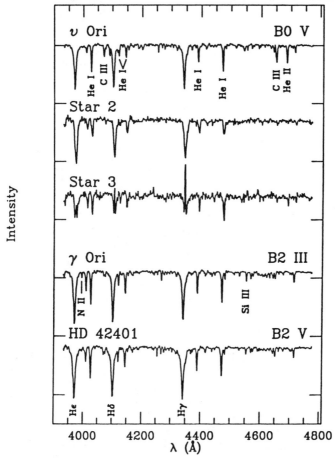

Figure 4: Spectra of Stars 2 and 3 with nebular background removed. Comparison spectra of standard stars are also shown.

A second line of evidence that Star 3 is a Be star is provided by its photometry (Figure 5), which shows it to be variable. One might be concerned that this might be some artifact of contamination by the reflection nebula around the supernova, but it is noteworthy that the visible data show an initial brightening, followed by a decline; this variation is difficult to explain as a contamination effect.

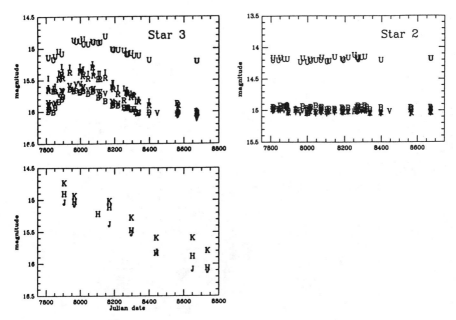

Figure 5: Photometry of Stars 2 and 3. The UBVRI data for Star 2 show no evidence of variability, while those for Star 3 show a slow variation of over half a magnitude. The near-infrared data for Star were referenced to Star 2 for many of the observations, so not much independent data exists for Star 2.

LIGHT ECHO EFFECTS

The well-known visible reflection nebulosity around the supernova (see, for example, Wang 1992) also appeared in the near-infrared (Suntzeff et al 1991). Also like the visible nebula, it has slowly faded with time, as shown in Figure 6, which is a montage of images taken at J on the CTIO 4-m telescope. While the nebulsoity is apparent in the top three images, taken in 1990 and 1991 (less so in the 1991 image), the bottom two, taken in 1992, show little or no nebulosity.

Images taken in the He I line at 1.083 μm also show structure similar to that seen in the reflection nebulosity (Figure 7). The infrared images start at in 1990, roughly 1000 days after outburst, but extended He I emission was detected almost a year earlier by Allen et al (1989).

The structure seen in Figure 7 is similar to that seen in both the visible images and in Figure 6, but perhaps somewhat sharper. In the broad-band images, which are true light echoes, one is observing light from the supernova at peak brightness, which has been scattered by dust, after a time delay due to the additional distance traversed.

This cannot be the case for the He I images (see Allen et al 1989), and it is more likely that what one is seeing is a similar effect, except that the line emission is from gas ionized by the initial ultraviolet peak rather than simple scattered, but with the same time delays.

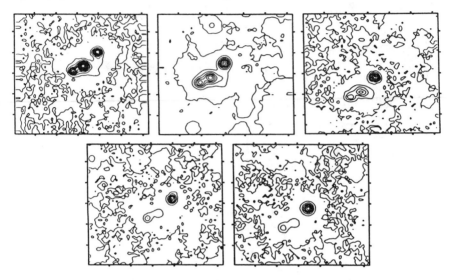

Figure 6: J (1.25 μm) images of SN1987A. All images were taken on the CTIO 4-m telescope, and cover a field of roughly 15 by 18 arcsec. The tic-marks along the ends are 3 arcsec apart. The contour levels have been set to the same level for all images. The dates of the images are: top row, left to right, 1990 March, 1990 November, 1991 April; bottom row, left to right, 1992 January, 1992 April.

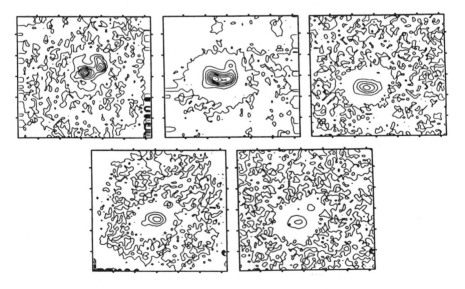

Figure 7: He I 1.083 μm images of SN1987A, as in Figure 6.

While traces of the extended emission were still present as late as the April 1991 image, most of the development of the phenomenon must have occurred prior to then, but after the time of the Allen et al (1989) observations, since these also showed the emission to be limited to a region close to the supernova itself.

ACKNOWLEDGEMENTS

Our work and this review would not be possible without the sharing of results by others working on SN1987A; we would like to especially thank Patrice Bouchet and Lifan Wang. NOAO is operated by AURA, Inc, under a cooperative agreement with the National Science Foundation.

REFERENCES

Allen, D. A., Meikle, W. P. S., and Spyromilio, J. 1989, Nature, 342, 403.

Cummings, R. 1992 in Massive Stars: Theirs Lives in the Interstellar Medium, ed. J. Cassinelli (San Francisco, ASP), in press.

Elias, J. H., DePoy, D. L., Gregory, B., and Suntzeff, N. B. 1991, in SN 1987A and Other Supernovae, ed. I. J. Danzigar and K. Kjar (Garching, ESO), p. 293.

Panagia, N. 1992 in Massive Stars: Theirs Lives in the Interstellar Medium, ed. J. Cassinelli (San Francisco, ASP), in press.

Suntzeff, N. B., and Bouchet, P. 1990, AJ, 99,650.

Suntzeff, N. B., Phillips, M. M., DePoy D. L., Elias, J. H., and Walker, A. R. 1991, AJ, 102, 1118.

Suntzeff, N. B., Phillips, M. M., DePoy, D. L., Elias, J. H., and Walker, A. R. 1992, ApJ, 384, L36.

Wang, L. 1992 in Massive Stars: Theirs Lives in the Interstellar Medium, ed. J. Cassinelli (San Francisco, ASP), in press.

Wang, L., D'Odorico, S., Gouiffes, C., and Wampler, J. 1992, IAU Circ. 5449.

Wang, L., and Mazzali, P. 1991 A&A, , 241, L17.

Massive Stars: Their Lives in the Interstellar Medium
ASP Conference Series, Vol. 35, 1993
Joseph P. Cassinelli and Edward B. Churchwell (eds.)

A POSSIBLE LBV IN NGC 1313?

ERIC M. SCHLEGEL
NASA-GSFC, Code 668, Greenbelt, MD 20771; also Universities Space
Research Association

EDWARD COLBERT[1], ROBERT PETRE
NASA-GSFC; [1]also, University of Maryland

ABSTRACT   A new x-ray source has turned on in the barred spiral
galaxy NGC 1313. The source was not seen with EINSTEIN, but is a
strong ROSAT source. A brief summary of the known properties to date
suggest the object may be a LBV, although the possibility that the object
is an unusual supernova is also explored.

OBSERVATIONS

NGC 1313 was observed with the Position Sensitive Proportional Counter
(PSPC) of the ROSAT x-ray satellite between 1991 April 24 and 1991 May
11, for a total exposure time of 11,183 seconds. The spectral bandpass of the
ROSAT PSPC is approximately 0.3 to 2.2 keV, a band conducive to soft x-
ray emitting sources. In the NGC 1313 field, three sources were found to be
present. Background-subtracted spectra of the three objects revealed different
spectral behaviors, suggesting three different types of objects. The central
source has an x-ray spectral behavior somewhat similar to a starburst galaxy
nucleus. The source to the south appears to be an x-ray binary. The third
source, southwest of the nucleus of the galaxy, has a spectrum best fitted by
a coronal plasma model (Raymond-Smith model), although the χ^2 values are
such that a bremsstrahlung or power-law model cannot be ruled out. Figure
1 (next page) shows the extracted ROSAT x-ray spectrum of the third source.
The measured flux is 4.6×10^{-13} ergs s^{-1} cm^{-2}, or a luminosity of 1.1×10^{39} ergs
s^{-1}, assuming a distance of 4.5 Mpc (de Vaucouleurs 1963)
    An examination of the EINSTEIN database showed that NGC 1313
was observed with the Imaging Proportional Counter (IPC) on 1980 January
2, with an exposure of 8293 seconds (Fabbiano et al. 1984, Fabbiano and
Trinchieri 1987). A quick comparison of the field showed that the third source
was not present in 1980, whereas the central and southern sources were
present, at about the same ratio in brightness. There was no detection of the
third source above the background level of the IPC. An upper limit on the
flux was assigned by convolving the EINSTEIN IPC response function with
the Raymond-Smith plasma model, using the known background counts as
the assumed count rate from the source. This produced an upper limit of

1.1×10^{-13} ergs s^{-1} cm^{-2}, or a luminosity of 3×10^{38} ergs s^{-1}. These values are approximately four times lower than the corresponding ROSAT values.

data and folded model

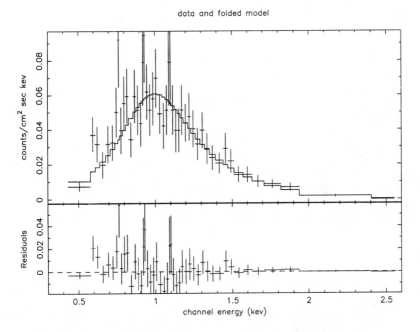

Fig.1. Extracted spectrum of the source in NGC 1313. A Raymond-Smith plasma model has been used to fit the data, with a temperature of 0.5 keV. The confidence limits on this temperature are about 0.3-0.8 keV.

DISCUSSION

By itself, the above x-ray behavior is not particularly useful, as the optical identification of the third source is required to connect it with the astrophysical database of known sources. As it so happens, a group in Australia (Ryder, Staveley-Smith, and Malin 1992) has been examining this object with optical and radio instruments. A spectrum obtained with the 2.3m telescope at Siding Spring Observatory revealed a narrow-line emission spectrum, with the spectrum dominated by the presence of Hα and Hβ. Little continuum emission was present, suggesting a nebular source. The spectrum was originally classified as a nova (Dopita and Ryder 1990) as the Balmer emission lines were narrow (velocity widths < 1000 km s^{-1}), and the ratio of the [O III] 4363Å to [O III] 5007Å lines implied an electron density of approximately 10^{6-7} cm^{-3}.

A more likely explanation for this object followed from an examination of the archival plates of the UK Schmidt telescope. A crude light curve was compiled, based upon 22 plates of the field of NGC 1313 taken in the photographic J band. Magnitudes were estimated by comparison with field stars which were later measured with CCD photometry to establish a reasonable absolute scale. The resulting light curve shows the object to be below the plate limit until about 1980, when it erupted by at least 6 magnitudes. The object faded by about 2 magnitudes within about a month, and returned to its pre-outburst brightness. Recent observations (1986, 1987, 1992) show the object undergoing a very slow brightening at a rate of about a magnitude per five years. The object has also been detected at radio wavelengths, where it appears to be a non-thermal source.

The apparent re-appearance of the object suggests that it is not a supernova, and the distance precludes a nova. A possible explanation for an object with supernova-like radio power, a non-thermal radio spectrum, narrow emission lines in the optical, an outburst, and an x-ray source is a luminous blue variable (LBV) (compare with SN 1961V in Goodrich et al. 1989). The apparent return is the key point in arguing for a classification of the object as an LBV. It should be pointed out that there do appear to be supernovae with similar properties, albeit not well-understood due to a lack of observations. SN 1986J in NGC 891 (Rupen et al. 1986) is also an x-ray source (Bregman and Pildis 1992), and exhibits similar properties to the NGC 1313 object.

More observations are clearly needed to establish the complete behavior of this object. A more extended discussion is in preparation.

REFERENCES

Bregman, J. and Pildis, R. 1992, *ApJ (Letters)* in press.
de Vaucouleurs, G. 1963. *ApJ*, **259**, 302.
Dopita, M. and Ryder, S. 1990, *IAU Circular* 4950.
Fabbiano, G., Feigelson, E., and Zamorani, G. 1982. *ApJ*, **256**, 397.
Fabbiano, G. and Trinchieri, G. 1987. *ApJ*, **315**, 46.
Goodrich, R. W., Stringfellow, G. S., Penrod, G. D., and Filippenko, A. V. 1989. *ApJ*, **342**, 908.
Rupen, M. P., van Gorkom, J. H., Knapp, G. R., Gunn, J. E., and Schneider, D. P. 1987. *AJ*, **94**, 61.
Ryder, S., Staveley-Smith, L., and Malin, D. 1992, in preparation.

Massive Stars: Their Lives in the Interstellar Medium
ASP Conference Series, Vol. 35, 1993
Joseph P. Cassinelli and Edward B. Churchwell (eds.)

AN OLD SUPERNOVA-PULSAR ASSOCIATION

J.A. PHILLIPS
Owens Valley Radio Observatory, Caltech 105-24, Pasadena, CA 91106

J.S. ONELLO
Dept. of Physics, State University of New York, Cortland, NY, 13045

ABSTRACT   PSR 1855+02 is located less than 6' from a supernova remnant (SNR) in the SNR-HII complex G35.6-0.5. The kinematic distance to the complex has two possible values, 4 and 12 kpc. The larger value is in good agreement with the dispersion-measure distance to the pulsar (13 kpc) and is consistent with the absence of visible nebulosity on the Palomar Sky Survey prints. The apparent separation between 1855+02 and the SNR implies a mean tangential velocity of 125 km/s for the pulsar and a proper motion of $\sim$ 2 mas/yr. The age of the proposed association is 1.6×10^5 yr. If confirmed by HI absorption measurements of the nonthermal remnant, the association would be the oldest known pulsar-SNR pair.

INTRODUCTION

Clifton et al (1992) have published a list of newly-discovered pulsars in the northern Galactic plane. We noticed that one of the pulsars from their survey, PSR 1855+02 ($P = 0.416$s, characteristic age $\simeq 1.6 \times 10^5$ yr), is nearly coincident with the non-thermal radio source, G35.6-0.5 (Fig. 1). G35.6 exhibits the classical characteristics of a radio supernova remnant: it is extended ($\Delta\theta \sim 10'$), exhibits $\sim$5% linear polarization at 5 GHz, and has a steep non-thermal spectrum ($\alpha \simeq -0.96$). The source was originally identified as a supernova remnant (SNR) by Milne (1970), but was later classified as an HII region based on the detection of thermal recombination lines (Dickle & Milne 1972). Subsequent observations (Angerhofer et al 1977) have established the non-thermal spectrum of G35.6, and we believe it to be an SNR in a star-forming complex.

To prove an association between PSR 1855+02 and G35.6, it is necessary to show that they are at the same distance. The dispersion measure of 504 pc cm^{-3} for PSR 1855+02 is large, and implies a distance $\simeq$ 13 kpc (Lyne, Manchester & Taylor 1985). This is consistent with a lower limit of 8 kpc obtained from HI absorption measurements of the pulsar (Clifton, Frail & Kulkarni 1988). The distance to the remnant has been estimated from the '$\Sigma - D$' relation to be between 8 and 18 kpc (e.g., Milne 1979, Allakhverdiyev 1983). Such estimates are notoriously uncertain (Green 1984) and we have made radio recombination line measurements to obtain an independent kinematic distance.

H168α RECOMBINATION LINE MEASUREMENTS

Figure 1 shows a 5 GHz continuum map of the G35.6 complex. The northern-most source, G35.6-0.0, is a thermal HII region. There appears to be a bridge connecting it with G35.6-0.5 so we observed positions around both regions. We obtained spectra of the H168α recombination line (1374.6006 MHz) using a 2048-channel autocorrelation spectrometer at the Arecibo Observatory. Table 1 summarizes the data. The spectral observations were frequency switched, and continuum flux measurements were obtained at only three positions. The line-to-continuum ratios were near 1%, as expected for HII regions observed near 1 GHz. The line widths at position 5 and the toward the pulsar were ~ 2 times greater than expected from thermal broadening in a 10^4 K HII region. The broad lines are indicative of enhanced turbulence in the non-thermal source and could account for the lower line-to-continuum ratio at those positions.

TABLE I   H168α Recombination Line Observations

Position	Epoch	l deg	b deg	T_l mK	T_l/T_c	ΔV km/s	V_{lsr} km/s
1	Mar 1989	35.7	−0.0	160 ± 10	-	24 ± 3	52 ± 1
2	Mar 1989	35.6	−0.0	389 ± 15	0.023	20 ± 1	50 ± 1
3	Mar 1889	35.5	−0.0	184 ± 10	-	22 ± 3	56 ± 1
4	Mar 1989	35.4	−0.0	151 ± 10	-	20 ± 2	55 ± 1
PSR	Apr 1992	35.6	−0.4	51 ± 3	0.006	33 ± 2	53 ± 1
5	Apr 1992	35.6	−0.5	91 ± 5	0.007	39 ± 5	54 ± 1

The LTE line-to-continuum ratio expected for a 10^4 K HII region with line width 35 km/s is 0.005, in excellent agreement with that measured in the direction of PSR 1855+02. Taken at face value, this result suggests that most of the flux was thermal. The emission measure inferred from the continuum temperature (T_c = 8.5 K at the pulsar position) is $EM = 7 \times 10^3$ pc cm^{-6}. From the emission measure we estimate the dispersion measure through the nebula to be $DM \approx 540$ pc cm^{-3}. The nebular DM is larger than the pulsar dispersion measure. Clearly, either the HII region is behind the pulsar or we have overestimated the thermal flux from the ionized gas. A better estimate of the free-free emission could be obtained from high frequency ($\nu \gtrsim 10$ GHz) observations of the remnant.

From the velocity of the recombination lines ($\simeq 54$ km/s) we can derive a kinematic distance to the HII-SNR complex. The Schmidt model for galactic rotation yields $D = 4$ kpc or $D = 12$ kpc. We inspected the Palomar Sky Survey prints for an optical nebula at the location of G35.6-0.5 and none was visible. On that basis we believe the larger distance more likely to be correct.

CONCLUSIONS

The kinematic distance to G35.6-0.5 and the DM distance to PSR 1855+02 are in good agreement. Nevertheless it would be valuable to resolve the kine-

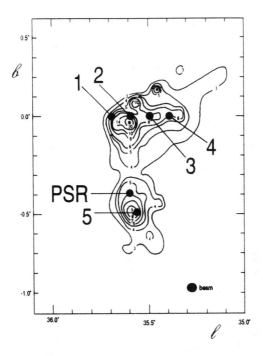

FIGURE I   PSR 1855+02 and G35.6-0.5. The contours are from the 2.'6-resolution, 5 GHz continuum map of Altenhoff et al (1979). We obtained H168α recombination line spectra with 3' resolution at positions indicated by the dots.

matic distance ambiguity to the nebula using HI absorption measurements of the non-thermal SNR. Based on the current location of PSR 1855+02 and the continuum peak of G35.6-0.5 we predict a pulsar proper motion of 2 milliarcseconds per year away from G35.6-0.5 and toward the galactic plane. The corresponding tangential velocity is 125 km/s, a value that is typical of measured pulsar velocities. The age of the proposed association is 1.6×10^5 yr and the remnant is probably in the post-Sedov "snowplow" phase of expansion. Older remnants with subsonic expansion velocities are expected to disperse in the ambient ISM.

ACKNOWLEDGMENTS

We owe a special thanks to José Navarro for writing and maintaining the *pdb* pulsar database. We also thank Shri Kulkarni, José Navarro, and Gautam Vasisht for helpful discussions, and Donna Curtin of the Sperry Research Center at SUNY Cortland for her excellent work on Figure 1.

REFERENCES

Allakhverdiyev, A.O. et al. 1983, Ap. Space Sci., 97, 287.

Altenhoff, W.J., Downes, D., Pauls, T. & Schraml, J. 1978, A&A Supp, 35, 23.

Angerhofer, P.E., Becker, R.H. & Kundu, M.R. 1977, A&A, 55, 11.

Clifton, T.R., Frail, D.A. & Kulkarni, S.R. 1988, ApJ, 333, 332.

Clifton, T.R. et al. 1992, MNRAS, 254, 177.

Dickel, J.R. & Milne, D.K. 1972, Australian J. Phys, 25, 539.

Green, D.A. 1984, MNRAS, 209, 449.

Lyne, A.G, Manchester, R.N. & Taylor, J.H 1985, MNRAS, 213, 613.

Milne, D.K. 1970, Australian J. Phys., 23, 425.

Milne, D.K. 1979, Australian J. Phys., 32, 83.

Massive Stars: Their Lives in the Interstellar Medium
ASP Conference Series, Vol. 35, 1993
Joseph P. Cassinelli and Edward B. Churchwell (eds.)

RECENT Hα ECHELLE OBSERVATIONS OF CYGNUS X-1

L. A. SHANLEY, J.S. GALLAGHER
University of Wisconsin, Department of Astronomy,
475 N.Charter St., Madison, WI 53706

G. KOENIGSBERGER, E. RUIZ, D. VERA
UNAM, Instituto de Astrinomia, Apartado Postal 70-264,
DF04510 Mexico City, Mexico

INTRODUCTION

Cygnus X-1 (HDE 226868) remains a prime example of a high mass,
luminous x-ray binary that is likely to contain a black hole. Through optical
observations of time dependent emission line strengths and profiles, e.g., from
the He I, He II, and Hα lines, the gas flows of this system can be probed.
Optical emission line profiles in Cyg X-1 have been observed since the early
1970s (e.g. Brucato and Zappala 1974; Hutchings *et al.* 1974). Both the He II
λ4686 and Hα lines of Cyg X-1 display complex line profiles, which vary
systematically with orbital phase. An intriguing question is then whether the
Cyg X-1 optical line profiles display long term stability which might indicate
constancy of mass motions within the system.

OBSERVATIONS AND REDUCTION

During an exploratory program to measure time variations in stellar emission
line profiles, five echelle spectra of Cygnus X-1 were obtained in August of
1991 at San Pedro Martir Observatory (SPMO). A journal of observations is
provided in Table I. For these observations, the SPMO 2-m telescope was
equipped with a Cassegrain echelle spectrograph and a 516 x 516 pixel CCD
detector. The CCD was retrofitted into the existing spectrograph camera using
a fiber bundle as an optical coupler. The slit was set at a width of 1.5 arcsec
and a ThAr lamp provided a wavelength calibration. This arrangement yielded
a spectral resolution of R = 18,000. Flat field observations were made twice a
night with the telescope at the zenith position.
   Handled in the standard manner, the data were reduced at UNAM and at
Wisconsin with IRAF's spectral reduction packages. We extracted the orders
cointaining the Hα and He I λ5876 lines, made wavelength calibrations, and
then corrected for the blaze response of the spectrograph. Because of a flexure
problem with the dewar, each spectrum had to be fit individually for the blaze

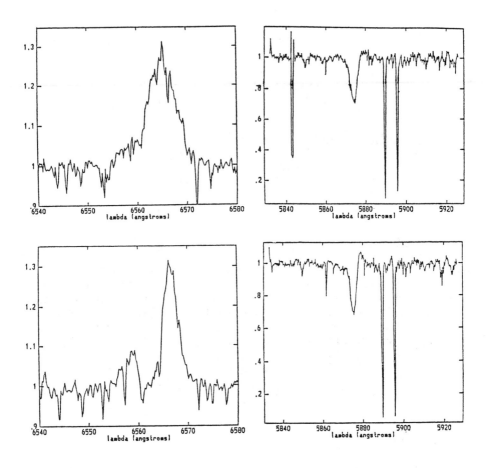

Figs. 1 through 4. The Cygnus X-1 spectra for the He I and Hα
regions, shown in Figures 1 through 4 (numbered clockwise from the
left), are averages of each night. The August 19th data were used to
check for short term variability in the line profiles. No differences were
seen between the three observations at the 2-3% level that is set by the
signal-to-noise. Figures 1& 2 are averages of two spectra taken on Aug
18; figures 3 & 4 are averages of three spectra taken on Aug 19.The left
column contians the spectra of the Hα line. The right contains spectra of
the He I line and the Na doublet.

function. Therefore, the final spectra are insensitvie to the prescience of any very broad velocity features. The signal-to-noise is approximately 30:1. The ephemeris used to compute the orbital phase is that of Batten *et al.* (1989); the primary minimum is JD 2443045.068 + 5.5996.

Properties of the lines, such as the equivalent widths, were measured with the IRAF SPLOT task. Using the IRAF program RVCORRECT, we determined heliocentric corrections for the raidal velocities of the major line components.

TABLE I   Journal of Observations

UT Date	UT Time	JD 2440000+	Exp Time seconds	Phase
8-18-91	8:02	8486.837	600	.814
8-18-91	8:17	8486.845	1800	.815
8-19-91	9:11	8487.882	1800	.001
8-19-91	9:45	8487.906	1800	.005
8-19-91	10:21	8487.931	1800	.009

RESULTS

During phase 0.8, the Hα line displays a symmetrical core centered at v=100 km/s. The emission core is superimposed on a broader emissin feature that we trace over a width of ±400 km/s in radial velocity. Reaching a peak normalized intensity of 1.05 times the continuum, the blue wing of this feature is enhanced. The emission equivalent width is 1.8 Å. The phase 0.0 data, on the other hand, exhibit absorption at v=-185 km/s which has split the line into blue and red components. A sharp peak is now seen at v=155 km/s and an increase in emission intensity for the peak redwards to about v=200 km/s is present. The broad component, however, appears to be unchanged between these two phases except in the region of the v=-185 km/s absorption. The emission equivalent width at phase 0.0 is 1.5 Å.

The He I λ5876 absorption center is at -83 km/s at phase 0.8 and a weak P-Cygni emission component begins at about 70 km/s. The absorption core has shifted redward to -18 km/s and the P Cygni component appears at about 140 km/s. No change is found in the depths of the line cores or in the equivalent widths, although weak absorption may extend further to the blue at phase 0.8 than at phase 0.0.

The Hα line profiles for Cyg X-1 published by Hutchings *et al.*. (1974; 1977) provide a convenient point of comparision with our observations. The generic types of line profiles seen in our data were also observed in 1973 and 1977, although the phasing of the two classes of profiles that we have seen may have been different in the earlier data. In addition, the mean emission intensity of the redward Hα peak was measured to be 0.24 times the continuum, which is slightly lower than what we have observed. Further-more, the equivalent width may have been slightly lower than in the 1970s, where Hutchings *et al.* (1977) found a mean value of 1.8 Å *after* correcting for an underlying absorption feature.

A more recent data set has been presented by Ninkov *et al.* (1987) based on observations made in 1982 with the Dominion Astrophysical Observatory 1.83 m telescope. These measurements give profiles very similar to those in our data but occurring with phase differences of up to 0.2 and with lower peak emission intensities of < 20%.

DISCUSSION

These data are perhaps most readily understood in terms of the focused stellar wind model (Friend and Castor 1982; Gies and Bolton 1986). The Hα spectrum of Cyg X-1 is interpreted by Ninkov *et al.* (1987) in terms of three major components: an underlying photospheric absorption line from the primary star, a broad emission component that remains roughly constant with orbital phase, and a narrow component associated witht the focused stellar wind. Most of the profile variation with orbital phase appears to occur in the focused wind component of Hα. Similarly, changes in the shape of He I in our data seem to be associated with low velocity material. These results are consistent with conclusions from the earlier Hutchings *et al.* and Ninkov *et al.* surveys regarding long term Hα profile variability.

Understanding fully the source and implications of longer term narrow line changes might give us better insight into the gas flow which is responsible for the accretion process. To this end, it would be especially interesting to look for correlations between the X-ray state of Cygnus X-1 and the form of its Hα emission line. In addition, it would be worthwhile to extend atmospheric models in the spirit of Gies and Bolton to the Hα line.

ACKNOWLEDGEMENTS

We would like to extend our thanks to C.-W. [Rick] Lee for his invaluable help with IRAF. Also, we would like to acknowledge the use of SIMBAD.

REFERENCES

Batten, A. H., Fletcher, J. M., and MacCarthy, D. G. 1989, *Publ. D. A.O.*, **17**, 104.
Brucato, R. J., and Zappala, R. R. 1974, *Ap. J.* (Letters), **189**, L71.
Friend, D. B. and Castor, J. I. 1982, *Ap. J.*, **261**, 293.
Gies, D. R. and Bolton, C. T. 1982, *Ap. J.*, **260**, 240.
Gies, D. R. and Bolton, C. T. 1986, *Ap. J.*, **304**, 371.
Gies, D. R. and Bolton, C. T. 1986, *Ap. J,*. **304**, 389.
Hutchings, J. B. and Crampton, D. 1979, *Publ. A.S.P.*, **91**, 796.
Hutchings, J. B., Crampton, D., Glaspey, J., and Walker, G.A.H. 1973, *Ap. J,* **182**, 549.
Hutchings, J. B. , Cowely, A. P., Crampton, D., Fahlmann, J., Glaspey, J. W., and Walker, G. A. H. 1974, *Ap. J,* **191**, 743.
McCray, R. and Hatchett, S. 1975, *Ap. J.*, **199**, 196.
Ninkov, Z., Walker, G. A. H., and Yang, S. 1987, *Ap. J.*, **321**, 438.

Massive Stars: Their Lives in the Interstellar Medium
ASP Conference Series, Vol. 35, 1993
Joseph P. Cassinelli and Edward B. Churchwell (eds.)

FORMATION OF THE CIRCUMSTELLAR SHELL AROUND SN 1987A

JOHN M. BLONDIN
Deparment of Physics and Astronomy, University of North Carolina,
CB# 3255 Phillips Hall, Chapel Hill, NC 27599

PETER LUNDQVIST
Deparment of Astronomy, University of Virginia, PO Box 3818,
Charlottesville, VA 22903

ABSTRACT   We have modeled the nebulosity around SN 1987A in the
context of the presupernova's BSG wind interacting with an asymmetric
RSG wind, using 2D time-dependent hydrodynamics. We find that a
model with an average $\dot{M}_r \sim 2 \times 10^{-5} M_\odot \text{ yr}^{-1}$ and an equator-to-
pole density ratio in the RSG wind of ~20 can adequately explain the
observations. The large asymmetry required in the RSG wind appears
to suggest the presence of a binary companion at least during the RSG
stage.

1. INTRODUCTION

Observations with the New Technology Telescope (NTT, Wampler et al.
1990), and to higher spatial resolution with the Hubble Space Telescope (HST,
Jakobsen et al. 1991) show that the line emission from the region closest
to the supernova (SN) is dominated by an elliptical structure. Because the
HST pictures show that the inside of the ellipse is nearly empty of emission
(Jakobsen et al. 1991), the observed structure is most likely physical and
not just a limb brightening effect of an elliptical nebula that is complete
around the SN. The idea of a ring geometry is supported by the velocity
structure of the nebulosity which gives a larger velocity shear across the
ellipse's minor axis than along its major axis (Crotts & Heathcote 1991). If the
ellipse is a projection effect of a circular ring tilted by ~43° to the line of sight
towards Earth, the observed velocity structure translates into an expansion or
contraction velocity of 10.3 km s^{-1}. Further support for a circumstellar ring is
provided by models for the UV line light curves (Lundqvist 1991), which give
better fits to the observations for a ring than for a spherical shell (Lundqvist &
Fransson 1991).

The NTT observations show that the ring is connected to an outer
structure which forms loops with largest extensions (roughly 2"5) along
the ellipse's minor axis. The loops are seen both in lines as well as in the

continuum, indicating that the radiating gas is mixed with dust. Wampler et al. (1990) note that a bipolar nebula similar to what is seen for many planetary nebulae (PNe) could also explain the observed structure around SN 1987A .

The evolution of the progenitor to SN 1987A from a red supergiant (RSG) star emitting a slow, dense wind to a blue supergiant (BSG) star emitting a relatively fast wind is expected to produce a circumstellar shell of wind material as the fast blue wind overtakes and sweeps up the slow red wind. Given this history of the SN 1987A progenitor, it is natural to expect the presence of a circumstellar shell surrounding the SN. Indeed, Chevalier & Fransson (1987) went so far as to predict such a structure. It is not surprising, therefore, that a shell has in fact been observed. Rather, it is the asymmetry of the circumstellar shell that has drawn so much attention. A bipolar nebula is most naturally explained by asymmetric mass loss during the star's RSG stage (Luo & McCray 1991, Wang & Mazzali 1992). Because of the higher density in the RSG wind on the equator than along the poles, the stellar wind bubble pushes out faster along the low density poles forming the shape of a peanut. The high density gas of the shocked RSG wind in the "waist" of the peanut-shaped shell is suggested to produce the observed ring seen by HST, while limb-brightening of the two lobes of the peanut-shaped shell creates the loops seen with the NTT. The cause of asymmetry is not known, but may be due to a binary companion.

2. A WORKING MODEL

The observed properties of the circumstellar shell around SN 1987A can be used to put stringent limits on the progenitor wind properties in the context of the colliding winds model (see Blondin & Lundqvist 1992). These properties include the expansion velocity of the ring ($10.3\,\mathrm{km\,s^{-1}}$; Crotts and Heathcote 1991), the density of the gas within the ring inferred from the UV and optical line emission ($\sim 10^4\,\mathrm{cm^{-3}}$; Lundqvist 1992), and the observed shell geometry, including the compactness of the ring and the ratio of the lobe radius to the ring radius.

To arrive at a working colliding winds model for the formation of the circumstellar shell surrounding SN 1987A we are forced to assume relatively extreme values for the progenitor wind parameters. The low expansion velocity of the ring requires a very low momentum flux in the BSG wind and a massive, slow RSG wind. A second constraint is implied by the low expansion velocity coupled with the observed high gas density in the ring; the RSG wind must be very dense along the equator. In order to satisfy these requirements and at the same time avoid the strong theoretical limits on the average mass loss rate of the RSG wind, we are forced to propose a very large asymmetry in the RSG wind. A very asymmetric RSG wind is nonetheless consistent with the shape of the bipolar lobes seen in the NTT images.

Blondin & Lundqvist (1992) applied the results of 2D hydrodynamic simulations and analytic arguments to derive the following set of parameters: $\dot{M}_b = 3 \times 10^{-7}\,M_\odot\,\mathrm{yr^{-1}}$, $v_b = 300\,\mathrm{km\,s^{-1}}$, $\langle \dot{M}_r \rangle = 2 \times 10^{-5}\,M_\odot\,\mathrm{yr^{-1}}$, $v_r = 5\,\mathrm{km\,s^{-1}}$, and an equator-to-pole density ratio of 20 in the RSG wind. The resulting circumstellar shell is shown in Figure 1, and has ring velocity of $\sim 8\,\mathrm{km\,s^{-1}}$. A similar model with a slightly higher value of v_b (say $350\,\mathrm{km\,s^{-1}}$) would probably give a very good match to the observed properties of the circumstellar shell around SN 1987A.

Fig. 1.   Circumstellar shell formed from colliding winds with parameters as described in the text. Density contours are spaced in logrithmic intervals of 2.

ACKNOWLEDGEMENTS

We thank Roger Chevalier, Ding Luo and Lifan Wang for useful discussions. We thank the North Carolina Supercomputing Center for time on their Cray Y-MP/464. P.L. acknowledges a Hubble Space Telescope Fellowship under contract HF-1015.01-90A. J.M.B. acknowledges support by NSF grants PHY90-57865 and PHY90-01645.

REFERENCES

Blondin, J.M., Lundqvist, P. 1992, submitted to ApJ.
Chevalier, R.A. 1988, Nature, 332, 514.
Chevalier, R.A., & Fransson, C. 1987, Nature, 328, 44.
Crotts, A.P.S., & Heathcote, S.R. 1991, Nature, 350, 683.
Jakobsen et al. 1991, ApJL, 369, L63.
Lundqvist, P. 1991, in Proc. ESO/EIPC Workshop, SN 1987A and Other Supernovae, ed. I. J. Danziger & K. Kjär (Garching:ESO), 607.
Lundqvist, P. 1992, PASP, in press.
Lundqvist, P., & Fransson, C. 1991, ApJ, 380, 575.
Luo, D. &, McCray, R. 1991, ApJ, 379, 659.
Wampler, E.J., Wang, L., Baade, D., Banse, K., D'Odorico, S., Gouiffes, C., & Tarenghi, M. 1990, ApJL, 362, L13.
Wang, L., & Mazzali, P.A. 1992, Nature, 355, 58.

Massive Stars: Their Lives in the Interstellar Medium
ASP Conference Series, Vol. 35, 1993
Joseph P. Cassinelli and Edward B. Churchwell (eds.)

CIRCUMSTELLAR INTERACTIONS: KEPLER'S SUPERNOVA REMNANT

KAZIMIERZ J. BORKOWSKI
Astronomy Program, University of Maryland, College Park, MD 20742

JOHN M. BLONDIN
Deparment of Physics and Astronomy, University of North Carolina,
CB# 3255 Phillips Hall, Chapel Hill, NC 27599

ABSTRACT   Observations of Kepler's supernova remnant have revealed a strong interaction with the ambient medium. We explore the dynamics and emission properties of this SNR based on the bowshock model of Bandiera, in which the supernova ejecta is interacting with a bowshock of swept up stellar wind from the progenitor star moving through a tenuous interstellar medium.

1. BOWSHOCK MODEL

Bandiera (1987) has proposed a model for Kepler's SNR in which a massive SN progenitor was ejected from the galactic plane $\sim$ 4 x 10^6 yrs ago, and lost significant amounts of matter shortly before the explosion. This mass loss presumably took place during the red supergiant stage of the stellar evolution, in the form of a dense, slow stellar wind. The ram pressure of the ISM due to the motion of the progenitor swept up and compressed this wind into a dense bowshock shell. The observed strong asymmetry of the remnant is thus a reflection of this asymmetric circumstellar material. Strong mass loss in the evolutionary phases prior to the SN explosion might have exposed the stellar core, whose explosion would produce a light curve characteristic of Type I SNe.

The CSM of the SN progenitor is thus a region of undisturbed stellar wind from the red supergiant progenitor in the immediate vicinity of the SN. Surrounding this wind is a bowshock-shaped shell of dense wind gas swept up by the ram pressure of the tenuous ISM. Outside of the bowshock shell is the low-density, hot ISM.

The blastwave generated by the SN is still relatively young, and has not yet entered the Sedov phase. The *Einstein* SSS detected strong X-ray lines from silicon, sulfur, argon, and calcium, which are also over-abundant (Becker et al. 1980). This emission probably originates in the shocked ejecta which contains Fe synthesized during the SN explosion. This implies that the remnant is in an early evolutionary stage, with the reverse shock still propagating into the ejecta. Until the outer blastwave reaches the bowshock shell we can use the self-similar solutions of (Chevalier 1982) for a blastwave driven by freely expanding ejecta with a power-law density distribution, propagating through a steady-state stellar wind. Once the blastwave reaches the bowshock shell, this spherically symmetric solution will no longer apply. It is this subsequent asymmetric evolution that we have attempted to model here. A complete description of this work is given in Borkowski *et al.* (1992).

2. HYDRODYNAMICS OF SHOCK/SHELL INTERACTION

We have evolved dynamical models of the interaction of the SN blastwave with the bowshock material using the hydrodynamics code VH-1. The initial solution for the expanding shock region is a self-similar driven wave with a fully developed instability region (taken from Chevalier *et al.* 1992). This shock solution is placed just inside of a bowshock of dense gas. As the evolution begins, the shock first reaches the shell at the apex of the bowshock, and later at larger distances from the explosion center. Upon hitting the shell, the blast wave splits into transmitted and reflected shocks. We identify the transmitted shock with nonradiative filaments found in Kepler's SNR (Fesen *et al.* 1989).

These numerical simulations show that the shocked shell is subject to the Richtmyer-Meshkov instability, because of the impulsive acceleration of the interface between the stellar wind and the bowshock shell by the blast wave. Even before the blastwave reaches the bowshock, the shell is expected to be inhomogeneous as a result of dynamical instabilities. Under these conditions, the RM instability is likely to disrupt the shocked shell early on in the shock/shell interaction, leaving behind a highly turbulent remnant.

3. EMISSION CALCULATIONS

The next stage of this work is to calculate the photon emission expected from these simulations, and to compare the results with observations. Here we briefly describe our current methods for producing synthetic observations from the hydrodynamic simulations. It is important to point out that presence of dynamical instabilities leading to mixing and clumping of the shock-heated gas will result in substantially different emission characteristics derived from multi-dimensional simulations as compared to one-dimensional models. We hope to exploit this fact when comparing results with observations in order to better understand the role of dynamical instabilities in SNR.

3.1 X-Ray Emission
Accurate calculation of the X-ray emission requires a knowledge of the history-dependent ionization state of the gas. We have devised a scheme to generate the time-history of each parcel of gas on the numerical grid by labeling each parcel of gas on the initial grid with a Lagrangian marker (i.e., the original grid coordinates). As the flow evolves, these markers are advected with the gas, and at specified time intervals the (advected) Lagrangian coordinates are output to a data file. Each Eulerian zone contains the Lagrangian coordinates corresponding to the location of that parcel of gas at the beginning of the calculation. Similarly, at the end of the calculation each grid zone contains the (advected) Lagrangian coordinates corresponding to the original location of that parcel of gas. We can then use these final Lagrangian coordinates to go

back to each previous time interval and interpolate on that grid of (advected) Lagrangian coordinates to find the density and pressure corresponding to the gas in each final grid zone (but at a previous time interval). In this way a time history can be built up for each grid zone in the final output. This data can then be used in a time-dependent ionization code to calculate the current ionization state and X-ray emissivity of that gas (Hamilton *et al.* 1983).

3.2 Radio Emission

The radio emission from SNR is much less understood than other wavebands. The observed synchrotron emission is generated by an unknown non-thermal distribution of electrons in the unknown magnetic field of the interior. In light of these uncertainties, we have employed a simple prescription as outlined by Fulbright and Reynolds (1990): The relativistic electrons are assumed to be accelerated at the shock front with a given efficiency and spectrum, and subsequently leak into the interior with some energy-dependent diffusion coefficient. The magnetic field is prespecified, allowing only for compression behind the shock.

3.3 Infrared Emission

The IR emission is calculated using the dust cooling function from Dwek (1986), assuming a constant dust/gas ratio. A more refined calculation will incorporate the effects of grain sputtering by including a time-dependent (using the same history file as generated for the X-ray emission) size distribution of the dust grains. Dwek has pointed out that as the grain size is shifted to smaller grains through sputtering in the hot postshock gas, the IR emission spectrum will change. Spatially resolved IR colors can therefore be compared with these simulations to provide further constraints on the model.

ACKNOWLEDGEMENTS

K.J.B. acknowledges support by the Center for Theoretical Physics and the Department of Astronomy at the University of Maryland. J.M.B. acknowledges support by NSF grants AST90-16687, PHY90-57865 and PHY90-01645. Support from the Pittsburgh Supercomputing Center is gratefully acknowledged.

REFERENCES

Borkowski, K. J., Blondin, J.M., and Sarazin, C. L. 1992, ApJ, 400, in press.
Becker, R. H., Boldt, E. A., Host, S. S., Serlemitsos, P. J., and White, N. E. 1980, ApJ, 237, L77.
Bandiera, R. 1987, ApJ, 319, 885.
Chevalier, R. A. 1982, ApJ, 258, 790.
Chevalier, R. A., Blondin, J. M., and Emmering, R. T. 1992, ApJ, 392, 118.
Dwek, E. 1986, ApJ, 302, 363.
Fesen, R. A., Becker, R. H., Blair, W. P., and Long, K. S. 1989, ApJ, 338, L13.
Fulbright, M. S., and Reynolds, S. P. 1990, ApJ, 357, 591.
Hamilton, A. J., Sarazin, C. L., and Chevalier, R. A. 1983, ApJS, 51, 115.

Massive Stars: Their Lives in the Interstellar Medium
ASP Conference Series, Vol. 35, 1993
Joseph P. Cassinelli and Edward B. Churchwell (eds.)

COLD BRIGHT MATTER NEAR SUPERNOVA 1987A

ROBERT J. CUMMING
Imperial College of Science, Technology and Medicine, Blackett Laboratory,
Prince Consort Road, London SW7 2BZ, UK. (present address: Royal
Greenwich Observatory, Madingley Road, Cambridge CB3 0EZ, UK.)

W. PETER S. MEIKLE
Royal Greenwich Observatory, Madingley Road, Cambridge CB3 0EZ, UK.

ABSTRACT   We report the observation of a short-lived, remarkably
narrow component in Hα and Hβ, found in high-resolution spectra of the
circumstellar medium of SN 1987A. The width of the Hα narrow
component indicates T <1000 K, much lower than the Hα temperature we
find for the bright ring nebula (13750±100 K). The narrow emission is
spatially unresolved and is not part of the ring. We investigate the cooling
and recombination of the emission source, and comment on its likely origin.

OBSERVATIONS

By high resolution spectroscopy of the strongest circumstellar lines, we are
investigating the detailed structure, dynamics and evolution of the circumstellar
medium of SN 1987A. The spectra are obtained using the échelle spectrograph
(UCLES) at the Anglo-Australian Telescope (Meikle et al. 1991). We have
observed a remarkably narrow component in the Hα and Hβ emission. A
corresponding component may be present in [O III] $\lambda\lambda$4959, 5007, but not in any
other lines. A similar effect has also been reported in Hα by other observers
(Hanuschik 1990, Crotts & Heathcote 1991) but at lower spectral resolution. In
this poster paper we briefly describe our observations of this phenomenon. We
present a fuller account in Cumming & Meikle (1992).

We present here data from CCD échelle spectra collected on 1990 Oct 29 (day
1344 after the SN explosion) at spectral resolution 6 km s^{-1}; and 1″.4 seeing.
We observed with a 0″.7 slit at p.a. 110–290° and 20–200°. Fig. 1 shows the Hα
position-velocity plots for 20–200°. Analysis of the H I line profiles shows that
the emission is made up of two components. A broad, extended component arises
from the expanding ring, while a second, completely unexpected component lies
redshifted by 11 km s^{-1} — it is spatially unresolved and extremely narrow.

We find that the narrow component is located 0.3±0.1 arcsec from the SN at
p.a. 234±13° and therefore cannot be part of the ring nebula. Taking into
account all sources of line broadening (including the intrinsic multiplet structure

433

of the line), we modelled the likely profiles for both components. We found that at all temperatures, the broad and narrow components are approximately Gaussian, so we used two-Gaussian fits to the data to determine their temperatures. We find $T_{H\alpha} = 13750\pm100$ K for the ring. Over the detector rows where it emission is apparent, the narrow component however has a width of 8.72 ± 0.24 km s^{-1} FWHM, and is *narrower than would be expected for T approaching 0 K*. For the region along the 20–200° slit where the narrow component is strongest, we exclude temperatures >1000 K with confidence limits of >91%. We believe that the temperature of the gas is <1000 K. Fig. 2 shows the best fits to these profiles.

In conjunction with contemporary low-resolution spectra, we examined the luminosity evolution of the $H\alpha$ narrow component, estimating its luminosity from our high-resolution spectra on days 1254 (when it was weakly detected) and 1344. On days 1105, 1397 and 1415, we adopt upper limits for the undetected narrow component (Fig. 3).

DISCUSSION

In Cumming & Meikle (1992), we discuss models in which the gas producing the narrow emission was ionised by the EUV flash from the SN explosion, and subsequently recombines and cools. We find that atomic gas cooling cannot reproduce both the observed light curve and the low temperature. Cooling by collisions with grains typical of red giant winds fails to supply the necessary extra cooling for the gas. It appears that other cooling mechanisms are needed, e.g. cooling by molecules such as CO. If these mechanisms exist, then to reproduce the observed low temperature, luminosity and evolution, the cloud density must have been about $(1–2)\times10^5$ cm^{-3}, and would have had a radius of 0.6–0.8 light days. The observations place the flash-ionised cloud $\sim1.5\times10^{18}$ cm (1.6 light yr) behind the SN.

In the circumstellar shell models of Blondin & Lundqvist (1992), dynamical instabilities develop and cause blobs of dense gas to be torn from the edge of the shell and be drawn into the shell's interior. These blobs may be the sources of observed narrow H I emission. Our cloud appears to be close the edge of the northern lobe of Blondin & Lundqvist's best model for the hourglass-shaped nebula (their model [h]) though its line-of-sight velocity, 11.2 ± 0.7 km s^{-1}, is half of that expected for the shell at this position. It seems that the cold cloud was located just inside the shell, but travelling (as Blondin & Lundqvist predict) more slowly outwards.

REFERENCES

Blondin J. M., Lundqvist P., 1992, ApJ (submitted)
Crotts A. P. S., Heathcote S. R., 1991, Nat, 350, 683

Cumming R. J., Meikle W. P. S., 1992, MNRAS (submitted)

Hanuschik R. W., 1990, A&A, 237, 12

Meikle W. P. S., Cumming R. J., Spyromilio J., Allen D. A., Mobasher B., 1991, in Danziger I. J., Kjär K., eds, Proc. ESO/EIPC Workshop on SN 1987A and Other Supernovae, ESO, Garching, p. 595

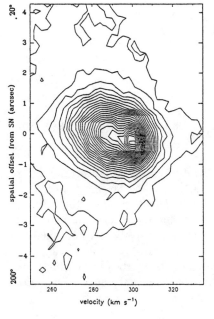

◀ FIGURE 1.   Velocity-position map of Hα intensity at p.a. 20–200° (p.a. 20° at top).

▼ FIGURE 3. Evolution of the narrow component luminosity

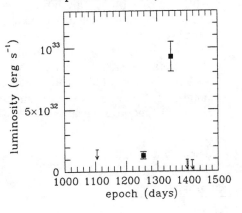

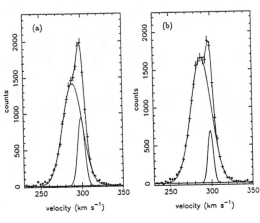

◀FIGURE 2. Best Gaussian fits to Hα at p.a. 20–200° for (a) distance 0″.42 from the SN towards 200°, (b) 0″.04 towards 20°.

Massive Stars: Their Lives in the Interstellar Medium
ASP Conference Series, Vol. 35, 1993
Joseph P. Cassinelli and Edward B. Churchwell (eds.)

RADIO SUPERNOVAE & MASSIVE STELLAR WINDS

KURT W. WEILER AND SCHUYLER D. VAN DYK
Naval Research Lab, Code 4215, Washington, DC, 20375-5320

NINO PANAGIA
Space Telescope Science Inst., 3700 San Martin Dr., Baltimore, MD 21218

RICHARD A. SRAMEK
National Radio Astronomy Observatory, P.O. Box O, Socorro, NM 87801

ABSTRACT   Many supernovae have been found to be powerful emitters
of radio emission which arises from interaction of the SN shock with a
dense cocoon of matter surrounding the SN. This cocoon was established
by mass loss in a dense stellar wind from the massive stellar progenitor of
the supernova itself or from a massive companion to the SN in a binary
system. Detailed study of this radio emission is now available for a
number of SNe and provides estimates for the properties of and mass loss
rates from the presupernova stellar systems in the final stages of evolution.

INTRODUCTION

Detailed study (see Weiler et al. 1986) of the radio emission from supernovae
(SNe) has revealed that they all have the properties of : (1) nonthermal emission with high brightness temperature, (2) turn-on first at shorter wavelengths
and later at longer wavelengths, (3) initial rapid increase of flux density at
each wavelength, (4) power law decline at each wavelength after maximum is
reached, (5) decrease in spectral index α between any two wavelengths as the
longer wavelength goes from optically thick to optically thin, and (6) final
asymptotic approach of α to an optically thin, nonthermal, constant negative
value ($S \propto \nu^{+\alpha}$). This behavior is well described by the relations:

$$S \text{ (mJy)} = K_1 \, (\nu/5 \text{ GHz})^\alpha \, ([t-t_0]/1 \text{ day})^\beta \, e^{-\tau} \tag{1}$$
$$\tau = K_2 \, (\nu/5 \text{ GHz})^\alpha \, ([t-t_0]/1 \text{ day}))^\delta \tag{2}$$
$$\delta \equiv \alpha - \beta - 3 \tag{3}$$

This formulation assumes that both the flux density S and the optical depth τ
are power-law functions of the supernova age $(t - t_0)$, with powers β and δ,

respectively; that the absorption is purely thermal, free-free in an ionized, external medium (frequency dependence $v^{-2.1}$) with radial dependence of r^{-2} from a constant speed, red supergiant wind; and that the radio emission is nonthermal synchrotron with an optically thin spectral index α. K_1 and K_2 are scaling factors for the units of choice, and correspond formally to the flux density (K_1) and optical depth (K_2) at 5 GHz 1 day after the explosion.

TYPES OF RSNe

Radio supernovae (RSNe) can be grouped similarly to the optical classifications, e.g., Type I and Type II. Type Ia SNe do not appear to be radio
emitters to the sensitivity limit of the VLA, but Type Ib and Type Ic SNe have
been detected. Type II SNe appear to all be radio emitters at some level, but
show a variation of properties. The most extreme case is the Type II SN1986J,
which has significant internal absorption not described by Eqs. 1, 2, and 3.

Type Ib/c RSNe
Type Ib/c RSNe: (1) turn-on very early and rapidly (smaller K_2) followed by a
very rapid turn-off (larger $/\beta/$), and (2) have a steeper optically thin radio
spectral index (larger $/\alpha/$).

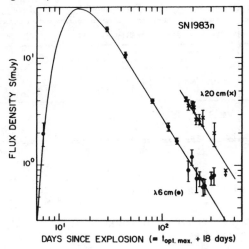

Fig. 1. Typical radio "light curve" for a Type Ib/c RSN (SN1983N in
NGC5236). Both λ6 cm (filled circles) and λ20 cm (crosses) are shown
with the solid curves from a minimum χ^2 fit to Eqs. 1, 2, and 3.

Type II RSNe
Type II RSNe: (1) turn-on rapidly, although often delayed (larger K_2), followed by a slower turn-off (smaller $/\beta/$), and (2) have a flatter optically thin
radio spectral index (smaller $/\alpha/$).

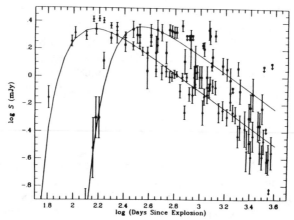

Fig. 2. Typical radio "light curves" for a Type II RSN (SN1980K in
NGC6946). Both λ6 cm (filled triangles) and λ20 cm (filled squares) are
shown with the solid curves from a minimum χ^2 fit to Eqs. 1, 2, and 3.

MASS LOSS RATES AND ZAMS MASSES

TABLE I   Mass Loss Rates and Estimated ZAMS Masses

	'70G	'79C	'80K	'81K	'83N	'84L	'86J	'90B
SN Type	II	II	II	II	Ib	Ib	II	Ic
Mass loss ($M_\odot$ yr^{-1})	2×10^{-5}	1×10^{-4}	3×10^{-5}	$<5\times10^{-6}$	2×10^{-6}	2×10^{-6}	2×10^{-4}	3×10^{-6}
Est. ZAMS Mass ($M_\odot$)	≥10	≥13	≥7	$\sim8?$	~6.5	~6.5	~25	6-8

Mass Loss Rates
The mass loss rates in Table I are estimated using the Chevalier (1981a,b,
1984a,b) model and Eq. 16 of Weiler et al. (1986):

$$\dot{M}/w = 3.02 \times 10^{-6} \, \tau_{5\,GHz}^{0.5} \, m^{-1.5} \, (v_i \, /[10^4 \text{ km s}^{-1}])^{1.5} \, (t_i \, /[45 \text{ days}])^{1.5} \qquad (4)$$
$$(t/t_i)^{1.5m} \, (T/10^4 \text{ K})^{0.68} \, M_\odot \text{ yr}^{-1},$$

where w is the RSG wind velocity, τ is the optical depth at t days after
explosion date t_0, m ($\equiv -\delta/3$) is the deceleration rate of the ejecta (R $\propto t^m$), v_i
is the ejecta velocity at t_i, and T is the electron temperature in the wind (w =
10 km s^{-1}, t = 1 day, τ(day 1) = K_2, v_i = 10^4 km s^{-1}, t_i = 45 days, T = 10^4 K).

ZAMS Masses
If the presupernova mass loss originates in an RSG wind, the models of
Maeder & Meynet (1988) yield estimates of the ZAMS masses in Table I.

PHYSICAL MODELS

Type Ib/c
Sramek et al. (1984) and Van Dyk et al (1993) suggest that the progenitor systems for Type Ib/c SNe involve a star with ZAMS $\sim$ 6.5 $M_{\odot}$ and a slightly less massive companion. The more massive star evolves faster to become a $\sim$ 1.35 $M_{\odot}$ white dwarf and, during its RSG phase, transfers part of its mass to its previously less massive companion. When the companion reaches the RSG phase with a high mass loss rate, it establishes the high density stellar cocoon excited by the SN shock to produce the radio emission. It also supplies $\sim$ 0.1 $M_{\odot}$ to the white dwarf SN progenitor, pushing it over the Chandrasekhar limit and causing it to explode. With this concept, it is thus the companion to the Type I SN which generates the dense circumstellar cocoon.

Type II
Type II SNe are thought to originate in isolated, massive stars with M $\geq$ 8 $M_{\odot}$ which follow normal stellar evolution through a RSG phase and then suffer core collapse and explode. Thus, the presupernova star establishes its own dense circumstellar cocoon with which the SN shock wave interacts to produce the radio emission. Two deviations from this scenario have been noted, however: SN1979C (Weiler et al. 1992) shows periodic modulation of its radio light curve, possibly caused by a non-interacting binary companion, and SN1986J (Weiler et al. 1990) implies that, above some ZAMS mass limit, internal, thermal matter mixed with the supernova ejecta is significant for the development of a RSN's radio light curve.

REFERENCES

Chevalier, R.A. 1981a, ApJ **246**, 267
_____ 1981b, ApJ **251**, 259
_____ 1984a, Ann NY Acad Sci **422**, 215.
_____ 1984b, ApJ **285**, L63.
Maeder, A. & Meynet, G. 1988, A&AS **76**, 411.
Sramek, R.A., Panagia, N., & Weiler, K.W. 1984, ApJ **285**, L59.
Van Dyk, S.D., Sramek, R.A., Weiler, K.W., & Panagia, N. 1993, ApJ, in press.
Weiler, K.W.. Sramek, R.A., Panagia, N., van der Hulst, J.M., & Salvati, M. 1986, ApJ **301**, 790.
Weiler, K.W., Panagia, N., & Sramek, R.A. 1990, Ap.J. **364**, 611.
Weiler, K.W., Van Dyk, S.D., Pringle, J.E., & Panagia, N. 1992, ApJ, in press.

Massive Stars: Their Lives in the Interstellar Medium
ASP Conference Series, Vol. 35, 1993
Joseph P. Cassinelli and Edward B. Churchwell (eds.)

SUPERNOVAE AND MASSIVE STAR FORMATION REGIONS IN LATE-TYPE GALAXIES

SCHUYLER D. VAN DYK
Naval Research Lab, Code 4215, Washington, DC   20375-5320

MARIO HAMUY
CTIO, Casilla 603, La Serena, Chile

ABSTRACT   From the association of Type Ib/c and Type II supernovae with giant H II regions in parent late-type galaxies, we conclude that both SNe types have massive stellar progenitors which are possibly very similar in mass range.

INTRODUCTION

We are updating a recent study (Van Dyk 1992) of the association of supernovae (SNe) with massive star formation regions in their parent late-type galaxies. We concentrate here only on the Type II SNe (SNe II) and Type Ib/c SNe (SNe Ib/c), since their progenitors have been modelled by the explosions of massive short-lived stars ($M_{\mathrm{ZAMS}} \geq 8\ M_\odot$; e. g., Woosley & Weaver 1986). Late-type galaxies are more often SNe parents than other galaxy types (e. g., Evans et al. 1989). Massive star formation is traced in the parent galaxies through narrow-band Hα imaging of giant H II regions (GHRs). Kennicutt et al. (1989) found that in late-type galaxies the GHRs are sites of the majority of massive star formation. Therefore SNe associated with GHRs possess a greater chance of having had massive stellar progenitors than SNe which occurred well outside GHRs. Association of SNe with GHRs is established by comparing the projected angular separation of each SN from its nearest GHR (Δ) with the maximum angular radial extent (r_{max}) of the GHR measured toward the SN position.

PROCEDURE

Our sample now includes 15 SNe Ib/c and 31 SNe II, all spectroscopically classified. Continuum-subtracted, flux-calibrated CCD Hα data were acquired either by us (at KPNO, CTIO and Lowell Observatory) or by other investigators. We have reduced the level of positional uncertainty for many SNe by using absolute positions or precise nuclear offset positions. We have transformed many of the CCD images to absolute coordinates (errors $\sim 1''$ – $2''$) from astrometry using GASP at STScI. For lack of a better discriminator, those SNe with absolute positions or accurate nuclear offsets are given three times the weight in our statistics than those SNe before ~ 1980 with only less accurate offsets (errors $\sim 5'' - 10''$) and twice the weight of those SNe after $\sim$ 1980 with less accurate offsets.

Detected H II regions nearest SNe were generally more luminous than $\sim 5 \times 10^{37}$ erg s^{-1}. The quantities r_{max} for GHRs were generally measured to a limit of $\sim 2 \times 10^{-17}$ erg cm^{-2} sec^{-1} arcsec^{-2} for SNe with $\Delta \lesssim 40''$. For each SN the ratio R is defined as $R \equiv \Delta/r_{max}$.

RESULTS

The cumulative proportions of SNe associated with GHRs are given in Table I. Ideally, if $R \leq 1$, then a SN is associated with a GHR. Since an indeterminable uncertainty in R exists for many SNe (arising from generally unknown uncertainties in Δ), we computed the proportions for $R>1$ as well. Also given in Table I are the statistical errors in the proportions, $\sqrt{N_{assoc}}/N_{type}$, where N_{assoc} is the number of SNe of a given type associated with GHRs and N_{type} is the total number of SNe of that type. The results here for a larger sample of both SNe II and SNe Ib/c are similar to the results of Van Dyk (1992), but the additional SNe Ib/c here have substantially reduced the effects due to small-number statistics.

Comparing these proportions to the range in the probability of chance superposition (i. e., the ratio of the fraction of the galactic disk covered by GHRs to the proportion of associated SNe) from Van Dyk (1992) for the SNe II with $R \leq 1$, which is $0.09 - 0.29$, we can be confident that the associations of both SNe II and SNe Ib/c with GHRs are quite likely to be real.

TABLE I   Association of SNe with GHRs

| SN Type | Proportion of SNe with | | |
	$R \leq 1.0$	$R \leq 2.0$	$R \leq 3.0$
Ib/c	0.35 ($\pm$0.10)	0.75 ($\pm$0.15)	0.86 ($\pm$0.16)
II	0.52 ($\pm$0.10)	0.79 ($\pm$0.12)	0.79 ($\pm$0.12)

CONCLUSIONS

We conclude that both SNe II and SNe Ib/c have massive stellar progenitors. Any differences between the proportions of SNe II and SNe Ib/c associated with GHRs given in Table I are statistically indistinguishable, which suggests that the progenitors for both types may be very similar in mass range, as is also suggested by observations of some SNe Ib/c with spectral characteristics of SNe II (e. g., Filippenko 1988, 1991a, 1991b, 1992). We can safely discount the Wolf-Rayet star progenitor model (with $M_{ZAMS} \geq 30\ M_\odot$) for SNe Ib/c (e. g., Ensman & Woosley 1988; Fransson & Chevalier 1989; Swartz & Wheeler 1991), since if correct, then the proportion of SNe Ib/c associated with GHRs would be greater than the proportions of SNe II. It is more likely that SNe Ib/c arise from massive interacting binary systems, while SNe II arise from single massive stars.

For both SNe Ib/c and SNe II, $\sim$ 20 % must have occurred in H II regions below our detection limit (although some may have had runaway OB stars as progenitors). Many of the SNe, particularly SNe II, occurred in faint H II regions near this limit.

ACKNOWLEDGEMENTS

We appreciate the generosities of Rob Kennicutt, Bill Keel, Jeff Kenney, and Florence Durret for lending us several CCD images, and of Mike Rupen for some accurate SNe positions. Astrometry of the CCD images was obtained using the Guide Stars Selection System Astrometric Support Program developed at STScI (operated by AURA, Inc., for NASA). KPNO and CTIO, National Optical Astronomy Observatories, are operated by AURA, Inc., under contract with the National Science Foundation.

REFERENCES

Ensman, L. M., & Woosley, S.E., ApJ, 333, 754.
Evans, R., van den Bergh, S., & McClure, R.D. 1989, ApJ, 345, 752.
Filippenko, A. V. 1988, AJ, 96, 1941.
Filippenko, A. V. 1991a, IAU Circ., No. 5169.
Filippenko, A. V. 1991b, in SN 1987A and Other Supernovae, edited by
	I. J. Danziger and K. Kjär (ESO, Garching bei München), p. 343.
Filippenko, A. V. 1992, ApJ, 384, L37.
Fransson, C., & Chevalier, R. A. 1989, ApJ, 343, 323.
Kennicutt, Jr., R.C., Edgar, B.K., & Hodge, P.W. 1988, ApJ, 337, 761.
Swartz, D. A., & Wheeler, J. C. 1991, ApJ, 379, L13.
Van Dyk, S. D. 1992, AJ, 103, 1788.
Woosley, S. E., & Weaver, T. A. 1986, ARA&A, 24, 205.

Massive Stars: Their Lives in the Interstellar Medium
ASP Conference Series, Vol. 35, 1993
Joseph P. Cassinelli and Edward B. Churchwell (eds.)

A KINEMATIC SEARCH FOR SUPERNOVA REMNANTS IN GIANT EXTRAGALACTIC H II REGIONS

HUI YANG and EVAN SKILLMAN
Astronomy Dept., University of Minnesota, Minneapolis, Minnesota

ABSTRACT   We have used the TAURUS imaging Fabry-Perot interferometer to obtain the $H\alpha$ emission velocity fields of the giant H II complexes NGC 5471 in M101, NGC 2363 in NGC 2366, and the largest H II region in NGC 2403. Sources were detected whose velocity profiles have unusually large widths when compared with the surrounding H II region. Radio continuum data from the VLA show nonthermal emission in NGC 5471 and NGC 2363 which coincides with these large-velocity-width-sources (LVWSs).

I INTRODUCTION

Since Giant Extragalactic H II Regions (GEHRs) are the sites of intense bursts of massive star formation, it is highly likely that they might be accompanied by supernova remnants (SNRs). Conventional searches for SNRs make use of radio continuum surveys, optical emission line surveys, and X-ray imaging. Because of the brightness of the nebular background in optical and radio emission, very few SNRs are found in GEHRs by these methods. Supernova rates based on SNR detections are therefore incomplete since a significant fraction of more massive stars are born in GEHRs and their SNRs are buried in nebula (van den Bergh & Tammann 1991). In principle, SNRs can also be detected by the use of spectroscopic searches for high velocity gas (Chu & Kennicutt 1986; Casteñeda et al. 1990). Chu & Kennicutt employed long-slit echelle spectroscopy and discovered four large-velocity-width sources inside NGC 5471 A,B,C and NGC 5461 in M101. They proposed that the LVWSs might be due to supernova remnants embedded in the GEHRs.

II OBSERVATIONS

We have used the TAURUS imaging Fabry-Perot interferometer with the 2.5m Isaac Newton Telescope at the Roque de los Muchachos Observatory on La Palma (Canary Islands). Three GEHRs were observed in the light of $H\alpha$. The data format was 240×240 pixels of 0.6" size by 100 channels of 6.8 km s^{-1} with a resolution of 18 km s^{-1}(FWHM). NGC 5471 and NGC 2363 were each observed for a total of 50 minutes and NGC 2403 # 1 was observed for 33 minutes. In the GEHR, where the H II radiation is dominant, if there are supernova remnants, we would expect to see a profile which is the combination of a relatively narrow Gaussian profile and a broad component at low intensity. In order to increase our sensitivity to low surface brightness, broad components, the data were convolved with a broad Gaussian profile (40 km s^{-1} FWHM). In ordinary GEHRs, the emission line profile is close to a Gaussian shape with FWHM smaller than 50 km s^{-1}.

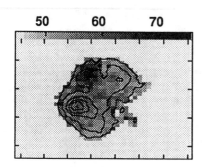 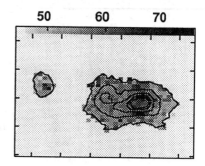

Fig. 1 A contour representation of the Hα image overlaid on a greyscale of the FWHM field of NGC 5471 (left) & NGC 2363 (right). The gray scale ranges from 45 km s^{-1} to 75 km s^{-1}. The spacing in declination is 5 arcsecond.

Figure 1 shows the velocity dispersions (second moments) for two of the Hα data cubes. In all three of the observed GEHRs there are areas with anomalously large dispersions. In figure 1 left, the LVWS at NGC 5471 B discovered by Chu & Kennicutt is recovered. In Figure 1 right, there are two LVWSs symmetrically located at the sides of the brightest nucleus where Roy et al. (1991) detected [O III] λ 5007Å line splitting. An expanding shell structure with a velocity of 45 km s^{-1} has been proposed to explain this feature. Two more LVWSs are found near the nucleus of NGC 2403 #1.

Radio continuum observations were taken at 1465 MHz and 4885 MHz for NGC 5471 and NGC 2363. The A-configuration data at 1465 MHz and B-configuration data at 4885 MHz were chosen to match the same uv coverage of 4.5 to 130 kilowavelengths. Region B in NGC 5471 clearly shows nonthermal emission while regions A and C are dominated by thermal emission, confirming Skillman's (1985) earlier observation. NGC 2363 region A also shows nonthermal emission.

III PHYSICAL NATURE OF NGC 5471 B

NGC 5471 is the most luminous H II complex in the giant spiral galaxy M101. The high velocity width of NGC 5471 B could be caused by either stellar winds or supernova remnants. The large velocity dispersion, the high mass, the large size and the lack of evidence for Wolf-Rayet stars make stellar winds an unlikely cause. Nonthermal radio continuum emission and enhanced [O I] and [S II] emission point to the presence of shocks due to SNRs. However, if NGC 5471 B is a single SNR, it is a very unusual one. Its flux density at 408 MHz is 4.8 mJy which is about 6 times the luminosity of the brightest SNR in the LMC (corrected for distance). Its Hα flux is 15×10^{-14} ergs cm^{-2} s^{-1} which is 30 times more luminous than the most luminous SNR in the LMC.

The reason for this extremely high Hα luminosity is not clear. We can think of two possibilities. First, for a SNR inside an already extant H II region, all material in the remnant is subject to the ionizing radiation from nearby massive stars. This should significantly add to the effect of ionization by UV photons from the inner side of the SNR shock fronts. Since the ionization of the shocked gas in the SNR is dominated by photoionization from the nearby stars, its Hα luminosity should be comparable with an H II region not an isolated SNR. The second possibility is that NGC 5471 B might be a superbubble formed by the coalescence of many SNRs. It has many similar properties to the superbubble found in IC 10 (Skillman 1988, Yang & Skillman in prep.). If interpreted as a single SNR, the IC 10 source would be quite spectacular exceeding 200 pc in size. However, the IC 10 source is roughly 7 times closer and is resolved into a collection of nonthermal sources which have properties comparable to the SNRs in the Magellanic Clouds. Similarly, NGC 5471 B may be an unresolved collection of supernova remnants although its size is less than 60 pc.

IV CONCLUSIONS

Within our Local Group, supernova remnants are most likely detected through nonthermal radio emission and optical emission lines. At distances greater than 1 Mpc, individual SNRs are no longer spatially resolved, and their detectability drops rapidly with distance. Ordinary SNRs, like those in M33, are difficult to detect at the distance of M101. The discovery of the SNRs in the giant H II regions of M101 and NGC 2366 prove that the properties of these SNRs are different from isolated ones. An interpretation of these objects as multiple supernova remnants results in an evolutionary scenario linking these broad line sources observed in giant H II regions to large nonthermal sources (e.g., as seen in IC 10), and finally to the supershell structure in galaxies.

We conclude that LVWSs are easily found in GEHRs using relatively short integrations with an imaging Fabry-Perot. The fact that we found LVWSs in all 3 GEHRs studied may indicate that they are a common phenomenon in GEHRs. The LVWSs are probably SNRs embeded in H II regions.

REFERENCES

Castañeda, H.O., Vilchez, J.M., Copetti, M.V.F. 1990 ApJ, 365, 164.

Chu, Y.-H., & Kennicutt, R.C. 1986 ApJ, 311, 85.

Roy, Boulesteix, Joncas, & Grundseth 1991 ApJ, 367, 141.

Skillman, E.D. 1985 ApJ, 290, 449.

Skillman, E.D. 1988 in *Supernova Remnants and the Interstellar Medium*, ed. Roger, R.S. & Landecker, T.L. (Cambridge: Cambridge University Press), p.465.

van den Bergh, S., & Tammann, G.A. 1991 Ann. Rev. Astron. Ap., 29, 371

SECTION 5

THE EFFECTS ON GALAXIES AND BROADER ISSUES

Massive Stars: Their Lives in the Interstellar Medium
ASP Conference Series, Vol. 35, 1993
Joseph P. Cassinelli and Edward B. Churchwell (eds.)

WOLF-RAYET STARS AND WOLF-RAYET GALAXIES

PETER S. CONTI

Joint Institute for Laboratory Astrophysics and
Department of Astrophysical, Planetary, and Atmospheric Sciences,
University of Colorado, Boulder CO 80309-0440

ABSTRACT   Wolf-Rayet stars are the hot, luminous, helium
burning descendants of the most massive stellar objects; their emission line spectra are due to strong stellar winds. These stars live only
a few hundred thousand years, following normal O star lifetimes of
a few million years after which stellar wind mass loss has enabled
chemically altered material to reach the surface. Wolf-Rayet galaxies are those starburst objects, photoionized by hot stars, in which
broad emission features due to Wolf-Rayet stars have been detected
in their integrated spectra. I shall review the overall properties of
Wolf-Rayet galaxies, many of which show evidence for mergers or
collisions but some appear to be isolated. Following recent work by
Vacca (1991; Vacca & Conti 1992) I estimate the numbers of Wolf-
Rayet stars and other exciting objects in a sample of mostly metal
deficient star-forming galaxies from their optical emission line spectra. The inferred Wolf-Rayet star fraction in many of these galaxies
seems to be sufficiently large that it can have come about only in
a relatively recent starburst of relatively short duration: perhaps
lasting only a million years.

INTRODUCTION

Emission-line galaxies are a subset of starburst galaxies whose integrated spectra
have relatively blue continua and are dominated by nebular emission lines due
to photoionization from the presence of significant numbers of O-type stars. It
has been suggested by Sargent & Searle (1970) that these galaxies represent
relatively recent, potentially sporadic, episodes of massive star formation. *Wolf-
Rayet galaxies* are a subset of emission line galaxies that contain a large number
of Wolf-Rayet (W-R) stars, as inferred from the presence of a broad He II λ4686
emission line (Conti 1991). *W-R stars* are the highly evolved descendants of
the most massive O stars (i.e., stars with initial masses $M_O \gtrsim 35 M_\odot$; Conti

et al. 1983*a*; Humpheys *et al.* 1985). These stars are hot, extremely luminous and exhibit spectra characterized by strong, broad emission lines resulting from dense, high velocity stellar winds.

WOLF-RAYET STARS

There are two main types of W-R stars, WN and WC, for which the optical emission lines serve as the basis for the current one-dimensional classification scheme. In Fig. 1, I show examples of the spectra of WN subtypes, in which the dominant emission lines are those from helium and nitrogen ions. Note that on this *log-log* plot, the extinction corrected continuum fluxes can be closely approximated by a power law over most of the observed spectral range from the far-UV through the optical to the near IR (Morris *et al.* 1992). Numerical subtypes have been assigned to WN stars based on the relative strengths of the He II $\lambda4686$, N III $\lambda\lambda4634,4640$, N IV $\lambda4057$, and N V $\lambda\lambda4604,4620$ emission lines (see e.g., van der Hucht *et al.* 1981; Conti *et al.* 1983*b*). One of the most noteworthy aspects of these spectra is the strong emission feature at 4686 Å due to He II. In the latest type WN star (WN8), a N III $\lambda\lambda4634,4640$ emission line is present, with a strength almost comparable to $\lambda4686$; in other WN stars, this line, and all others of various ions, are weaker.

Lines of helium, carbon, and oxygen are the principal emission features in the spectra of WC stars. Examples of the WC spectral sequence are shown in Fig. 2. Again note the power law nature of the continua. WC stars have been classified into numerical subtypes based upon the relative strengths of the C III $\lambda5696$, C IV $\lambda\lambda5801,5812$, and O V $\lambda5592$ emission lines (see e.g., Torres *et al.* 1986). In the latest type WC9 star, the C III $\lambda5696$ line is much stronger than C IV $\lambda\lambda5801,5812$. Note the strong C III emission at $\lambda4650$, comparable in strength to the C IV lines in the yellow region of the spectrum.

The classification system is primarily an ionization/excitation sequence; the coarser designations of "early" (E) and "late" (L) (i.e., WNE, WNL, WCE, and WCL) have been defined by Conti & Massey (1989). The strong emission lines in W-R spectra indicate the presence of substantial, dense, optically thick stellar winds in which mass flows out from the stars at typical rates of $\dot{M} \sim 2 \times 10^{-5}$ $M_\odot$ yr^{-1} and with terminal velocities of $V_\infty \sim 2500 - 3000$ km s^{-1} (Abbott & Conti 1987). The WN stars exhibit the results of core hydrogen burning (in the CNO cycle) which have been brought to the stellar surface through the action of mass loss and mixing (convective overshooting) according to current thinking (e.g. Maeder & Meynet 1987, 1989). The spectra of WC stars reveal the presence of substantial quantities of the products of core helium burning. Observations and modeling lead to the conclusion that W-R stars represent highly evolved objects, most of which are burning helium in their cores. In this scenario, the W-R stage lasts approximately 10% of the main sequence lifetime of the O star predecessors.

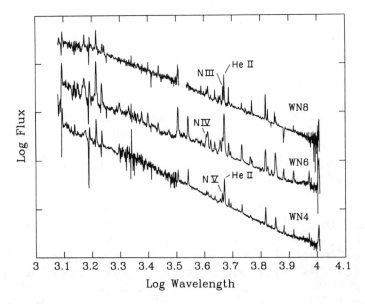

Fig. 1. Extinction corrected spectra of three WN stars (WR128:WN4; WR134:WN6; Br13:WN8) from Morris (private communication). The optical classification lines are indicated.

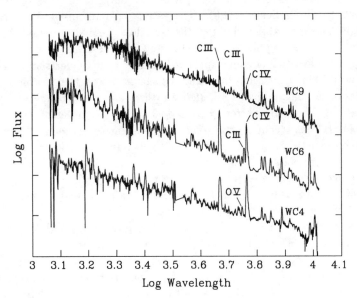

Fig. 2. Extinction corrected spectra of three WC stars (Br7:WC4; WR23:WC6; WR69:WC9) from Brownsberger (private communication). The optical classification lines are indicated.

Evolutionary models may be used to predict the ratio of the number of W-R stars to the number of O star progenitors (i.e., O stars with initial masses large enough such that the star may eventually evolve into a W-R star), N_{W-R}/N_O. This ratio should vary with metallicity and reflect the relative lifetimes of the two evolutionary stages (e.g. Maeder 1991). His models predict a number ratio of about 0.1 for the solar neighborhood, in reasonable agreement with the observationally determined value. Other quantities, such as the predicted ratio of the number of WC stars to WN stars, N_{WC}/N_{WN}, are also found to be in good agreement with the observed values. These evolution models confirm the widely accepted belief that W-R stars are extreme Population I objects which are the descendants of only the (initially) most massive O stars.

WOLF-RAYET GALAXIES

The first W-R galaxy so identified was He2-10 (Allen *et al.* 1976). As can be seen from Fig. 1, He II $\lambda 4686$ is one of the most prominent emission features in the spectra of WN stars; similarly from Fig. 2 C III $\lambda 4650$ and C IV $\lambda 5801, 5812$ are strong in WC stars. The presence of hundreds to thousands of W-R (mostly WN) stars in these galaxies has been inferred from a comparison of the equivalent width of the broad features with those of the corresponding emission lines in the spectra of Galactic W-R stars (e.g., Kunth & Sargent 1981; Osterbrock & Cohen 1982; Kunth & Schild 1986). The estimated number of W-R stars in these galaxies is remarkable in light of the fact that there are only ~ 175 W-R stars known in our Galaxy and only ~ 110 known in the LMC.

W-R galaxies are distinguished spectroscopically from Seyferts and AGN by their narrower Balmer emission lines and, as shown below, their forbidden line intensity ratios are also indicative of *stellar* rather than *nonthermal* (i.e., power law) excitation. There are currently nearly 50 known examples of W-R galaxies. Conti (1991) has recently catalogued these very heterogeneous objects and reviewed their general properties. Many W-R galaxies are Markarian or Zwicky objects; several are classified as Blue Compact Dwarf Galaxies. Many exhibit disturbed morphologies which may be the result of interactions or mergers. At least one W-R galaxy is an extremely powerful far-infrared source (Armus *et al.* 1988).

Previous studies of some W-R galaxies have not always had the spectral resolution to distinguish the He II $\lambda 4686$ feature from other nearby broad (i.e., stellar) lines. In several galaxies a broad feature near 4650 Å (the "W-R bump"; e.g. Kunth & Joubert 1985; Smith 1991) has been detected and attributed to the blends of N III, C III, or C IV as well as He II $\lambda 4686$. The N III blend is characteristic of the spectra of WNL stars; the C III/IV features are characteristic of the spectra of WC stars (Conti & Massey 1989; Conti *et al.* 1990). Although a broad C IV $\lambda 5808$ emission line, indicative of WC stars, has been detected in the spectra of a few W-R galaxies (Sargent & Filippenko 1991; Sargent, private communication), most W-R galaxies are found to contain primarily WNL stars. It has been suggested that the lack of WC stars is the result of the low metallicities in these galaxies (Smith 1991).

One of the most surprising results of studies of W-R galaxies, in addition to the sheer numbers of W-R stars present, is the large inferred value of N_{W-R}/N_O. Values as large as unity have been reported in the literature (e.g., Osterbrock & Cohen 1982; Armus $et\,al.$ 1988; Sargent & Filippenko 1991). The current models of massive star evolution, as calculated by Maeder (1990, 1991), for example, do not yield such large number ratios even if the metallicity of the star-forming region is much greater than that found in the solar vicinity. The Maeder (1991) predictions can be used if the star formation processes are in equilibrium with star deaths; for massive stars with relatively short evolution times this means time scales larger than 10^7 years ("continuous" star formation). Arnault $et\,al.$ (1989) have synthesized W-R star populations based on the earlier massive star evolutionary tracks calculated by Maeder & Meynet (1987). These models, which had no mass loss - mass dependence in the W-R phase, have since been superceded by the newer Maeder (1991) models. Arnault $et\,al.$ give predictions based upon an assumption of an "instantaneous" burst of star formation, with a time scale less than 10^6 years. (Their predictions for "continuous" star formation are much lower than the Maeder 1991 calculations, as expected, since the latter W-R models have longer lifetimes.) The "burst" models give N_{W-R}/N_O ratios about four times larger than in the "continuous" case, according to Arnault $et\,al.$, but the values never approach unity.

OBSERVATIONS AND ANALYSIS

Vacca & Conti (1992) have obtained moderate resolution ($\sim$ 3.5 Å) and S/N ($\sim$ 15 − 30) spectra of 10 known W-R galaxies with the 2D-Frutti detector on the 4-m telescope at CTIO in order to investigate the general properties of W-R galaxies and study their stellar populations. The spectra cover the wavelength ranges $\sim$ 3120 to $\sim$ 7000 Å. In our galaxies there is often a bright core, and one or sometimes two other star forming knots; our long slits, with width 1''5, did not always fully cover the regions of interest. This work is more fully described in Vacca (1991) and Vacca & Conti (1992).

In Fig. 3, I show a spectrum of the well-known W-R galaxy He2-10. Note the strong nebular emission line spectrum due to photoionization by stars. Broad emission lines can be seen at He II $\lambda4686$ and C IV $\lambda\lambda5801, 5812$, showing that both WN and WC stars are present. Similarly, in Fig. 4 is the spectrum of BCDG IIZw40. A broad emission line is found at He II $\lambda4686$, indicating the presence of WN stars.

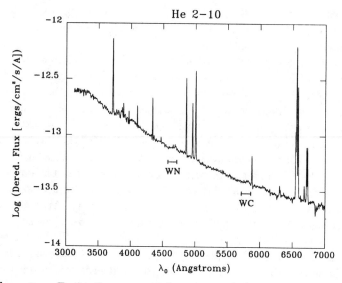

Fig. 3. Extinction corrected spectra of the W-R galaxy He2-10 (from Vacca, private communication). Note the relatively blue continuum and the weak Balmer Jump. The presence of WN stars is indicated by the broad feature at He II λ4686; similarly WC stars appear to be present due to the broad emission at C IV $\lambda\lambda$5801,5812.

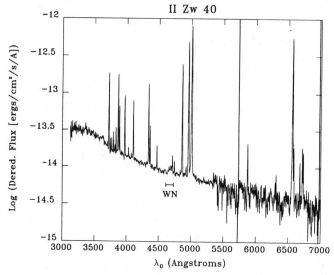

Fig. 4. Extinction corrected spectra of the W-R galaxy IIZw40 (from Vacca, private communication). Note the relatively blue continuum and the nearly absent Balmer jump. The presence of WN stars is indicated by the broad emission feature at He II λ4686. The narrow emission lines just longward are due to [Ar IV].

The nebular emission lines have been used to determine reddenings, electron temperatures, densities, and oxygen abundances in these galaxies by the usual methods of analysis of H II regions (e.g. Osterbrock 1989). Distances were derived from heliocentric redshifts ($H_0 = 75$ km/s/Mpc). Our spectra were corrected for Galactic foreground reddening, as given by Burstein & Heiles (1984), as well as internal reddening. Internal extinctions, in terms of color excesses E_{B-V}, were estimated from the observed $F(H\alpha)/F(H\beta)$ flux ratios, an assumed intrinsic flux ratio from Hummer & Storey (1987), and the extinction curve appropriate for the LMC H II region 30 Doradus (Vacca 1991). Electron temperatures were derived from the reddening-corrected [O III] $F(\lambda 4959)/F(\lambda 4363)$ flux ratios and are about 10^4 K for all galaxies. Electron densities were derived from the corrected [S II] $F(\lambda 6716)/F(\lambda 6731)$ flux ratios and are of the order of 10^2 cm^{-3}. Total oxygen abundances were derived from the various oxygen emission lines.

In Fig. 5, I plot the reddening corrected emission line ratios utilized by Veilleux & Osterbrock (1987) for our galaxies. For these objects, the ratios consistently indicate that the emission line excitation is similar to that of H II regions, that is, from hot stars. We are thus justified in using the strength of the Hβ to determine the numbers of hot stars present.

The analysis procedure was as follows: The number of equivalent WNL stars was calculated by dividing the estimated luminosity in the He II $\lambda 4686$ emission feature in each galaxy by the average value derived for single WNL stars in the LMC (Vacca 1992), $\langle L_{WNL}(\lambda 4686) \rangle \approx 1.7 \times 10^{36}$ ergs s^{-1}. We then assumed each WNL star contributed $\langle Q_0^{WNL} \rangle \approx 1.7 \times 10^{49}$ photons s^{-1}, an average value derived by Vacca (1991) from application of the models of W-R stars of Schmutz $et\,al.$ (1992) to a sample of LMC WNL stars. We now have an estimate of the number of WNL stars and their contribution to the number of Lyman continuum photons in these galaxies.

The hot star content can be expressed in terms of 'equivalent' O7V stars. We assumed each equivalent O7V star contributes $\langle Q_0^{O7V} \rangle \approx 1.0 \times 10^{49}$ photons s^{-1} to the ionizing flux (Leitherer 1990). Each of these stars has an $H\beta$ luminosity of $\langle L_{O7V}(H\beta) \rangle \approx 4.8 \times 10^{36}$ ergs s^{-1}. We then use the observed $H\beta$ luminosity to estimate the total number of exciting stars (O7V $plus$ WNL) and, with the above numerical values, and some algebra, the number ratio of WNL to O7V stars, N_{WNL}/N'_{O7V}, can be calculated from the equation

$$N_{WNL}/N'_{O7V} = [0.35\, L(H\beta)/L(\lambda 4686) - 1.68]^{-1} \quad ,$$

which is appropriate for densities $n_e \approx 10^2$ cm^{-3} and temperatures $T_e \approx 10^4$ K. Note that this equation predicts a maximum value of the ratio $L(\lambda 4686)/L(H\beta) = 0.21$ for Case B ionization balance by a population of W-R stars.

The total number of O stars present, N_{Otot}, was determined from the number of equivalent O7V stars, N'_{O7V}, with the relation $\eta \equiv N'_{O7V}/N_{Otot}$. The mass dependency factor η was introduced by Vacca (1991) and depends upon the

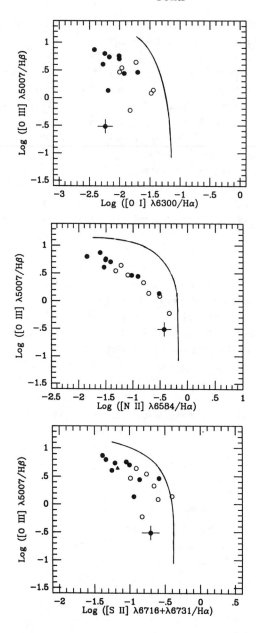

Fig. 5. Nebular excitation diagrams for W-R galaxies showing the relationships of [S II], [N II], and [O I] to Hα *versus* [OIII] to Hβ (following Veilleux & Osterbrock 1987). All of the galaxies illustrated show H II region characteristics.

slope of the Present Day Zero Age Mass Function (PDZAMF), the metallicity (which affects the lower mass limit for the definition of an O star), and the upper mass cutoff. For each galaxy, η was determined from our oxygen measurement, Maeder's (1991) models, Leitherer's (1990) calibration of the Lyman continuum flux from O stars of various spectral types, a PDZAMF with a slope $x = 1.5$ as found for 30 Dor (Parker 1992), and an upper mass limit $M_{upper} = 130\ M_\odot$. I present in Table I the various properties of the W-R galaxies derived from our data (Vacca & Conti 1992).

According to Moffat *et al.* (1987) there are some 20 W-R stars in the inner regions of 30 Dor. All of our galaxies contain starburst regions with *at least* this many and up to 30 times more WN stars! The numbers of O stars in 30 Dor, about 150 from the stellar census of Parker (1992), may also be compared with the numbers derived for our W-R galaxies. From this comparison it is easily seen that these starbursts are similar to, or greatly scaled up versions of 30 Dor.

Vacca (1991) has confirmed that the method used above is approximately correct through a similar spectral line analysis of an $8' \times 8'$ area spectral scan of 30 Dor. This spectrum was obtained by Mark Phillips on the 1.5-m at CTIO using a long $8'$ slit and by adjusting the diurnal rate so that at the end of the predetermined 15 min exposure time the telescope had moved $8'$ in R.A. The spectral analysis of the 30 Dor region revealed approximately the same numbers of O and W-R stars as found by Parker and by Moffat *et al.* We have subsequently obtained better spectral resolution coverage of several different areas centered on 30 Dor and analysis is underway (Vacca, Phillips, & Conti 1992).

From N_{Otot} and the model atmospheres of Kurucz (1979), we computed the fraction f of galaxy light at ~ 4900 Å contributed by O stars, by comparing it to the observed galaxy continuum through the slit. This fraction is shown in the last column of Table I; the median value is about 10%, but in several cases closer to half of the light comes from the O stars. The origin of the remainder is presumably due to previous generations of stars in the underlying galaxy.

NGC1365

Drew Phillips serendipitously observed the GHII region L4 in the galaxy NGC 1365 with high resolution but limited spectral coverage as shown in Fig. 6. In this object a broad emission feature is found at the correctly redshifted position of C III $\lambda5696$; the C IV $\lambda5808$ is weakly present, if at all. As can be seen from Fig. 2, these spectral features are characteristic of WC9 stars. Phillips & Conti (1992) have analyzed this metal rich region by a procedure similar to that above. They found at least 300 WC9 stars, and over 1000 O7V stars, but there are uncertainties in the reddening. This GHII region is located at the end of a bar in NGC 1365; the galaxy core has a Seyfert spectrum.

TABLE I - Properties of Wolf-Rayet Galaxies

Galaxy	D (Mpc)	$-W_\lambda(H\beta)$ (Å)	$L(H\beta)$ $\times10^{39}$ (ergs/s)	$L_\lambda(\lambda4861)$ $\times10^{38}$ (ergs/s/Å)	$-W_\lambda(\lambda4686)$ (Å)	$L(\lambda4686)$ $\times10^{38}$ (ergs/s)	N_{WNL}	$\dfrac{L(\lambda4686)}{L(H\beta)}$	$\dfrac{N_{WNL}}{N_{O7V}}$	(O/H) $\times10^{-4}$	η	$\dfrac{N_{WNL}}{N_{Otot}}$	N_{Otot}	f
He 2-10	9.0	39.6	25	6.3	0.7	5.0	290	0.020	0.07	1.15	1.06	0.07	4400	0.05
Mrk 1094	36	18.4	5.7	3.1	0.6	2.2	130	0.038	0.14	2.63	1.22	0.17	780	0.02
Mrk 1236	23	131	4.7	0.36	0.8	0.4	20	0.008	0.03	1.23	1.07	0.03	890	0.18
NGC 1741	51	86.0	94	10.9	0.9	12	710	0.013	0.04	2.02	1.07	0.05	16000	0.11
NGC 3049	18	35.1	8.4	2.4	2.3	6.5	380	0.077	0.36	12.2	1.59	0.57	680	0.03
NGC 3125 A	12	96.2	13	1.4	4.2	6.6	390	0.050	0.19	2.52	1.21	0.23	1700	0.10
NGC 3125 B	12	64.4	18	2.8	2.7	8.9	530	0.048	0.18	2.41	1.20	0.22	2400	0.07
POX 4	45	250	75	3.0	2.6	8.9	520	0.012	0.04	1.10	1.05	0.04	14000	0.33
POX 139	27	409	2.7	0.07	12.5	1.1	70	0.041	0.14	1.10	1.05	0.15	430	0.53
Tol 35	24	89.8	14	1.6	2.6	4.6	270	0.033	0.11	1.91	1.15	0.13	2100	0.10
II Zw 40	9.2	245	19	0.78	4.2	3.4	200	0.018	0.06	1.33	1.08	0.06	3300	0.31

DISCUSSION

Before comparing the observed N_{W-R}/N_O in W-R galaxies with the predictions of Maeder (1991) let me first consider the uncertainties. Several factors could lead to *larger* derived N_{W-R}/N_O values. I have assumed all stars are of WNL subtype, but WNE stars might be present in appreciable numbers. The $\langle L_{WNE}(\lambda 4686) \rangle$ is smaller in WNE stars; their single star $\langle Q_0^{WNE} \rangle$ are larger. Both effects drive the N_{W-R}/N_O to larger values, which may be as large as a factor of 2 effect. The contribution of WC stars has also been ignored. Since these objects have not been detected in most W-R galaxies of Table I, this assumption is probably a good one, aside from He2-10.

Next, what factors could lead to *smaller* derived N_{W-R}/N_O values? There are several possible effects which could lead to an underestimate of the N_O stars. Some Lyman continuum photons are undoubtedly outside the slit dimension, and therefore have not been accounted for in the derivation of the number of O stars from the Hβ luminosity, despite the fact that they are produced by O stars within the region. Salzer (private communication) has investigated this geometric issue from CCD Hα photometry of some emission line galaxies and finds it to be important. Following Salzer, I believe this aperture problem could drive the inferred N_{W-R}/N_O down by a factor 2-3, depending on the starburst geometry, thus more than counterbalancing the neglect of WNE stars (see above). Next, some of the star forming regions *could* be *density bounded* to Lyman continuum radiation; that is, there might be insufficient hydrogen to completely absorb all the Lyman continuum photons, which are otherwise leaking out of the galaxy. While there is as yet no way to quantify this effect for our galaxies, in other emission line galaxies the H Could a I radio maps typically extend to much larger dimensions than the starburst cores. Thus this problem might not be too important for the N_{W-R}/N_O determinations. Similarly, extensive dust could absorb Lyman continuum photons. This is probably *not* important for most of these W-R galaxies, but might play a role in those heavily reddened galaxies such as IRAS 01003-2238 (Armus *et al.* 1988).

Could a nebular contribution to He II $\lambda 4686$ be present in the W-R galaxy spectra? The spectral resolution of the Vacca & Conti (1992) analysis was sufficiently large such that the broad nature of this line was readily apparent. In POX 4, for example, a narrow nebular component centered on a broader stellar feature in He II $\lambda 4686$ was detected; the values in Table I refer to the broad portion only. There are, of course, uncertainties in the measurements, amounting to 30% for the He II $\lambda 4686$ line, but probably only 10% for H β. There are larger uncertainties in the $\langle L_{WNL}(\lambda 4686) \rangle$ and $\langle Q_0^{WNL} \rangle$ calibration. Both of these could be different from our adopted values by a factor 2. However, these kinds of uncertainties could affect the N_{W-R}/N_O in *either* direction.

There are also several factors which could lead to *larger predicted* N_{W-R}/N_O values. Some W-R stars could be hydrogen burning (Conti *et al.* 1983a); these are *not* accounted for in the Maeder (1991) models. This is a 30% effect, at

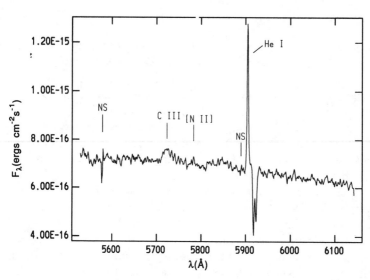

Fig. 6. Spectrum of the giant HII region L4 in the barred galaxy NGC 1365 (from Phillips & and Conti 1992).

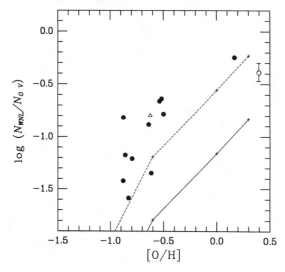

Fig. 7. The number ratio of WNL to O V stars in W-R galaxies, as a function of the oxygen abundances ($[O/H] \equiv \log(O/H) - \log(O/H)_\odot$ and $\log(O/H)_\odot = -3.08$: *Filled circles*, galaxies of Table I; *triangle*, $8' \times 8'$ region centered on the core of 30 Doradus (from Vacca 1991); *open circle* with error bars, region L4 of NGC 1365 (from Phillips & Conti 1992). The solid line denotes the predictions of Maeder (1991) in the "continuous" star formation mode. The dashed line denotes the values estimated from "instantaneous burst" models, following Arnault *et al.* (1989).

most. Similarly, some W-R stars are close binaries which are not taken into account in the single star evolution models. This effect is also about 30% in the N_{W-R}/N_O predictions.

In summary, in a comparison of the *derived* with the *predicted* N_{W-R}/N_O values, the inferred values could be *too large* by about a factor 2 to 3. It is now appropriate to compare the observed N_{W-R}/N_O in W-R galaxies with the predictions of Maeder (1991). The N_{W-R}/N_O are plotted *vs.* relative oxygen abundance in Fig. 7. The lower line is the prediction (Maeder 1991) for "continuous" star formation; that is, star formation proceeding on time scales longer than 10^7 yr. The upper line is our estimate, following Arnault *et al.* (1989), of the predicted ratio under the assumption that the star formation is a short "burst," lasting less than 10^6 yr. Maeder, who is making detailed calculations of the predictions for a "burst" scenario, agrees with this appraisal but the reader is cautioned that my estimate is uncertain. From Fig. 7, it can be seen that all the N_{W-R}/N_O values for the galaxies are on or above the "burst" predictions, and well away from the "continuous" star formation scenario. Thus it appears that in the W-R galaxies discussed here the *star formation processes for the massive stars seem to be operating on short time scales,* $\leq 10^6$ yr, rather than longer ones, $\geq 10^7$ yr. These starbursts can thus be thought of as greatly scaled up versions of nearby giant HII regions such as 30 Dor.

The unanticipated finding that the starburst time scales are "short" may have more general applicability in considerations of the numbers of massive stars in other galactic systems. In certain starburst galaxies the current absence of the most massive stars has sometimes been ascribed to differences in the slope of the IMF, or "upper mass cutoffs." But if the massive star formation time scales are short, as in a"burst," then the lack of the hottest O type stars might be due merely to an evolution effect as the most massive stars age and disappear.

W-R galaxies are examples of *extreme* starburst phenomena, where hundreds to thousands of massive stars have recently formed. The understanding of the initiation of such energetic events is a question for future work.

I am especially indebted to Bill Vacca for his substantial contributions to much of what is reported in this review, our continuing fruitful interactions, and for providing me with Figures 3 and 4. Likewise Pat Morris and Ken Brownsberger have given me Figures 1 and 2, respectively. I appreciate continuing support from the NSF under grant AST 90-15240. Some of this review is taken from a paper presented at a conference in Madrid (Conti & Vacca 1992).

REFERENCES

Abbott, D. C., & Conti, P. S. 1987, ARA&A, 25, 113
Allen, D. A., Wright, A. E., & Goss, W. M. 1976, MNRAS, 177, 91
Armus, L., Heckman, T., & Miley, G. 1988, ApJ, 326, L45
Arnault, Ph., Kunth, D., & Schild, H. 1989, A&A, 224, 73
Burstein, D., & Heiles, C. 1984, ApJS, 54, 33

Conti, P. S. 1991, ApJ, 377, 115

Conti, P. S., Garmany, C. D., de Loore, C., & Van Beveren, D. 1983a, ApJ, 274, 302

Conti, P. S., Leep, E. M., & Perry, D. N. 1983b, ApJ, 268, 228

Conti, P. S., & Massey, P. 1989, ApJ, 337, 251

Conti, P. S., Massey, P., & Vreux, J.-M. 1990, ApJ, 354, 359

Conti, P. S. & Vacca, W. D. 1992, in The Nearest Active Galaxies, ed. J. Beckman and L. Colina (in press)

Hummer, D. G., & Storey, P. J. 1987, MNRAS, 224, 801

Humphreys, R. M., Nichols, M., & Massey, P. 1985, AJ, 90, 101

Kunth, D., & Joubert, M. 1985, A&A, 142, 411

Kunth, D., & Sargent, W. L. W. 1981, A&A, 101, L5

Kunth, D., & Schild, H. 1986, A&A, 169 71

Kurucz, R. 1979, ApJS, 40, 1

Leitherer, C. 1990, ApJS, 73, 1

Maeder, A. 1990, A&AS, 84, 139

Maeder, A. 1991, A&A, 242, 93

Maeder, A., & Meynet, G. 1987, A&A, 182, 243

Maeder, A., & Meynet, G. 1989, A&A, 210, 155

Moffat, A. F. J., Niemela, V. S., Phillips, M. M., Chu, Y.-H., & Seggewiss, W. 1987, ApJ, 312, 612

Morris, P., Brownsberger, K., Conti, P. S., Massey, P., & Vacca, W. D. 1992 (in preparation)

Osterbrock, D. E. 1989, Astrophysics of Gaseous Nebulae and Active Galactic Nuclei (Mill Valley: University Science Books)

Osterbrock, D. E., & Cohen, R. D. 1982, ApJ, 261, 64

Parker, J. Wm. 1992, Ph.D. thesis, University of Colorado, Boulder

Phillips, A. C. & Conti, P. S. 1992, ApJL (in press)

Sargent, W. L. W., & Filippenko, A. V. 1991, AJ, 102, 107

Sargent, W. L. W., & Searle, L. 1970, ApJ, 162, 155

Schmutz, W., Vogel, M., Hamann, W.-R., & Wessolowski, U. 1992 (in preparation)

Smith, L. F. 1991, in IAU Symposium 143, Wolf-Rayet Stars and Interrelations with Other Massive Stars in Galaxies, ed. K. A. van der Hucht & B. Hidayat (Dordrecht: Kluwer), p. 601

Torres, A. V., Conti, P. S., & Massey, P. 1986, ApJ, 300, 379

Vacca, W. D. 1991, Ph. D. thesis, University of Colorado, Boulder

Vacca, W. D. 1992 (in preparation)

Vacca, W. D., & Conti, P. S., 1992, ApJ (in press)

Vacca, W. D., Phillips, M., & Conti, P. S. 1992 (in preperation)

van der Hucht, K. A., Conti, P. S., Lundström, I., & Stenholm, B. 1981, Space Sci. Rev., 28, 227

Veilleux, S., & Osterbrock, D. E. 1987, ApJS, 63, 295

Massive Stars: Their Lives in the Interstellar Medium
ASP Conference Series, Vol. 35, 1993
Joseph P. Cassinelli and Edward B. Churchwell (eds.)

STARBURSTS IN THE NEARBY UNIVERSE

John S. Gallagher, III, Department of Astronomy, Washburn Observatory,
475 North Charter St., University of Wisconsin, Madison, WI 53706

ABSTRACT Starbursts result from the rapid conversion of interstellar
gas into large numbers of OB stars. The physical causes of galactic
starbursts however, are not fully understood. Luminous starbursts
can be produced by collisions between galaxies, but not all luminous
starburst systems show evidence of recent interactions with other galaxies.
Furthermore, smaller starburst galaxies often are relatively isolated. The
possibility therefore exists that starbursts can be caused by instabilities
within galaxies. This paper briefly examines the case for interactions
between OB stars and interstellar matter as a potential mechanism for
starbursts in non-interacting galaxies.

THE CONCEPT OF GALACTIC STARBURSTS

It is often assumed that galactic star formation rates (SFRs) vary smoothly
with cosmic time. Thus models for the star formation histories of galaxies such
as those derived by Searle et al. (1973; SSB), or more recently by Sandage
(1986), represent average galactic SFRs with simple functions which for most
structural classes of galaxies monotonically decline over time. This type of
model has the dual advantage of being both consistent with what is known
about the star formation history of the Milky Way (but see Barry 1988 for an
alternative view) and of yielding relatively simple predictions for the evolution
of galaxy populations as a function of lookback time.

On the contrary, if the smooth models for galactic star formation histories
turn out to be incorrect, then the task of interpreting the evolutionary properties
of galaxies could become even more challenging. Unfortunately, measurements
of faint galaxies already provide hints that all is not well with the application of
the smooth model to the evolution of distant galaxies (Colless *et al.* 1989; Lilly
et al. 1991). The issue of what causes some galaxies to depart from a simple
evolutionary path and to experience episodes of enhanced SFRs therefore is
becoming increasingly important. We need to understand starbursts not only
in their own rights, but also as potentially critical evolutionary mechanisms in
galaxies.

In this paper I first comment on the development of the idea that some
galaxies experience episodic phases of rapid evolution, during which they are
seen as "starbursts." In the second part of the paper I will consider how we

might determine whether the inevitable interaction between massive stars and the interstellar medium from which they form is a significant source of galactic starbursts. More details about starbursts can be found in the proceedings of recent conferences on this topic (e.g. Thuan *et al.* 1987; Persson 1988; Pudritz and Fich 1988; Leitherer *et al.* 1991).

First we must understand what is meant by a "starburst." Since star formation is locally an episodic process, almost *all* stars could be said to have been born in "starbursts". Instead the starburst term should apply only to special circumstances occurring over large regions of galaxies. I will therefore define a starburst in the spirit of the SSB and Weedman (1988) discussions to to be when the SFR is at an enhanced level relative to what would be expected from the normal statistical fluctuations in the numbers and the sizes of sites of stellar births in a large region. In other words, a starburst region has a more than a 3σ upwards fluctuation in the SFR, and will occur over a small fraction of the cosmic time scale ($\leq 10^8$ yr).

HISTORICAL PERSPECTIVE ON THE STARBURST IDEA

Interacting Galaxies and Violent Star Formation

In a back-to-back pair of papers, Burbidge, Burbidge, and Hoyle (1963; BBH) and Sandage (1963) presented opposing views about the possibilitiy that peculiar galaxies, such as those cataloged by Vorontsov-Velyaminov, are cosmologically young objects. BBH supported the affirmative by arguing that it is consistent with the continuing galaxy formation predicted by the steady-state theory. Sandage countered by demonstrating that a range of ages exists for stellar populations within two interacting galaxy systems, VV117 and VV123. Furthermore, Sandage emphasized the ability of young stellar populations to "overpower" older stellar populations as an explanation for why demonstrably old galaxies (e.g., the LMC or M33) can have the blue optical colors of young (1 Gyr) star clusters. Sandage therefore showed that optical colors of galaxies can reflect star formation histories more strongly than ages.

The broader evolutionary implications of galaxy-galaxy interactions began to be recognized in the 1960s. Toomre and Toomre (1972) described these effects in their prescient discussion of the physics of interactions, which left little doubt that evolution in *some* galaxies has been anything but smooth (see also Toomre 1977). Larson and Tinsley (1978) made a further critical step by beautifully demonstrating the connection between dynamical interactions and the "burst" mode of star formation. The interacting galaxy starburst model now is confirmed and is widely accepted (e.g., Bushouse 1987; and from the *IRAS* results, see Soifer *et al.* 1987).

Interactions between galaxies can produce large-scale starbursts of evolutionary importance in luminous galaxies. But are they the only mechanism for luminous starbursts? The answer to this question is probably a complicated no. A minority of nearby, luminous starburst galaxy candidates show no overt indications of having been victims of interactions with other galaxies; other starburst mechanisms in luminous galaxies are consistent with and may even be required by the data (e.g., the "heretical" view of van den Bergh 1972). An example of an outwardly unperturbed luminous starburst galaxy is shown in Figure 1.

Several alternative explanations have been proposed for "non-interacting" luminous starbursts (e.g., Keel 1991): (1) an internal instability in the SFR is involved (e.g., Ikeuchi 1988; Hensler and Burkert 1990); (2) long (≈ 1 Gyr) delays can occur between a triggering interaction with another galaxy and the resultant starburst so that the interacting companion is now long gone (e.g., Noguchi and Ishibashi 1986, Noguchi 1988a,b); (3) some bursts are caused by intergalactic gas rather than other galaxies (e.g., Silk *et al.* 1987); or (4) "stealth" galactic interactors exist that are hard to observe (such as gas rich dwarfs; Taylor *et al.* 1992). Past experience suggests that nature is likely to have used all four of these reaction channels (and probably others) in making luminous, "non-interacting" starburst galaxies.

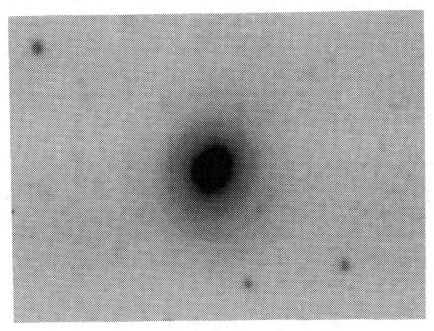

Fig. 1. A gray scale version of an R-band image of the starburst galaxy II Zw 168 is shown here to emphasize the regular structure of its outer regions. There are no indications in the outer light distribution of this galaxy for a recent external gravitational perturbation. This image was taken with a TI CCD on the KPNO 2.1-m telescope.

Starbursts in Small Galaxies

Evolution of stellar populations in large and in small galaxies may differ because of both physics and statistics. The discovery of the extragalactic HII regions in surveys of Zwicky objects prompted SSB to consider how such apparently young systems could exist among small galaxies. SSB first appealed to the statistics of star formation. The near simultaneous presence of several star-forming complexes in a small galaxy could make that galaxy appear to be flashing into optical prominence. Thus SSB argued that the statistics of normal star formation will favor bursty evolution of small galaxies.

To see that something more than the statistical noise of normal star formation process was required in extraglactic HII regions, it was necessary to make measurements of the masses of young stars and thus of recent SFRs. These data were derived for the HII components of starburst dwarfs from radio and optical observations, and showed that the numbers of OB stars required to make extragalactic HII region galaxies exceed even the OB stellar content of a supergiant HII region like 30 Doradus (e.g, O'Connell *et al.* 1978). Statistical effects do not readily yield such large concentrations of young stars, so additional physics is required to make some extragalactic HII regions and related blue dwarfs.

In more modern terminology, the extragalactic HII regions are folded into the broader category of blue compact dwarfs (BCDs). Although the BCDs are an inhomogeneous structural class, a significant fraction of these systems have starburst characteristics (see Thuan 1991). A phenomenological model for BCDs was developed by Seiden and his collaborators who considered how star formation might propagate in galaxies. The time behavior of these models is sensitive to galactic dimensions and percolation parameters, with smaller galaxies showing more of a tendancy to form stars in episodic bursts (see Seiden and Schulman 1990). The stochastic self-propagating star formation (SSPF) models suggest that star bursts in small galaxies could reflect internal physics through the ability of young stars to affect the interstellar medium and thereby influence future star formation, rather than purely reflecting statistical effects.

SSPSF however, has the difficulty of being a macroscopic description whose relationship to the detailed physics of star formation is still unclear. Thus the types of external mechanisms that are listed above for producing luminous starbursts in non-interacting systems might also apply to the smaller BCDs. And again there are indications that multiple processes are likely to be occuring in nature to produce the observed suite of violently star-forming small galaxies.

STARBURSTS AND MASSIVE STARS

High Mass Stars in Starburst Galaxies

Starburst galaxies are rich in high mass ($M \geq 8\ M_\odot$) stars. These stars make their presence known both directly through their unique spectral features and indirectly via the photoionization of gas and production of HII regions. Since for a wide range of initial mass functions (IMFs) photoionization is dominated by stars with initial masses of $\geq 20\ M_\odot$, emission from ionized gas is a favorite way to trace the numbers of massive stars (see Kennicutt 1983; Gallagher 1987). However, with the recognition of spectral features that stem uniquely from even more massive stars, as in Wolf-Rayet spectra (e.g., Massey 1985; Conti 1990), direct methods are now also being employed to confirm the presence of vast assemblages of OB stars in starburst galaxies.

The densities of massive stars in starburst galaxies should be compared over a variety of spatial scales with the fiducial provided by normal spirals. While this comparison is yet to be made in a completely systematic way, it is clear that galaxies exist with more intense star formation over large regions than is found in normal late-type Sc spirals. This fact is perhaps most extensively demonstrated by the high *IRAS* far infrared luminosities and large ratios of L(FIR)/L(B) that are found for many luminous blue galaxies (LBGs; e.g. Belfort *et al.* 1987; Hunter *et al.* 1989). In a more limited number of cases Hα images of LBGs display the presence of multiple supergiant HII regions–signatures of what J. Heidmann appropriately called "hyperactive" star formation. The Hα imaging data are more complete for the BCDs, which often have the further advantage of low internal extinction. For example, Gibson *et al.* (1992) have used digital Hα imaging as a means to explore star formation in three BCDs that could be in different phases of a starburst cycle.

Feedback from Massive Stars on Star Formation
Once massive OB stars have formed they will influence the surrounding interstellar medium in a variety of ways.   On the destructive side are photoionization, stellar winds, and supernovae, all of which will inhibit the ability of a region to form new stars. However, these processes will also produce an overpressure region that can create interstellar clouds and possibly also trigger pre-existing interstellar clouds to form stars (e.g., McCray and Kafatos 1987) This combination of processes lie at the heart of SSPSF and related models for propagating star formation.

Starbursts, in addition to requiring large amounts of star formation further require that the star formation process operate with a high degree of coherence over large regions with sizes of $\geq$ 1 kpc. If we take 10^8 yr for the maximum duration of a starburst, then a mean velocity of $\geq$ 10 km/s is needed for star formation to propagate over 1 kpc. This may be difficult to achieve with Oort cycle models based on cloud-cloud interactions (e.g., Vazquez and Scalo 1989) since the HI dispersion velocities are typically $\approx$ 10 km/s while dissipational radial flow velocities are perhaps ten times slower. The short coherence time required to make observed starbursts is a significant issue for propagating star formation starburst models (see also Larson 1988; Gibson *et al.* 1992).

Another limitation for this class of model is set by the ability of overpressure to stimulate star formation within galactic disks. For example, models for the development of superbubbles of hot gas around OB stellar clusters show that the radius of the superbubble in the disk plane will be less than about two gas scale heights (e.g., Tenorio-Tagle and Bodenheimer 1988; Igumentschev *et al.* 1990). Once this size is reached, hot gas begins to preferentially move upwards, and the superbubble will depressurize. Further expansion then must occur at low velocities of $\leq$ 20 km/s as the superbubble enters a snowplow phase. Mean propagation velocities averaged over the lifetime of the star forming complex are less than the terminal shell velocity, so coherence is also an issue for this class of model.

Initial conditions that are necessary for the outbreak of rapid, stimulated star formation have not been widely explored. A rich supply of raw materials for star formation–such as dense molecular clouds–would surely help. This idea leads us to another facet of the starburst problem: large scale, explosive star formation evidently requires large masses of star-forming molecular clouds. How do huge arrays of interstellar clouds assemble and wait to be triggered to explode into star formation and make a starburst (see Gibson *et al.* 1992)? Why don't the clouds instead simply turn into stars slowly via the types of processes that we see locally within the Galactic disk? This is Keel's "fizzle problem." How do we collect enough fuel for rapid star formation without first losing our supply to the "slow burn" of normal star formation processes?

The gas transport problem is exacerbated by the tendancy for starbursts to be located in the centers of all but the least massive non-interacting starburst galaxy candidates. An example of this effect is shown in Figure 2. The centers of galaxies are on the one hand good locations for starbursts because densities are high, the gravitational potential is deep, and nuclear region gas cannot be readily moved to another part of the galaxy (e.g., Loose *et al.* 1982). However, on the other hand, centers are not such logical places for starbursts since any disk gas must traverse large radial distances to arrive in the nucleus of a non-interacting galaxy, and while in transist it must avoid being destroyed by star formation. The effectiveness of such processes in preventing the build-up of nuclear region gas are demonstrated by the central holes in the cool ISM which are seen in many spirals (e.g., M31).

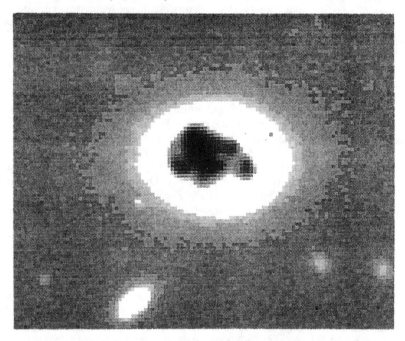

Fig. 2. R band continuum and Hα line CCD images of the starburst candidate galaxy III Zw 12 are combined to show locations of star-forming sites. The continuum image is displayed as a positive picture while the Hα image is printed as a negative. These data were obtained with a TI CCD on the KPNO 2.1-m telescope. As in other starburst systems without strong evidence for recent external dynamical perturbations, star formation in this system is centrally concentrated as shown by the the Hα emission.

Rapid transport of dense gas is one way around this apparent conundrum. Thus collisional stimulation of starbursts via gas flows again appears as a natural process for accumulating gas near the centers of galaxies over short time scales (e.g., Norman 1991). Stellar bars may also play a critical role by fostering rapid radial gas inflows into galactic nuclei, and the presence of nuclear starbursts in non-interacting barred galaxies such as NGC 253 or NGC 4736 is probably not coincidental. Yet there are also starbursts in systems without bars, and so we should continue to look for other types of internal SFR instabilities.

We therefore turn to the ISM phase instability models which place their emphasis on local physics rather than gas transport. These models consider the flow of gas between various phases of the ISM, such as the hot ionized, warm HI/HII, and cold molecular phases. The time evolution of each phase can be complex, and starbursts can result, for example, from a rapid build-up of cold phase gas on a time scale that is short compared with the disruption time set by massive stars. As these classes of models are unstable, starbursts are possibly a natural feature of the multi-phase ISM (Shore 1981).

Sophisticated numerical models now exist for multi-phase ISM models of galactic evolution in which star formation is but one step in a complex evolutionary cycle (see Ikeuchi 1988; Larson 1988). One-dimensional models such as those illustrated by Hensler and Burkert (1990) demonstrate the connection between episodic cooling of the hot ISM phase and galactic starbursts. Two-dimensional ISM models by Chiang and Prendergast (1985; see also Chiang and Bregman 1988) similarly show fluctuating SFRs, although these models are highly idealized. Struck-Marcell and Scalo (1986, 1987) have presented a related class of unstable SFR model derived from cycles in the physical state of the ISM, and again burst behavior is found.

Based on the current generation of theoretical models for galactic SFRs, highly time-dependent SFRs should be a feature of many galaxies. We therefore might expect to see extreme examples of this type of process as non-interacting starburst galaxies, although we should also be aware that comparisons between models and observations are still difficult to make. A question for the future is then whether we can make the mapping between the observed characteristics of nearby starbursts and the necessarily simplified unstable SFR models. Once this is accomplished, we will gain better insights into the role of internal SFR instabilities in fostering unsteady evolution in galaxies.

BRIEF COMMENTS ON NEARBY, NON-INTERACTING STARBURSTS

Where Starbursts Occur. Starbursts are centrally concentrated in all but the smallest galaxies (e.g., Telesco 1988; Condon *et al.* 1991; Dressel and Gallagher 1992). We therefore need to be certain that we understand why the centers of galaxies are primary starburst sites. Is this a signature of external influences on galactic star formation processes, or does central concentration of star formation also result from internal instabilities?

Gas Supplies. Not surprisingly, many starbursts take place in gas-rich galaxies. How the gas gets into position to support a starburst is not clear. For example, I am not aware of any obvious candidates for nearby galaxies in pre-starburst states; i.e., with lots of dense gas and little ongoing star formation within that gas. We also should be sensitive to the implications of the huge cold gas reservoirs seen in some starbursts. Where did this gas come from? Are such galaxies then in an abnormal evolutionary state, perhaps resulting from recent gas captures? If so, do starbursts then signify evolutionary sea-changes in the lives of galaxies?

Starburst Stellar Population Evolution Rieke *et al.* (1988) presented a useful evolutionary scheme for starbursts that is based on models for the evolution of stellar populations. Generally starbursts would be expected to die by blowing most of the cool ISM out of the burst region, thereby shutting down star formation. The galaxy would then enter a post-burst phase in which the dying intermediate mass stars from the burst would give rise to an "E+A" spectrum (Dressler and Gunn 1983). Unfortunately, few E+A spectra have been found (see Oegerle *et al.* 1991). There are two obvious solutions to this problem: (1) Starbursts are not "clean" events which end with a total turn-off of all star formation in the affected region. (2) The IMF in starbursts are such that few intermediate mass stars are made, and as a result such galaxies rarely show E+A postburst spectra simply because the requisite large numbers of intermediate mass, main-sequence stars do not exist.

ISM Cycles in Starbursts Using NGC 253 as an example, we see that multiple phases of the ISM are present within a typical starburst. These include hot, x-ray emitting gas, HII, HI, and molecular clouds. It is obviously critical to understand the interrelationships between these various ISM phases and the young stellar populations (Watson *et al.* 1992). Since interactions bewteen ISM phases are central to internal starburst physics, detailed probes of the state of the ISM within starbursts must remain a high priority.

Structure of Starbursts The only patterns of star formation that are known in any detail are those in the low density disks of galaxies. Here we see the OB associations and stellar clusters that reflect the cloudy state of the prenatal ISM. These processes are quiescent in the sense that interstellar clouds are unlikely to suffer substantial external perturbations during the starbirth process. We do not know if this is true within starbursts. For example, in the BCDs we find evidence for a preference for making relatively dense (and sometimes luminous) star clusters (Caldwell and Phillips 1989; Waller and Dracobly 1992) that may indicate a change in the structure of star-forming clouds. With the advent of high angular resolution from space (e.g., with the HST) and the ground in the infrared, it is now possible to observe the structure of star formation within starbursts, which can then be compared with two- and three-dimensional model predictions.

Starburst Frequencies Do all galaxies experience one or more starbursts, even in the absence of strong external perturbations? Are there specific galactic evolutionary phases where starbursts are most likely to occur, as suggested by theoretical models? Complete answers to these questions are not available. However, approximately 5-10% of nearby galaxies with $M_B \leq$ -18 are potentially in starburst phases at the present epoch (Huchra 1977), which is larger than the fraction of known interacting systems. Since the lifetimes of starbursts are probably on the order of 0.01 of a Hubble time, these statistics suggest that many galaxies could have experienced one or more starburst episodes during their lifetimes. Nature seems to be giving us some interesting hints about how galaxies evolve! The more luminous of the nearby starburst galaxies furthermore resemble distant blue galaxies, and therefore may eventually provide a direct link to conditions in younger galaxies (Gallagher 1990).

ACKNOWLEDGEMENTS

It is a pleasure to thank my many colleagues who have collaborated with me during the past decade in studying blue galaxies, including Howard Bushouse, Linda Dressel, Claudio Firmani, Deidre Hunter, Susan Lamb, Jean Keppel, Mark Phillips, Sahsa Tutukov, and Don York. My present students, Steve Gibson, Pam Marcum, Lynn Matthews, and Alan Watson, also have contributed to this research program. Their help is appreciated, with special thanks to Lynn Matthews for a critical reading of this paper. Support for this project at Wisconsin has been provided by the University of Wisconsin Graduate School and by NASA.

REFERENCES

Barry, D. C. 1988, ApJ, 334, 436.
Belfort, P., Mochkovitch, R., and Dennefield, M. 1987, A&A, 176, 1.
Burbidge, E. M., Burbidge, G. R., and Hoyle, F. 1963, ApJ, 138, 873.
Bushouse, H. A. 1987, ApJ, 320, 49.
Caldwell, N. and Phillips, M. M. 1989, ApJ, 338, 789.
Chiang, W.-H. and Prendergast, K. H. 1985, ApJ, 297, 507.
Chiang, W.-H. and Bregman, J. N. 1988, ApJ, 328, 427.
Colless, M. M., Ellis, R. S., Taylor, K., and Hook, R. N. 1989, MNRAS, 223, 811.
Condon, J. J., Huang, Z.-P., Yin, Q. F., and Thuan, T. X. 1991, ApJ, 378, 65.
Conti, P. 1990, ApJ, 377, 115.
Dressel, L. L. and Gallagher, J. S. 1992, in preparation.
Dressler, A. and Gunn, J. 1983, ApJ, 270, 7.
Gallagher, J. S. 1987, Mitt. Ast. Gesell., 70, 126.
Gallagher, J. S. 1990, in *Evolution of the Universe of Galaxies*, ed. R. Kron, ASP Conf. Series 10, p157.
Gibson, S. J., Gallagher, J. S., and Hunter, D. A. 1992, *this workshop*.
Hensler, G. and Burkert, A. 1990, in *Windows on Galaxies*, ed. G. Fabbiano, J. S. Gallagher, and A. Renzini, (Dordrecht: Kluwer Academic), p321.
Huchra, J. P. 1977, ApJS, 35, 171.
Hunter, D. A., Gallagher, J. S., Rice, W. L., and Gillett, F. C. 1989, ApJ, 336, 152.
Igumentschev, I. V., Shustov, B. M., and Tutukov, A. V. 1990, A&A, 234, 396.
Ikeuchi, S. 1988, Fund. Cos. Phys., 12, 255.
Keel, W. C. 1991, in *Dynamics of Galaxies and Their Molecular Cloud Distributions*, ed. F. Combes and F. Casoli (Dordrecht: Kluwer Academic), p243.
Kennicutt, R. C. 1983, ApJ, 272, 54.
Larson, R. B. and Tinsley, B. M. 1978, ApJ, 219, 46.
Larson, R. B. 1988, in *Galactic and Extragalactic Star Formation*, ed. R. E. Pudritz and M. Fich, (Dordrecht: Kluwer Academic), p459.
Leitherer, C., Walborn, N. R., Heckman, T. M., and Norman, C. A. (eds.) 1991, *Massive Stars in Starbursts*, Space Telescope Science Inst. Symp. Ser. 5, (Cambridge: Cambridge Univ. Press).
Lilly, S. J., Cowie, L. L., and Gardner, J. P. 1991, ApJ, 369, 79.

Loose, H. H., Krügel, E., and Tutukov, A. 1982, A&A, 105, 342.

Massey, P. 1985, PASP, 97, 5.

Noguchi, M. and Ishibashi, S. 1986, MNRAS, 219, 305.

Noguchi, M. 1988a, A&A, 201, 37.

Noguchi, M. 1988b, A&A, 203, 259.

Norman, C. N. 1991, in *Massive Stars in Starbursts*, ed. C. Leitherer *et al.*, (Cambridge: Cambridge Univ. Press), p271.

O'Connell, R. W., Thuan, T. X., and Goldstein, S. J. 1978, ApJ, 226, L11.

Oegerle, W. R., Hill, J. M., and Hoessel, J. G. 1991, ApJ, 381, L9.

Persson, C. L. (ed.) 1988, *Star Formation in Galaxies*, NASA Conference Pub. 2466.

Pudritz, R. E. and Fich, M. (eds.) 1988, *Galactic and Extragalactic Star Formation*, (Dordrecht: Kluwer Academic).

Rieke, G. H., Lebofsky, M. J., and Walker, C. E. 1988, ApJ, 325, 629.

Sandage, A. 1963, ApJ, 138, 863.

Sandage, A. 1986, A&A, 161, 89.

Scalo, J. M. and Struck-Marcell, C. 1986, ApJ, 301, 77.

Searle, L., Sargent, W. L. W., and Bagnuollo, W. 1973, ApJ, 179, 427.

Seiden, P. E. and Schulman, L. S. 1990, Adv. Phys., 39, 1.

Shore, S. N. 1981, ApJ, 249, 93.

Silk, J., Wyse, R. F. G., and Shields, G. A. 1987, ApJ, 322, L59.

Soifer, B. T., Houck, J. R., and Neugebauer, G. 1987, ARAA, 25, 187.

Struck-Marcell, C. and Scalo, J. N. 1987, ApJS, 64, 39.

Taylor, C. L., Brinks, E., and Skillman, E. D. 1992, Preprint.

Telesco, C. M. 1988, ARAA, 26, 343

Tenorio-Tagle, G. and Bodenheimer, P. 1988, ARAA, 26, 145.

Thuan, T. X. 1991, in *Massive Stars in Starbursts*, ed. C. Leitherer *et al.*, (Cambridge: Cambridge Univ. Press), p183.

Thuan, T. X., Montmerle, T., and Tran Thanh Van, J. (eds.) 1987, *Starbursts and Galaxy Evolution*, (Gif sur Yvette: Editions Frontières).

Toomre, A. 1977, in *The Evolution of Galaxies and Stellar Populations*, ed. B. M. Tinsley and R. B. Larson, (New Haven: Yale Univ. Obs.), p401.

Toomre, A. and Toomre, J. 1972, ApJ, 178, 623.

van den Bergh, S. 1972, JRASC, 66, 237.

Vázquez, E. C. and Scalo, J. N. 1989, ApJ, 343, 644.

Waller, W. H. and Dracobly, C. L. 1992, *this workshop.*

Watson, A., Gallagher, J. S., Merrill, K., Gatley, I., Phillips, M. and Keppel, J. 1992, *this workshop.*

Weedman, D. 1988, in *Star Formation in Galaxies*, ed. C. J. L. Persson, NASA Conference Pub. 2466, p351.

Massive Stars: Their Lives in the Interstellar Medium
ASP Conference Series, Vol. 35, 1993
Joseph P. Cassinelli and Edward B. Churchwell (eds.)

GIANT HII REGIONS

PAUL HODGE
Astronomy Department
University of Washington, Seattle, Washington 98195

ABSTRACT   The various studies of giant HII regions in our Galaxy
and in other (normal) galaxies are reviewed. The properties of
these objects are illustrated by examples, which are contrasted with
normal HII regions. For our Galaxy, the giant HII region W49 is
described as an example of those objects that are observed only at
radio wavelengths. In the LMC, the well-observed 30 Doradus is
contrasted with a normal HII region, DEM 8. The SMC is represented
by the supercluster NGC 346 and its enveloping HII region, which are
compared to the normal HII region DEM 161. NGC 604 and NGC
5471 in M33 and M101, respectively, are other nearby, well-studied
giant HII regions. For M31, however, it is shown that no giant HII
regions are present; this fact is related to the general situation in
early Hubble types of galaxies. Finally, the question of whether the
giant HII regions are a physically distinct class of objects is addressed;
the answer seems to be that some of them, at least, are, while other
objects so-called are probably just the largest examples in the range
of star-forming regions. Irregular galaxies, especially small ones with
simple, quiet dynamics, appear to be able to generate larger HII
regions than can earlier type galaxies.

INTRODUCTION

Giant HII regions are relatively well-observed objects in other galaxies,
but for our Galaxy they are relatively incompletely studied. Two reasons
explain this odd fact: (1) almost all giant HII regions in the Galaxy are
so heavily obscured that they are detectable only at radio and infrared
wavelengths, and (2) the giant HII regions in other galaxies are among
the brightest objects in the galaxies and so are easy to observe. For these
reasons, most of our knowledge of this class of object comes from examples
in other galaxies.

Properties of giant HII regions have been reviewed by Shields (1990).
A pioneering study was published by Kennicutt (1984), who made several
important points, including:

1. morphologies vary, with the high surface brightness objects being the
most frequently observed; other types are huge diffuse regions, multiple
complexes, and ringlike objects,

2. sizes range up to about 1 kpc, with core radii having values of 50-100 pc,

3. the mass of ionized gas can be several $\times 10^6$ solar masses,

4. the ionizing luminosity is the equivalent of several thousand O5 V stars,

5. giant HII regions are essentially absent in galaxies of Hubble type Sb and earlier.

Other recent papers that illuminate the nature of these objects include many extensive spectroscopic surveys, such as that of McCall, Rybski and Shields (1985), and the many papers of Rosa, D'Odorico and their collaborators (*e.g.*, Rosa and D'Odorico 1986). An earlier review of the abundance questions for giant HII regions is given by Pagel and Edmunds (1981).

Figure 1 shows a typical, normal Sc galaxy, NGC 628. The brightest HII region in NGC 628 is the object indicated in the figure, No. 627 in the Hodge-Kennicutt (Hodge 1976) survey. Its total Hα flux is 5×10^{-13} erg cm^2 sec^{-1} (Kennicutt and Hodge 1980) and its abundances based on emission lines are given by McCall *et al.* (1985), who find that it has a line ratio of [OIII]/Hβ that is unusually large. Its diameter is 1300 pc and it is a single object, rather than being a complex of HII regions. Its Hα luminosity is 10^{40} ergs s^{-1} and the mass of excited gas is 2×10^5 solar masses. It would require 5000 O9 V stars to excite the gas. At its position in the 1520 Angstrom UV image obtained with the UIT on the ASTRO-1 mission (Chen *et al.* 1992) is an extremely bright OB association, which apparently fires HII region #627.

Fig. 1. NGC 628, an Sc galaxy, with the giant HII region #627 indicated by an arrow (from Hodge 1966).

GIANT HII REGIONS IN THE MILKY WAY

The Galaxy has a number of giant HII regions, but most of them lie in heavily-obscured regions of the Milky Way, where they were first detected by radio surveys. An example is the object W49, which has a diameter of 150 pc and an equivalent Hα luminosity of 2×10^{39} ergs s^{-1} (Kennicutt 1984). Using a Salpeter IMF down to 10 solar masses, Kennicutt (1984) calculates that the presumed OB association involved in W49 has a mass of 7000 solar masses. This object, perhaps the most luminous giant HII region in the Galaxy, has only 1/5th the luminosity of NGC 628 No. 627, which is in a more luminous galaxy; the difference may be a population effect, but is probably also a result of the fact that our Galaxy is an Sbc galaxy, earlier in type than NGC 628.

The properties of W49 can be contrasted with those of the more average object, the Orion Nebula, which is 5 pc in diameter, has an Hα luminosity of 10^{37} ergs cm^2 sec^{-1}, and which is involved with only a couple of dozen massive stars.

LARGE MAGELLANIC CLOUD HII REGIONS

The nearest galaxy, the Large Magellanic Cloud (LMC), has an excellent example of a giant HII region (GHR): 30 Doradus. This is now the most thoroughly-studied example of its class, better understood than GHRs of the MWG. With a diameter of nearly 500 pc and an Hα luminosity of 1.5×10^{40} ergs s^{-1} (Kennicutt and Hodge 1986), 30 Dor is a remarkable object to be found in such a small galaxy, though we find other examples in a few other nearby irregular galaxies (*e.g.*, NGC 2363; see below). The mass of excited gas is 6×10^5 $M_\odot$ and the estimated total stellar mass is 5×10^4 $M_\odot$ (Kennicutt 1984).

A great deal of spectroscopy and photometry of the nebulosity and embedded stars in 30 Dor has been carried out. Examples of important contributions are: Savage *et al.* (1983) on luminosities and ionization; Melnick (1985) on spectra and photometry; Walborn (1986) on spectra; Walborn and Blades (1986) on spectra; Kennicutt (1991) on massive star formation; Heap *et al.* (1991) on HST spectra; Wang and Helfand (1991) on the X-ray image; Weigelt *et al.* (1991) on the HST image of R136; Parker (1992) on spectra and photometry; and Campbell *et al.* (1992) on HST photometry. The Hubble Space Telescope has been especially useful for separating the various images that make up the very small core image of the nebula, which had previously been resolved only by interferometry from the ground. Within the core of the image, known as R136a, with a radius of only 0.5 pc, there are more than 47 stars detectable. The composite spectral type is O3If and there are at least 6 WR stars. Masses of these stars are estimated to range from 115 down to < 10 solar masses.

As a contrast to the properties of the GHR in the LMC, we can note the properties of the "normal" LMC HII region, DEM 8 (Wilcots 1992). It has a diameter of 6 pc and a total Hα luminosity of 1.5×10^{37} ergs s^{-1}

(Kennicutt and Hodge 1986). Wilcots finds that there probably is only one exciting star, but its spectral type is not yet known.

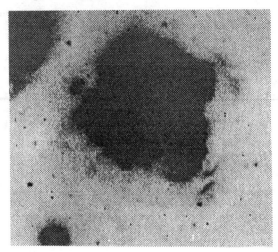

Fig. 2. The Orion-like HII region DEM 8 in the LMC (from Wilcots 1992).

GIANT HII REGIONS IN THE SMC

The Small Magellanic Cloud (SMC) does not have as extreme a giant HII region as does its Magellanic companion. The brightest HII region, NGC 346, has 1/15th the luminosity of 30 Dor (the Hα luminosity is 9×10^{38} ergs s^{-1}; Kennicutt and Hodge 1986). It is 250 pc in diameter and has a total mass of excited gas of 8×10^4 $M_\odot$ (Kennicutt 1984). Out of a total stellar mass of 2×10^3 $M_\odot$, there are 33 O-type stars, 11 of which are type O6.5 or earlier (Massey et $al.$ 1992).

Fig. 3. The large HII region NGC 346 in the SMC (from Hodge and Wright 1977).

By contrast, the "normal" HII region DEM 161 in the SMC, as studied by Wilcots (1992), is more like the Orion Nebula. It has a diameter of 3 pc, an Hα luminosity of 2.3×10^{37} ergs s^{-1}, and only one exciting star. A young star cluster, HW 81, is embedded within the nebula, although the majority of this cluster's stars are probably older than the HII region and its central star (Wilcots 1992).

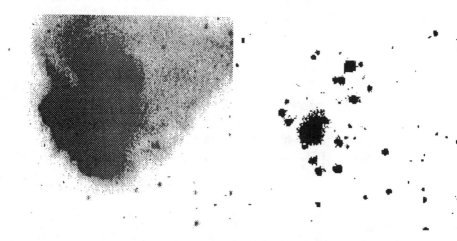

Fig. 4. The SMC Orion-like HII region DEM 161. At left is the Hα image and at right is the continuum image, which shows the stars of the cluster HW 81 (from Wilcots 1992).

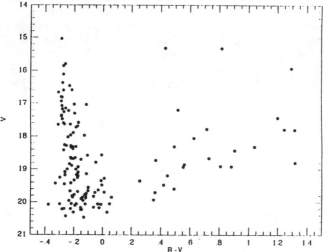

Fig. 5. A color-magnitude diagram for the cluster HW 81, the presumed exciting cluster in DEM 161 (from Wilcots). This can be contrasted with that of 30 Dor (e.g., in Parker 1992), which has brightest stars that are four magnitudes brighter than found here.

SOME OTHER WELL-STUDIED GHRs

There are many other well-studied GHRs in the galaxies near enough
for detailed studies. In fact, perhaps as much because they are so bright
as because they are especially compelling astrophysically, the giant HII
regions of a galaxy are usually the best studied, if not the only studied,
HII regions in it. Three examples of the GHRs that are so bright that
they have their own entries in the NGC are listed and briefly described
in Table I.

TABLE I   Some Well-Studied GHRs

Galaxy	HII Region	Diameter (pc)	Mass (young stars)
M33	NGC 604	350	$10^4 \ M_\odot$
NGC 2366	NGC 2363	500	$5 \times 10^5 \ M_\odot$
M101	NGC 5471	700	$2 \times 10^5 \ M_\odot$

THE GIANT HII REGIONS OF M31

Four recent studies of the HII regions of the Sb galaxy M31 are the global
catalog of Pellet *et al.* (1978), the photometric study and luminosity
function of Kennicutt *et al.* (1989), the comprehensive study of Walterbos
and Braun (1991) and the detailed photometry of selected areas of Hodge
and Lee (1992). All agree that there are no giant HII regions in M31, even
though it is probably a more massive galaxy than the MWG and at least
twenty times more massive than the LMC. The lack of GHRs in M31 is
consistent with the situation for other Sb and earlier galaxies (Kennicutt
et al. 1989) and probably is connected to the star-formation history of
these galaxies (Hodge 1989) and to their dynamics (Waller and Hodge
1990).

ARE GIANTS REALLY DIFFERENT?

The fact that some small galaxies (*e.g.*, the LMC, NGC 2366, as well as
the extreme examples of HII and star-burst galaxies) have GHRs and
other, more massive galaxies do not, has raised the question of whether
these objects are actually a different sort of beast from the normal run
of HII regions. This question was first raised by van den Bergh (1981),
who noticed that HII region sizes occasionally showed an apparently
anomalously-large object that did not fit the well-established exponential
distribution of sizes that was the general case. Further evidence was
put forward for irregular galaxies (Hodge 1983), but was questioned, at

least for their large sample of mostly Sc galaxies, by Kennicutt *et al.*
(1989). More recently, Ye (1992) used maximum likelihood methods to
examine the size distribution of HII regions in irregular galaxies and found
that certain unusually-large objects had to be removed from the size
distributions in order to get a satisfactory fit.

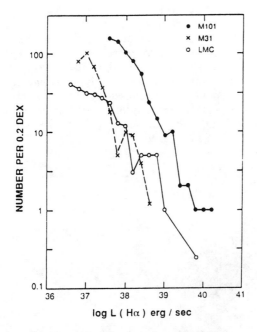

Fig. 6. The HII region Hα luminosity functions for the galaxies M31,
LMC and NGC 5457, showing the conspicuous absence of luminous
objects in the case of M31 (from Kennicutt *et al.* 1989).

Reviewing the data given above, approximately 40% of the Irr
and Sc galaxies surveyed seem to have anomalously-large HII regions.
The question is: are these giant HII regions a separate class of object
with separate astrophysical properties and/or a separate formation
mechanism, or are they merely statistical fluctuations in the tail end of
the mass distribution? Arguing from present evidence and from more
theoretical (perhaps "speculative" is a better word) considerations
mentioned below, it appears that the giant HII regions of some galaxies
really are anomalous. The carefully-argued statistical treatment of Ye
(1992) suggests that the exponential distribution of sizes is, though still
empirical, a good representation of the data; it is therefore a small step
to extrapolate the argument to say that the large examples that do not
fit are a separate class of object (though it would be much preferable to
have a theoretical underpinning for this statement). It might be suggested
that NGC 2363 in NGC 2366 is anomalous, while NGC 346 of the SMC is
merely at the large end of the normal HII region mass distribution.

If the GHRs are really anomalous, then there are two possibilities: (1) they may be large because they have been added to the population of HII regions by a mechanism that prefers to make very massive star-forming regions, or (2) they may be the massive end of the HII region spectrum, but they are only able to form so massively in certain, relatively rare circumstances.

It is suggested that **both** of these possibilities are true. Certain irregular galaxies, like NGC 1140, which has a largest HII region that is 4200 pc in diameter, or NGC 4861, the largest of which is 2400 pc in diameter, are in general peculiar objects in morphology; they may be related to the extreme cases of star-burst and HII galaxies, such as NGC 1569 (Waller 1992). These objects are apparently experiencing an unusual period of intense star formation, caused by (possibly) an outside interference, such as gas-accretion or tidal interaction. Perhaps less-extreme cases of this mechanism explain some of the GHRs in more normal-looking galaxies.

Fig. 7. The irregular galaxy NGC 1140 in Hα light. The largest HII region in this galaxy has a diameter of 4200 pc (from Hodge and Kennicutt 1983).

On the other hand, Figure 6 is very suggestive. The Sc (M101) and the Irr galaxy (LMC) show nearly parallel luminosity functions that extend to high luminosities, while M31 appears to have a large-luminosity cut-off. If this is the case, then there must be some physical mechanism that prevents large HII regions from forming in galaxies like M31. Such a mechanism, the shearing of GMCs by differential rotation and the break-up of them by tidal effects, has been suggested to explain the lack of large HII regions in the central regions of M101 (Waller and Hodge 1991); it is suggested that it also operates for spiral galaxies, especially of Sb type and earlier, where the disk is massive and rapidly rotating. To combine this with the previous argument above, it must be that the HII regions of Sb and earlier galaxies do not fit the empirical exponential size distribution rule, a fact that can not yet be checked satisfactorily because of the lack of enough size distribution measurements for galaxies of early Hubble types (*e.g.*, see Hodge 1987).

In conclusion, it appears that giant HII regions are often, but not always, cases of special circumstances, not members of the normal population of HII regions in a galaxy. They can be unusually large either because of particular mechanisms, possibly externally-triggered, that promote large scale star-formation, or because they exist in an environment that particularly encourages it. More light on the subject will no doubt be shed when a thorough theoretical understanding of the luminosity function and other parameters describing HII regions is reached (*e.g.*, McKee 1992).

ACKNOWLEDGEMENTS

I am in the debt of many astronomers for their help, ideas, data, and gentle criticisms, particularly Eric Wilcots, Bryan Miller, Robert Kennicutt and Evan Skillman.

REFERENCES

Campbell, B. *et al.* 1992, *A. J.*, in press.

Chen, P. *et al.* 1992, *Ap. J.*, in press.

Heap, S., Ebbets, D. and Malamuth, E. 1991, *BAAS*, **23**, 1330.

Hodge, P. 1966, *An Atlas and Catalog of HII Regions in Galaxies*, (Univ. Washington: Seattle).

Hodge, P. 1976, *Ap. J.*, **205**, 728.

Hodge, P. 1987, *PASP*, **99**, 915.

Hodge, P. 1989, *Ann. Rev. Astr. Ap.*, **27**, 139.

Hodge, P. and Lee, M. G. 1992, submitted.

Hodge, P. and Wright, F. 1977, *The Small Magellanic Cloud*, (Univ. Washington Press: Seattle).

Kennicutt, R. 1984, *Ap. J.*, **287**, 116.

Kennicutt, R. 1991, in *The Magellanic Clouds*, R. Haynes and D. Milne, eds. (Kluwer: Dordrecht), p. 139.

Kennicutt, R., Edgar, B. and Hodge, P. 1989, *Ap. J.*, **337**, 761.

Kennicutt, R. and Hodge, P. 1980, *Ap. J.*, **241**, 573.

Kennicutt, R. and Hodge, P. 1986, *Ap. J.*, **306**, 130.

Massey, P., Garmany, K. and Parker, J. 1992, *A. J.*, in press.

McCall, M., Rybski, P. and Shields, G. 1985, *Ap. J. S.*, **57**, 1.

McKee, C. 1992, in *Star Forming Galaxies*, J. Franco and F. Ferrini, eds. (Cambridge Univ. Press: Cambridge), in press.

Melnick, J. 1985, *A. & A.*, **153**, 235.

Pagel, B. and Edmunds, M. 1981, *ARAA*, **19**, 77.

Parker, J. 1992, Ph.D. Thesis, Univ. Colorado.

Pellet, A., Astier, N., Viale, A., Courtes, G., Maucherat, A., Monnet, G. and Simien, F. 1978, *A. Ap. S.*, **63**, 441.

Rosa, M. and D'Odorico, S. 1986, in *Luminous Stars and Associations in Galaxies*, C. De Loore, A. Willis, and P. Laskarides, eds. (Reidel: Dordrecht).

Savage, B., Fitzpatrick, E., Cassinelli, J. and Ebbets, D. 1983, *Ap. J.*, **273**, 597.

Shields, G. 1990, *ARAA*, **28**, 525.

van den Bergh, S. 1981, *A. J.*, **86**, 1464.

Walborn, N. 1986, in *Luminous Stars and Associations in Galaxies*, C. De Loore, A. Willis, and P. Laskarides, eds. (Reidel: Dordrecht).

Walborn, N. and Blades, J. 1987, *Ap. J.*, **323**, L65.

Waller, W. 1992, this volume.

Waller, W. and Hodge, P. 1990, in *Dynamics of Galaxies and their Molecular Cloud Distributions*, F. Combes and F. Casoli, eds. (Kluwer: Dordrecht), p. 187.

Walterbos, R. and Braun, R. 1991, *A. Ap. Suppl.*, in press.

Wang, Q. and Helfand, D. 1991, *Ap. J.*, **370**, 541.

Weigelt, G. *et al.* 1991, *Ap. J.*, **378**, L21.

Wilcots, E. 1992, Ph.D. Thesis, Univ. Washington.

Ye, T. 1992, *MNRAS*, **255**, 32.

Massive Stars: Their Lives in the Interstellar Medium
ASP Conference Series, Vol. 35, 1993
Joseph P. Cassinelli and Edward B. Churchwell (eds.)

UNRESOLVED QUESTIONS REGARDING MASSIVE STARS AND THEIR INTERACTION WITH THE ISM

KRIS DAVIDSON
University of Minnesota, Department of Astronomy,
116 Church St., Minneapolis, MN 55455

ABSTRACT Some potentially important astrophysical questions receive little observational or theoretical attention. A few "orphan problems" pertaining to mass-loss by luminous stars are presented as examples.

Isn't that an awful title? -- Intended to finish off this meeting with ponderous solemnity, no doubt. Anyway, it would be presumptuous here in this last talk to review the others, because they're good reviews themselves. So let me be presumptuous in a different way: Let's reflect on the problems and ideas that haven't gotten much attention yet!

Someone has remarked that astrophysical topics fall into three categories: Bandwagon, Fancy That, and No Stone Unturned. With a slightly different projection in Sentiment Space we can invent another category, Orphan Problems. Some of these were once on bandwagons but somehow never got solved; occasionally a whole bandwagon has been abandoned long before it completed its parade, not because it was a bad vehicle but because fashion is fickle. Let's mix a few more metaphors: Recall how many of us, at one time or another, have self-consciously labeled some object a Rosetta Stone of astrophysics, or how often we declare intentions to Confront Theory with Observation. My point here is that some of those Rosetta Stones haven't been examined past the end of the Greek part; or sometimes Theory Confronts Observation, but they don't recognize each other and they just go separate ways. Examples in this talk are some of my own personal favorites; of course many astronomers have others of their own.

Just to illustrate this kind of thing, take the Crab Nebula, formerly known as The Rosetta Stone of Astrophysics. (It genuinely is related to the topic of this meeting, but I choose it here because it's a good example of an Orphan Topic.) In the 1950's, and then for a couple of years around 1970, this was a small bandwagon itself. Today, because of its textbook-example status ("the best-known SNR"?), it seems natural to assume that the Crab must be pretty well understood by now, or at least that it doesn't seriously conflict with theory. After all, this is the most observable young SNR, arguably the most critical example for some parts of SN and pre-SN theory. We shouldn't claim to understand SN's and young SNR's if the Crab is not understood! But in fact it is NOT understood, and the extent to which we don't understand it would probably surprise many non-specialists.

Craig Wheeler (1978) was probably the first to notice the theoretically or observationally embarrassing "nitrogen paradox" in the Crab Nebula. This takes some explanation. The observed thermal gas (SN ejecta + swept-up stellar ejecta) is mostly helium, seemingly without much extra carbon and oxygen made by helium burning. The usual interpretation is that most of the helium-burning-and-subsequent products ended up in the pulsar, probably indicating that the pre-SN star was initially about 10 solar masses. [For reviews of these matters see Davidson and Fesen (1985) and Kafatos and Henry (1985).] If most of the helium was made by the CNO cycle as it should have been, then most of the original carbon and oxygen should have converted to nitrogen. But there isn't that much nitrogen in the Crab; a very basic prediction seems to have failed. Until we understand this, our most observable young supernova remnant does not offer any support to nucleosynthesis theory! -- But the rush to solve this problem has so far been underwhelming, and most astronomers are probably not even aware of it. The most plausible explanations need to be tested by UV spectroscopy, an obvious task for the HST. The HST has not yet obtained any useful relevant observations of this famous object.

Some other aspects of the Crab, though perhaps less fundamental, are also surprisingly little understood. Its unusual-looking "northern jet" remains mysterious (see, e.g., Fesen and Gull 1986), the orientation of its likely bipolar morphology still hasn't been ironed out even after all these years (Fesen et al. 1992), some Ni/Fe emission line ratios are unexpected (Henry 1990), etc. Yet, for about 20 years there has been little general interest in further work on the Crab -- it lost its fashionable bandwagon status before its problems were truly solved.

Next, a few words about bipolar morphology in general, and something in the "Fancy That" category in particular. last year Dennis Ebbets obtained a short-exposure HST/PC image of another of my favorite objects, the most extroverted star in the Galaxy, Eta Carinae. (For a review of this very massive star see, e.g., Davidson 1989.) We've processed the picture to show the whole structure over several orders of magnitude in surface brightness, exhibited in a poster at this meeting. The version in Fig. 1 here is subtly disturbing, at least with two circles drawn in it, especially when I shortly mention one or two other particular objects.

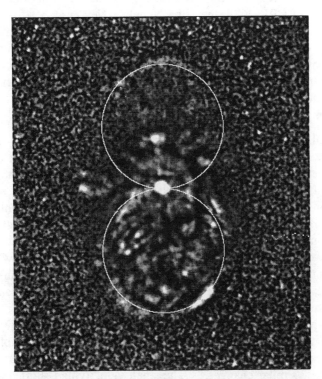

Fig. 1   HST/PC image of Eta Carinae at violet wavelengths. Northwest is upward in this picture and northeast is to the left. This looks like an image of a faint object because an unsharp-masking process has been applied, in order to show both the bright inner part and the much fainter outer edges. The two circles are exactly the same size and they are mutually tangent exactly at the central star.

This is clearly bipolar, generally consistent with models proposed, e.g., by Meaburn et al. (1987) [see Hester et al. (1991) for an unconventional interpretation, though]. However, some of the details are not what we should expect from existing discussions of bipolar structure (e.g. Morris 1981). First, it's a little surprising that the projected lobes are so circular. More surprisingly, both circles are exactly the same size. But what really bothers me is that the two circles are mutually tangent, almost precisely at the central star. Statistically this seems very unlikely for the projection of a three-dimensional bipolar structure, especially considering that the southeast lobe is thought to be closer to us. Moreover, here's another strange observation: An image of He2-104, a.k.a. "the Southern Crab", also matches the osculatory circles in Figure 1; see Schwartz et al. (1989) and Lutz et al. (1989). He2-104 and Eta Car are nothing like each other as stars, of course, differing enormously in mass and nature. For what it's worth, a picture of CRL 2688, the "Egg Nebula" (Ney et al. 1975) can also be superimposed on the same figure so that its two lobes nestle comfortably within the two circles. Apparently some basic physical or geometric characteristic of the ejection process can make widely different objects look this way. My main point here is that bipolar morphology may deserve geometrically detailed examination, more than just qualitatively noting the presence of an axis and two lobes.

Now, another question about bipolar structures: Do they nearly all involve binary stars, or is rapid rotation a feasible alternative? The only easy way to get an axis of symmetry is from angular momentum, making binaries and rapid rotators the obvious alternatives. Since many of us dislike both of these because they are complicated, not enough attention has yet been given this question. Let me offer some informal, anecdotal evidence, again concerning Eta Carinae. Hydrogen and helium emission lines from the central object and its envelope usually include narrow components, only about 40 km/s wide (e.g. Ruiz et al. 1984). I think that sometimes these narrow components change or even temporarily disappear within a few months; see Bandiera et al. (1989). [A few years ago I was shown a high-resolution spectrum made in 1987, wherein the narrow component of H-alpha was absent; but this was probably never published. Other spectra, made years earlier and also never published, are also rumored to show similar changes.] The small velocity and short time-scale together suggest that this gas is in a region at most several astronomical units across. However, if the eruption phenomenon in Eta Car depends on a close companion star then its orbital speed has to be several hundred km/s. Such a star might be expected to stir the observed envelope gas so that its line widths should become far larger than 40 km/s. Possible conclusion: The bipolar axis of symmetry in Eta Carinae seems likely to be determined by rapid rotation, not a binary orbit. But this leaves us with the familiar physical question: Why should rapid rotation become more (rather than less) influential as the star evolves and expands? Maybe the modified Eddington limit with temperature-dependent opacity, not just electron scattering, can help here, but anyway the question hasn't yet been discussed very often. [Cf. Sreenivasan and Wilson (1989).]

At this point, while preparing this talk another point occurred to me, concerning the modified Eddington limit in a close binary system. Radiation pressure can decrease the effective surface gravity of a star but it has no effect on orbital velocities in a binary system. This can dramatically change the sizes and even the topology of Roche surfaces. If the L/M ratio is more than perhaps 70 percent of the Eddington Limit (depending on assumptions), then the vertex where the primary star loses mass can be on the side opposite its secondary companion! In a sense the main effect of the companion star in this situation is to enforce rapid rotation. Now Dany Vanbeveren has mentioned some references to pertinent calculations done many years ago: Schuerman (1972) and Vanbeveren (1978). Since this phenomenon is very sensitive to opacity, while opacity is sensitive to the potential surfaces (because they affect the surface area), here we have a potential cause of chaotic mass-loss, maybe applicable to LBV's.

To some extent the cause of gigantic LBV eruptions is another Orphan Problem -- it has received amazingly little theoretical analysis -- but I'd better not discuss it here, having done so at some length before (Davidson 1989). And here's a riddle: How are very massive stars like bosons? Answer: They're gregarious. Even though only one new star in a million is very massive (say 50 solar masses or more), if you find one there are usually others nearby -- consider R136 or the Carina Nebula, for examples. This generality was noticed long ago but maybe it's not widely known; has anyone attempted to develop an explanation?

Let me close with a possible observational project that has not been discussed much. A few LBV eruptions reach prodigious absolute magnitudes of -14 or so: Eta Car did so in the nineteenth century and "SN 1961V" in NGC 1058 may have been an even more extreme LBV eruption (Goodrich et al. 1989). This is so bright that we can almost contemplate a patrol for such events in nearby galaxies; it wouldn't be very much different from looking for supernovae of the SN 1987A type. Extreme LBV eruptions are so dramatic and mysterious that it would be fun to detect one or two of them.

So that's one roster of hobbyhorses, "orphan problems" involving mass loss by massive stars; other astronomers can cite others. I suspect that modern styles of both theoretical and observational research tend to inhibit work on certain special types of problems. Many theoretical groups now develop big computational codes, which get passed along and evolve through successive graduate students; such a research group is less likely to consider any problem that is not a natural subject for the group's code. Mainstream observers today seem to feel most comfortable when they have big lists of targets for each telescope run; but sometimes intensive work on some unique object is more valuable than the same amount of telescope time expended on a set of objects. Anyway, the march of observational and theoretical progress sometimes leaves unexplored spaces behind; one of these may embarrass us someday.

REFERENCES

Bandiera, R., Focardi, P., Altamore, A., Rossi, C., and Stahl, O., 1989, in Physics of Luminous Blue Variables (Davidson et al., eds.) p. 279.

Davidson, K., 1989, in Physics of Luminous Blue Variables (Davidson et al., eds.), 101.

Davidson, K. and Fesen, R.A., 1985, Ann. Rev. A. Ap. 23, 119.

Fesen, R.A. and Gull, T.R., 1986, Ap. J., 306, 259.

Fesen, R.A., R.A., Martin, C.L., and Shull, J.M., 1992, Ap. J. in press.

Goodrich, R.W., Stringfellow, G.S., Penrod, G.D., and Filippenko, A.V., 1989, Ap. J. 342, 908.

Henry, R.B.C., 1990, Publ. A.S.P. 98, 1044.

Hester, J.J., Light, R.M., Westphal, J.A., Currie, D.G., Groth, E.J., Holtzman, J.A., Lauer, T.R., and O'Neil, E.J., 1991, A. J. 102, 654.

Kafatos, M. and Henry, R.B.C. (eds.), 1985, The Crab Nebula and Related Supernova Remnants. (Cambridge University Press)

Lutz, J.H., Kaler, J.B., Shaw, R.A., Schwarz, H.E., and Aspin, C., 1989, Publ. A.S.P. 101, 966.

Meaburn, J., Wolstencroft, R.D., and Walsh, J. R., 1987, A&A, 181, 342.

Morris, M., 1981, Ap. J. 249, 572.

Ruiz, M.T., Melnick, J., and Ortiz, P., 1984, Ap. J. (Letters), 285, L19.

Schuerman, D.W., 1972, Ap. Sp. Sci., 19, 351.

Schwartz, H.E., Aspin, C., and Lutz, J.H., 1989, Ap. J. (Letters) 344, L29.

Sreenivasan, S.R., and Wilson, W.J.F., 1989, in Physics of Luminous Blue Variables (Davidson et al., eds.), P. 205.

Vanbeveren, D., 1978, Ap. Sp. Sci. 225, 212.

Massive Stars: Their Lives in the Interstellar Medium
ASP Conference Series, Vol. 35, 1993
Joseph P. Cassinelli and Edward B. Churchwell (eds.)

POWERING THE SUPERWIND IN NGC 253

ALAN WATSON, JAY GALLAGHER
University of Wisconsin – Madison, 475 North Charter Street, Madison,
WI 53706.

MICHAEL MERRILL
K.P.N.O., PO Box 26732, Tucson, AZ 85726.

JEAN KEPPEL
603 Pauley Drive, Prescott, AZ 86303.

MARK PHILLIPS
C.T.I.O., Casilla 603, La Serena, Chile 1353.

NGC 253 AS A NUCLEAR STARBURST GALAXY

NGC 253 is a prototypical moderate nuclear starburst galaxy. It is a barred SBc
spiral galaxy at a distance of approximately 3 Mpc and can be studied on scales
down to 15 pc in the optical and near IR.

It is a bright IRAS source with a flux of 1000 Jy at 60 μm (Rice et al.
1988) and a FIR luminosity of $3 \times 10^{10} L_\odot$ (Telesco and Harper 1979, Rice et
al. 1988). It has a strong Brγ emission line, a signature of ongoing massive star
formation (Rieke, Lebofsky, and Walker 1988), and deep CO absorption bands,
indicative of the dominance of red supergiants in the near IR. It contains a
population of compact radio sources, similar to those seen in M82 (Turner and
Ho 1985, Antonucci and Ulvestad 1988, Ulvestad and Antonucci 1991). Optical
spectra show that the nucleus is heavily reddened, with a Balmer decrement of
approximately 30.

NGC 253 possesses a 'superwind,' seen both in X-ray emission (Fabbiano
1988) and in optical line emission (McCarthy, Heckman, and van Breugel 1987).
Nuclear ejection was first suggested by Demoulin and Burbidge (1970) to explain
the kinematics of the nuclear region.

NEAR IR OBSERVATIONS

We have obtained J, H, and K images of the entire galaxy at 1.3 arcsec/pixel
(18 pc/pixel) using SQIID on the KPNO 1.3m. We have constructed a mosaic
of 180 s exposures which traces the galaxy over much of its optical extent. The
data were shifted, rotated, magnified, and calibrated following normal practice.

The PtSi detectors in SQIID have a low quantum efficiency (of order 5%)
and excellent stability, linearity, and cosmetic quality. To check the quality
of the calibration, we compared our images to single-channel measurements (de
Vaucouleurs and Longo 1988) in apertures with diameters ranging from 10 arcsec
to 100 arcsec and found excellent agreement to around 10%.

Rieke, Lebofsky, and Walker (1988) published images of the nucleus with a pixel size of 1.2 arcsec. They report that the nucleus was significantly bluer than the circumnuclear region (of order 0.5 mag in $H-K$); we find no trace of this in our data, nor in data taken in October 1991 with SQIID in its current configuration, nor in the data of Forbes et al. (1992). It is possible that seeing or focus effects could blur our colors or that Rieke, Lebofsky, and Walker caught a transitory event, such as a supernova.

We have constructed the $J-H$ and $H-K$ colors for each pixel in our image. We show the color-color diagram for all of the pixels in a region 14 × 14 arcsec centered on the nucleus. The area of the symbol is proportional the the K flux in that pixel. The points clearly form a locus leading from $J-H \approx 0.8$ and $H-K \approx 0.25$ up to much redder colors in the nucleus. We have indicated the standard Milky Way extinction law by an arrow which corresponds to $A_K \approx 0.5$ or $A_V \approx 4.5$. In agreement with Scoville et al. (1985), we interpret the near IR color-color diagram of the nucleus to be due to a dust-shrouded population of red supergiants plus a contribution at K from hot (1000 K) dust.

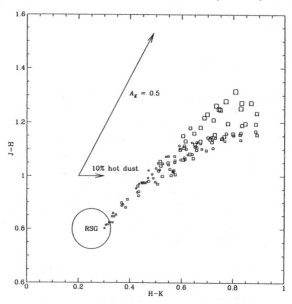

ESTIMATING THE SUPERNOVA RATE

We have obtained the 'dereddened' H magnitude within a radius of 10 arcsec (180 pc) of the nucleus by correcting each pixel for reddening, assuming an intrinsic $J-H$ color of 0.8 for red supergiants. The 'dereddened' H magnitude within 10 arcsec is 7.28, compared the the observed H magnitude of 7.78. (Corrections for extinction are complex, but we consider our correction to be conservative.)

The luminosity of the red supergiants in the nucleus corresponds to about 2% of the bolometric luminosity of NGC 253. The luminosity of massive stars remains roughly constant during their lifetime and they spend a few percent of their life as red supergiants. Assuming a typical red supergiant absolute magnitude at H of −11.0 and a distance of 3 Mpc, the H magnitude of the nucleus corresponds to 4400 red supergiants within 180 pc of the nucleus.

A typical lifetime of a massive star in the red supergiant phase is approximately 10^5 yr, and if we assume continuity over this lifetime we derive a supernova rate of 0.04 yr^{-1} in a 180 pc radius aperture. This is somewhat lower than the rate of 0.1 yr^{-1} to 0.25 yr^{-1} in a 300 pc radius aperture derived from VLA observations of compact sources by Antonucci and Ulvestad (1988). Although our apertures differ in size, our results are directly comparable as most of the compact sources are in the central 10 arcsec. Possible systematic differences between our results may be due to confusion with H II regions and the possibility that the evolution of supernovae in dense environments is not straightforward.

POWERING THE SUPERWIND

From Einstein X-ray observations, Fabbiano (1988) estimates that the temperature and mass of the superwind are of order 10^7 K and about 2×10^7 $M_\odot$. The extent of the wind is at least 4.5 kpc. The wind is highly over-pressurized and is likely to be flowing out at close to the sound speed. This is confirmed by the presence of 300 km s^{-1} bulk velocities at the base of the wind seen in long-slit echelle spectra obtained with the CTIO 4m. With these properties, the superwind is being supplied with hot gas on a timescale comparable to the sound crossing time of 10^7 yr. Mass is being supplied to the superwind at a rate of 1.5 $M_\odot$ yr^{-1}. Energy is flowing into the wind at a rate of about 2×10^{41} erg s^{-1} or 6×10^7 $L_\odot$. This corresponds to about 15% of the energy released by supernovae.

This figure should not be over-interpreted. The extinction correction used to derive the supernova rate ignores complications from nebular lines and patchy extinction which will leads us to believe that the true supernova rate wil be somewhat higher than our estimate. Also, the physical properties of the superwind, especially its temperature, are uncertain.

Nevertheless, we can draw two conclusions. It seems inevitable that energy is being efficiently transferred from supernovae in the nucleus into the base of the superwind. The high supernova rate in the nuclear region must lead to an ISM with a global structure vastly different to the ISM in quiescent disks.

We thank Ian Gatley for his help in taking the SQIID images.

REFERENCES

Antonucci, R.J., and Ulvestad, J.S. 1988, ApJ 330, L97.
Demoulin, M.-H., and Burbidge, E.M. 1970, ApJ 159, 799.
Fabbiano, G. 1988, ApJ 330, 672.
Forbes et al. 1992., MNRAS 254, 509.
McCarthy, P.J., Heckman, T.M., and van Breugel, W. 1987, AJ 93, 264.
Scoville, N.Z., et al. 1985, ApJ 289, 129.
Rice, W., et al. 1988, ApJS 68, 91.
Rieke, G.H., Lebofsky, M.J., and Walker, C.E. 1988, ApJ 325, 679.
Telesco, C.M., and Harper, D.A. 1980, ApJ 235, 392.
Turner, J.L., Ho, P.T.P. 1985, ApJ 299, L77.
Ulvestad., J.S., and Antonucci, R.J. 1991, AJ 102, 875.
de Vaucouleurs, A., and Longo, G. 1988, Catalogue of Visual and IR Photometry
    of Galaxies (University of Texas: Austin).

Massive Stars: Their Lives in the Interstellar Medium
ASP Conference Series, Vol. 35, 1993
Joseph P. Cassinelli and Edward B. Churchwell (eds.)

SUPERGIANT STARS AND STAR CLUSTERS IN THE STARBURST IRREGULAR GALAXY, NGC 1569

WILLIAM H. WALLER
NASA Goddard Space Flight Center, Code 681, Greenbelt, MD 20771

CHERYL L. DRACOBLY
Peace Corps; and University of Washington, Seattle, WA 98195

ABSTRACT   NGC 1569 is a Magellanic-type irregular galaxy which
has recently undergone a global burst of star formation. At a distance
of 2.2 Mpc, it is close enough for ground-based resolution of its brightest
stars and star clusters. We report the results of a CCD imaging study
conducted with the KPNO 0.9-m telescope. Using DAOPHOT to identify
and measure discrete sources in the R and I-band images, we identify 92
sources belonging to the galaxy. These span a wide and continuous range
of magnitudes at R-band — from 14 mags for the brightest "supergiant
star cluster" known as "object A" to 21 mags for the faintest resolved
sources. Given the adopted distance and reddening along with the
observed colors, these latter sources are consistent with being B–K-type
supergiant stars that were formed during the recent starburst episode.

OBSERVATIONS

R, I, and Hα-band images were obtained with the KPNO #1 0.9-m telescope
and RCA-3 CCD camera during clear weather but mediocre seeing (FWHM
$\approx 2.2''$). The large-scale morphology of the continuum emission — and of the
"eruptive" Hα emission — was reported in (Waller 1991). Figure 1 shows 3′
× 3′ (1.9 kpc × 1.9 kpc) fields of the continuum-subtracted Hα emission and
of the underlying red continuum emission. Star-like sources are evident along
the perimeter of the R-band image. Other discrete sources can be found in the
bright interior, most especially the two bright nuclear objects "A" and "B" (cf.
Arp and Sandage 1985; Waller 1991).

PHOTOMETRY

Photometry of the discrete continuum sources in NGC 1569 was carried out
with DAOPHOT, a computer program designed for crowded-field stellar
photometry (Stetson 1987). After trimming the R and I-band images to 3′ ×
2′ fields, we used DAOPHOT's 'FIND' algorithm to locate discrete sources in
both images. Cosmic ray hits and other glitches were eliminated by demanding

spatial matchups of sources in both images. A total of 128 matching sources were found. The spatial distribution of sources was then compared with median filtered versions of the images to verify the discrete nature of the sources. Bright foreground sources were eliminated by visual inspection of the images. Of the remainder, 92 sources had standard errors in their $(R - I)$ colors of less than 0.25 mags. We estimate that at least 90% of these sources are actual members of the galaxy.

RESULTS

The resulting R-band magnitudes and $(R - I)$ colors are shown in Figure 2a. The errorbars denote the standard errors in magnitude and color, based on DAOPHOT's PSF-fitting photometry. This diagram reveals a wide and continuous range of magnitudes at R-band — from 14 mags for the brightest source known as "object A" (Arp and Sandage 1985) to 21 mags for the faintest resolved sources. Figure 2b shows the *absolute* color-magnitude distribution, after correcting for reddening $(E(B - V) = 0.56)$ and adjusting for distance (d = 2.2 Mpc), as determined by Israel (1988). The loci of supergiant stars are also plotted for comparison (Schmidt-Kaler 1982; Johnson 1966). This diagram shows that the faintest sources are consistent with being B–K-type supergiant stars with progenitor masses exceeding 20 $M_\odot$ and hence ages of no more than 10 Myr. A significant population of even brighter sources is evident which can only be attributed to clusters of luminous stars. The brightest of these (objects "A" and "B") qualify as "supergiant star clusters", having luminosities in between that of 30 Doradus and NGC 5461 (the "hypergiant" HII region in M101).

CALL FOR MORE OBSERVATIONS!

These findings indicate that NGC 1569 is a rich source of extremely luminous stars and star clusters — vestiges of the starburst activity which has wracked this tiny galaxy during the past ~50 million years. However, the actual composition of the stellar populations remains to be determined. The results presented here are based on imagery with mediocre resolution (~2″). Much higher resolution imaging at more wavebands (using the Hubble Space Telescope or the HRCAM on the Canada-France-Hawaii Telescope) would go far in evaluating a particular source's true membership in the galaxy and whether it is stellar or clustered. Jay Gallagher reports (this conference) that HST/WFPC images of NGC 1569 have been recently obtained which — when fully analyzed — should provide well-resolved photometric measurements of the supergiant stars and cluster candidates. The high probability of there being Cepheid supergiants could also be exploited (by imaging over several days) to derive a more accurate distance to the galaxy. Lastly, high-resolution spectroscopy using small entrance apertures would provide the essential spectral classification of the stars. This could be easily done at visible wavelengths with HST's Faint Object Spectrograph but may be more difficult in the UV due to the strong Galactic reddening.

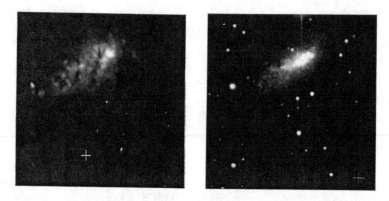

Fig. 1.   NGC 1569 in the light of Hα (left) and R-band (right). Field of view is 3′ × 3′ (1.9 kpc × 1.9 kpc at a distance of 2.2 Mpc).

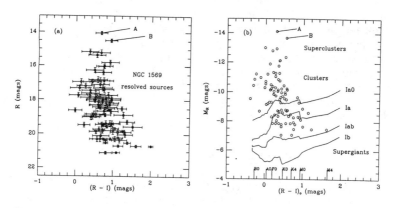

Fig. 2.   Color-Magnitude diagrams of the discrete sources in NGC 1569, before (a) and after (b) adjusting for distance and reddening.

REFERENCES

Arp, H. C., and Sandage, A. R. 1985, *A.J.*, **90**, 1163.
Israel, F. P. 1988, *Astr. Ap.*, **194**, 24.
Johnson, H. L. 1966, *Ann. Rev. Astr. Ap.*, **37**, 193.
Schmidt-Kaler, Th. 1982, in *Landolt-Bornstein*, New Series, Vol. **2c**, p. 46.
Stetson, P. B. 1987, *Pub. A.S.P.*, **99**, 191.
Waller, W. H. 1991, *Ap. J.*, **370**, 144.

Massive Stars: Their Lives in the Interstellar Medium
ASP Conference Series, Vol. 35, 1993
Joseph P. Cassinelli and Edward B. Churchwell (eds.)

KINEMATICS OF STARS IN THE NUCLEUS OF M82:
THE NUCLEAR MASS

DAN F. LESTER and NIALL I. GAFFNEY
Department of Astronomy and McDonald Observatory
University of Texas, Austin TX 87812

CHARLES M. TELESCO
Space Sciences Lab, ES-63, NASA Marshall Space Flight Center
Huntsville AL 35812

ABSTRACT   We have measured the velocity dispersion of the red stellar
population in the heavily obscured nucleus of the starburst galaxy M82.
This was achieved by measuring the degradation of the otherwise very
sharp ^{12}CO 2-0 bandhead at 2.3μm in the stellar continuum. We find an
enclosed mass of 5×10^7 M$_\odot$ within 7.5pc of the center, close to that for
the same region in our own galaxy. The M/L ratio in the nucleus of M82
is similar to that in our own galactic nucleus too, where the red light is
dominated by red giants.

INTRODUCTION

M82 is the closest (3 Mpc) and best observed example of extragalactic
organized star formation on a galactic scale. As such, it has been used as a
prototype for more distant, luminous starburst systems. In order to understand
the dynamical processes that have been responsible for the starburst, and the
stellar population that is extant there, measurements of the total mass are
fundamentally important.

In heavily obscured starburst systems, the only information about the
kinematics has been from measurements of the gas. Such velocities can be
measured very accurately, and on a small spatial scale, but nongravitational
processes may play an important role in the kinematics. This is seen in M82 in
the bipolar outflow that is produced by the supernova activity in the core.
Furthermore, dissipative mechanisms will have a particularly important effect in
bar driven kinematics, which probably best represents M82. Such processes
will be troublesome in galaxies that are the most violent examples of starbursts.

Measurements of the *stellar* rotation curve and velocity dispersion are, however,
unaffected by these processes that distort the kinematical information from the
gas. Such measurements are much more difficult to do, because they require
high S/N in the stellar continuum which, per resolution element, is usually

much fainter than in the emission lines that characterize the starburst. Furthermore, the optical absorption lines that are used for this kind of work are simply not accessible in these usually heavily obscured systems. We have used the degradation of the 2.3μm [12]CO bandhead in the red stellar continuum from M82 to derive the velocity dispersion in that heavily obscured nucleus. This bandhead is nearly infinitely sharp in the spectra of the individual late-type stars that dominate the continuum at this wavelength. The observed shift and broadening of what would otherwise be a sharp bandhead can be interpreted reliably as the beam averaged velocity and velocity dispersion.

OBSERVATIONS

The observations were made on 27 April 1992 with the IRTF CSHELL echelle spectrometer on a shared-risk basis during its commissioning period. This instrument is described in detail by Tokunaga et al. (1992). We used two quadrants of the Rockwell HgCdTe NICMOS 256x256 array to make the observations. The slit was oriented at PA~69°, through the nucleus, and along the starburst ridge, as defined by Telesco et al. (1991). We offset from BD+70°857 to this position using the astrometry of Lester et al. (1990). The slit was 1" wide and 14" long, on a monochromatic scale of 0.25"/pixel. The dispersion was 3.3 km/s/pixel. In order to improve the S/N, the spectra were coadded in four pixel bins, giving a final velocity resolution of 13 km/s and a spatial resolution of 1". Good wavelength calibration and instrumental stability are important to the success of this experiment, and we have determined the velocity scale to better than 0.2 pixel (<1 km/s). Observations of the hot blue star BS 4295 were used to cancel out telluric features.

ANALYSIS

Figure 1 shows the resulting spectrum of the nucleus of M82 (solid line) with the spectrum of a late type model stellar atmosphere appropriate to the mean spectral type of M82 convolved to the instrumental resolution (thin dashed line). This model has been shifted to the systemic velocity of M82, for comparison. It reproduces actual stellar spectra almost perfectly, and it is used here for computational convenience. It is important to understand that the results from our analysis are completely insensitive of the particular model that is used. This is because while the CO band *strength* changes with spectral type, the *shape* changes very little, and we fit the strength to the observed spectrum directly. The washed-out appearance of the bandhead in the M82 spectrum is due entirely to kinematics.

Figure 1 also shows a simple fit to the observed spectrum of the central 1" (r<7.5 pc). The heavy dotted line is the convolution of the model template shown in Figure 1 with a simple Gaussian profile. The fit indicates a systemic velocity $V_r = 240 \pm 20$ km/s and line-of-sight velocity dispersion $\sigma = 100 \pm 20$ km/s. A better fit may be possible with multiple components, but will not be explored here. This σ is about two times larger than that derived from [NeII] in the same area by Achtermann (private communication), suggesting that the ionized gas is not kinematically representative of the gravitational potential.

We derive a mass enclosed within the $r<7.5$ pc beam of $M_{enc} \sim 3r\,\sigma^2/G = 5 \times 10^7$ $M_\odot$. This is similar to that measured, in the same size region, in our own galactic nucleus (see McGinn et al. 1989). With K band continuum flux $F_K = 32$ mJy in this beam and an assumed K extinction of $A_K = 0.5$ (see Telesco et al. 1991), we derive $M/L_K \sim 0.5$ for the nucleus of M82. Using the large beam measurements of the galactic center, and a conventionally assumed $A_K = 2.7$, we derive $M/L_K \sim 0.7$. Thus, the M/L ratio of the stellar population in M82 is similar to that in our own galactic nuclear bulge. This supports the suggestion by Lester et al. (1990) that the red starlight from the nucleus arises in a normal bulge population that predates the starburst that now envelops it.

REFERENCES

Lester, D.F., Carr, J.S., Joy, M., and Gaffney, N.I. 1990 Ap. J., 352, 544.
McGinn, M.T., Sellgren, K., Becklin, E.E., and Hall, D.N.B. 1989 Ap. J., 338, 824.
Tokunaga, A.T., Toomey, D.W., Carr, J., and Hall, D.N.B. 1992 in SPIE#1235 "Instrumentation in Astronomy VII", p.131.
Telesco, C.M., Campins, H., Joy, M., Dietz, K. and Decher, R. 1991 Ap. J., 369, 135.

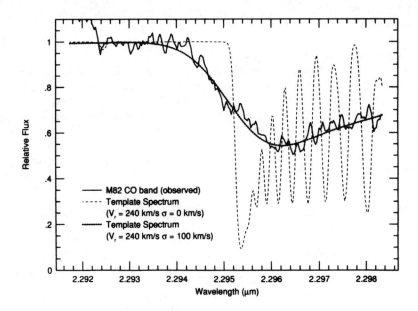

Figure 1: The observed 2.3μm spectrum of M82 is compared with that of a template star spectrum, and that of a synthetic population with σ=100 km/s.

Massive Stars: Their Lives in the Interstellar Medium
ASP Conference Series, Vol. 35, 1993
Joseph P. Cassinelli and Edward B. Churchwell (eds.)

OBSCURATION EFFECTS IN STARBURST GALACTIC NUCLEI

WILLIAM C. KEEL
Dept. of Physics and Astronomy, University of Alabama, Box 870324,
Tuscaloosa, AL 35487

INTRODUCTION

Starburst nuclei have been observed throughout the accessible spectrum, with an overall goal of understanding the stellar populations and interstellar media of these most extreme star-forming environments. Many of the observed properties resemble those of scaled-up individual H II regions as seen in our own neighborhood - similar emission-line ratios, blue and mostly stellar continua, emission from heated dust, and in some cases spectral features of W-R stars. However, there are troubling inconsistencies between the results of some kinds of observations, especially when interpreted as simple scaled-up single regions. These are of two kinds. First, estimates of obtained at various wavelengths disagree substantially, such that IR indicators give higher extinctions than those at shorter wavelengths (as cogently discussed by Waller et al. 1992). This indicates mixing of emitting and absorbing sources, and that there is no unique value of extinction to be applied at a given wavelength independent of knowledge of the source structure. Second, measurements of thermal emission confirm that many starbursts have a large amount of associated dust, but strong optical emission lines and ultraviolet continua are observed.

In its starkest form, the problem may be stated as follows: for objects so dusty and compact, why do we see anything at all in the optical and UV? The pat answer is that the dust is patchy, and this is almost certainly correct. However, the effects of patchy obscuration may well be different for stars and ionized gas, leading to a profound influence on the observed properties and our interpretation. This paper shows two ways of demonstrating that such selective obscuration goes on, and shows the implications with a simple two-dimensional model for a starburst nucleus.

EXAMPLES: LYMAN-CONTINUUM EXCESS AND OBSCURED CLUSTERS

A set of consistent spectral energy distributions has been generated for a sample of starbursts in galactic nuclei and disks, using the IUE archives and new ground-based optical and K-band data. The sample consists of 12 nuclei listed by Balzano (1983) or Devereaux (1989) for which observations were available in the IUE archives. Photometry from B through K bands, and in Hα, was performed on CCD and IR Imager data from Lowell and Kitt Peak, using matched numerical apertures. This allowed production of spectral energy distributions covering exactly matched parts of each galaxy, so that, at the expense of possible added light from older bulge stars in the redder bands, the resulting spectral shape refers to one spatially limited stellar population.

The far-UV intensities calculated from the Hα fluxes and recombination theory fall well above any extrapolation of the directly observed stellar continuum longward of Lyman α. An example is shown in Fig. 1, for two nuclei with dramatically different Balmer decrements. The ionizing continuum is almost ten times stronger than the starlight longward of Lyman α in each case.. This is a general property of starburst nuclei: the ionized gas is illuminated by more stellar UV radiation than emerges along the direct line of sight. Furthermore, the reddenings derived by techniques using optical, UV, and IR indicators may disagree dramatically.

Selective obscuration is demonstrated perhaps more vividly by comparison of optical continuum, emission-line, and near-infrared images of the popular colliding starburst system NGC 3690/IC 694 (Mkn 171, Arp 299...). Such a set of images is shown by Keel (1989). The K image shows several knots, mostly regions of intense star formation. Only one of these is at all prominent even in the red continuum, while each has a strong Hα counterpart. We can see emission lines powered by star clusters that we can't see directly. While this bodes well for using optical emission lines to probe dusty objects, it means that diagnostics involving emission lines versus the continuum (such as equivalent widths for determination of cluster ages or initial-mass functions) are meaningless unless a useful model can be constructed. Hence, the short burst ages inferred from large Hα equivalent widths may not pose a physical problem. Furthermore, indicators of IMF slope must be viewed with great caution; even for nearly monochromatic indices, there may well be bias if stars of different ages have different distributions with respect to the obscuring matter.

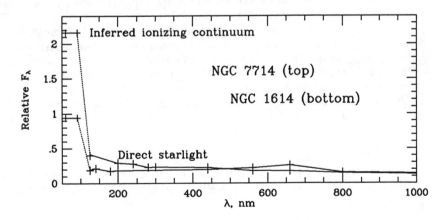

Figure 1. Spectral energy distributions for the nuclei of NGC 1614 and 7714. The Lyman-continuum flux is shown schematically as flat in F_λ; its level will vary somewhat depending on effective stellar temperature.

A TOY MODEL: OBSERVABLES AND STARBURST STRUCTURE

A simple picture of star formation on distinct clouds illuminates many issues in the structure of starbursts. The density and typical filling factor are estimated from recent CO interferometry. The amount of starlight absorbed by dust in these clouds and the surrounding medium are constrained by IRAS flux measures. Finally, the extent and dust content of gas surrounding the star-forming zone are given by emission-line images, which in many cases resolve the Strömgren volume, and by Balmer emission decrements. These have been incorporated first into a two-dimensional model which allows quick examination of the impact of such factors as compactness of the cloud configuration or typical shape of clouds on such observables as total IR or line emission, or escaping stellar UV continuum, for a fixed total SFR. Full 3D modelling is in progress.

In these models, a set of molecular clouds is distributed in a Gaussian configuration, and star clusters are formed near the cloud surfaces. Emergent rays are traced until they either leave the grid (at low intensity) or intercept another cloud. Grid points outside the clouds maintain values of local UV intensity. This is related to surface brightness in recombination lines; the medium surrounding the clouds has measurable dust content, so the details need to be fine-tuned to match specific objects. Models have been generated for fixed cloud number and SFR, varying the size of the whole configuration relative to the cloud size and the aspect ratio of individual clouds. In the two-dimensional limit, thermal IR emission from clouds heated by the young clusters increases for smaller configurations, while observed line emission decreases; note that the ratio of these may vary by a factor 10 as the size of the starburst region changes by a factor 5.

These results begin to resolve some long-standing puzzles in understanding starburst regions. We see some of the ionizing star clusters directly (giving UV radiation with only slight reddening by the diffuse surrounding material), and we see emission-line material powered by the many "searchlight beams" escaping between more or less distinct molecular clouds from within the configuration, as well as by clusters we can see directly. Diagnostics often used for the age or stellar population in H II regions, such as Balmer-line equivalent widths or UV/infrared flux ratios, will be quite misleading in such an environment. The continuum and emission lines refer to different fractions of the overall stellar populations.

This project has been supported by NASA ADP grant NAG5-1386, and by EPSCoR. Observing time at Lowell Observatory and Kitt Peak National Observatory is gratefully acknowledged.

REFERENCES

Balzano, V.A. 1983, ApJ, 268, 602

Devereaux, N. 1989, ApJ, 346, 121

Keel, W.C. 1989, S&T, 77, 18

Waller, W.H., Gurwell, M., & Tamura, M. 1992, AJ, 104, 63

Massive Stars: Their Lives in the Interstellar Medium
ASP Conference Series, Vol. 35, 1993
Joseph P. Cassinelli and Edward B. Churchwell (eds.)

CONSTRAINTS ON STARBURST MODELS FOR DWARF GALAXIES USING HI OBSERVATIONS IN THE LSC

LYNN D. MATTHEWS AND JOHN S. GALLAGHER, III
Astronomy Department, University of Wisconsin, Madison, WI 53706

JOHN E. LITTLETON
Physics Department, West Virginia University, Morgantown, WV 26506

INTRODUCTION

Starbursts are a common feature of many theoretical pictures of the star-forming histories of dwarf irregular galaxies (dIrrs). In some scenarios (e.g., Silk, Wyse, and Shields 1987; SWS), these bursts are governed by *external* factors. In this picture, only dwarfs located in moderate density regions should be able to undergo starbursts due to the recapture of natal gas lost during an early, intense epoch of star formation. In other models (e.g., Tyson and Scalo 1988; TS), *internal* processes bring about intermittant episodes of intense star formation, interspersed with much longer quiescent stages. Both of these types of theories make specific predictions regarding the spatial and internal properties of dIrrs. Because dIrrs tend to be rich in neutral hydrogen (HI), they are easy targets for 21-cm line observations. Properties obtained this way can provide an empirical test for star formation models involving starburst phenomena.

THE SAMPLE

The availability of high-quality, deep-sky UK Schmidt photographic plates for the Southern skies has allowed optical selections of large numbers of previously unobserved, low surface brightness dIrrs (de Vaucouleurs types $T = 8 - 11$) as targets for 21-cm line observations. Data from HI surveys undertaken by Littleton *et al.* (1992) have been combined with the catalogued data of Fisher and Tully (1981) and Fouqué *et al.* (1990) to form a reasonably complete sample of dIrr galaxies within a region of the Local Supercluster (LSC) spanning approximately a sterardian in solid angle and defined by the following coordinates: $22^h < \alpha < 15^h$; $-40^\circ < \delta < -18^\circ$; $|\, b \,| > 25^\circ$; $v_o \leq 3000 \ km/s$.

EXTERNALLY TRIGGERED STARBURST MODELS

SWS have proposed a mechanism to externally trigger starbursts in dwarf galaxies. They suggest that early, intense star formation in dwarfs created a metal-enriched, diffuse, hot intergalactic medium which intially had no effect on subsequent star formation. However, during the present epoch, the commencement of nonlinear evolution on the scale of groups of galaxies would be sufficient to initiate cooling of this gas and to bring about subsequent infall of HI onto gas-stripped dwarfs, thus provoking further starburst episodes.

The model of SWS predicts that at present there should exist a large population of gas-rich dIrrs within the local universe. Since cooling of the enriched gas can occur only in galaxy groups, present-day dIrrs are expected to be found in moderately high-density group environments. None should be found in large-scale voids. In order to investigate these hypotheses, it is necessary to examine the environments in which dIrrs are found in order to assess whether extremes in HI properties could be the result of external influences.

Fig. 1 is a plot of the log of the HI mass (M_{HI}) versus the log of the HI velocity profile width (W_{20}). W_{20} was chosen as the ordinate since it is useful as a measure of the gravitational potential of the galaxy. Inclination corrections for both W_{20} and M_{HI} were ignored due to the uncertainty involved in making such corrections for dIrrs. Using Fig. 1, a subset of galaxies with extreme HI properties could be chosen (indicated by large dots or circles). Included were the extreme high- and low-HI-mass dwarfs ($\log M_{HI} > 9.4 M_{\odot}$ or $\log M_{HI} < 8.4 M_{\odot}$), as well as those galaxies displaying unusual values of W_{20}. The density of the regions surrounding the dIrrs with extreme HI properties were estimated from Tully (1988) and are indicated on Fig. 1.

As predicted by SWS, we found that there exist large numbers of gas-rich dIrrs within our region of study. These tend to be located in regions delineated by brighter galaxies. None were found in large voids. Moreover, Fig. 1 reveals an apparent correlation between HI mass and environment; most of the low-HI-mass dIrrs at a given W_{20} are located in low-density regions, while the highest-HI-mass dIrrs are found in more densely populated environments. This could be indicative of the predicted ongoing HI accretion onto dwarfs, the efficiency of which is governed at least in part by the local density of galaxies.

INTERNALLY TRIGGERED STARBURST MODELS

The starburst model of TS invokes a stochiastic, self-propagating cycle to internally trigger starbursts in dwarf galaxies. These burst phases are predicted to be interspersed with much longer quiescent phases whose onset is caused by large-scale flucuations in the low-mass dwarfs. This scenario requires that the mean dwarf density should be at least $10 \ Mpc^{-3}$, a value much greater than observations indicate. However, TS have argued that since the discovery of galaxies is limited by optical biases, there may exist a large number of optically dim, quiescent galaxies which have eluded detection.

Fortuitously, HI surveys aimed at detection of optically selected targets afford an opportunity to sample a reasonably large portion of the local universe, both because the beam size is generally much larger than the extent of the HI in the dIrr, and because observations are generally done in total power mode, whereby regions of "empty" sky are integrated during each set of observations, hence faint, uncatalogued galaxies may be serendipitously discovered in these "blind", off-source measurements. This can serve as an indirect means of searching for optically invisible galaxies.

The difference between the least squares fits on Fig. 2 suggests that there is a correlation between surface brightness and HI mass; brighter dIrrs also tend to be richer in HI. It is also not apparent that the lowest-HI-mass, lowest surface brightness dIrrs have optically bright (i.e., bursting) counterparts at comparable masses. Both of these results are inconsistent with the TS model unless considerable HI mass is expended during the starburst phase, leaving quiescent dwarfs in a gas-depleted state.

Despite reasonable sensitivity limits and optical selection of galaxies as faint as 26 mag/arcsec^2, few new galaxies were detected in the low-HI-mass regime. Undetected objects are thus either larger, more-distant galaxies beyond our velocity search range, or are nearby, gas-poor systems rather than the hypothetical TS dwarfs. Finally, previously uncatalogued galaxies were discovered in the off beams in less than 1% of all new observations. The TS model predicts a chance detection probability of nearly 100% for our beam size (21 arcmin) and low HI-mass sensitivity threshold ($v \approx 2000\ km/s$).

These results imply that large numbers of undetected low-mass, gas-rich galaxies do not exist unless they are satellites of much brighter galaxies which in turn inhibits our ability to optically select them as targets. The apparent absence of a population of very low-mass galaxies reaffirms the conclusions of Briggs (1990) and Weinberg et al. (1991) which argued that the true volume density of HI-rich dwarfs is far below that required by the TS model.

SUMMARY

New 21-cm HI observations of dwarf irregular galaxies have been combined with existing catalogued data to form a reasonably complete sample of galaxies with known HI properties in a region of the LSC spanning approximately a steradian in solid angle. Both spatial and individual HI properties of the sample galaxies were utilized to place constraints on two classes of starburst models for dwarf galaxies. Despite the simplistic nature of our tests, several inferences were made:

(1) The data are consistent with the SWS starburst scenario in which low-mass galaxies preferentially recapture gas in denser environments.
(2) A substantial deficiency of low optical brightness, low-to-moderate-HI-mass galaxies was found compared with the predicitons of the TS starburst model. This reaffirms earlier studies (e.g., Briggs 1990) which suggested that this model is too extreme.
(3) The combination of the trend toward higher HI-mass in optically brighter galaxies and the tendency for these more HI-massive galaxies to be found in denser environments is consistent with the suggestion made by Schneider (1992) that low surface brightness is a manifestation of a lack of external triggering needed to invoke active star formation.
(4) Optical follow-up observations are needed to compliment our HI data for more detailed assessments. These data would allow us to determine for example, whether or not the SWS mechanism is actually responsible for starbursts and if indeed starbursts are an important mode of star formation in dwarf galaxies or if such galaxies instead evolve relatively smoothly over time (cf. Hunter and Gallagher 1985).

REFERENCES

Briggs, F.H. 1990, *AJ*, **100**, 999.

Fisher, J.R. and Tully, R.B. 1981, *ApJS*, **47**, 139.

Fouqué, P., Bottinelli, L., Durand, N., Gougenheim, L., and Paturel, G. 1990, *A&AS*, **86**, 473.

Hunter, D.A. and Gallagher, J.S. 1985, *ApJS*, **58**, 533.

Littleton, J.E., Gallagher J.S., Hunter D.A., and Matthews, L.D. 1992, unpublished.

Silk, J., Wyse, R.F.G., and Shields, G.A. 1987, *ApJ*, **322**, *L59* (SWS).

Schneider, S.E. 1992, private communication.

Tully, R.B. 1987, *ApJ*, **321**, 280.

—— 1988, *Nearby Galaxies Catalog*, (Cambridge Press: Cambridge).

Tyson, N.D. and Scalo, J.M. 1988, *ApJ*, **329**, 618 (TS).

Weinberg, D.H., Szomoru, A., Guhathakurta, G., and van Gorkom, J.H. 1991, *ApJ*, **372**, *L13*.

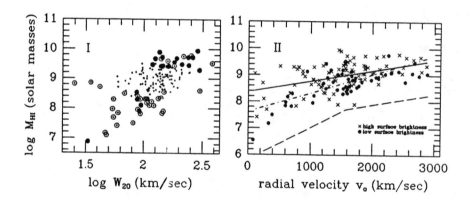

FIGURE I $\log M_{HI}$ vs. $\log W_{20}$. Tiny dots: dIrrs with intermediate HI properties; filled circles: galaxies located in high-density environments ($\rho > 0.20$ galaxies/Mpc^3); open circles: galaxies in low-density regions ($\rho \leq 0.20$).

FIGURE II $\log M_{HI}$ vs. radial velocity (v_o). Velocities have been corrected for motion with respect to the Local Group and $H_o = 75\ km\ s^{-1}Mpc^{-1}$ was assumed. Sample completeness begins to drop off significantly for $v_o \geq 2000\ km/s$. The dashed line represents the estimated detection threshold. The solid and dotted lines represent least squares fits to the data points denoting high and low surface brightness dIrrs respectively.

Massive Stars: Their Lives in the Interstellar Medium
ASP Conference Series, Vol. 35, 1993
Joseph P. Cassinelli and Edward B. Churchwell (eds.)

STARBURST PROPAGATION IN DWARF GALAXIES

Steven J. Gibson and John. S. Gallagher, III.
Department of Astronomy, University of Wisconsin, 475 North Charter
Street, Madison, WI 53706

Deidre A. Hunter
Lowell Observatory, 1400 West Mars Hill Road, Flagstaff, AZ 86001

INTRODUCTION: DWARF STAR-FORMING GALAXIES

Dwarf galaxies offer an excellent opportunity to study the processes of star formation and galactic evolution. They are very common objects. They show an impressively wide range in levels of star formation, as evidenced by considerable variation in the optical dominance of young stars over old stars. They are often very rich in H-I gas and poor in metals; hence they are generally unevolved galaxies (Hunter & Gallagher, 1986; Skillman, Kennicutt, & Hodge, 1989).

Do objects with high star formation rates represent a particular, repetitive evolutionary phase in the lives of small, gas-rich galaxies? How important are starburst cycles? One aspect of this question concerns how a starburst might evolve in a small galaxy (Searle, Sargent, & Bagnuolo, 1973; Tyson & Scalo, 1988).

We investigate three dwarf galaxies which may represent different stages of starburst evolution. We have constructed a model which defines physically and observationally what a starburst is and how it may evolve. This model, which will be presented in detail in an upcoming paper (Gibson & Gallagher, in preparation), is applied to the galaxies discussed below.

APPEARANCES OF THREE DWARF GALAXIES

Knowledge of galactic and extragalactic H-II regions can be applied to dwarf star-forming galaxies to study their evolution. II Zwicky 40, NGC 1569, and NGC 1800 all have H-I rich, low-mass disks with slow rotation velocities, visible stellar populations of comparable size (3-5 kpc for $\mu_B \leq 25$ mag/arcsec^2), and large Population I components, which yield strong Hα emission, high L_{FIR}, and warm FIR colors (Joy & Lester, 1988; Waller, 1991). However, the broadband colors, especially the Population I tracers, are different, as are the extents of star formation, shown by the distribution of H-II regions within each galaxy. The Hα morphologies vary widely: II Zwicky 40 looks very compact, while the other two show extended filaments (Figure 1). These differences suggest that the three galaxies, though perhaps generally alike, are not in the same evolutionary state.

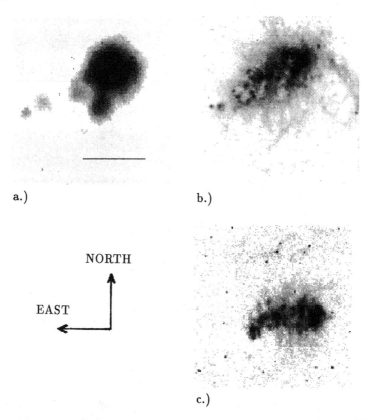

a.)   b.)

NORTH

EAST

c.)

Fig. 1.   Each of these Hα images were taken through narrow band interference filters and corrected for continuum fluxes from off-band observations. The image scales, all originally 0.4 arcsec/pixel, have been corrected to the same physical scale for comparison. a.) II Zwicky 40 in Hα; 2400 second exposure taken with the Perkins 1.8-m telescope at Lowell Observartory, using a 4:1 focal reducer and a TI 800×800 CCD. The length of the bar is $\sim$ 1 kpc. b.) NGC 1569 Hα; 900 second exposure taken with the KPNO 2.1-m telescope and a TI 800×800 CCD, binned to 400×400. c.) NGC 1800 Hα; 2700 second exposure taken with the CTIO 1.5-m telescope, and a Tektronics 1024×1024 CCD.

We interpret the data as follows: II Zwicky 40 is undergoing a local starburst, which may or may not propagate along its H-I distribution. NGC 1569 is experiencing a global starburst, which may or may not have been triggered globally. NGC 1800 had a global burst on the order of 30 Myr ago and is lapsing into quiescence. This interpretation is based on the repetitive starburst model. An alternative not explored here is that Population I dominated galaxies represent a true transition between evolutionary phases. Using our model of starburst behavior, we investigate the likelihood that these three objects represent different phases of the same type of starburst galaxy, specifically, whether or not a compact burst such as that in II Zwicky 40 could expand into a more global phenomenon like that of NGC 1569.

IMPLICATIONS FOR THREE DWARF GALAXIES

II Zwicky 40 contains a single, localized starburst, which appears to have plenty of H-I fuel around it (Brinks & Klein, 1988), although the 3-D distribution remains unknown. The possibility of future propagation cannot be ruled out. A slow propagation would essentially relocate the starburst, whereas a fast one would engulf the majority of the fuel at the same time, perhaps in the fashion of NGC 1569. If the II Zwicky 40 starburst does not propagate, it may end up as a super version of NGC 1705, which had a "1-shot" local starburst (Meurer, Freeman, Dopita, & Cacciari, 1992). NGC 1569 displays a weak age gradient across its global starburst. This is consistent with either fast propagation from an initial local burst or a rapid series of triggers. The latter case could be due to a past collision with another galaxy (Noguchi & Ishibashi, 1986). NGC 1800 appears to be done with a global starburst and on its way back to quiescence. This is the way NGC 1569 may appear in the future.

REFERENCES

Brinks, E. & Klein, U., 1988. *M.N.R.A.S.* 231, 63
Hunter, D. A. & Gallager, J. S., 1986. *P.A.S.P.* 98, 5
Joy, M. & Lester, D. F., 1988. *Ap.J.* 331, 145
Meurer, G. R., Freeman, K. C., Dopita, M. A., & Cacciari, C., 1992.
     Ap.J. 103 60
Noguchi, M. & Ishibashi, S., 1986. *M.N.R.A.S.* 219, 305
Searle, L., Sargent, W. L. W., & Bagnuolo, W. G., 1973. *Ap.J.* 179, 427
Skillman, E. D., Kennicutt, R. C., & Hodge, P. W., 1989. *Ap.J.* 347, 875
Tyson, N. D. & Scalo, J. M., 1988. *Ap.J.* 329, 618
Waller, W. H., 1991. *Ap.J.* 370, 144

Massive Stars: Their Lives in the Interstellar Medium
ASP Conference Series, Vol. 35, 1993
Joseph P. Cassinelli and Edward B. Churchwell (eds.)

MASSIVE STAR FORMATION IN DWARF IRREGULAR GALAXIES

ROBERT RUOTSALAINEN
Physics Department, Eastern Washington University, Cheney, WA 99004

Image-tube exposures of the galaxies Sextans B (= DDO 70) and Leo A
(= DDO 69) were obtained at the 2.2 m telescope of Mauna Kea Observatory.
The resolved stellar populations of these galaxies are dominated by luminous
blue stars. Such hot luminous stars are understood to be short-lived, and
indicative of regions of recent star formation. Color-magnitude diagrams were
produced from the resolved images of some 280 stars in Sextans B, and 450
stars in Leo A. These observations were compared with synthetic color-
magnitude diagrams based upon stellar evolutionary models of Brunish and
Truran (1982a, b), and of Bertelli, Bressan, Chiosi and Angerer (1986).
Matches, of the modeled distributions of points in the synthetic diagrams to the
observations, are consistent with a picture of migratory "bursts" of activity in
the formation of massive and intermediate-mass stars within these galaxies.

Point-spread functions were fit to the digitized photographic data for the
measurement of apparent blue magnitudes and U-B colors. The color index
U-B, more sensitive than B-V to the effective temperature for early-type stars,
was used to model intervals of recent star formation in these two galaxies.
Figures 2 and 4 illustrate models adjusted to match observations, seen in
Figures 1 and 3, of stars in two fields of Sextans B. The broken lines seen near
the bottoms of these figures represent assigned modeled plate limits and
estimated observational plate limits. The average difference in apparent blue
magnitude, for 92 stars in Sextans B measured here and also by Tosi, Greggio,
Marconi and Focardi (1991), is less than 0.1, with a dispersion about this
average of 0.4 magnitudes.

The modeled distributions depend on values assigned for intervals of star
formation over ages **t**; and also on values for the initial mass function index α
(of the form: $N \propto M^{-\alpha}$ for **N** number of stars of mass **M**), the metallicity **Z**,
and the distance modulus μ. The modeled interval of star formation matched to
observations of stars in the western field of Sextans B is older than the model
matched to observations of stars in the central field. The Salpeter (1955) value
of $\alpha = 2.35$ was used. Better matches of the modeled distributions to the
observations were noted for metal-poor (Z = 0.001) models than for those of
solar (Z = 0.02) composition, consistent with values of the oxygen abundance
reported by Skillman, Kennicutt and Hodge (1989) for H II regions in Sextans
B. A distance modulus of $\mu = 26.0$ is reflected in the plate limits assigned to
the modeled distributions. Magnitude-dependent smoothing was applied to
these models for the simulation of random photometric errors.

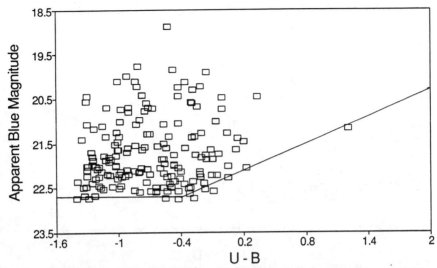

Fig. 1.   Color-magnitude diagram for 174 stars in Sextans B: Central Field.

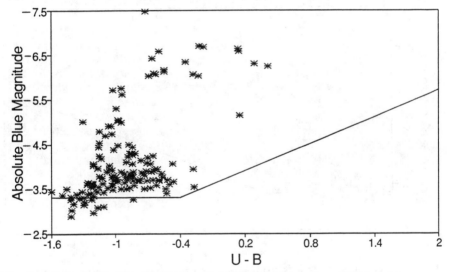

Fig. 2.   Metal-poor (Z = 0.001) model of 150 points: 0 < t(Myr) < 40.

ACKNOWLEDGEMENTS

Observations, undertaken in collaboration with R. Brent Tully, were supported
by the National Science Foundation. The Dominion Astrophysical
Observatory, National Research Council of Canada, provided facilities and
most generous assistance for the digitization of the image-tube exposures.
Images were processed with the use of routines written by James N. Heasley
and Kenneth A. Janes, and with PCVISTA, developed by Michael W.
Richmond and Richard R. Treffers.

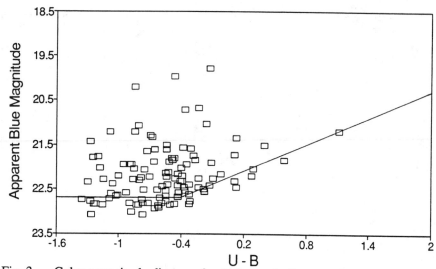

Fig. 3.   Color-magnitude diagram for 110 stars in Sextans B: Western Field.

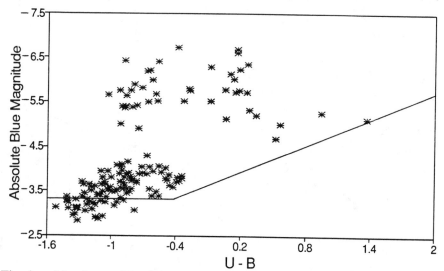

Fig. 4.   Metal-poor (Z = 0.001) model of 150 points: 10 < t(Myr) < 50.

REFERENCES

Bertelli, G., Bressan, A., Chiosi, C., and Angerer, K. 1986, *Astr. Ap. Suppl.*, **66**, 191.

Brunish, W. M., and Truran, J. W. 1982a, *Ap. J.*, **256**, 247.

Brunish, W. M., and Truran, J. W. 1982b, *Ap. J. Suppl.*, **49**, 447.

Salpeter, E. E. 1955, *Ap. J.*, **121**, 161.

Skillman, E. D., Kennicutt, R. C., and Hodge, P. W. 1989, *Ap. J.*, **347**, 875.

Tosi, M., Greggio, L., Marconi, G., and Focardi, P. 1991, *A. J.*, **102**, 951.

Massive Stars: Their Lives in the Interstellar Medium
ASP Conference Series, Vol. 35, 1993
Joseph P. Cassinelli and Edward B. Churchwell (eds.)

THE EFFECTS OF ENERGY INPUT FROM MASSIVE STARS ON THE INTERSTELLAR MEDIUM OF STARBURST GALAXIES

MATTHEW D. LEHNERT

STScI and The Johns Hopkins University, Baltimore, MD 21218

ABSTRACT

Optical spectroscopy and narrow-band Hα+[NII] imagery of a sample of IR-selected, edge-on galaxies has revealed that extra-planar emission-line gas extended along the minor axis is a very common feature of these galaxies. The properties of this extended gas (high ratios of [NII]/Hα, [SII]/Hα, [OI]/Hα compared to the major axis and disk HII-regions, large line widths, bubble- and shell-like morphologies, the relative brightness of the extended emission) and the correlation between these properties and the IR properties of the galaxies suggest that we are seeing outflows driven by massive stars in the nuclei of these galaxies and the strength of these flows depends strongly on the rate and distribution of star-formation.

INTRODUCTION

The structure and ionization of the interstellar medium (ISM) is thought to be governed by energy and momentum injection by stars, particularly massive stars. Models of the interaction between energy input from massive stars through stellar winds and supernovae and the ISM suggest that the strength of the interaction should be a function of the rate and distribution (concentrated versus dispersed) of star-formation. Such interaction should manifest itself by heating and energizing the ISM thereby increasing its scale height and forming structures such as chimneys, bubbles, and shells around the most intense star-forming activity. Since starburst galaxies are forming many OB associations over a short time period (t $< 10^{7-8}$ yrs) and in a small volume (r $<$ few x 10^{2-3} pc), energy is being injected at very high rates compared to "normal" galaxies. Thus the effects of massive stars on the ISM might be more easily gauged in starburst galaxies. This suggests that by studying starburst galaxies, we can gain new insight into the role played by energy injection from massive stars to the structure and energy balance of the ISM. Such insight would not only have important consequences for our understanding of the formation and nature of the structure of the ISM in our galaxy and other "normal" galaxies, but indeed, could, for example, lead to fundamental insight into the physical processes occurring during galaxy formation, give clues on the enrichment and heating of the intra-cluster medium, might provide insight into the processes that lead to the metallicity-radius relation in galaxies and the mass-metallicity relation among galaxies, provide a mechanism whereby proto-ellipticals are able to clear away their remaining interstellar medium, etc.

SAMPLE, DATA, AND ANALYSIS

To select galaxies with a range of star-formation rates and distributions, we selected galaxies that are infrared bright ($S_{60\mu m} > 5.4$ Jy), edge-on (b/a > 2), infrared warm ($S_{60\mu m}/S_{100\mu m} > 0.4$), and to avoid the confusing infrared cirrus and galactic extinction, only galaxies with high galactic latitude ($|b| > 30°$). The edge-on orientation is crucial because the relevant properties of the emission line gas can be far more easily measured when the gas is seen in isolation against the sky rather than projected onto the much brighter gas associated with the starburst itself. Our final sample for which we have some data consists of about 50 galaxies that have infrared luminosities from 10^9 to 10^{12} $L_\odot$ which implies star-formation rates from 0.3 to 300 $M_\odot$ yr^{-1}.

Our data consists of longslit spectra that include the lines [OI]λ6300, [NII]$\lambda\lambda$6548, 6583, and [SII]$\lambda\lambda$6716, 6731 taken along both the minor and major axes. In addition, we have narrow-band Hα+[NII] and broad-band R images of most of the galaxies. The spectra allow us to determine the densities, temperatures, pressures, velocity field, and ionization state of the gas. From these results we can derive the relative importance of various energy input mechanisms (such as shock heating, photoionization by massive stars and/or AGN), the mass, kinetic energy, momentum, and filling factor of the optical emission-line gas. The images allow us to determine the fluxes, luminosities and morphologies that are crucial in assessing the total energy input and how the input is distributed.

Since each longslit spectrum contains a great deal of spectral and spatial information, one could easily get lost in the data reduction and large amount of information for each galaxy. Thus we tried to adopt a strategy for binning the data based on some physically meaningful scale that will allow us to characterize the the emission-line properties of the galaxies and make meaningful comparisons among galaxies. In this poster, we will describe and give the results of one such strategy. We summed the columns of the CCD spectra into 5 bins whose sizes were based on the half-light radius of the line emission. We determined the half-light radius, R_e, for each galaxy using the relationship between R_e of the emission line gas and the infrared luminosity ($R_e \propto$ {IR-luminosity}$^{\frac{1}{2}}$). The nuclear bin contains the light from the central half-light diameter of the galaxy, two near-nuclear bins contain the light within one to two R_e on each side of the nucleus, and two off-nuclear bins contain the light for radii greater than two R_e. Adopting this scheme allows us to judge the relative importance of the starburst to the ionization and structure of the gas in three regions where we might naively expect different dominant sources of emission (the starburst in the nuclear bin, combination of disk and starburst emission in the near-nuclear bins, and emission from disk HII-regions in the off-nuclear bins). By comparing the results for the bins along the minor and major axes and by comparing these results with the IR (*e.g.,* infrared luminosity, infrared color, ratio of IR-to-optical luminosity, etc.) and physical (*e.g.,* rotation speed and morphological type) properties of the galaxies, we should be able to gain new insight into the physical processes that are dominating the interaction between the massive stars and the ISM of the "host" galaxy.

RESULTS AND CONCLUSIONS

Statistically speaking, we find that the strength of the low ionization lines of [SII] and [OI] relative to Hα increase as a function of distance from the nucleus along the minor axis. We also find that the distributions of strength of these lines relative to Hα (now including [NII]/Hα) of the extended gas along the minor axis are statistically different from those in the disk along the major axis. The ratios of [NII]/Hα, [SII]/Hα, and [OI]/Hα are higher along the minor axis (being shock/AGN-like on average) than along the major axis (being almost exclusively HII-region-like). We also find that the line widths along the minor axis are on average broader than the line widths in the disk along the major axis. The widths out along the minor axis however are not statistically broader than in the nucleus. Moreover, we find a strong relationship between the line ratios ([NII]/Hα and [SII]/Hα) and line width.

Another important finding of this project is that we see very strong correlations between the infrared properties (infrared luminosity, infrared color, and ratio of IR-to-optical luminosity) and the line ratios and line widths of the near-nuclear and off-nuclear bins along the minor axis and see no correlation between these properties and rotation speed. Specifically, we find that the line ratios ([NII]/Hα and [SII]/Hα) in the extended emission line gas are strongly related to the ratio of IR-to-optical luminosity and, in the case of [NII]/Hα, strongly related to the IR-luminosity (more extreme IR properties implies higher line ratios). The widths of the lines also strongly correlate with the IR-luminosity and the ratio of IR-to-optical luminosity (again, more extreme IR properties implies larger line widths).

In many cases, the emission-line gas along the minor axis has a bubble- and/or shell-like morphology and its brightness is in excess of what we expect emission from disk alone. In order to make this statement more quantitative, we have attempted to determine the amount of excess Hα+[NII] emission along the minor axis above what would be expected for "normal" disk emission. While we do not have enough space to go into the details of this determination, we find that amount of excess, extended Hα+[NII] emission is again correlated with the IR-properties of the galaxies (being strongly related to the IR-color and weakly related to the IR-luminosity and ratio of IR-to-optical luminosity – again, more extreme IR properties implies a larger excess).

All of the above results taken together suggest that extended emission line gas is a common feature of these galaxies. The properties of this extended gas (high ratios of [NII]/Hα, [SII]/Hα, [OI]/Hα compared to the major axis and disk HII-regions, large line widths, bubble- and shell-like morphologies, the relative brightness of the extended gas) and the correlation between these properties and the IR properties of the galaxies suggest that we are seeing outflows driven by massive stars in the nuclei of these galaxies and the strength of these flows depends strongly on the rate and distribution of star-formation.

I would like to thank my thesis advisor Dr. Tim Heckman for suggesting such an interesting project and for putting great intellectual effort towards its *hopefully* successful conclusion.

Massive Stars: Their Lives in the Interstellar Medium
ASP Conference Series, Vol. 35, 1993
Joseph P. Cassinelli and Edward B. Churchwell (eds.)

30 DORADUS: THE STARS, THE ISM

JOEL WM. PARKER
CASA, Box 389, University of Colorado, Boulder, CO 80309

ABSTRACT   I present *UBV* CCD photometry and spectroscopic classifications of stars in 30 Doradus to study their intrinsic properties: colors, effective temperatures, luminosities, and masses. I also investigate the ISM in 30 Doradus, including its distribution, the mean extinction, and the slope of the reddening law.

OVERVIEW

The *UBV* photometry in this study is complete to $\sim 18^{\mathrm{m}}$ for 2400 stars within a 40 square arcminute field of 30 Doradus in the Large Magellanic Cloud. Figure 1 shows the main region involved in this analysis. New spectroscopic observations of 54 stars (mostly OB types) were obtained for this study. The spectrograms have 3Å resolution over the wavelength region 3800–4800Å. Spectral types and luminosity classes determined from these data were combined with published and unpublished classifications of over 100 additional OB stars.

The Johnson Q-parameter, $Q = (U - B) - \frac{E(U-B)}{E(B-V)}(B - V)$ (Johnson & Morgan 1953), is used to deredden the OB stars and to determine the mean reddening for the other stars in 30 Doradus. From the intrinsic colors of the stars, I calculate the effective temperatures. As discussed by Massey (1985) the colors alone are not sufficient to determine the temperatures of O and early B stars, so the photometry is used to select candidate OB stars for spectroscopic observations. Effective temperatures and bolometric magnitudes are then determined for all stars using photometry and spectral types (if known).

These data are used to plot the stars on the H-R diagram shown in Figure 2. Using evolutionary mass tracks, the stars are binned by mass to determine the IMF. However, since there clearly have been more than one epoch of star formation (*e.g.*, see McGregor & Hyland 1981; Lortet & Testor 1991) the "initial" mass function is not well defined in these observations.

Additionally, these data can be used to calculate the Lyman continuum luminosity for each star. The sum of these luminosities allows one to estimate the nebula's expected Balmer Hα luminosity, which then can be compared to observed values.

Full details of this study are given by Parker (1992).

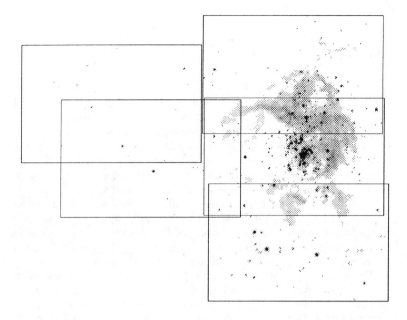

Figure 1: A mosaic of the V-filter images of 30 Doradus. Each CCD frame is $4.1' \times 2.6'$. North is up and east is to the left.

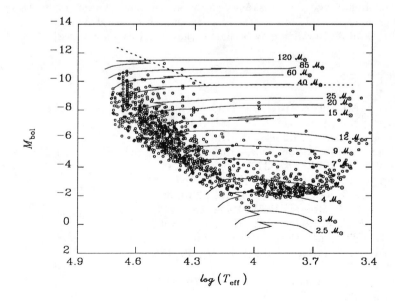

Figure 2: This is the H-R diagram for the stars with $V < 18^{\mathrm{m}}$. Evolutionary mass tracks are from Maeder & Meynet (1988) and Maeder (1990). The dotted line shows the location of the Humphreys & Davidson (1979) luminosity limit.

SUMMARY OF RESULTS

A few of the results from this study are:

⋆ The mean total reddening (including Galactic foreground) of the stars in 30 Doradus is $\overline{E(B-V)} = 0.44 \pm 0.20$, in agreement with values from other studies (Israel & Koornneef 1979; Fitzpatrick & Savage 1984; Lee 1990; Vacca 1991; Heap *et al.* 1991). The mean reddening slope is $\frac{E(U-B)}{E(B-V)} = 0.75$. Using *IUE* observations, Fitzpatrick (1985) finds a considerably steeper slope of 0.89 for 30 Doradus and a slope of 0.76 for other regions in the LMC.

⋆ The intrinsic colors of OB stars in 30 Doradus do not differ significantly from their Galactic counterparts.

⋆ The best-fitting overall IMF slope in 30 Doradus is in the range of $\Gamma = -1.3$ to -1.5 (where the Salpeter slope is $\Gamma = -1.35$).

⋆ The total calculated Lyman continuum luminosity for the stars in the region is $N_{\rm Ly} \gtrsim 3.2 \times 10^{51}$ photons s^{-1}. This is a lower limit since the effects of W-R stars and the unresolved components of R 136 have not been included.

⋆ The calculated Balmer luminosity, $L_{\rm H\alpha} \approx 4.5 \times 10^{39}$ erg s^{-1} agrees well with the observed flux (Kennicutt & Hodge 1986) for this central region.

⋆ The shape of the IMF as well as calculations of the Lyman continuum luminosity indicate that a large number of hot, massive stars still have not been observed spectroscopically and classified. Most of these stars probably reside in the central cluster of R 136 (unresolved in these observations). Since the effective temperatures of O and early B stars as calculated from their photometry alone tend to be too cool, the resulting estimated masses tend to be too low. Also, the initial masses of the W-R stars are unknown, and other OB stars already may have disappeared as supernovae. Accounting for these effects probably would increase the relative number of massive stars, so the true overall IMF of 30 Doradus may be flatter than the IMF shown in this study.

REFERENCES

Fitzpatrick, E.L. 1985, *ApJ*, **299**, 219
Fitzpatrick, E.L., & Savage, B.D. 1984, *ApJ*, **279**, 578
Heap, S.R., Altner, B., Ebbets, D., Hubeny, I., Hutchings, J.B.,
        Kudritzki, R.P., Voels, S.A., Haser, S., Pauldrach, A., Puls, J., &
        Butler, K. 1991, *ApJ*, **377**, L29
Humphreys, R.M., & Davidson, K. 1979, *ApJ*, **232**, 409
Israel, F.P., & Koornneef, J. 1979, *ApJ*, **230**, 390
Johnson, H.L., & Morgan, W.W. 1953, *ApJ*, **117**, 313
Kennicutt, R.C., & Hodge, P.W. 1986, *ApJ*, **306**, 130
Lee, M.G. 1990, Ph.D. thesis, University of Washington
Lortet, M.-C., & Testor, G. 1991, *A&AS*, **89**, 185
Maeder, A. 1990, *A&AS*, **84**, 139
Maeder, A., & Meynet, G. 1988, *A&AS*, **76**, 411
Massey, P. 1985, *PASP*, **97**, 5
McGregor, P.J., & Hyland, A.R. 1981, *ApJ*, **250**, 116
Parker, J.Wm. 1992, Ph.D. thesis, University of Colorado
Vacca, W.D. 1991, Ph.D. thesis, University of Colorado

Massive Stars: Their Lives in the Interstellar Medium
ASP Conference Series, Vol. 35, 1993
Joseph P. Cassinelli and Edward B. Churchwell (eds.)

THE EDDINGTON-LIMIT IN GALAXIES
OF DIFFERENT METALLICITIES

Henny J.G.L.M. LAMERS[1,2] and Rene NOORDHOEK[1]

[1]SRON Laboratory for Space Research and Astronomical Institute,
Sorbonnelaan 2, 3584 CA, Utrecht, NL

[2]Joint Institute for Laboratory Astrophysics and
Department of Astrophysical, Planetary, and Atmospheric Sciences,
University of Colorado, Boulder CO 80309-0440

ABSTRACT   We calculated the location of the Eddington-limit in galaxies of different metallicities with $Z = 0.002$ ($\sim$SMC), $Z = 0.01$ ($\sim$LMC), $Z = 0.02$ (solar), $Z = 0.04$ ($\sim$M31). The Eddington-limit is higher for lower metallicities. This has important consequences for the predicted luminosity upperlimit of blue and red supergiants, and for the luminosity of the Luminous Blue Variables.

INTRODUCTION

The most massive stars with $M > 50\ M_\odot$ reach their photospheric Eddington limit when they evolve away from the main sequence. This is due to the fact that the flux-mean opacity in the photosphere *increases* with decreasing $T_{\rm eff}$ in the range of 10 000 $< T_{\rm eff} <$ 50 000 K. For $T_{\rm eff} <$ 10 000 K the opacity *decreases* with decreasing $T_{\rm eff}$. This implies that the location of the photospheric Eddington-limit in the Hertszsprung-Russell Diagram (HRD) is not a horizontal line at constant luminosity, but that it decreases in L from $T_{\rm eff} = 50\ 000$ K to about 10 000 K, reaches a minimum at $T_{\rm eff} \simeq$ 10 000 K, and increases again towards lower $T_{\rm eff}$. We will call this "the Eddington-trough." Lamers and Fitzpatrick (1988), hereafter called L&F, have calculated the location of the Eddington-trough in stars of solar metallicity.

In this paper we report the first results of a study of the location of the Eddington-trough for galaxies of different metallicities of $Z = 0.002$ ($\sim$ SMC), $Z = 0.01$ ($\sim$ LMC), $Z = 0.02$ (solar) and $Z = 0.04$ ($\sim$ M31).

The Eddington-trough has important consequences for the evolution of massive stars:

1. There is a strip in the HRD close to the left side of the trough in the range of T_{eff} from 50 000 K to about 10 000 K, where stars have very unstable photospheres because the radiation pressure almost balances the gravity. Appenzeller (1986) and Lamers (1986) suggested that the Luminous Blue Variables (LBV's) which are located in this strip are unstable because of radiation pressure.

2. Massive stars will evolve from the ZAMS onto the left side of the Eddington-trough during or after their core-hydrogen burning. The resulting high mass loss and instability will prevent the stars from evolving further to the red. This explains the observed luminosity upper limit (Humphreys-Davidson limit) for luminous hot stars in the Galaxy, LMC and M31 (Humphreys and Davidson, 1979; L&F).

3. The luminosity limit for Red Supergiants is determined by the lowest point of the Eddington-trough which can be passed by stars on their horizontal redward motion in the HRD (L&F).

THE EDDINGTON LIMIT IN THE (g, T_{eff}) DIAGRAM

The method used to derive the photospheric Eddington-limit is the same as described in L&F. We used the new Kurucz (1991) model atmospheres in the range of $8500 \leq T_{\text{eff}} \leq 50\ 000$K with metallicities of $Z = 0.002, 0.01, 0.02$ (solar), and 0.04. These models have the new opacities including about 10^7 lines. For each model, characterized by Z, T_{eff} and $\log g_N$ (with $g_N = GM/R^2$), we derived the maximum value of the radiation pressure force, $g_{\text{rad}}(max)$, in the optical depth range of $10^{-2} < \tau_R < 10^2$. This optical depth range was chosen for the following reason. If $g_{\text{rad}} > g_N$ at $\tau_R > 10^2$ and $g_{\text{rad}} < g_N$ at $\tau_R < 10^2$ the layers of $\tau_R < 10^2$ may stabilize the photosphere. (The layers with $g_{\text{rad}} > g_N$ may be Rayleigh-Taylor unstable.) If $g_{\text{rad}} > g_N$ at layers with $\tau < 10^{-2}$ the star can develop a stationary stellar wind as in normal luminous OB stars. However, if $g_{\text{rad}} > g_N$ at $\tau_R \simeq 1$, the star cannot have a stable photosphere and will suffer very severe and possibly variable mass loss.

For each value of Z and T_{eff} we plotted the ratio $g_{\text{rad}}(max)/g_N$ versus $\log g_N$. This ratio increases rapidly with decreasing $\log g_N$. (See Figure 3 of L&F for solar metallicity.) The extrapolations of these relations to $g_{\text{rad}}(max)/g_N = 1$ gives the value of $g_N = g_{\text{Edd}}(Z, T_{\text{eff}})$ where the radiative force in the photosphere balances the gravitational force.

The values of $\log(g_{\text{Edd}})$ are plotted versus $\log(T_{\text{eff}})$ for the four metallicites in Figure 1. Stars below these lines in the (g, T_{eff}) diagram cannot have stable photospheres. Stars *just* above these lines will have very high mass loss rates because the effective gravity in their photospheres will be small.

Figure 1 shows that g_{Edd} increases with metallicity, as expected, because the contribution to g_{rad} by spectral lines increases with metallicity. The increase in g_{Edd} with Z is only small with a difference of about 0.15 dex between $Z = 0.002$ and 0.04. We will show below that this small difference in g_{Edd} translates into a large difference in luminosity.

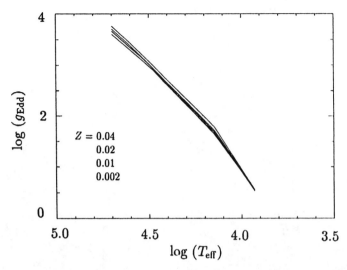

Figure 1.   The relation between g_{Edd} and T_{eff} for different metallicities of $Z = 0.04$ (top), $Z = 0.02$, $Z = 0.01$, $Z = 0.002$ (bottom). The small differences in g_{Edd} due to metallicity translate into substantial differences in L_{Edd}.

THE EDDINGTON-LIMIT IN THE HR-DIAGRAM

The location of the Eddington-limit in the (g, T_{eff})-diagram can be converted into the location in the HRD if the M/L ratio of the stars are known. For a given M/L ratio the luminosity at the Eddington-limit (L_{Edd}) can be derived from the g_{Edd} (Z, T_{eff}) relations by substituting $R^2 = L/4\pi\sigma\,T_{\text{eff}}^4$ into the expression for the gravity. L_{Edd} is the solution of the equation

$$g_{\text{Edd}}(Z, T_{\text{eff}}) = 4\pi G\sigma T_{\text{eff}}^4 M(Z, L, T_{\text{eff}})/L \ . \tag{1}$$

We derived the M/L ratios from the evolutionary models of Maeder (1991) for $Z = 0.002, 0.01$ (interpolated between 0.005 and 0.02), 0.02 and 0.04, i.e., the same values as used in the model atmospheres.

For the first approximation we used the M/L ratio after the core-hydrogen burning (CHB) phase for stars with $L > 10^5 L_\odot$. This relation can be written as

$$\log(M/M_\odot) = m_0 + m_1 \log(L/10^5 L_\odot) \tag{2}$$

with coefficients: $m_0 = 1.228$, $m_1 = 0.5340$ for $Z = 0.002$; $m_0 = 1.246$, $m_1 = 0.4889$ for $Z = 0.01$; $m_0 = 1.252$, $m_1 = 0.4341$ for $Z = 0.02$; $m_0 = 1.229$, $m_1 = 0.3975$ for $Z = 0.04$.

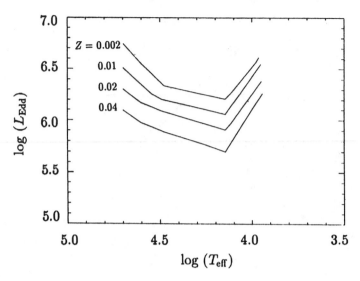

Figure 2. The location of the photospheric Eddington-limit in the HRD for different metallicities. The limit is higher for metal poor galaxies. This implies that the predicted luminosity upper limit for red and blue supergiants will depend on the metallicity of the galaxy (see text).

The resulting Eddington-limits are shown in Figure 2.

Notice that all four curves show the characteristic shape of a trough with the deepest point reached at $T_{eff} \simeq 12000$ K. The values of L_{Edd} depend on Z with the lowest metallicities corresponding to the highest L. This implies that the Humphreys-Davidson-limit in galaxies with different metallicities will have the same general shape, and that the LBV's at minimum brightness will occupy a strip to the left of the Eddington-trough in the HRD with a similar shape in different galaxies. Both these predictions are confirmed by observations (Humphreys, 1987; Fitzpatrick and Garmany, 1990; Wolf, 1989). Figure 2 also shows that the bump in the Eddington-limit, which appeared in the calculations by L&F around $T_{eff} \simeq 13000$ K due to the underestimate of the opacity from twice ionized metals in the Kurucz (1979) models, has now disappeared by the inclusion of many more metal lines in the Kurucz (1991) models.

The results mentioned above should be considered only on a qualitative basis, because we assumed that the stars reach their Eddington-limit *after* the CHB-phase. For the most luminous stars with initial masses larger than $60 M_\odot$ the Eddington limit may already be reached *during* the CHB-phase. In that case Eq. (2) should be replaced by the actual M/L ratio and the location of the Eddington-limit in the HRD will change accordingly. Therefore it is not possible at this moment to make a quantitative comparison between predictions and observations.

CONCLUSIONS

1. The Eddington-limit of luminous stars has the shape of a trough with a minimum near $T_{eff} \simeq 12000$ K. Massive stars that evolve onto the left side of the trough will have a very high mass loss rate which may prevent the further evolution to the right of the HRD. The LBV's are located just to the left of the Eddington-trough.

2. The region inside the trough will be empty except for LBV's during outbursts.

3. The region on the right side of the trough will be empty because single stars cannot evolve into this region.

4. The luminosity upper limit of red supergiants is determined by the stars that can just pass under the Eddington-trough to the right.

5. The luminosity of the Eddington-trough depends on the metallicity of the galaxy and is higher for low metallicity galaxies. The differences with respect to the solar metallicity is 0.3 dex for $Z = 0.002$, 0.15 dex for $Z = 0.01$, and -0.2 dex for $Z = 0.04$.

6. The theoretical luminosity upper limit for red supergiants will depend on Z by the same factors.

We thank Robert Kurucz and Andre Maeder for sending us their new model atmospheres and new evolutionary tracks prior to publication and the JILA Scientific Reports Office for preparing this manuscript. H.J.G.L.M.L. is grateful to Peter Conti for hospitality at JILA and for support by his National Science Foundation grant AST-9015240.

REFERENCES

Appenzeller, I., 1986, in Luminous Stars and Associations in Galaxies, eds. C. de Loore, P. Laskarides, A. Willis (Dordrecht:Reidel) p. 139

Fitzpatrick, E.L., Garmany, C.D. 1990, Ap.J. 363, 119

Humphreys, R.M., 1987, in Instabilities in Luminous Early Type Stars, ed C. de Loore and H.J.G.L.M. Lamers (Dordrecht, Reidel)

Humphreys, R.M., Davidson, K., 1979, Ap.J. 232, 409

Kurucz, R.L., 1979, Ap. J. Suppl. 40, 1

Kurucz, R.L., 1991, Private Communication

Lamers, H.J.G.L.M., Fitzpatrick, E.L., 1988, Ap.J. 324, 279 (L&F)

Lamers, H.J.G.L.M., 1986 in Luminous Stars and Associations in Galaxies, eds. C. de Loore, P. Laskarides, A. Willis (Dordrecht: Reidel) p. 157

Maeder, A. , 1991, Private Communication

Wolf, B. 1989 Astr. Ap. 217, 87

Massive Stars: Their Lives in the Interstellar Medium
ASP Conference Series, Vol. 35, 1993
Joseph P. Cassinelli and Edward B. Churchwell (eds.)

DISTANT MASSIVE STARS AT LOW GALACTIC LONGITUDES: PROBES OF THE INTERSTELLAR MEDIUM OF THE INNER GALAXY

Todd M. Tripp, Kenneth R. Sembach, and Blair D. Savage
Department of Astronomy, University of Wisconsin - Madison,
475 N. Charter St., Madison, WI 53706

Massive stars affect the disk ISM through the action of ionizing radiation, winds, and supernova explosions. According to Galactic fountain theory, massive stars also affect the ISM at large distances from the plane when supernova remnants blowout and spew hot collisionally ionized gas into the halo. We use distant massive stars as background targets for the study of interstellar absorption in the inner Galaxy. The most intriguing feature in the interstellar profiles is a significant amount of absorption at negative (i.e. forbidden) velocities.

OVERVIEW

We analyze *International Ultraviolet Explorer* high-dispersion observations of interstellar absorption towards four distant inner Galaxy lower halo stars in order to test kinematic predictions of interstellar medium models. We concentrate on inner Galaxy stars in order to minimize the effects of Galactic rotation and thereby improve our ability to search for radial inflow or outflow that is predicted by some theories. For example, in the hot ($T_0 = 10^6$ °K at $z = 0$ kpc) Galactic fountain model explored by Bregman (1980), supernova-heated disk gas buoyantly rises into the halo and moves radially outward before cooling into clouds that ballistically return to the galactocentric radius from which the fountain started. The rising gas may be too hot to be detected in absorption by the *IUE*, but the returning clouds could be detected and should show radial inflow. However, if the fountain is initially cooler ($T_0 = 3 \times 10^5$ °K), then it might generate radial outflow detectable in UV absorption. Other potential sources of inflow/outflow are expanding superstructures, spiral structure, a bar in the Galactic center, or a Galactic wind. A full description of this work will appear in the *Astronomical Journal*.

OBSERVATIONS

The inner Galaxy stars we have studied are HD167402, HD168941, HD172140, and HD173502. All of the stars are between $l = 2.3°$ and $l = 5.8°$ and are 6-7 kpc distant. We have retrieved all observations of these stars from the *IUE* data archives, and we have obtained new observations of HD167402 and HD168941. Following the procedure employed by Sembach, Savage, and Massa (1991), we have acquired new exposures with the star offset along the long axis of the large (10" x 20") *IUE* aperture. By co-adding these spectra after correcting for the offset, the *IUE* fixed pattern noise is smeared out and the signal-to-noise is improved from ~ 5-10 (typical single exposure S/N) to ~ 15-25. The data are reduced and co-added as described in Sembach and Savage (1992).

SOME ANALYSIS DETAILS

We use the *apparent optical depth method* (Savage and Sembach 1991) to check for unresolved saturation in the high ion profiles and to determine their column densities.

We have reviewed the stellar classifications. On the basis of UV stellar photospheric and wind lines, we have assigned spectral/luminosity classifications. Based upon these classifications, we have derived the distances to the stars using the O and B star absolute magnitude calibration of Walborn (1972).

INTERSTELLAR ABSORPTION AT "FORBIDDEN" VELOCITIES

The most peculiar and intriguing feature in the interstellar absorption profiles is the significant amount of absorption at negative velocities. We have used the Galactic rotation curve obtained by Clemens (1985) to calculate the span of LSR velocities expected for the four sight lines of interest, assuming that the halo corotates with the underlying disk. If the rotation of gas in the disk and halo is circular and follows standard rotation curves, then *absorption should only be detected at positive LSR velocities towards these stars.* The instrumental point spread function and the velocity dispersion of the gas must be accounted for. The FWHM of the *IUE* is ~ 25 km s^{-1}, and if the gas has a velocity dispersion of 25-30 km s^{-1}, then the absorption profiles might be expected to extend from ~ −30 to 100 km s^{-1}. However, *towards three out of the four stars, interstellar absorption by low as well as high ionization stages is detected from ~ −100 to 100 km s^{-1}.*

Evidence of Radial Outflow from the Inner Galaxy?

What is the source of the negative velocity absorption? This absorption cannot be explained by placing the stars farther away. We estimate that the 1σ uncertainties in the stellar distances are $\pm 25\%$. Placing the stars 25% farther will only extend the allowed velocity ranges to greater positive velocities. The negative velocity absorption cannot be explained by decoupling halo corotation either. If the halo does not rotate at all, for example, inside of a galactocentric radius of 4 kpc, then the Solar motion towards HD168941 will carry the absorption profile to only ~ -20 km s^{-1} (assuming $R_0 = 8.5$ kpc and $\Theta_0 = 220$ km s^{-1}).

One viable interpretation is that we are detecting gas flowing radially away from the Galactic center. The existence of expanding features in the inner Galaxy at moderate to high negative velocities has been known for many years (e.g. Oort 1977 and references therein), but these have mostly been detected in H I studies. If the negative velocity absorption presented here can be shown to be associated with some of these expanding features, then the *IUE* data provide important new information about the ionization, spatial extent, and physical conditions in the expanding features.

Another interesting possibility is that the absorption occurs in outflowing and/or turbulent gas associated with spiral arms, especially the 3 kpc arm. Recent detailed calculations by Martos and Cox (1992) show that gas flowing at ~ 30 km s^{-1} into the small gravitational potential perturbation of a spiral density wave will lead to a large density enhancement inside of a spiral arm. The density jump will cause hydrostatic equilibrium to break down and the resultant pressure gradients will lead to gas flow away from the plane *and radial outflow* with velocities on the order of 100 km s^{-1}.

REFERENCES

Bregman, J. N. 1980, *Ap.J.*, **236**, 577.
Clemens, D. P. 1985, *Ap.J.*, **295**, 422.
Martos, M. and Cox, D. P. 1992, in preparation.
Oort, J. H. 1977, *A.Rev.A.str.&A.p*, **15**, 295.
Savage, B. D., and Sembach, K. R. 1991, *Ap.J.*, **379**, 245.
Sembach, K. R., Savage, B. D., and Massa, D. 1991, *Ap.J.*, **372**, 81.
Sembach, K. R., and Savage, B. D. 1992, *Ap.J.Supp.*, in press.
Walborn, N. R. 1972, *A. J.*, **77**, 312.
Walborn, N. R. 1973, *A. J.*, **78**, 1067.

Massive Stars: Their Lives in the Interstellar Medium
ASP Conference Series, Vol. 35, 1993
Joseph P. Cassinelli and Edward B. Churchwell (eds.)

A UV STUDY THROUGH THE COMPLETE GALACTIC HALO BY
THE ANALYSIS OF HST-FOS SPECTRA OF QSOS AND AGNS

G. S. BURKS, F. BARTKO, M. SHULL, J. STOCKE, E. SACHS,
University of Colorado, CASA Campus Box 389, Boulder CO 80309

M. BURBIDGE, R. COHEN, V. JUNKKARINEN,
University of California, San Diego, CASS, La Jolla, CA 92093

R. HARMS, D. MASSA
Applied Research Corp., 8201 Corporate Drive, Landover, MD 20715

ABSTRACT   The Spectra of a number of QSOs and AGNs observed with
the *Hubble Space Telescope-* Faint Object Spectrograph have been studied.
They include 3C273, PKS 0454, PG 1211, CSO 251, TON 951, PG 1351.
The interstellar line strengths of certain ions have been measured. Special
attention is paid to the C IV/C II ratios, and the Si IV/Si II ratios. These are
compared with these ratios in the 3C273 sight line. 3C273 has the highest C
IV/C II and Si IV/Si II ratios of the six. 3C273's strong high ionization
absorption may be explained more straight forwardly by a nearby supernova
remnant than by highly ionized gas higher up in the galactic halo.

METHOD AND REDUCTIONS

The *HST* Faint Object Spectrograph (FOS) is well suited to study complete
sight lines through the galactic halo. The local Galactic interstellar absorption in
the spectra of six QSOs and AGNs were studied. Such a study can allow comparison of our Galactic Halo with QSO absorption line systems.

The spectra used are from the FOS using its high resolution gratings. The
errors for equivalent widths are determined from the maximum and minimum
equivalent widths were determined using the most extreme plausible continua. is an
over estimate, because of the way the SN is determined. The results are given in
Tables I-II. Table I contains equivalent widths in A, for Galactic Halo absorption
lines of Si. Table II contains equivalent widths for Carbon. In the tables, 's' means
the equivalent width is somewhat uncertain because the line is on a slope of an
emission line, 'p' means that the line is near a QSO emission peak. 'b' means that
the line is uncertain due to a blend.

DISCUSSION

We show plots of Si II(1526.7 A) Galactic Halo absorption vs. Si IV(1393.8 A) + Si IV(1402.8) absorption in Figure 1. We show plots of C II(1334.5 A) + C II*(1335.7 A) Galactic Halo absorption vs. C IV(1548.2 A) + C IV(1550.7 A) absorption in Figure 2. From these two plots, we see that the Galactic absorption in the 3C273 sightline is atypical, relative to the others. For silicon, we see that the Si IV line strength is stronger than one would expect for the corresponding Si II strength. Table II contains coordinates, and the CIV (tot)/C II(tot) and Si IV(tot)/Si II(1526.7) ratios. These ratios are less than one in all cases except 3C273. The C IV/C II and Si IV/Si II ratios for 3C273 are about 1.25. C IV/C II is less than one in the preponderance of Galactic and low halo sight lines (Pettini & West 1982, Danly 1989). C IV/C II is greater than 1 in most QSO absorption line systems (Wu et al. 1990). How can we explain the 3C273 absorption? 3C273 is in the direction of one of the brightest soft X-ray regions in the sky as seen in a resent ROSAT image (Dyer 1991). This X-ray feature corresponds to continuum radio Loops I and IV (Berkhuijsen 1971), which have been shown to be associated with supernovae events. The high C IV and Si IV may therefore be due to shock effects near a Supernova shell (Burks 1991, Burks et al. 1991). This absorption is probably not due to highly ionized gas high in the halo; since it is not seen in our other halo sight lines, none of which show both a high X-ray and radio continuum emission.

From our data, it appears that Galactic Halo absorption is not homogeneous. In five sight lines in our sample low, ionization species dominate over high ionization species. In most QSO metal absorption line systems the opposite is true. Therefore, our Halo does not absorb like most QSO absorption line systems, with the exception to the 3C273 sight line. 3C273's strong absorption in highly ionized species may be most easily explained as being a result of a nearby Supernova shell. Absorption in highly ionized species through our Halo could be dominated by nearby supernovae events. Our work supports the theory that the QSO absorption line systems may be associated with high Supernova rates, because of our explanation for the 3C273 sightline (Yanny, York and Gallagher 1989).

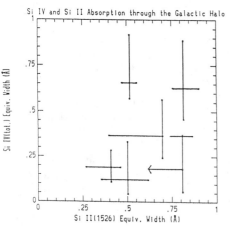

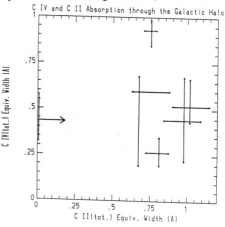

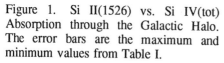

Figure 1. Si II(1526) vs. Si IV(tot) Absorption through the Galactic Halo. The error bars are the maximum and minimum values from Table I.

Figure 2. C II(tot.) vs. C IV(tot) Absorption through the Galactic Halo. The error bars are the maximum and minimum values from Table I.

TABLE I   Interstellar Absorption Equivalent Widths(in A)

Object	Qual	Si II 1526.7	Si IV(tot) 1393 + 1402	CII + CII* 1334 + 1335	CIV$_{total}$ 1458 + 1450
PG1211	best	0.502	0.123	0.978s	0.442
	max.	0.618	0.331	1.089	0.676
	min.	0.357	0.044	0.843	0.221
PKS 1351+640	best	0.817	0.628	1.017s	0.519
	max.	0.903	0.866	0.903	0.667
	min.	0.757	0.459	1.146	0.432
3C273	best	0.516	0.656s	0.753	0.929
	max.	0.556	0.917	0.788	0.985
	min.	0.473	0.570	0.708	0.846
CSO 251	best	0.698	0.366	0.673	0.598
	max.	0.865	0.565	0.878	0.679
	min.	0.399	0.244	0.629	0.195
TON 951	best	0.409	0.191	0.811	0.264
	max.	0.461	0.284	0.875	0.346
	min.	0.269	0.111	0.725	0.195
PKS 0454	best	<0.81	0.183	-	0.432
	max.		0.371		0.577
	min.		0.057		0.147

TABLE II   QSO Coordinates and Interstellar Absorption Ratios

Object	$l°$	$b°$	HVC	C IV/C II	
PG1211	267.55	+74.315	N	0.45	0.25
3C273	289.94	+64.359	N	1.23	1.27
CSO 251	188.32	+55.378	Y	0.89	0.52
PKS 1351+640	111.89	+52.020	Y	0.51	0.77
TON 951	188.56	+37.965	N	0.33	0.47
PKS 0454-220	221.98	-34.650	N	-	>0.23

REFERENCES

Burks, G. S., 1991, Ph.D. thesis, The University of Chicago

Burks, G. S., York, D. G., Blades, J. C., Bohlin, R. C., Wamsteker, W. 1991, ApJ, 381, 55

Berkhuijsen, E. M., 1971, A&A, 14, 260

Danly, L. 1989, ApJ, 342, 785

Dyer, A. 1991, Astronomy, 19, no. 6, 43

Pettini, M., and West, K. A., 1982, ApJ, 260, 561

Wu, C. C., Caulet, A., York, D. G., Blades, J. C., Boggess, A., Morton, D. C 1991, ApJ, 379, 66

Yanny, B., York, D. G., Gallager, J. S., 1989, ApJ, 338, 735

York, D. G., Burks, G. S., and Gibney, T. B., 1986, AJ, 91 354

Massive Stars: Their Lives in the Interstellar Medium
ASP Conference Series, Vol. 35, 1993
Joseph P. Cassinelli and Edward B. Churchwell (eds.)

THE WOLF-RAYET CONTENT OF NGC 595 AND NGC 604: A VIEW FROM HST

LAURENT DRISSEN
Space Telescope Science Institute,
3700 San Martin Drive, Baltimore, Maryland 21218

ANTHONY F. J. MOFFAT
Département de Physique, Université de Montréal
C.P. 6128, Succ "A", Montréal (Qué), Canada, H3C 3J7

MICHAEL M. SHARA
Space Telescope Science Institute,
3700 San Martin Drive, Baltimore, Maryland 21218

ABSTRACT   We have obtained HST *Planetary Camera* images of NGC 595 and NGC 604, the most massive giant HII regions in M33 in order to study their Wolf-Rayet (W-R) population. A dozen W-R stars are detected in each region. All previously claimed "superluminous" W-R stars are found to be tight (diameter $\leq$ 3pc) stellar aggregates. The W-R/O number ratio is significantly (3 times) higher in NGC 595 than in NGC 604. Although their contribution to the total blue light is not dominant, Wolf-Rayet stars are major contributors to the present-day output of momentum and energy into the interstellar medium in these two regions.

INTRODUCTION

Giant HII regions (GHRs) are the spectacular signatures of recent bursts of star formation in galaxies. A typical GHR contains $10^5 - 10^7 M_\odot$ of hydrogen ionized by a young ($\leq 10^7$ yrs) cluster rich in massive, O-type stars (Kennicutt 1984; Hodge, these proceedings).

NGC 604 and NGC 595 are the most massive GHRs in the nearby spiral galaxy M33; in the Local Group, only 30 Dor surpasses them. Wolf-Rayet stars in these two clusters were first discovered via spectroscopy by Conti & Massey (1981) and D'Odorico & Rosa (1981). D'Odorico & Rosa concluded that there must be a very high number ($\sim$ 50) of W-R stars in NGC 604 (W-R/O number ratio $\sim$ 1), a possibility that is difficult to explain if our current understanding of massive star evolution (and the W-R phase) is correct. More recently, Drissen *et al.* (1990, hereafter DMS90) have shown, on the basis of interference filter imagery, that some ten stellar knots in each of these two GHR contain excess HeII 4686 emission. The actual number and luminosity of the W-R and O stars could not be determined precisely however, because of severe crowding problems.

These uncertainties prompted us to re-observe these two regions with the high spatial resolution provided by the HST.

OBSERVATIONS AND REDUCTIONS

The images were obtained with the HST Planetary Camera (1 pixel=0.043″ = 0.15 pc at the distance of M33). For both targets, three exposures (140s, 600s, 600s) were taken with the wide-band blue filter F439W, as well as four exposures (1700s each) with the narrow-band filter F469N centered on the strong HeII 4686 Å W-R emission line.

Although photometry on HST images is complicated by spherical aberration, it is still possible to obtain results reliable enough for our purpose, i.e. identify W-R stars and allow a global comparison of the stellar population between the two clusters. In order to supplement the photometry, image subtraction has been used: after proper scaling, the continuum image is subtracted from the "on-line" image, and regions having an excess of light in the line of HeII 4686 (e.g. W-R or Of stars) show up in the resultant image. This technique has been successfully applied to ground-based data (see DMS90) and is particularly powerful in very crowded regions.

RESULTS

About a dozen stars in each of NGC 595 and NGC 604 show some excess at λ4686. Most of them are W-R stars (as inferred from ground-based spectroscopy), but some stars with a low λ4686 excess might be Of stars.

All previously claimed overluminous W-R stars are now resolved into multiple components. This is illustrated in Figure 1, which shows the core of NGC 595. The brightest W-R stars in both regions have $M_B \sim$ -7.3. This places them in the upper tail of the distribution of visual magnitudes for WNL stars (van der Hucht $et\ al.$ 1990). The remaining W-Rs have M_B ranging from -4.8 to -7.1, i.e. well within the limits of Galactic W-Rs, and hence are not unusual in this respect. The brightest unresolved stars ($M_B \sim$ -8.7 in NGC 604 and -8.1 in NGC 595) do not show W-R characteristics. They are probably evolved supergiants and deserve further spectroscopic work.

At a given magnitude up to the completeness limit ($M_B \sim -4.0$, corresponding to $M \sim 15\ M_\odot$), NGC 604 contains about three times as many stars as NGC 595. Moreover, assuming that all non-WR stars having $M_B \leq$ -4.8 are O stars, we obtain that WR/O $\sim$ 0.1 in NGC 604 and 0.3 in NGC 595. Obviously, spectroscopic classification of all the blue stars is needed to derive more reliable values, but these numbers are consistent with estimates based on global properties (i.e. Hα flux, total B magnitude) of these regions.

The mechanical power of the W-R wind greatly exceeds that of an O star of the same luminosity. The energetics of a starburst region may thus be strongly influenced, if not dominated, by W-R stars for a few million years before supernovae take over (Robert $et\ al.$, these proceedings). Both NGC 604 and NGC 595 seem to be in this W-R wind-dominated phase. The influence of the W-R stars on the interstellar medium are obvious in NGC 604 (Rosa & Solf 1983, Clayton 1988). Much less attention has been paid to NGC 595 which has, however, a proportionately higher W-R content. High resolution Fabry-Perot observations might prove very interesting.

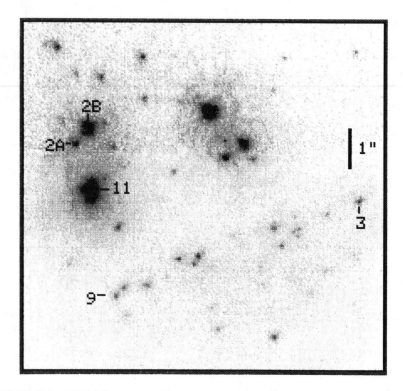

FIGURE 1. HST/Planetary Camera image (broad-band B filter) of the central part of NGC 595. North is at the top, east to the left. Wolf-Rayet candidates are identified. The W-R candidate no 11 is a member of a trapezium-like cluster (diameter = 1.4 pc).

L. D. gratefully acknowledges CRSNG (Canada) for a Post-Doctoral Fellowship and STScI's Director's Research Fund for travel support.

REFERENCES

Clayton, C. A. 1988, MNRAS, 231, 191.
Conti, P. S., & Massey, P. 1981, ApJ, 249, 471.
D'Odorico, S., & Rosa, M. 1981a, ApJ, 248, 1015.
Drissen, L., Moffat, A. F. J., & Shara, M. M. 1990, ApJ, 364, 496 (DMS90).
Kennicutt, R. C. 1984, ApJ, 287, 116.
Rosa, M., & Solf, J. 1983, A&A, 130,29.

Massive Stars: Their Lives in the Interstellar Medium
ASP Conference Series, Vol. 35, 1993
Joseph P. Cassinelli and Edward B. Churchwell (eds.)

THE *CUBIC* COSMIC X-RAY BACKGROUND EXPERIMENT

MARK A. SKINNER, DAVID N. BURROWS, GORDON P. GARMIRE,
DAVID H. LUMB, JOHN A. NOUSEK.
Department of Astronomy and Astrophysics, Penn State University,
525 Davey Laboratory, University Park, PA 16802 USA

ABSTRACT   *CUBIC*, the *Cosmic Unresolved X-ray Background
Instrument Using CCDs,* is instrumented to make non-dispersive
spectral observations, with moderate energy resolution over the
energy range 200 eV - 10 keV. Using mechanically collimated CCDs,
we plan to make one to four day pointed observations of the Diffuse X-
ray Background (DXRB). Our aperture allows 5°x5° spatial resolution
of the galactic XRB (below 1 keV), and 10°x10° spatial resolution of the
cosmic XRB(CXRB). After its December 1994 launch aboard the joint
NASA/Argentine mission SAC-B, we anticipate up to 50% sky
coverage during the planned three year lifetime. We plan to use
CUBIC to study the thermal states and abundances of hot gas in the
galactic DXRB, to make accurate measurements of the intensity and
slope of the CXRB, and to search for line emission in the CXRB.

The *CUBIC* experiment, currently under construction by the Penn State X-ray
astronomy group, is an example of some 'Big' science being done on a very
small scale. The instrument is an outgrowth of our group's sounding rocket
program, and as such mixes the flexibility, quick response, and low cost of a
sounding rocket with the enhanced data return of a small satellite (~3 years).
    CUBIC will fly aboard the first Argentine engineering satellite, SAC-B.
This is a joint mission between NASA and the Argentine civilian space agency
CONAE. Our instrument will be the largest (30 kg, 30 W) of the four
experiments to fly aboard this small (<200 kg, 120 W) satellite. A December,
1994 launch from Wallops Island, VA is planned; a Pegasus XL should place
SAC-B into a 550 km, 38° inclination orbit. The main ground station and mission
support operations center will be in Buenos Aires. There will be five orbits per
day with ground station contact; up to 30 Megabits of science and engineering
data will be dumped per contact. Two momentum wheels and magnetic coils
will control the spacecraft attitude, which will be determined to within 2°
realtime, ≤ 1° post-facto.
    The main scientific goal of *CUBIC* will be to obtain high quality, non-
dispersive spectral observations of the DXRB. The *CUBIC* CCDs will
provide moderate energy resolution (E/ΔE ~10-60) over a continuous energy
range from ~200 eV to over 10 keV. This will allow accurate measurement of the
intensity and slope of the CXRB (>1 keV). Previous measurements with
proportional counters are inconsistent; counters with thin entrance windows,
sensitive to X-rays below 3 keV, found the CXRB to follow a powerlaw with
an intensity of 11.0 (McCammon *et al.* *1983,* Wu *et al.* 1991; Garmire *et al.*
1991). Thick window counters, sensitive to energies greater than 3 keV, found
a power law intensity of 8.0 (Marshall *et al.* 1980; Schwartz 1979). *CUBIC*,
continuously sensitive above and below 3 keV, should be able to determine if
there is an actual break in the spectrum, or if the discrepancy is due to
systematic instrumental effects.

Active Galactic Nuclei (AGN) may provide a continuum component to the extragalactic DXRB, due to redshift smearing of the 6.4 keV Fe line. After many months of observations, *CUBIC* may be able to help distinguish between various AGN turn-on and evolutionary models(Schwartz 1991).

X-rays of energy below 1 keV are generally from the galactic DXRB, and *CUBIC* should be able to distinguish between the strong line groups of oxygen, neon, iron, carbon, etc (fig. 2). At these energies we will be provided with 5°x5° spatial resolution. We will also be able to make good sensitivity measurements in the 300-500 eV range. This band pass has had limited observations, as the thin films necessary for proportional counters in this energy range are typically carbon-based and have low transmittance to these energies.

With its wide spatial resolution, *CUBIC* will be most valuable in studying large and diffuse objects, e.g., the ISM and SNRs, but we will also use it to study some bright X-ray sources (Sco X-1, the Crab, etc.) that are too bright for missions with large area X-ray mirrors. This will not be an all-sky survey, but with its three year expected useful observing lifetime, *CUBIC* can cover up to ~50% of the sky.

HARDWARE DESCRIPTION

The heart of the *CUBIC* X-ray camera are two 1024x1024 pixel CCDs, designed by PSU, Dick Bredthauer of Loral, and Jim Janesick of JPL, of a novel "thin-poly" design, currently under fabrication by Loral. These front-illuminated CCDs have 18μx18μ pixels, with an approximate 18μx12μ "open area" covered by a thin (400Å) layer of silicon, and oxide and nitride. The remaining pixel area is covered by the gate structure. The CCDs were designed to provide good sensitivity to soft X-rays (figure 1). We will cool the CCDs to ~-100°C with a Marlow thermoelectric cooler. Thick aluminum walls in the camera will provide radiation shielding. A 1200Å Al-Ti filter will block visible light and UV geocoronal lines from our CCDs. A rare-earth permanent magnetic broom will exclude electrons from the camera. The energy-dependent aperture of the camera is formed by cutting a square open area in a 30μ beryllium foil, bounded by thicker aluminum. Below one kilovolt, X-rays can only penetrate the open area, forming a 5°x5° field of view. Above 3 keV, the thin Be sheet readily transmits X-rays, and this forms our larger 10°x10° fov. As the DXRB spectrum follows a sharply falling power law, the larger fov at higher energies increases our sensitivity.

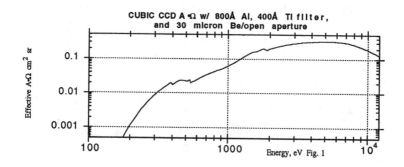

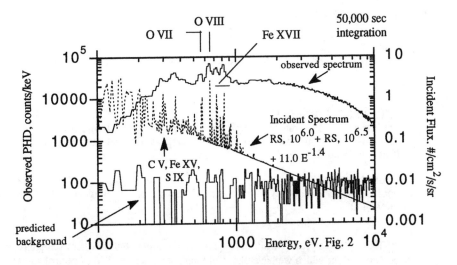

Fig. 2

The relatively high A•Ω of this experiment allows it to compare favorably with other past and future experiments for the study of the diffuse background, as seen in this table.

Mission	A•Ω (cm² sr)		Exposure	ΔE (eV)
	(<2 keV)	(~6 keV)	(m² s sr)	(@ 6 keV)
CUBIC	~0.03	~0.15	~100/year	~120
BBXRT [1]	~0.013	~0.0025	~0.05	160 [2]
ASTRO-D GIS [1]	~0.42	~0.19		~470
SIS [1]	~0.076	~0.035		~120
AXAF (ACIS) [1]	~0.01	~0.002		~120
XMM	~0.16	~0.06	~290/year	~120
UW DXS [3]	0.02		~0.1	10 [4]

[1] No significant response below 500 eV

[2] 155 eV in the central pixel, 165-170 eV in the outer pixels

[3] No significant response above 300 eV

[4] at 200 eV, not 6 keV

REFERENCES

Garmire, G.P., et al. 1991, Ap. J. submitted
Marshall, F.E., et al. 1980, Ap. J., **235**, 4.
McCammon, et al. 1983, Ap. J., **269**, 107.
Schwartz, D.A. 1979, in X-ray Astronomy, ed. W.A. Baity & L.E. Peterson (Pergamon Press), 453.
Schwartz, D.A. 1991, Ap. J. submitted
Wu, X., et al l. 1991, Ap. J., **379**, 564.

Massive Stars: Their Lives in the Interstellar Medium
ASP Conference Series, Vol. 35, 1993
Joseph P. Cassinelli and Edward B. Churchwell (eds.)

SEARCH FOR W-R FEATURES IN A SAMPLE OF 12 NORTHERN HII
GALAXIES

VICTOR ROBLEDO-RELLA
Joint Institute for Laboratory Astrophysics, Campus Box 440, Boulder, CO
80309-0440, and Instituto de Astronomía, Universidad Nacional Autónoma de
México, México, D.F.

PETER S. CONTI
Joint Institute for Laboratory Astrophysics, Campus Box 440, Boulder, CO
80309-0440

ABSTRACT   We present work in progress on a sample of 12 northern HII galaxies
(see Table I), for which we have obtained long slit "green" and "red" optical spectra,
of medium resolution (5Å), using the 2.1m telescope of the Observatorio
Astronómico Nacional de México, located at Sierra de San Pedro Mártir, Baja
California, México. We have partially reduced the spectra, and detected a narrow
(FWHM ~7Å) feature at HeII 4686 for only 1 galaxy (Case-gal 184), and barely
detected that feature for 2 others (Case-gal 845 and Mrk 496). Our observations show
no distinguishable HeII 4686 feature either for Mrk 33 or G0833+625. For the rest of
the galaxies, our signal to noise ratio did not allow us to confirm the presence (or
not) of this feature in their spectra. The ratios of the nebular forbidden lines
[OIII]5007/Hβ and [NII]6584/Hα for all our galaxies indicate that the photoionizing
flux has an stellar origin. Case-gal 184 is the lowest metal abundance galaxy in our
sample, as inferred from its [NII]6584/Hα ratio. This is consistent with the
occurrence of narrow nebular HeII 4686 in metal-deficient galaxies. We intend to
enlarge our observations of W-R galaxy candidates and to apply them population
synthesis techniques in order to study their stellar content in a more systematic way.

INTRODUCTION

Wolf-Rayet galaxies (e.g. Conti 1991) have been defined as emission line galaxies whose
integrated spectra show a "broad" (FWHM > 10Å) emission feature at HeII 4686, attributed to
W-R stars, but at the same time have narrow nebular emission lines. In this sense, W-R
galaxies may be considered systems with a relatively strong component of massive stars,
which suggests the existence of a "very recent" (a few 10^6 years) burst of star formation. W-R
galaxies can be used to study problems such as the relation between the inferred number of
massive stars (O and W-R stars) and the metallicity of the parent galaxy, or the relation
between the estimated age and strength of the burst with environmental parameters as, for
example, the local density of surrounding galaxies. Stellar evolution models can also be tested
by comparing their predictions with the derived ratios of massive stars (say the W-R/O ratio)
in these galaxies (Vacca and Conti 1992).
     The main goal of this work is to look for any possible "broad" emission feature
around λ4686 in a selected sample of northern HII galaxies (taken from Sanduleak and Pesh
1989 and references therein, and Wasilewsky 1983, among others) in order to increase the
number of known W-R galaxies.

OBSERVATIONS AND DATA REDUCTION

We obtained long slit optical spectra at mid-resolution of the galaxies listed in Table I, "green frames", around Hβ and "red frames, around Hα. We used a Thompson CCD of 384x576 pixels attached to a Boller & Chivens spectrograph, with a 600*l*/mm grating giving us a dispersion of 2.00 Å/pix in the green and 2.15 Å/pix in the red. The projected width of the slit on the sky was about 4.5". The "green" and "red" exposures were, respectively, of 20 min. and 15 min. each. Unfortunately the sky conditions were not completely photometric all the nights. The λ–calibration was done using a He-Ar comparison lamp, and for the flat-field correction we used an incandescent lamp inside the spectrograph.

All the "green" and "red" images have been partially reduced using the IRAF package: the images were bias subtracted and flat-field corrected; the spectra were extracted with an aperture that encompassed the complete galaxy (in both λ ranges); sky subtraction was done defining appropriate apertures at both sides of the galaxy. We removed the presence of cosmic rays as well as zones of bad sky subtraction, combining 2 or more frames. We also took spectra of the standard stars: EG 247, Feige 34, Feige 56 and Hiltner 600, but we have not yet completed the flux calibrations, and consequently we have not yet corrected the spectra for the λ-response of the CCD.

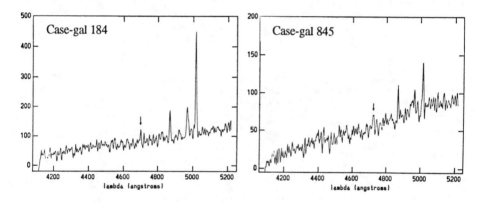

Figs. 1a and 1b. "Green" spectra (counts vs. lambda) of Case-gal 184 and Case-gal 845 showing the possible presence of a feature around λ4686. The λ-response of the CCD has not been applied yet.

RESULTS AND COMMENTS

Due to the non-photometric conditions during some of the "green" nights, we did not attain a very good S/N ratio for all the galaxies (< 10 for the "green" frames and < 20 for the "red" ones). Unfortunately this is a critical factor when trying to look for faint features in the spectra, and our results should not be considered definitive as to the absence of W-R stars in our galaxies. In Table I we present the list of observed galaxies along with a comment about any distinguishable feature found (or not) around λ4686. In Figures 1a and 1b we show the "green" frames (properly shifted in λ) for Case-gal 184 and Case-gal 845 respectively. We consider Case-gal 184, Case-gal 845 and Mrk 496 the only galaxies for which we found some evidence of a faint feature around λ4686. We neither found any other noticeable feature of W-R stars (NIII 4640, CIII 4650 or CIV 5808) in our spectra.

As the ratios [OIII]5007/Hβ and [NII]6584/Hα do not depend very much either on the adopted Galactic reddening law or on the λ-response of the CCD, we could verify the stellar nature of the photoionizing field in all our galaxies by plotting them on the "diagnostic diagram" 5007/Hβ vs. 6584/Hα. See Figure 2.

From the semi-empirical relation used by Robledo-Rella and Firmani (1990; derived from data given by Kunth and Joubert 1985 for a sample of 42 HII region-like objects), which relates the Hα/6584 ratio with the (O/H) oxygen abundance through the linear regression:

$$\log(H\alpha/6584) = 14.136 - 1.572 \times [12 + \log(O/H)] \qquad (1)$$

(correlation coefficient C = 0.93), we have assigned an approximate oxygen abundance to each galaxy (see Table I). We found that all our galaxies have an (O/H) value lower than solar, and that Case-gal 184 has the lowest (O/H) value (this being an upper limit), which corresponds well with the suggestion that galaxies showing narrow nebular λ4686 tend to be metal deficient systems.

TABLE I   List of observed galaxies

Galaxy	Features around λ4686	FWHM (Å) λ4686	(O/H)* / (O/H)⊙
Case 68	none/noise	- - - -	0.25
Case 184	detected/narrow	7	<0.15
Case 224	none/noise	- - - -	- - - -
Case 259	none/noise	- - - -	0.45
Case 310	none/noise	- - - -	0.20
Case 845	barely detected/'broad'	10	0.55
Mrk 33	none	- - - - -	0.40
Mrk 52	none/noise	- - - - -	0.80
Mrk 496	barely detected/noise	- - - - -	0.70
Wasi 12	none/noise	- - - - -	0.70
Wasi 67	none/noise	- - - - -	- - - -
G0833+652	none	- - - - -	0.50

(*) Approximated oxygen abundance derived using eq. 1. Here (O/H)⊙ ≈ 8.92 x 10⁻⁴
=====================================

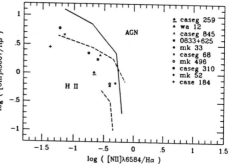

Fig. 2 "Diagnostic diagram" log(5007/Hβ) vs. log(6584/Hα) for our galaxies. The solid line shows the division between the loci occupied by AGNs and HII region-like objects. The dashed upper and lower lines correspond to the Dopita and Evans (1986) models for Teff = 56,000K and 37,000K respectively. Adapted from Veilleux and Osterbrock (1987).

REFERENCES

Conti, P.S. 1991, *Ap.J.*, **377**, 115.
Dopita, M.A. and Evans, I.N. 1986, *Ap.J.*, **307**, 431.
Kunth, D. and Joubert, M. 1985, *A.A.*, **142**, 411.
Robledo-Rella, V. and Firmani, C. 1990, *R.Mex.A.A.*, **21**, 236.
Sanduleak, N. and Pesh, P. 1989, *Ap.J.S.S.*, **70**, 173.
Vacca, W.D. and Conti, P.S. 1992, *Ap.J.*, in press.
Veilleux, S. and Osterbrock, D.E. 1987, *Ap.J.S.S.*, **63**, 295.
Wasilewsky, A.J., *Ap.J.*, **272**, 68.

Massive Stars: Their Lives in the Interstellar Medium
ASP Conference Series, Vol. 35, 1993
Joseph P. Cassinelli and Edward B. Churchwell (eds.)

MASSIVE STARS AND THE KINEMATICS OF GIANT HII REGIONS

Héctor O. Castañeda
Instituto de Astrofísica de Canarias and Isaac Newton Group of
Telescopes, La Palma Observatory.
E-38200 La Laguna. Tenerife. SPAIN.

ABSTRACT   Giant extragalactic HII regions are active centers of
star formation and by their nature are ideally suited to study the interaction
between massive stars and their environment. I discuss in the text the
preliminary results of an ongoing observational program designed to study the
kinematics and dynamics of the ionized gas in this type of objects.

INTRODUCTION

Giant extragalactic HII regions (GEHR) are characterized by the supersonic
line widths in their emission lines, first observed by Smith and Weedman
(1970). Several models have been advanced to explain the motions of the
gas, including virial equilibrium between the gas and the massive stars
(Terlevich and Melnick 1981), and the effect of the stellar winds of the ionizing
stars (Rosa and Solf 1984).   In collaboration with C. Muñoz-Tuñon and J.
Vilchez (IAC), R. Terlevich (RGO), and M. Copetti (U. Santa Maria) we are
conducting a long term program to study the kinematics of the gas in the main
complexes of nearby galaxies and to understand the source of the turbulent
motions in the gas. These observations should provide a new insight into the
interaction between the ionizing cluster and the interestellar medium. I discuss
in this paper some preliminary results of our program.

OBSERVATIONS

One-dimensional spectra were obtained with the Intermediate Dispersion
Spectrograph of the 2.5m Isaac Newton Telescope of the Observatory of the
Roque de los Muchachos. Two-dimensional imaging spectroscopy in the lines of
Hα and [OIII]λ5007 were obtained using the TAURUS-II Fabry-Perot imaging
spectrograph at the Cassegrain Focus of the 4.2m William Herschel Telescope
of the Observatorio del Roque de los Muchachos, with the Image Photon
Counting System (IPCS) as detector. The one-dimensional spectroscopy
program is designed to detect high-velocity, low intensity gas, thanks to the
high S/N ratio that can be achieved via the CCD; Fabry-Perot interferometry
allows the mapping of the velocity field over the entire extent of the HII
regions.

RESULTS

Most previous kinematical studies were based on line profiles integrated over
the whole of the region. A question not yet answered is whether the mechanism
of broadening is due only to random motions of filaments and clouds. In
that case we also should see chaotic radial velocities over the surface of the
nebulae. It is interesting to note that in fact systematic motions in the form of
velocity gradients with a well defined pattern have been observed, confirmed
by comparison of velocity maps in different emission lines. An example is
NGC 588, an "egg-shaped" nebula in M33, with a morphology very similar to a
planetary nebula. The velocity map in the [OIII]λ5007 line shows a gradient in
the velocity of the order of 25 km s^{-1} between the inner and outer zones, with
the more blueshifted velocities in the central part of the region. This is also the
area where the dispersion along the line of sight is larger. This pattern appears
to indicate that random motions, where important, are not the only source of
line broadening observed in the integrated emission profiles. Synthetic aperture
spectra with large aperture show gaussian line profiles, but spectra obtained
with small apertures reveal asymmetric line profiles and multiple components.

Shell-like structures are likely to be formed in GEHR by the stellar winds
from the embedded OB associations. Wind-blown bubbles have been studied
in our program, a typical case being the region Hubble III in NGC 6822. Very
peculiar features are observed in the region (Sabalisck et al. 1992): a large
fraction of the spectra shows non-symmetric line profiles, in some cases a line
splittings up to 70 km s^{-1} are measured, and radial velocity gradients of 20
km s^{-1} between the inner and outer zones of the zones of the filaments that
define the region have been observed.

The typical ages of the regions ($\gtrsim 10^6$ years) may indicate the presence of
Wolf-Rayet (WR) stars, that evolve from high mass O stars. WR stars have
been postulated as candidates for the mechanical energy source of turbulent
motions. The mechanical power injected into the interstellar medium during
their lifetime can be of the order of 2.10^{51} ergs, comparable to the internal
energy in the form of supersonic mass motions for a typical GEHR (Hunter
and Gallaguer 1985). To test this hypothesis, we have examined some of the
regions of our sample where WR stars have been detected by direct imaging,
and we have not observed a peculiar enhancement in the velocity dispersion in
the positions of the stars; the larger turbulent velocities seem to be associated
with the zones of low surface brightness. Considering also that recent surveys
reveal that the average number of WR/O stars in a GEHR is much lower than
unity (Drissen 1991), and that the efficiency of transfer of kinetic energy from
the stellar wind to the ionized gas could be of the order of 1% (Chu 1982),
suggests that the effect of WR stars on the overall kinematic structure of the
HII complexes could have been overestimated.

The effect of the massive stars on the kinematics of the regions even
extends to the end of their lifetimes. It is likely that an appreciable fraction
of the stars that ionize the regions have already evolved, making probable
the existence of supernova remnants within the complexes (stars with M $\gtrsim 7$
M$_\odot$ must end their lives as supernovae). The supernova remnants could then
produce high velocity gas at very low intensity levels (the same phenomenon
could be associated with the stellar winds of the massive stars). We have
conducted a search for this spectral feature in the Hα line over ten regions

associated with M101, M51, NGC 6822, and NGC 4861, with a succesful detection only in NGC 5471 (Castañeda, Vilchez, and Copetti 1990). A similar phenomenon has been reported recently in NGC 2363 (Roy *et al.* 1992). In both cases there is no conclusive evidence for the nature of the source of the high velocity gas, and follow-up observations will be necessary to clarify the issue.

ACKNOWLEDGEMENTS

I want to gratefully acknowledge the DGICYT (Spain) for a fellowship under which part of this work was carried out. The INT and WHT are operated on the island of La Palma by the Royal Greenwich Observatory at the Observatorio del Roque de los Muchachos of the Instituto de Astrofísica de Canarias.

REFERENCES

Castañeda, H.O., Vilchez, J.M., and Copetti, M.V.F. 1990, *Ap. J.*, 365, 164.
Chu, Y.-H. 1982, *A. J.*, 254, 578.
Drissen, L. 1991, *IAU Symposium 143: Wolf-Rayet Stars and Interrelations with Other Massive Stars in Galaxies*, K.A. van der Hucht and B. Hidoyat (eds.), p. 595.
Hunter, D.A, and Gallagher, J.S. 1985, *A. J.*, 90, 80.
Rosa, M., and Solf, J. 1984, *Astr. Ap.*, 130, 29.
Roy, J-R, Aubé, M., McCall, M.L., and Dufour, R.J. 1992, *Ap. J.*, 386, 498.
Sabalisck, N. Castañeda, H.O., Muñoz-Tuñon, C., Copetti, M.V.F., and Terlevich R. 1992, *Proc. EIPC Workshop: Star Forming Galaxies and Their Interestellar Medium*, F. Ferrini and J. Franco (eds.), in press.
Smith, M.G., and Weedman D. 1970, *Ap. J.*, 161, 33.
Terlevich R, Melnick J. 1981, *M.N.R.A.S.*, 195, 839.

Massive Stars: Their Lives in the Interstellar Medium
ASP Conference Series, Vol. 35, 1993
Joseph P. Cassinelli and Edward B. Churchwell (eds.)

MASSIVE STARS AND GALACTIC HALOS: PHOTOIONIZATION IN THE HALO OF NGC 891

James Sokolowski
Space Telescope Science Institute, Baltimore, Maryland

ABSTRACT We show that massive stars can produce the uniform and enhanced, low-ionization emission observed throughout the lower halo of NGC 891. Depletion of the gas phase coolants O, Si and Fe within dust grains and hardening of the composite stellar radiation field by absorption within the disk ISM are essential in producing the line ratios observed. Estimation of the photon flux yields a slowly varying ionization parameter throughout the lower halo, which naturally produces uniform excitation conditions, and suggests a filling factor of order 10^{-1}.

INTRODUCTION

Recent imaging and spectroscopic studies reveal a *diffuse ionized medium* (DIM) in the edge-on, spiral galaxy NGC 891[1,2] similar to that seen in the Galaxy[3] In NGC 891, Hα emission is observed throughout the range $0 \lesssim R \lesssim 10$ kpc, 250 pc $\lesssim |z| \lesssim 4$ kpc, is *relatively* smooth and yields an electron density well fitted[2] as $n_e(R, |z|) = \sum_{i=1}^{3} n_{e,i} e^{-R/R_{0,i}} e^{-|z|/z_{0,i}}$, with coefficients[4] $n_{e,i} = (0.17, 0.26, 0.05)\phi^{-1/2}$ cm^{-3}, $R_{0,i} = (8.5, 3, 12)$ kpc and $z_{0,i} = (800, 1400, 2800)$ pc and filling factor ϕ. The vertical distribution of these DIM is similar, while their emission properties differ. In NGC 891, low-ionization emission is uniformly enhanced, with [NII] $\lambda 6583$/H$\alpha = 0.5$ - 1.0 and [SII] $\lambda\lambda 6716,6731$/H$\alpha = 0.4$ - 0.6, and shows no systematic variation in either R or $|z|$[2,5], while at low latitude ($|b| \lesssim 30°$), the Galactic DIM emits [NII] $\lambda 6583$/H$\alpha \sim 0.3$ and [SII] $\lambda 6716$/H$\alpha \sim 0.3$[3]. However, direct comparison of these media is difficult because in NGC 891 the DIM is observed only at high-$|z|$, while the high-latitude properties of the Galactic DIM are completely unknown[3,4] It has been shown[3,4] that only OB stars and supernovae are sufficiently energetic to power and maintain ionization equilibrium within these media and photoionization models have had success in reproducing the low-latitude emission properties of the Galactic DIM[5] However, published models are incapable of obtaining the ratios observed in NGC 891 and the role of supernova generated shocks is uncertain[6,7] Therefore, we examine models of DIM photoionization appropriate for the formation of the enhanced, low-ionization emission observed at high-$|z|$ in NGC 891.

MODELING THE DIM IN NGC 891

The Ionization Parameter

We estimate the flux of ionizing photons throughout the lower halo of NGC 891 by adopting a 1-dimensional calculation and appealing to the ionization equilibrium of its DIM. In this approximation, the flux of ionizing photons at a point $(R, |z|)$ is simply that required ionize the DIM above it. That is;

$$\Phi(R, |z|) \ cm^{-2} \ sec^{-1} = \int_0^\infty \phi \, n_e^2(R, |z'|) \, \alpha_{rec} \, dz' - \int_0^{|z|} \phi \, n_e^2(R, |z'|) \, \alpha_{rec} \, dz',$$

where α_{rec} is the case-B hydrogen recombination coefficient. The ionization parameter is then $U(R, |z|) = \Phi(R, |z|) \, n_e^{-1}(R, |z|) \, c^{-1}$, where c is the speed of light. This calculation produces a slowly varying ionization parameter throughout the DIM ($1.5 \times 10^{-4} \lesssim U \lesssim 1 \times 10^{-3}$ for $\phi = 10^{-1}$) providing a natural mechanism for the production of the uniform DIM excitation conditions observed.

The Stellar Radiation Field

We simulate the massive stellar population within NGC 891 using a Salpeter IMF over the mass range $15 \lesssim M_*/M_\odot \lesssim 40$ and use model atmospheres[8] to represent the radiation field from each stellar type. OB stars in the Galaxy and NGC 891 produce ionizing photons at a rate ~ 6 times that required to ionize their DIM,[3,4] suggesting that attenuation of the composite stellar radiation field may be significant as it propagates into the lower halo. We therefore introduce an absorption feature in the composite radiation field of the Salpeter IMF to simulate the effects of radiative transfer through the multi-phase, disk ISM. This absorption feature has a low energy cut-off at 13.6 eV, scales with photon energy[9] as $E_\gamma^{-2.43}$ and is normalized to absorb ⅔ of the ionizing photons from the pure Salpeter IMF. The presence of this absorption feature raises the average, ionizing photon energy from ~ 21 eV to ~ 26 eV, which allows for the generation of higher model electron temperatures.

Photoionization Modeling

Using the photoionization code CLOUDY,[10] we examine the parameter space available to individual DIM condensations (clouds, filaments, slabs) using plane-parallel, low-density models. All models include the optical and thermal (heating and cooling) properties of standard dust mixtures[10] and are radiation bounded. The results of our calculations for the pure Salpeter IMF and solar abundance gas[11] are shown in Figure 1. The agreement with previous models[6] is excellent, specifically they fail to reproduce the line ratios observed in the DIM of NGC 891. Full examination of these models shows that the elements O, Si and Fe combine to produce ~ 10 - 30% of their total cooling, suggesting that the strong elemental depletions due to the presence of dust may play a significant role in enhancing the high-$|z|$, low-ionization emission in DIM. Adopting standard elemental depletion parameters[12] for an average gas density of order unity (e.g. N, O, Si and Fe abundances are 0.75, 0.43, 0.046 and 0.0145 times solar), we plot the results of depleted abundance models and the hardened continuum in Figure 1. The agreement with observations is excellent and summarized in Table 1.

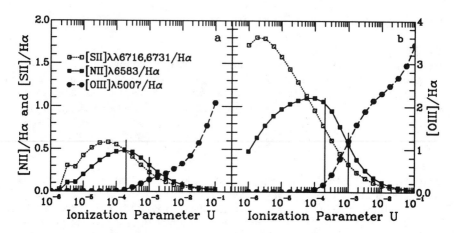

Figure 1. Photoionization model results for the pure Salpeter IMF and solar abundances (a) and the hardened Salpeter IMF with depleted abundances (b). The solid vertical lines delimit the ionization parameter regime for $\phi = 10^{-1}$.

TABLE I   Model Results ($\phi = 10^{-1}$)

Emission Property	Observed	Model
[NII] λ 6583/Hα	0.5 – 1.0	0.55 – 1.1
[SII] $\lambda\lambda$ 6716,6731/Hα	0.4 – 0.6	0.3 – 0.7
[OI] λ 6300/Hα	unknown	0.03 – 0.1
[OII] λ 3727/Hα	unknown	1.2 – 2.0
[OIII] λ 5007/Hα	unknown	0.2 – 1.2

REFERENCES

(1) Dettmar, R.J. 1990, A&A, 232, L15
(2) Rand, R.J., Kulkarni, S.R., & Hester, J.J. 1990, ApJ, 352, L1
(3) Reynolds, R.J. 1991, The Interstellar Disk-Halo Connection in Galaxies, ed. H. Bloemen, 67
(4) Sokolowski, J. 1991, Ph.D. Thesis, Rice University
(5) Keppel, J.W., Dettmar, R.J., Gallagher, J.S., & Roberts, M.S. 1991, ApJ, 374, 507
(6) Mathis, J.S. 1986, ApJ, 301, 423
(7) Sivan, J.P., Stasinska, G., & Lequeux J. 1986, A&A, 158, 279
(8) Kurucz, R.L. 1979, ApJS, 40, 1
(9) Cruddace, R., Paresce, F., Bowyer, S., & Lanpton, M. 1974, ApJ, 187, 497
(10) Ferland, G.J. 1991, OSU Internal Report 91-01
(11) Grevesse, N., & Anders, E. 1989, Cosmic Abundances of Matter, ed. C.J. Waddington
(12) Jenkins, E.B. 1987, Interstellar Processes, ed. D.J. Hollenbach and H.A. Thronson, 533

Author Index